Built in Niugini

Royal Anthropological Institute

The RAI Series

Series Editor: Jeremy MacClancy

The RAI Book Series is devoted to the publishing of major works in anthropology, both monographs and edited volumes. Enquiries as to future projects are welcome, and may be addressed to publications@therai.org.uk. Further details may be found at www.therai.org.uk/publications/rai-book-series.

Volume 1 *Built in Niugini: Constructions in the Highlands of Papua New Guinea*
Paul Sillitoe

Volume 2 *Made in Niugini: Technology in the Highlands of Papua New Guinea*
Paul Sillitoe

Volume 3 *Agnation, Alternative Structures, and the Individual in Chopi Society*
D.J. Webster

Built in Niugini

Constructions in the Highlands of Papua New Guinea

Paul Sillitoe

The RAI Series, Volume One

Sean Kingston Publishing
www.seankingston.co.uk
Canon Pyon

First published in 2017 by
Sean Kingston Publishing
www.seankingston.co.uk
Canon Pyon

British Library Cataloguing in Publication Data

A catalogue record for this book is available from the British Library.

Printed by Lightning Source

ISBN 978-1-907774-45-4

A one a one

4

J + R + H + 2A + P

I find nothing but praise for those householders who, in the confidence of learning, are emboldened to build for themselves.

Vitruvius, *Book VI* of *Ten books on Architecture*, p. 75

When a man has finished building his house, he finds that he has learned unawares something which he ought absolutely to have known before he – began to build.

Friedrich Nietzsche, *Beyond Good and Evil*, section 277

I do not want my house to be walled in on all sides and my windows to be stuffed. I want the cultures of all the lands to be blown about my house as freely as possible. But I refuse to be blown off my feet by any.

Mahatma Gandhi, *Young India* (1 June 1921), p. 170

Ceux qui faisaient des discours sur le travail manuel ne savaient pas de quoi ils parlaient.

Albert Camus, 'Les muets' in *L' exil et le royaume*, p. 76

Contents

FIGURES

MAPS

PLATES

TABLES

FOREWORD

It is a great pleasure that the RAI is able to publish this important volume, the first in our new series. Like all good books, this text can be read in a number of different ways. At one level, it is an appeal to a broad-based anthropology. At another, it is a superb traditional social-anthropological monograph, one based on an intimate knowledge of a particular people and their language. At another, it is a meticulous demonstration of how data can be collated and presented in ways that enhance a wider argument – a counterweight to generalized comment without sufficient empirical foundation. This last is backed up by a series of trenchant – and surely accurate – criticisms of the sometimes extravagant writing on materiality. It can also be used then, as a thoroughly up-to-date commentary on contemporary theorizing in anthropology.

When we look at the history of anthropology over the last century, it often appears that the subject is divided between its different parts. This division is something that the RAI has never wished to come about. The institute is certainly committed to a broad, pluralistic approach, but it is one that assumes that there is an underlying unity between anthropology, archaeology, biological anthropology and linguistic anthropology. Another way of stating this would be to say that, though sometimes great scholarly impetus can come through the pursuit of one or another sub-field, that impetus also has a cost. We should seek to be aware of where the balance lies – of those intellectual paths which are closed by sub-specialization as well as those that are opened.

Sillitoe, in this beautifully written and rich account, is profoundly aware of this. Reading his full and detailed exploration of the relationship between the humans and the built environment helps us realize once again, just how much anthropology lost when archaeology distanced itself from it. There were, it is clear, diverging methodologies which took them apart, and over time these methodologies have come to generate distinct bodies of knowledge. Yet, it is also possible to consider how the two may continue in tandem.

Perhaps, there will come a time when it is usual for anthropological investigations in the field to include an archaeologist working in partnership with them. What might be termed contemporary archaeology, in as much as this would imply using archaeology to look at living communities, could work in a multitude of ways. One obvious instance would be in peasant communities, where a stratigraphic building survey could potentially yield an accurate understanding of the different phases that a community has lived through, one that is just as revealing as the oral histories that the anthropologist collects, and which could be combined with it. There are doubtless many others, drawing in other specialisms too, perhaps leading to the possibility of social anthropologists more readily sharing their time in the field with other colleagues, in order to enjoy the cross-fertilisation of ideas and techniques from which large archaeological excavations have long benefitted.

Built in Niugini is the second major work that Sillitoe has written on this theme. The first, *Made in Niugini* ,was published first by the British Museum Press, and it is with equal pleasure that we reprint it as the second volume in our series. *Made in Niugini* was not only published by the British Museum, but also sponsored by it, and their holdings were greatly enriched by Sillitoe's collecting of artefacts in the field. Here, we see another example of the way that Sillitoe has crossed boundaries. Social anthropology and museums have always been intertwined at the RAI, and many of its Honorary Secretaries, most notably William Fagg, who was director of the Museum of Mankind, were drawn from the museum world. More broadly, however, thirty years ago it frequently appeared that anthropology and museums had separated, to their mutual detriment. Happily, Sillitoe's pioneering example here has presaged better times and today there is a lively dynamic between anthropology and museums that has been transformative. Let us hope that the equally pioneering path blazed by its successor *Built in Niugini* is met with a similarly propitious outcome.

David Shankland
Director, Royal Anthropological Institute
Honorary Professor, Anthropology, University College London

PREFACE

What can we understand about ourselves, others and the world we live in? This grand question has intrigued thinkers since ancient times, even possibly our proto-human ancestors as they became self-aware, although we only have a documentary record of thinking from the classical era onwards. The ensuing philosophical debate suggests perennial tail-chasing, to the bemusement of some, including Socrates apparently, who reputedly commented 'I know only one thing, and that is that I know nothing.' This book explores the implications of this apparently oxymoronic and postmodern-sounding comment for our presumed understanding of human behaviour, by focusing on tacit knowing.

We engage in many activities without being intentionally aware of doing so, about which we apparently know nothing consciously. We eat meals without focusing on the use of knife and fork, we walk without thinking about synchronizing our leg movements and we interact with other persons following social conventions automatically without reflection. Our minds are elsewhere, often engrossed in conversations, during which we talk without deliberating about grammar and the meaning of words. These everyday acts and things, which are intuitively familiar, do not necessarily involve mindful thought. Such a state of being is pre-reflective, informed by socio-cultural inculcation and experience. The way we are in the world, as the phenomenologists put it, presents these pre-reflective things and abilities to us. The implications for what we think we know about human actions are profound.

This book uses an in-depth ethnographic study of building practices in the New Guinea highlands as a vehicle to explore these issues, which may strike many as an unlikely, even unpromising context to do so. It affords an opportunity to found discussion of these elusive matters in a determinate context, mindful of the medieval Scholastics' logical distinction between different 'terms of intention' (Jacob 2003; Russell 1946:463–4); elaborated on subsequently in debates pitting empiricists against idealists. The 'terms of first intention' refer to specific tangible examples of phenomena, someone/thing that we can see, touch, hear, smell etc. – such as a person in a room reading this book – to which we can refer in our discussions and compare our ideas, whatever our socio-cultural background (assuming that all humans have the same senses). The 'terms of second intention', on the other hand, concern culturally relative abstractions – such as the idea of society for a group of socially interacting human beings – for which we have no empirical sense-derived evidence and which as intangible ideas afford no concrete referent (I cannot put my idea of English society in a room to investigate it further).

Second-order intellectual comprehension concerns abstract categories that are particularly difficult to access, translate and correlate cross-culturally, relating to concepts that are socio-historically informed and subject to wide cultural-linguistic variation. Yet discussions of contentious 'know-nothing' tacit or experiential issues in the social sciences, currently applying so-called practice or agent theories, depend on such elusive second-order abstractions in enquiring into the taken-for-granted values, responsibilities and obligations that inform the behaviour of persons socialized to hold these in common. The aim in this book is to direct discussion onto tangible first-order things, using the study of material constructions to further understanding of intellectual constructions, on the grounds that it affords us an opportunity to investigate more securely the significant implicit dimensions of human behaviour. The dependence of much materially related knowledge, like other domains of knowledge, on tacit understandings helps pose several questions regarding what we can aspire to know. We can rigorously examine the material aspects of life, in the empiricist tradition, which involve phenomena 'out there' against which we can test the veracity of arguments; unlike ideological issues, in the metaphysical tradition, favoured by many social theorists.

The individuality of experiential knowledge further affords an opportunity to enquire substantively into the variability that characterizes human behaviour. The social sciences are only now facing up to the implications of individual behavioural variation with the move from a structural to a processual focus, challenging assumptions about homogenous normative codes guiding social interaction, which have persons in the same social group sharing identical concepts, values etc. This is pertinent for the individual perspective that I controversially argue is prominent in Was valley behaviour, albeit set within a matrix of quite different, even opposed, cultural assumptions to those of Western individualism. The wider implications of the individual focus I follow up further here in respect of personal experience, which includes the implicit contexts of involuntary 'emotions' that are central to human social responses.

This materially focused discussion of tacit knowledge also affords an opportunity to illustrate and discuss further the connotations of what I have cheekily dubbed ethno-determinism, drawing not only on the classic ethnographic approach to enquiries but also currently in-vogue phenomenological philosophy. The ethno-deterministic perspective prioritizes lines of enquiry that emerge from the ethnography and places emphasis on evidence, circumscribing interpretations-cum-explanations of behaviour that depend overly on anthropologists' imaginations. It does not deny that we inevitably deploy a certain explanatory or interpretative framework to guide selection of what to include in any enquiry and how to make sense of it; but it advocates theoretical heterogeneity and flexibility in this respect, seeking to accommodate the idiosyncrasies of various cultural ways and uncertainties of social life. The study of construction activities affords what at first may appear an improbable opportunity to explore the philosophical underpinnings of the ethno-determinist approach, showing that it is serious about both ethnography and theory-cum-ideology, in addressing deep intellectual questions in incisive ways from different cultural directions. In short, the book is not just about building houses but also about building understanding of the limits of our comprehension of other ways of being-in-the-world, or to use the currently popular idiom, other ontologies.

On another tack, the book is a further contribution to my study of an 'economy' in the New Guinea highlands, so far as that category applies to subsistence livelihoods situated within a stateless exchange-oriented polity (Sillitoe 2010a). Being a study of property, it concerns the economy minimally defined, as activity relating to the material means of existence. But these peoples' property relations illustrate the incongruity of our economic assumptions, as they have no idea of private and public property but rather one of restricted common property. The builders of a house, for example, may construct it on almost any land in locales where they have active social relations with others, and while they enjoy prior rights of residence, other relatives or friends may live there too (so long as they are on good terms with other occupants). And the residents' have priority rights to occupy the site so long as they continue to maintain a house there. Similarly, labour arrangements queer the relevance of economic assumptions with a sexual division of labour structuring activities such as house building in a way that subverts commercial approaches to work, whether capitalist or socialist. These arrangements are integral to the independence of households, which as autarkic productive bodies render peripheral mercantile considerations of the sort that elsewhere afford hierarchical control of production and transaction to some actors, who use the ensuing power to govern the lives of others – the antithesis of Wola egalitarian relations and acephalous order.

This study of fixed capital assets is a companion to my earlier book *Made in Niugini* (Sillitoe 1988a) that deals with movable capital. The intention initially was that it would be a part of that volume, which, with its focus on artefacts, astutely foresaw a renewal of interest in material culture studies, although it was wide of the mark, until recently anyway, in assuming that this would include a focus on technology. This book continues to argue for attention to technology – in the context, for instance, of current interests in so-called materiality – and seeks to show what we can achieve by situating in-depth technological enquiry within a wide socio-cultural context. In this way it delves into topical issues pertaining to material issues, which are central to the domain marked out as the 'economy', and feature prominently in topical debates about the sociology of knowledge.

The book draws on archaeology, which has long had an interest in technological issues. It explores in greater depth relations between anthropology and archaeology, mentioned tentatively in my earlier volume on artefacts, and how their integrated study of technological and material issues may advance understanding of tacit aspects of knowledge, however improbable at first sight. While archaeologists have expended much effort in understanding buildings – the remains, largely foundations, of which comprise a significant part of the evidence of past lives – anthropologists have not taken much interest in them, which is strange as the building of shelters of various kinds, from igloos to skyscrapers, is one of the few universal cultural features of humankind. The gap is plainly evident, for instance, in New Guinea, where information on construction activities such as house building is scarce relative to other topics, such as exchange institutions, kin relationships and religious beliefs. It has fallen to vernacular architecture to fill the gap in part, albeit it approaches the many different construction practices and designs around the world by and large from a narrow architectural perspective. In this context, the book is a contribution to the local architecture literature, seeking to show how situating the study of construction activities with respect to wider issues, socially, culturally and intellectually, can further understanding.

Among the last paragraphs that the author of a book usually writes are those acknowledging the help of others. So, last but certainly by no means least, my heartfelt thanks to all those who have helped me in many and various ways. During my New Guinea odyssey, I have been lucky in having several persons generously assist me materially and intellectually. The list is long, as you might anticipate for a research project extending over forty plus years. I have thanked many individuals and institutions for their support in several previous publications, where their help was particularly central, and I shall avoid repeating myself here and producing a particularly long list, especially as I am likely to forget some persons and potentially cause offence. So anyone looking at this book who can see the reflection of some information or idea passed to me: thank you for your support and inspiration.

While it is invidious to select some persons from this very long list, there are a few I have to thank in the context of this current work. There are my Was valley friends, Mayka

Kot (sadly killed in an accident), Puwgael Piyp, Wenja Waebis and Maenget Zim, who built the house at Honael that features centrally in this book: my thanks to them for putting up with my constant observation of their activities and answering my sometimes to them naive questions. And all the other friends, who I name at appropriate junctures in the book, who also lent a hand with the Honael house and those who contributed, sometimes inadvertently, to my understanding of construction activities generally. Both Barry Craig and Tony Crawford have given valuable help with some images. I am grateful to several anonymous referees for helpful suggestions and the editorial staff of the Royal Anthropological Institute's newly restored book series for their assistance and for agreeing to take on a large manuscript at a time when presses increasingly eschew publishing such a work, which one commissioning editor described as a 'brick'. I owe a special thanks to Sean Kingston for his attention to detail and careful production of the book. Also, I have, as always, to thank Jackie, whose support and forbearance has been central to the success of these Was valley research endeavours, and who has always been ready to help me make sense of the ethnographic experience and data, and to encourage me when weary of what has turned out to be a far larger project than either of us ever imagined when I embarked on it.

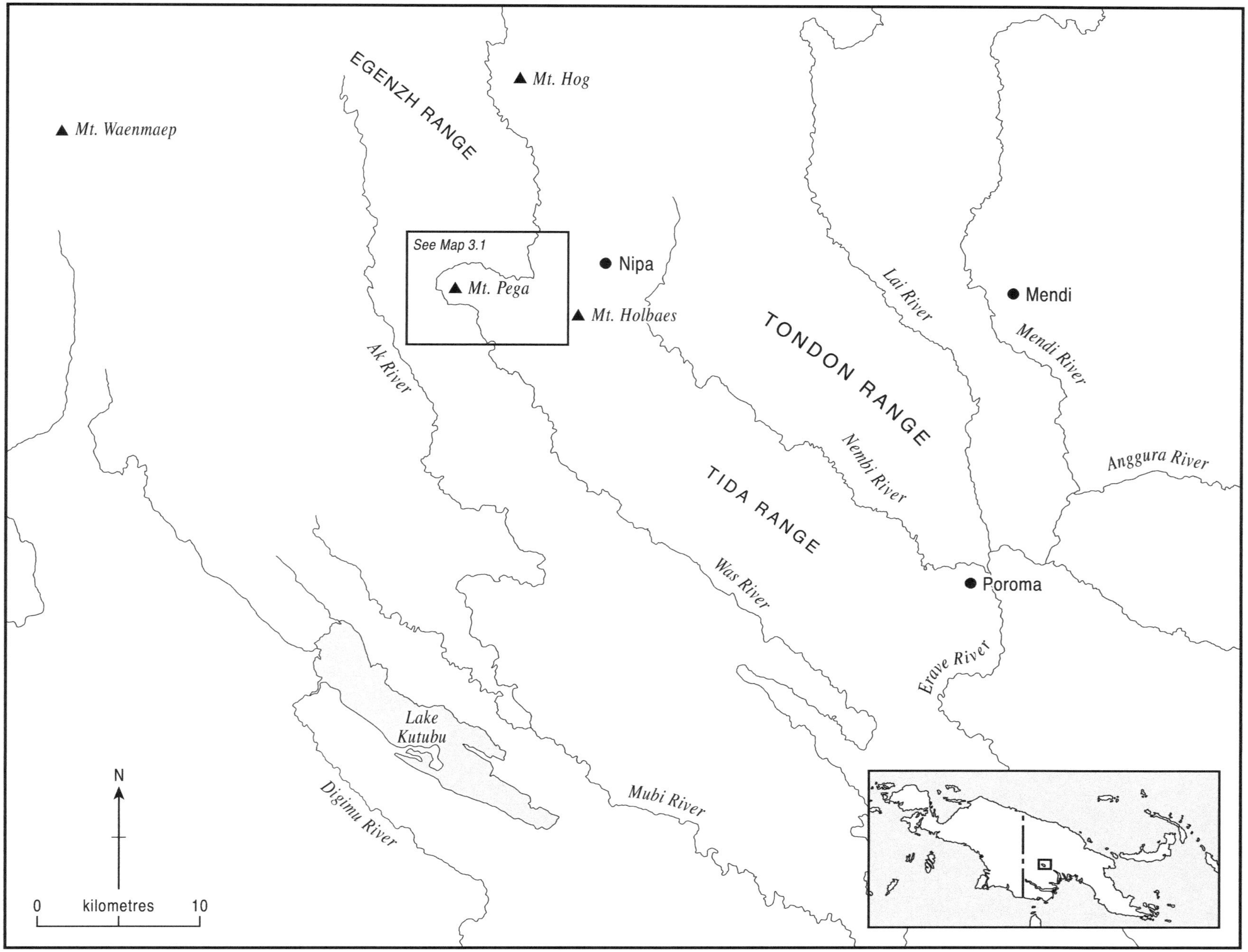

Map 1.1 The Wola speaking region.

1

Whys and wherefores of construction

What has menstruation to do with house construction? This question will doubtless strike any architects and builders, and probably most other readers too, as irrelevant, if not a downright eccentric way to open a book on building practices, even in such an 'exotic' place as the New Guinea mountains. It is not, I confess, a question that would have ever occurred to me before I knew my Wola friends. Yet it illustrates how taking our cue from the ethnography can take us off in unexpected directions, and also intimates how seeking an answer to such an unexpected question may take us deep into cultural issues framed according to locally relevant idioms.

We can approach the question in three ways. Firstly, we can ask those who associate menstruation with house construction about the connection. Persons living in the Was valley will explain that they build houses with unconnected rooms designated for men and women, or entirely separate men's and women's houses (sometimes only metres between them in the same house-yard enclosure, other times at some distance apart), to ensure men's well-being. They may go on to point out further that they believe that women can pollute men and make them sick, especially at certain times in their monthly cycle, and that these housing arrangements are a precautionary measure to protect men from 'women's poison'.

Secondly, we may observe aspects of associated behaviour for which the actors are seemingly unable to offer an explanation. We all do things without apparently thinking about them, using tacit knowledge, a topic that this book returns to repeatedly. If you ask people, for instance, why women's dwellings differ in design from men's, they will likely look puzzled and may reply that it is 'what our ancestors did'. Further enquiries along these lines, such as why they only construct pens for pigs in women's houses or quarters, and not in men's, will likely elicit similar perplexed looks and 'it's our custom' sort of responses. They may comment further that keeping pigs is women's responsibility, which it is largely, although men also contribute to pig herding (Sillitoe 2003:313–14).

Thirdly, we can draw on our intellectual precepts, which inform our being-in-the-world, for answers regarding why people say and do what they tell us and we observe. A question that leaps out, for instance, is why people believe that the natural emissions of one sex, which are patently non-toxic, are poisonous to the other, and what the implications are of this. Another, is why pig keeping should be women's responsibility, a question that locally brings forth repetition of the 'it's what our ancestors did' answer. These arrangements are an aspect of the sexual division of labour that assigns responsibility for various tasks to women and men. A separation underscored by men's horror of menses. The division of labour, integral to production arrangements, leads onto political-economic issues, with the gendered structuring of labour arguably being a key feature of the acephalous politics that persons live rather than reflect on (Sillitoe 2010a:353–4).

The series of interrelated questions sketched here illustrates how starting from the ethnography, rather than from some theoretical postulate, can result in the linking together of a disparate range of cultural features, extending from house architecture to stateless politics, which we are unlikely otherwise to connect. While it may take us off in unsuspected directions, these are locally relevant culturally, originating from autochthonous perspectives. And the approach adheres to the principle of holism, an issue returned to below. It can lead us in novel directions regarding social change too, the rapidity of which in the New Guinea highlands present particular challenges, for example with regards to identity, also explored further later. Over recent decades, for instance, attitudes to residence have gradually changed, resulting in more mixed dwellings, as a new generation, which knows that people elsewhere cohabit without apparent ill health, questions the dangers women supposedly pose to men – albeit many men remain wary of close contact with women, especially at their 'moon' times. This signals changes in gender relations that are occurring in parallel with wider socio-political change.

Ethno-determinism: holism to hermeneutics

Ethnography, not theory, drives enquiries, which is a tenet of the approach that I have dubbed ethnographic-determinism, abbreviated here to ethno-determinism. The term started out as a nose-thumbing gesture, after receiving referees' comments declaring that I needed to adhere to some theory or other (or as I prefer, ideological viewpoint – Sillitoe and Bicker 2004), preferably one they favoured of the moment. Some even declared my work atheoretical, a ridiculous charge because some view of the world informs any structured piece of writing, as I have pointed out (Sillitoe 2003:335–6), even in deciding what to record and describe let alone how to explain or interpret it. Irritation at the wide use and abuse of the term 'ethnography' currently across the social sciences furthered the irony intended in suggesting disparaged deterministic proclivities.[1]

Subsequently I found others responding positively to mention of ethno-determinism, somewhat to my surprise initially, even embarrassment; such as when I used the term with students, intending to lighten discussion. Following a postgraduate conference, for instance, where I was asked what advice I should give new researchers, several participants told me how useful they found my ethno-determinist-informed comments to follow where the ethnography leads and not worry about fitting it up to some theory, and wished to know more about the approach. It seemed that they were inadvertently joshing me in return, by appearing to take seriously what I was saying about being a determinist. After all, in illustrating ethno-determinist philosophy to them, I had light-heartedly recalled: 'If you don't know where a road leads, it sure as hell will take you there.' (Leo Rosten, cited by Koestler 1967:151). In part, I suspect that the approach arouses interest because it authorizes you not to worry overly much about understanding 'theories' – or, as I say, 'ideologies' – and their associated baffling jargon. Many students find these daunting, sometimes even impenetrable, and think they are somehow inadequate when struggling to get their work to square with one theory/ideology or another, preferably today's hot ideas, which in these pluralistic times comprise a bewildering array of alternatives, whereas previously we only had to conquer and apply the theory of the day, such as structuralism, Marxism or whatever, plus a smattering of previous theories sufficient to put them down.

The unexpected interest prompted me to consider more closely the implications of advocating this approach, with some surprising outcomes, notably seeing parallels with certain strands of phenomenology, as discussed in this book. It is somewhat ironic that intellectual endorsement of ethno-determinism, with its eschewal of the *théorie du jour*, should come from phenomenology, a philosophical approach that currently enjoys widespread favour in the social sciences. The irony is perhaps a justifiable upshot of ethno-determinism's sardonic origins. Indeed, irony permeates the approach. While it may come as a relief to students to not have to engage with theories in the way they suppose, they frequently have problems managing the contradictions associated with striving to work in the deductively constrained inductive-ethno-determinist way.

Although we can envisage working according to the previously mentioned three ways, the implication is not that we follow them sequentially as if steps. Extending outwards recursively from what people say and do, the approach seeks to further our understanding of issues from a local perspective, perhaps taking us in unexpected but culturally relevant directions: with house design taking us off on a uniquely Was-valley trajectory regarding perspectives on gender relations. But, as postmodernism stresses, what we think we already know, which is heavily culturally conditioned by our upbringing in a particular social tradition (here heightened by extended periods of formal education), inevitably intrudes deductively into our understanding of whatever we hear actors say and see them do. From the start, our experience and knowledge inform what we take an interest in (such as the gendered aspects of house design), which may also be the result of serendipity – some chance experience in the field or our intellectual inclinations informed by theoretical (e.g. feminist) interests. And furthermore, what we subsequently hear and encounter in the field informs and modifies our prior understandings. There is an unavoidable circularity. The aim is to strive to work inductively so far as possible, while acknowledging the limitations of doing so and seeking to control distortions where possible.

Another paradox of the ethnographic approach is adherence to the principle of holism, which, although little discussed, most anthropologists implicitly accept canonically, while overlooking its inconsistencies (Bubandt and Otto 2010). The approach requires us to frame whatever we investigate within wider socio-cultural context, on the grounds that we cannot foretell what issues may condition whatever it is that we think we know and experience – such as menstruation informing house design in the Was valley. The challenge of holistic practice, employing the 'logic of the all-over' (Ingold 2011b:226), is that it implies we have to know everything before we can really know anything (Sillitoe 1983:xiv–xv), which renders it an unachievable aspiration.[2] In seeking to balance between

1 The reaction of a reviewer of this book confirmed my provocative intentions, criticizing the term as 'clumsy' and suggesting 'emic-ethnography' instead.

2 Ingold (2011b:221–6) seems to think that it is achievable but criticizes this holistic approach for its 'conception of part-whole relations ... equated with totalisation', which results in 'integrative, all-over accounts that, in their very completion, establish a separation between ourselves .... and the others'. Instead of such 'studies *of* people' he advocates 'studies *with* people' (albeit with no reference to participatory methods and their implications – Sillitoe 2012b). The implication is open-ended engagement in peoples' lives, practising a 'holism that is anti-compositional, fluid,

such contradictory forces, ethno-determinism clearly involves more than ethnographic description, whether fashionably 'thick' (Geertz 1993 – after positivist philosopher Ryle) or otherwise. Although description plays a part, particularly during fieldwork – together with genealogical and survey information, historical narrative, quantitative data, indigenous rationalizations and so on – subsequent analysis of patterns and problems suggested by these data (albeit structured by our intellectual understanding of the world) entails far more than ethnographic narrative and interpretation.

While we may first encounter such perplexing issues as the holistic paradox as students, these are hoary intellectual teasers, which are discernible from the time when interest was first taken in communities elsewhere and ethnography became a major focus, from salvage to present day. The need to affirm an ethnographic *fons et origo*, such as featured centrally at the advent of the discipline, further adds to ethno-determinism's sense of irony (Sillitoe 2010a:4–5). And I am not alone in thinking it necessary to reassert the principles of such pioneering enquiries. Other approaches that argue for evidence not theory to drive research include 'grounded theory'[3] (Glaser and Strauss 1968) and recently 'causal analysis'[4] (Vayda 2009). The current focus on materiality, picked up later, somewhat supports the argument that lines of enquiry should emerge from the ethnography, demanding adequate evidence to limit overly imaginative, distorting explanations or interpretations of behaviour. In the context of his materially focused actor-network formulation, Latour (1991:129), for instance, maintains that, 'If we display a socio-technical network ... we have no need to look for additional causes. The explanation emerges once the description is saturated' or sufficiently thick. But his eschewal of any explanation or interpretation, which he thinks prevents 'Moderns' (i.e. Westerners) from accessing the ontology of 'preModerns' – and advocacy of 'saturated' description of the perpetual change that characterizes a 'collective' – is excessive from an ethno-deterministic perspective, which maintains that any description implies a certain explanatory or interpretative framework to make sense and guide selection of what to include; albeit it licenses radical accounts in seeking to accommodate the indeterminateness, even messiness of varying cultural ways of being.

The contradictions encountered in holistically framed enquiries date back considerably further and relate to classical philosophical questions, to which there appear to be no definitive answers, and which recur at intervals dressed up in new jargon that should not mislead us into thinking that they themselves are new. They bring to mind an experienced Manchester colleague's quip that there are only a few original ideas and our ancestors thought most of them while proverbial cave dwellers; since when all we have is their recycling through the generations, pondered over using differing lexicons in changing historical contexts. The comment perplexed me as a young fresh-faced academic but as I have in my turn achieved the status of experienced colleague, so I have come increasingly to appreciate the import of his admittedly somewhat deliberately exaggerated and provocative comment.

Medieval hermeneutics, for instance, invokes a principle that parallels the tenet of holism. In interpreting the Bible and ancient texts, scholars maintained that understanding the parts depended on understanding the whole, and vice versa. It gave rise to the idea of the hermeneutic circle, a process involving continual reciprocity between whole and parts. It is a concept to which a Melbourne social philosopher colleague introduced me, who, defending a seminar that I gave from criticism for teleology, argued that it is not possible to break out of the hermeneutic circle, which had me scurrying off for a dictionary to figure out what the defence was exactly. Hermeneutics underline the inherent circularity of understanding, arguing that we can only comprehend any text by having a tacit foreknowledge that alerts us to salient features that would otherwise escape our attention; albeit reading a text we may manage to pass beyond this backdrop of predetermined interpretative assumptions.

Subsequently, in the mid-nineteenth century, the circular perspective acquired a decidedly contemporary twist when Schleiermacher (1998) extended hermeneutic assumptions to linguistic meaning universally, giving a modern commentary on cultural relativity. He argued, long before today's postmodernists, that we cannot presume to understand others without closely scrutinizing our own hermeneutically held prejudices and remaining open to the possibility that what may on first acquaintance seem cogent from our position in the world, may be something quite unfamiliar; albeit such strict hermeneutic practice, he points out, cannot guarantee an entirely full understanding

---

processual and improvisatory'; a manifesto for vague unstructured enquiries bobbing along in the current of everyday life somewhere, at best it seems a version of the extended-case method or a diary of contemporary affairs that will subsequently become history.

3 The grounded theory of Glaser and Strauss (1968) derives inductively from the study of the phenomena represented. It does not begin with a theory, deploying it deductively to demonstrate some point. While ethno-determinism agrees with allowing what is relevant to the area of study to emerge rather than imposing it theoretically in advance, it does not necessarily agree with grounded theory's series of research protocols.

4 The causal analysis of Vayda (2009) argues for a pragmatic explanatory focused approach that investigates both human actions and environmental changes causally. It is critical of much recent scholarship for its indifference to evidence and methodology, vigorously opposing the idea that we can further understanding by deploying theories rather than seeking evidence for the causal forces behind actual actions and events.

but it keeps us alive to the implications of filtering another's communication through our own culturally informed experience. This applies not only to interaction with those who use another language but also with those of our own, because while speakers of the same tongue share vocabulary and grammar, they may use these symbolic resources in individually varied ways to express somewhat diverging views of the world, which is the premise that underpins current narrative analysis.

Later, in the mid-twentieth century, the phenomenological tradition extended the scope of hermeneutics further to a truly holistic perspective encompassing wider cultural and cognitive issues. According to Heidegger (1962), hermeneutics concern ontology, the nature of being in the world. An aspect of *Dasein* 'existence', they involve the placing of the actor with respect to understanding whatever occurs. A product of our socio-cultural conditioning and experiences, such being may be a pre-reflective condition that does not necessarily involve consciously considering interactions between persons. It is originally tacit, depending on experiential know-how not theoretical cogitation, an issue that is central to this book. We do not consciously reflect on everyday objects, for instance, such as a tap and a sink before running water. The way we are in the world, which is intuitively familiar, presents them to us. It is the holistic context of which actors are not consciously aware: it just is, which complicates matters further, suggesting again that holism is an unattainable aspiration, featuring inaccessible subconscious processes. In referring to hermeneutics in this phenomenological sense, I am using the term in a different, wider sense than in 'interpretative anthropology' that sees cultures as texts, comprising several layers of meanings to be interpreted in the manner of literature – which is to use hermeneutic in its original theological context regarding the interpretation of religious scripture. The lifeways of people only become texts when written down by social scientists, for instance, which are then fair game for critical literature analysis – as happened with the writing culture critique of the late-twentieth century – but before this, holistic-hermeneutic fieldwork is necessary to inform the writing.

Why bother?

The ethno-determinist commitment to holism partly accounts for the time it has taken to produce this companion study to *Made in Niugini* (Sillitoe 1988a), which marks the long overdue completion of my Was valley material-life enquiries, exploring the construction activities omitted from that work. I initially envisioned it as part of that book on material culture, in which I intended to deal with immovable property or fixed structures together with moveable property or portable artefacts. But the original inclusive plan came to nought as the British Museum, which commissioned *Made in Niugini* to complement a collection of Was valley artefacts made for the Museum, was only interested in the documentation of the moveable property that comprised the collection.[5] Anyway, to have included immovable property, as I had planned, would have made an already large book enormous and tried the publisher's patience to breaking point, as was evident by the expression on the then Keeper of Ethnography's face when I delivered the manuscript, overtopping a box that had held four reams of photocopier paper (McLeod 1993:73).

Another reason for tardiness in completing this study has been a nagging question, put to me sometimes about works such as *Made in Niugini*: 'who cares?' Will anyone be interested in a sequel on house building and other construction activities in the remote Was valley of the New Guinea highlands, no matter how ethnographically rich and intellectually challenging? The Was (or Wage) being one of five river valleys occupied by Wola speakers[6] in the Southern Highlands Province of Papua New Guinea, north-east of Lake Kutubu, where the majority of the population live between 1,600 and 2,000 metres a.s.l., from the Ak (or Augu) valley in the west to the Mend (or Mendi) valley in the east (Map 1.1).

The paucity of information on construction in the Melanesian region relative to other topics, such as 'ceremonial' exchange activities, may justify this work in the eyes of some, as an attempt to fill a gap in the ethnographic record.[7] Although paradoxically, it may nullify it for others, as proof that it is a humdrum topic and hence understandable gap, which it is absurd to bother to fill, regardless that building shelters of one sort or another is one of humanity's few universal activities. The rejection of papers based on my fixed structures research by journal editors who thought that it lacked scholarly merit exacerbated these doubts, combined with concerns expressed by some colleagues for the 'REFerability' of my

5 I did not manage to bring a Wola house back to London. Although I did mull over the possibility, a lack of resources prevented it!

6 They speak a language of the West-Central Family (part of the extensive East New Guinea Highlands Stock). See Sillitoe 1979a:xi–xiv and 2010b for further linguistic information and also the orthography used here. (Note: one sound omitted from the orthography is 'au', as pronounced in English 'cow' or Pidgin 'kaukau', and in the Wola *ninau* – 'my').

7 Among the few works that consider architecture and construction of buildings in the Melanesian region see: Bowden 1990; Clarke 1971:100–6; Coiffier 1990; Coudart 1994; Crawford 1981:169–73; Haddon 1912; Hauser-Schäublin 1990; Heider 1970:252–71; Landtman 1920; Lemonnier 2012; Loupis, Ruff and Ruff 1982; Malinowski 1935:240–72; Meggitt 1957; Oliver 1997 (vol. 2):1172–90; Pospisil 1963:257–66; Rensel and Rodman 1997; Saini 1965; Speiser 1996:99–114; Young 1993.

work.[8] This disappointing reception may be due not only to being out-of-step with the times, in eschewing today's theoretical fads in favour of ethnographic integrity, but also perplexingly a cost of subscribing to the doctrine of holism. It results in a corpus that demands a similar commitment to the holistic principle on the part of readers, who have to contend with a considerable body of material if they are going to engage with it at all – that is, connect with it through necessary in-depth awareness of socio-cultural context – which is unpopular in this age of butterfly-like expectations to twit or flit about rather than, beaver-like, building understanding in holistic depth; for it is easier to subscribe to the idea of holism in word than practice.

In responding to the 'why bother' question, one has to consider justifying one's work not only to colleagues, but also laypersons, which is one of the political reasons for the imposition of assessments such as the REF on university research. It focuses the mind, I tell students, who also sometimes ask sharp questions about the relevance of research, to take what I call the taxi test, where I imagine during a short cab trip having to defend the spending of some of the cabby's taxes on my work. One reason for doing so, as I argue in *Made in Niugini*, is that material-culture research, relatively ignored in anthropology for much of the twentieth century, has some lessons for us, despite our technological capabilities – the most advanced in humankind's history, witness motor vehicles such as taxi cabs. The fixed structures of the Was valley, like material possessions generally, are modest, yet effective, and advancing our understanding of such constructions adds not only to the sum of our knowledge academically, but can also illuminate practically current and historical issues. Such non-industrial technologies may, for instance, impact lightly on the natural environment, and teach us something regarding sustainable practices, as discussed in subsequent chapters, particularly with increasing concerns over the, evermore evident, globally deleterious environmental effects of industrial technology, such as the contribution of taxi-cab emissions and high-rise buildings to climate change.

It turns out that *Made in Niugini* was prescient in arguing that the time was right for a resurgence of interest in material culture studies, although it was askew in thinking that it would feature among others a focus on technology, until lately anyway. A structuralist-cum-post-structuralist agenda has prevailed instead, focusing on consumption, as an aspect of personhood and identity, and museum collections, which include innovative moves towards local participation in interpretation and display. This book seeks to further this renewed interest in material issues, by again promoting a technology focus and showing what we might achieve in situating the study of construction activities with respect to wider issues, ethnographically and intellectually. It understands technology to comprise culturally patterned knowledge that informs the conversion of materials into desired things, here constructing houses. Technical know-how and skills include familiarity with raw materials and practical processes, linguistically encoded or not, necessary to the making of things with particular uses. It includes task-based and conceptual knowledge acquired through spoken (and written) word and observation, which may be differentially distributed in a community along gender, age, class and other lines.

Furthermore, we might anticipate that archaeologists 'care about' the resurgence of interest in material culture, although the prediction in *Made in Niugini* that they would increasingly take custody of, and employ comparatively, the ethnographic record on communities such as that of the Was valley (with anthropologists' declining interest in their achievements -- exacerbated by the thinness of concerns with material issues), has proved somewhat wide of the mark, to date anyway, as also discussed in subsequent chapters. Nonetheless, archaeologists show an encouraging renewal of interest in technological issues, for

> a key issue in recent discussions ... has been about whether the focus of material culture studies should be on objects as 'end products' and their consumption or on the processes responsible for the formation of things and objects ... [it] calls into question this deeply embedded analytical divide and offers the building blocks for a cross-disciplinary unifying framework. This framework ascribes no analytic primacy and, when possible, blends production and consumption.
>
> (Malafouris and Renfrew 2010:9)

Moreover, this book is a further contribution, one of a series, to my long-term study of the 'economy' of the Was valley, which also partly accounts for the time that it has taken to appear. An ever-increasing burden of administration is to blame too, which is partly the result of career progression, growing with seniority, and of a burgeoning academic bureaucracy, much of which is unhelpful to scholarship. The series of books on the 'economy' grew out of my outrageously ambitious Ph.D. plans, which illustrate what adherence to the tenet of holism can result in, demanding, as it turns out, a large part of my career. The scare quotes around 'economy', by the way, indicate the ambiguity of the category regarding Was valley arrangements. It is defined minimally as activity relating to the material means of existence (Sillitoe 2010a:11–14). This definition encompasses the construction activities discussed here, while avoiding further engagement with the tangled debate about the relevance or not of economic

8 The REF 'Research Excellence Framework' (a revised version of the RAE 'Research Assessment Exercise') is the UK government's process for assessing university research and funding.

thought to the material conditions of Was valley life (for which see Sillitoe 2010a).

Further to the 'who cares' question, this book delves into issues pertaining to materiality, which are central to the domain marked out for investigation as the 'economy'. We can investigate the material aspects of life with a certain rigour, in the empiricist tradition, as they concern phenomena 'out there' against which we can assess arguments; unlike ideological issues of a metaphysical kind favoured by many social theorists.[9] But the dependence of much materially related knowledge on experiential understandings poses several questions regarding what we can aspire to know about it, drawing largely on philosophical ideas associated with the phenomenological tradition that problematize our assumptions. These tacit issues concern social behaviour generally too. The individuality of experiential knowledge puts the analytical spotlight on the individual, which is a key focus in economic thought with its assumptions of self-interest motivating human behaviour. It is also prominent in Was valley behaviour, albeit set within a matrix of completely different, even opposed, cultural assumptions to those of market economics, as I controversially argue (Sillitoe 2010a:30–44). This book follows through further on the wider implications of focusing on the individual, and idea that the 'emotions', or David Hume's (1896) 'passions', are central to human social interaction.[10] It draws on the Scottish philosopher's empirically anchored sceptical argument in delving into the knotty topic of tacit knowing. In this way, the investigation of construction activities affords what at first may appear an unlikely opportunity, to explore the philosophical underpinnings of the ethno-determinist approach and show that it is not only heavy on ethnography but also not light on theory-cum-ideology, addressing profound intellectual questions in insightful ways from novel cultural directions.

Was valley political-economy

The Was valley 'economy' is embedded in a stateless order, with intriguing political-economic implications. The engine of the 'economy' is subsistence farming, with livelihoods depending largely on swidden cultivation, which supplies the population with most of its food; hunting and gathering contributing little (Sillitoe 2003). The diet largely comprises vegetables, with sweet potato the staple, typically cultivated in composted mounds; other crops include various cucurbits, bananas, taro and green leafy plants. People are shifting and fallow horticulturalists, farming sites for varying periods before abandonment. They clear and cultivate some once or twice, before abandoning to tree and cane regrowth for many years to restore productivity, while they maintain others semi-permanently, managing fertility through composting and occasional brief grassy-herbaceous fallows (Sillitoe 1996). This results in two kinds of garden: those cleared and planted once or twice with a wide variety of crops, and those cultivated repeatedly, sometimes for decades, which support fewer crops, largely sweet potato. Compost and soil tillage feature in the latter, together with drainage control using mounded beds. Land cultivation is labour intensive, employing hand tools, such as axes and digging sticks. It features a sexual division of the labour, which as noted in the opening, is an aspect of the gender separation that characterizes Wola life: men and women observing various conventions to keep them apart, as in occupying separate houses. The initial heavy axe work, clearing vegetation and enclosing sites, falls to men. After the burning of vegetation when dry, women take over routine cultivation, tilling the soil, planting and harvesting many crops.

A similar division of labour characterizes activities with animals: hunting is generally a male domain and pig herding a female one. These arrangements are a significant aspect of gender relations and are central to their understanding. The Wola, like other Highlanders, keep pig herds of considerable size (Boyd 1984; Feacham 1973; Hide 1981; Pospisil 1963; Rappaport 1968). The work involved in herding animals seems ostensibly to be large, with them consuming about half the sweet potato tubers harvested (Sillitoe 1983:228–30), although it is not overly onerous. The daily routine is for women to release animals in their charge of a morning, to forage during the day in nearby forest and grassland;[11] unless they are unruly, when they may keep them penned or tethered up. The animals are conditioned to return to homesteads in the evening, when they are fed their tuber ration, and spend the night in stalls, traditionally located in women's houses, although today many are kept in separate lean-to shelters in response to warnings that living with pigs is unhealthy.

Access to garden land underwrites social arrangements, with the land-tenure system determining the areas where families can claim rights, through either consanguineal or

9 The ideological value-loaded inflection of theories in the social sciences facilitates aforementioned fashion-driven changes, with new (or resuscitated under a new name) theories every decade or so, unlike the cumulative building of a body of integrated theory in natural science.

10 While I draw on Hume's philosophy for advancing a point of view that helps me make some sense of my Was valley experiences and observations, the idea that the passions (what we feel, which does not equate with today's meaning of overwhelming feelings but refers to personal emotional responses generally) motivate much behaviour has a long history, extending from Aristotelian notions of the passions and the appetites through the *passiones animae* of Augustine and Aquinas to current day thinking on the emotional dimensions of the human psyche.

11 This free-ranging arrangement differs from that found in some more densely populated Highlands regions such as Simbu (Hide 1981:328) and Enga (Feacham 1973:27).

Plate 1.1 Homestead on valley side with garden and casuarina trees.

affinal links to the original, mythically legitimated settlers. The tenure arrangements afford considerable choice, with both kinship and personal circumstances playing a part, such that the resulting local groups have a bilateral complexion, regardless of many men saying they prefer to reside where their father lived (many do so each generation, obliging their wives to move to their place after marriage, husbands knowing the area well and its residents). The resulting territorial groupings of kin, called *sem* 'families', accordingly comprise loosely constituted patrifilially biased bilateral kin corporations (Lederman 1986; Ryan 1959; Sillitoe 1979a). The term *sem* 'family' has a wide scope depending on the context; it may refer to a nuclear or extended family, or larger groups of kin (such as are often glossed clan and lineage using descent idioms). These larger *sem* bodies may be distinguished between according to size and genealogical status, by the suffixes *gənk*[12] 'small' or *onda* 'large'. The Was region comprises many neighbourhoods identified with *semonda* 'large families' (average population *c.*300 persons), named, local groups that result from the territorial aggregation of several related families. They are internally subdivided into *semgənk* 'small families' that structure rights to parcels of land. These 'small families' comprise the descendants of the mythical *semonda* founders who sub-divided territory and resources between them as they grew in numbers over the generations; such fission continuing to occur gradually today with population growth. The *sem* groups exist through generations, have proper names (e.g. Aenda or Mayka), comprise several related families and are ideally exogamous.

Families live in squat houses scattered along the sides of the valley, where their neat cultivations are interspersed with extensive cane grassland and the grassy clearings of fallowed or recently abandoned gardens, the forested watersheds left largely unpopulated. Their homesteads comprise variably composed family groups; ranging from a man and his wife or wives and their children, to three or four related men together with their nuclear families and other relatives (such as unmarried sisters, widowed parents and so forth). The map of house locations and the associated genealogies in Chapter 3 give a detailed snapshot of the spatial pattern of residence and the composition of households at one period of time.[13]

The generic term for houses is *aenda*, and these are built according to a limited number of designs. They are mostly low, smoky structures with bark-lined wooden stake walls and thatched gable roofs. As the chapter's opening comments intimate, the various kinds of house are distinguished primarily by the gender of those who occupy them, reflected in relatively minor differences in construction. There is the *enza* 'men's house' that features an open-fronted foyer and separate sleeping room at the rear, occupied by adult men, frequently two or more relatives and their adolescent sons. The *tenda* 'women's house' is usually a single-roomed structure occupied by a woman and her children; larger structures may be divided into two rooms with two women and their children in residence, sometimes together with other female relatives such as unmarried sisters or widowed mothers. The *enzatenda* 'men and women's house' comprises three rooms, occupied, as the name suggests, by men, women and children, where men and their adolescent sons occupy the front two rooms and women and children the rear one. In contrast, the *momaenda* 'family house' is the same, usually single-roomed, design as the *tenda* 'women's house', but differs in having men, women and children – usually a man, his wife and their offspring – living together in it.

More recently there is the 'modern' style *mbolin* (or *kiapon*) *aenda* 'white persons' house' of introduced design and construction, which may be occupied by a family, while others serve as nominal trade stores. Other structures comprise various sorts of shelters that do not serve as residences, such as *pəgaenda* 'day shelters', often found in gardens, and the *wilaenda* 'men's day house', which is a small stand-alone version of the open-fronted foyer of an *enza* men's house. Other structures include *showmay aenda* 'pig sties', mentioned above, and various ritual buildings, of which the only extant version is another newcomer, the *horaenda* 'church' (*lotu* in Pidgin).

Houses are largely occupied from dusk until the following morning. During the daytime most houses are empty, unless the weather is particularly inclement, when occupants may stay at home, for instance sheltering from heavy rain (people do not let everyday rainfall stop them getting on with their activities, and rain often starts in the afternoon when they are away from home – Sillitoe 1996:54). The only persons who spend considerable periods of the day indoors are the elderly and the sick. And sometimes children may be found playing around homesteads. Other occasions when houses may be occupied in the daytime are during special events, notably socio-political exchanges when wealth may be displayed inside men's day rooms or under house eaves, such as at marriages or funerals. People also use houses as places to store their possessions, often wrapped up in stout leaf parcels, and to cook their meals, either in the central fireplaces or adjacent earth ovens in house yards, mainly in the late afternoon and evening.

The several related families residing in the same region do not recognize any obligations to support the territorial group as a corporation representing their

12 Schwa (ə) has been represented in previous publications of mine on the Wola by a superscript e (^{e}) due to limitations in the character set available.

13 Note: the numbers next to individuals on the genealogies correspond to those by the houses they occupy.

Plate 1.2 *Homestead of men's house (left) and women's houses (right and background).*

Plate 1.3 *An enza men's house.*

Plate 1.4 A tenda women's house.

Plate 1.5 A momaenda family house.

Plate 1.6 A mbolin aenda 'modern' style house.

Plate 1.7 A lean-to style pəgaenda shelter.

Plate 1.8 A showmay aenda pig sty.

collective interests.[14] They do not unite in this way to pursue political ends. The action groups are more akin to temporary coalitions of related individuals who come together to pursue an activity jointly that is in all their interests. It is socio-political exchange institutions, not corporate-clan group obligations, which are fundamental to this stateless, fiercely egalitarian polity, and which afford individuals a degree of political freedom that it is difficult for those of us living in states to appreciate, the degree of sovereignty and equality individuals expect seemingly posing problems for social order (Sillitoe 1979a). Pigs feature in exchanges, not only occasionally supplementing a largely vegetable diet with meat, but also importantly serving as wealth, together with other transactables such as seashells previously and money today. They pass to and fro in the interminable series of exchanges between defined categories of kin that accompany all significant social events, such as marriages and following deaths, sometimes bringing together considerable numbers of people. These transactions continue today as a significant force for ordering relations in the context of weak to non-existent central government authority and lawless 'rascal' activity, with violence possible when persons defend their individual rights. Kin usually support one another in disputes, which can erupt into armed confrontations, with 'tribal wars' continuing as a feature of life.

14 The reader should not take this account of Was valley arrangements as necessarily paradigmatic for the New Guinea Highlands region. There is considerable variation across the region regarding the constitution of local groups, the weight of influential men, intermarriage and the deployment of resources in wealth transactions.

Exchange transactions have a competitive edge. It is an aspect of the feisty egalitarianism that is a central feature of social life, in which men constantly struggle to show their worth and maintain their standing within a community of equals made up of both their competitors and judges. Some individuals excel in exchange at any time. Valuing these activities highly, people accord them high status, using the appellation *ol howma*,[15] 'men of the grass-clearing' for them.[16] Women who regularly manage several pigs achieve reputations as *ten howma* for their admired ability. It is possible that such admired persons may on occasion exert limited influence over decisions reached by a network-coalesced group united for some end. But they do not have the status of leader with any authority to direct the actions of others, as reported in some parts of New Guinea; any attempt to exert control, however subtle, over the actions of others would offend against the ethos of equality prominent in everyday life. Autonomy and equality are isomorphic. No one is beholden to anyone else. Consequently, there is no person in charge when several individuals decide to co-ordinate their actions in some undertaking; they reach decisions and organize themselves by mutual agreement.

It is arguable that if men with influence could use it in a political sense to improve their positions, they would employ it to increase their productive capacity (Sillitoe 2010a). They are not able to, and the implications are notable. While women assume most responsibility for herding pigs, for instance, men take over during public transactions. Some commentators interpret this arrangement as exploitative: women contributing considerably to pig keeping, while men enjoy the success that comes with exchanging the animals at esteemed public events (Josephides 1985:112, 208; Modjeska 1982:95–6). This purportedly controverts egalitarianism. However, I consider it a key aspect of the stateless order, thwarting access to power by control of production, and fundamental to the exchange-focused acephalous political-economy (Sillitoe 2010a:445–53) It also corresponds with attitudes to pig management, which are otherwise perplexing from a formal economic perspective (Sillitoe 2003).

Previously, supernatural beliefs centred on ancestor spirits that maliciously caused illness and death (by 'eating' vital organs), as could enemies' powers as sorcerers and 'poisoners', and various malevolent forest spectres. Sometimes people offered pigs to restrain these nasty supernatural forces (Sillitoe and Sillitoe 2009). The majority today have converted to Christianity and regularly attend mission services, albeit many continue to fear that the old spirit forces still inhabit their region. The station at Nipa in the Nembi valley nominally represents government administration, with police station, school and health centre, and some trade stores, although as noted, national government jurisdiction has all but collapsed locally, with central and provincial government malfeasance and corruption. Although the Highlands Highway runs along a neighbouring valley, the region is marginal regarding development, with little prospect of change presently, potential investors (such as companies that might further exploit the Southern Highlands' gas and oil reserves, on which the local population's hopes of significant material advancement rest) being wary of the failing state (seeing what has happened in the neighbouring new Hela Province to the west and Lake Kutubu region to the south). Other opportunities, such as farming cash crops, are few. Individuals seeking to enter their country's slowly expanding market economy migrate as labour elsewhere, having little chance locally. Today, they largely go to the Western Highlands region and some to cities such as Lae and Port Moresby; previously, a few experienced the brief colonial-era migrant-labour scheme.

Vernacular architecture

The discussion has focused so far on one aspect of the anthropological endeavour, namely the furthering of our understanding of particular socio-cultural orders, here that of Wola speakers who live in the Was valley. The other aspect is the comparative one, which seeks to advance our understanding of humanity by looking at a variety of cultural arrangements – such as those across New Guinea or the Pacific or even the world. It focuses on certain topics with a view to finding patterns, even universals, in socio-cultural arrangements and associated behaviour – such as in kin classification and behaviours, conflicts and their resolution, and so on. The scholarly concerns and questions of the researcher explicitly drive such enquiries, in contrast to enquiries at one locale that offer, as ethno-determinism endorses, the opportunity for culture-bearers to do so. The intellectual bias, though unavoidable – as someone has to set the agenda – nonetheless may open comparative studies to challenge, with the Euro-American academic tradition determining the topics identified and categories employed, whereas from another cultural perspective these terms of reference may be inappropriate for comparing the selected human pursuits.

The construction of dwellings and other structures is not only a pan-human activity – all populations erecting shelters of one kind or another – but it is also a category that appears universal, unlike many other human concepts that show bewildering variety. Furthermore, it is readily observable and reliably documentable, and so suitable for more objective comparative analysis. Yet curiously anthropology

15 The *howma* is a communal clearing – village green equivalent – where many exchange events occur, often featuring pigs. See Sillitoe 1979a:48n.1 for a description of a *howma* clearing.

16 They are the equivalent of big-men described elsewhere (Berndt and Lawrence 1973; Godelier and Strathern 1991; Lederman 1990).

has, until recently, shown relatively little interest in the field, and those anthropologists (e.g. Humphrey 1988) who do draw attention to construction and architecture usually fail – until recently – to follow through. The subject has passed by default to vernacular architecture studies, which is a vibrant sub-field of architecture that has recently enjoyed a resurgence, as evident in burgeoning activity, including publications (e.g. see the online Vernacular Architecture Forum Bibliography[17] at www.vernaculararchitectureforum.org/resources/bibliography.html).

On one level this book is a contribution to this local architecture literature. Vernacular architecture is an open and wide-ranging field that encompasses many different construction practices and ideologies (Asquith and Vellinga 2006; Carter and Cromley 2005; Glassie 2000; Rapoport 1969:4–5; Ryan 2011:1–53). There are several other terms that are largely synonymous with it and indicate the scope of the field, including indigenous, aboriginal and ethnic architecture, sometimes abbreviated to ethnoarchitecture (ethnoarchitecture.org); customary and ancestral architecture; popular, informal or rural architecture; and more colourfully, anonymous or spontaneous architecture, and buildings without architects (May 2010; Rudofsky 1987:58); or more derogatively, primitive and non-pedigree architecture (Guidoni 1978; White 1985). Sometimes also referred to as 'folk architecture', this varied field, concerning the 'ordinary' or 'traditional' built environment, involves contributions from several disciplines, including architecture, building construction, anthropology, archaeology, material culture, human geography, landscape studies, folklore, history and historical-building preservation.[18] The term 'traditional architecture' is used on occasion for buildings that consciously seek to maintain certain architectural practices unchanged through time, for heritage, romantic or aesthetic reasons; associated occasionally with efforts to preserve certain historical structures such as churches and palaces.

Some of these terms suggest that vernacular architecture concerns geographically or historically remote, even non-Western or exotic, building styles, which is incorrect. There are long-established societies in Europe and North America devoted to the study and preservation of local architectural traditions. In England, these range from grand medieval and later buildings to anonymously built country cottages, modest farm outbuildings such as barns, and craft workshops, even factories (Brunskill 2006; Clifton-Taylor 1987). And in the United States there is the 'American vernacular' tradition that studies the gamut of rural to urban buildings (Upton and Vlach 1986). These two regions are better represented, with respect to research and publications, than any others, particularly Oceania, the focus of this book, where as noted our knowledge is sparse, which is a matter of some concern in view of the rapid cultural change that is equally affecting building as other aspects of life (Rensel and Rodman 1997). Indeed, the field of vernacular architecture generally is poorly served regarding research and publications compared to its big brother, architect-designed architecture. This emphasis is hugely disproportionate, given the global significance of vernacular architecture, which accounts for an estimated 90 per cent or more of all buildings worldwide (Rapoport 1969:2).

Vernacular architecture features local designs that have evolved in harmony with the surrounding landscape, climatic conditions and raw materials. It concerns buildings constructed largely, if not exclusively from locally available resources. It follows common and well-tried regional forms and depends on local building customs and techniques as practised at a certain place and time. A review of physical circumstances, such as local ecology and climate, and associated practical matters, such as availability of raw materials and livelihood strategies (Rapoport 1969), shows that, as with other socio-cultural activities, these may constrain but do not determine architectural traditions. They depend equally on culturally informed ideas about basic needs, comfort, privacy and so on, such that populations living in similar environments construct quite different buildings. Vernacular architecture may have culturally specific symbolic significance and sometimes signals a strong ethnic identity. It often features owner-builders, maybe assisted by kin and neighbours. The resulting structures are often modest, built by common people, usually for their everyday use. Some commentators, considering functionality a defining feature, dubiously distinguish vernacular architecture from 'polite architecture', which includes vernacular aspects as aesthetic design features. But many, perhaps most, locally constructed buildings also include decorative elements that serve no functional purpose, albeit vernacular architecture is frequently strongly functional in style to meet inhabitants' needs. According to one authority, a vernacular builder is 'guided by a series of conventions built up in his locality' with the 'function of the building … dominant' and uses 'local materials … as a matter of course' with 'other materials … imported quite exceptionally' (Brunskill 2000:27–8).[19]

17 This searchable bibliographic database covers all aspects of vernacular architecture. It includes all publications listed in the *Vernacular Architecture Forum Newsletter* dating back to 1979. For earlier work, see Hall 1972; Wodehouse 1980; Vance 1983.

18 While the applicability and accuracy of some of these terms is open to question – for instance, if people use imported materials to build houses, are they indigenous or traditional (Bourdier and Al Sayyad 1989:6)? – their general drift is clear enough.

19 Elsewhere, in defining vernacular architecture, Brunskill (2006 [1981]:24) refers to its traditional foundations, its relation to ordinary folk's everyday lives, its connection with particular

While vernacular architecture draws on local building traditions, this does not imply that it is changeless (Rensel and Rodman 1997). It may continue to evolve in response to local environmental, cultural and historical circumstances and needs. Although the building designs and know-how are customary and they largely, though not exclusively, comprise knowledge and skills built up by trial and error, and inherited through the generations, a would-be builder may innovate on them, often as spur-of-the-moment inspirations that usually result in incremental change to traditional ways; in contrast to architects, who consciously plan and seek to innovate and come up with new designs incorporating new materials. Vernacular buildings may feature a mix of traditional and modern styles and materials, and even evidence hybridity of several styles. According to some, the defining feature of the vernacular is that its design and construction features no specialists (May 2010; Rudofsky 1987). Any buildings 'designed by professional architects or commercial builders' do 'not come within the compass of the vernacular' (Oliver 2003:15). While such architecture concerns 'structures made by empirical builders without the intervention of professional architects' (Arboleda 2006:1), that is without need of draughtsman's plans or formal architectural designs, the implication is not that design is absent. Rather, it employs shared design, where all those involved may contribute over time, as an 'architecture of the people, and by the people' (Oliver 2003:14).

Vernacular variation

A cursory glance through any illustrated book on vernacular architecture shows the fascinating diversity of traditions (Duly 1979; Oliver 1997). Every region of the world has its architectural code; the many ways we contrive to provide shelter for ourselves standing as edifices to human cultural variety (Crouch and Johnson 2001).[20] These vary widely from Inuit snow-built igloos to Arab mud-brick houses, English thatched cottages to stilted Malaysian kampong houses, Plains Indian tepees to Bedouin tents, Korowai Papuan tree houses to Shaanxi Chinese cave dwellings. While varying environmental and cultural factors influence the myriad local building traditions, there are regional features that give a certain generality to structural forms (Heath 2009; Noble 2007; Vellinga *et al.* 2007).

places, as reflected in the use of local materials, and its affective associations in addition to functional uses – a list of characteristics used by others (such as Bourdier and Al Sayyed 1989:6).

20 There are many publications on the architecture of various regions and for some idea of the variety see, for instance, Morgan 1881 and Nabokov and Easton 1989 on North America; Nas 2006 and Waterson 1990 on Southeast Asia; Berliner 2003 and Knapp 1990 on China; Hockings 1989 and Jowitt and Shaw 1999 on the Pacific; Elleh 1997 and Oliver 1971 on Africa; Hawker 2008 and Ragette 2003 on Arabia; and Andrews (1999) on Central Asia.

The climate is such an influence (Givoni 1969). Buildings in hot climates are usually open, designed with apertures to facilitate ventilation, and may be built of light airy materials, whereas those in cold climates have fewer or no openings to reduce heat loss and feature heavier thermally insulating materials (although this is not invariable, for such insulating materials combined with small or no windows can protect against heat too, as in parts of North Africa). And in hot regions there are design features to promote cool ventilation – such as the wind towers of Arabia; and to increase air humidity through evaporation – such as Mediterranean courtyard fountains (Fathy 1986). In cold climes we find the reverse, with arrangements to heat buildings ranging from chimneys for indoor fires to elaborate hypocaust underfloor and central heating systems. People who experience marked seasonal changes may alter their dwellings with the seasons. For instance, the Sami of Scandinavia move from tent to house. Pitched roofs are common where rainfall is high, and sometimes dwellings are built on raised platforms to lift them above flood water. Buildings in windy locations are oriented to reduce features presented to prevailing winds, and so on. The availability of local construction materials is another influence. Wood will likely feature where suitable timber occurs; otherwise people may use stone, mud (sometimes as baked-clay bricks) or animal dung and bones. Bamboo is a versatile building material in the tropics, as are palm fronds and various grasses, particularly in thatch.

The lifestyle of users will also influence structures. Some nomadic populations may use disposable materials such as saplings, leaves and snow to make readily erected structures – such as the Australian Aboriginal humpy and Canadian Inuit igloo; while others may use lightweight, reusable and pack-animal-transportable materials including woven coverings and poles to pitch more substantial shelters – such as Mongolian yurts and Bedouin tents. Sedentary populations, on the other hand, may use heavier and more durable materials, including timber and stone to construct permanent buildings, which may demand large labour inputs and represent sizeable capital assets indicative of social status. These trends are not invariable. Settled people may erect readily replaceable buildings – such as Japanese wood frame and paper houses –where severe natural events occur (such as floods, high winds or earthquakes), knowing that whatever structures they erect are unlikely to withstand these forces and in the event of collapse impermanent materials are less likely to result in serious injuries.

Other wholly cultural considerations may contribute to variation in building styles, resulting in different solutions to similar problems (Noble 2007). Architectural works do not have much to say about these issues, as Carsten and Hugh-Jones point out (1995:4), they usually focus 'on the more material aspects of dwellings, including environmental conditions, resource, technology, techniques of construction and types of building' and may discuss 'spatial organisation, symbolism and aesthetic values of buildings, but they often say relatively little about the social organisation of the people who live inside'. It is a gap that anthropological studies fill to some extent (Buchli 2013; Lawrence and Low 1990; Waterson 1990), albeit these may be criticized for the reverse, namely saying relatively little about the material aspects of buildings. And strangely enough there has to date been relatively little interdisciplinary mixing of these approaches (Schefold 1997:8).[21] Patterns of social interaction between occupants, for instance, influence building styles. The organization and size of families – how members interact, prepare and eat food, their sleeping arrangements, who shares what spaces – all influence the size and layout of dwellings. Structures may promote social interaction between inhabitants or the reverse, they may ensure segregation; as seen in the Was valley with separate male and female spaces. Wives in polygynous families may share facilities or they may occupy separate houses in the same compound; again as in the Was valley. Beliefs and values may inform decoration of buildings, particular designs having symbolic import, and the layout of buildings may have meanings too.

History of vernacular architecture

The concept of vernacular architecture emerged explicitly in the nineteenth century, although it is implicit earlier in notions of regional styles. The term first occurs in 1818 (Arboleda 2006:1) and features in the Picturesque, and Arts and Crafts movements in Britain, with their interest in gothic and rural revivals that sought to recreate past architectural styles, and the Stick and Shingle styles in the United States that drew on colonial clapboard architecture. The initial interest was stimulated by the search at the time for a national English architecture, with the Gothic pointed-arch style proposed as appropriate above the Romanesque, which as of foreign Norman origin was not thought vernacular. A prominent proponent was John Ruskin, who expressed strong views on architectural heritage, such as in the following commentary on rural Westmoreland architecture, which illustrates various points about the vernacular movement:

> The clay-slate and greywacke, as it is shattered by frost, and carried down by the torrents, of course forms itself into irregular flattish masses... These stones, thus shaped to his hand, are the most convenient building materials the peasant can obtain. He lays his foundation and strengthens his angles with large masses, filling up the intervals with pieces of more moderate size; and using here and there a little cement to bind the whole together, and to keep the wind from getting through the interstices; but never enough to fill them altogether up, or to render the face of the wall smooth. At intervals for from 4 feet to 6 feet a horizontal line of flat and broad fragments is introduced projecting about a foot from the wall. Whether this is supposed to give strength, I know not; but, as it is invariably covered by luxuriant stonecrop, it is always a delightful object... The roof is rather flat, and covered with heavy fragments of the stone of which the walls are built, originally very loose; but generally cemented by accumulated soil, and bound together by houseleek, moss, and stonecrop: brilliant in colour and singular in appearance... Such are the prevailing peculiarities of the Westmoreland cottage. 'Is this all?' someone will exclaim: 'a hovel, built of what first comes to hand, and the most simple and convenient form; not one thought of architectural beauty ever coming into the builder's head!' Even so, to this illustration of an excellent rule, I wished particularly to direct attention; that the material which nature furnishes, in any given country, and the form which she suggests, will always render the building the most beautiful, because the most appropriate.
>
> (Ruskin 1907:50–3).

An unrelated strand of interest, which emerged in the latter part of the nineteenth century and from which this study descends, concerns travellers' accounts of faraway places. Their book and magazine narratives often included descriptions and sometimes illustrations of buildings from various exotic places to titillate their curious readership. Some took the images of 'rude' dwellings as evidence of the builders' intellectual inferiority and confirmation of popular social evolutionary thought (Rykwert 1981). The argument also conveniently supported the colonial agenda of the time.

During the twentieth century, interest in vernacular architecture was largely kept alive in the intellectual backwater of folklore studies, together with research undertaken during the preservation of ancient buildings, which included some archaeological contributions. For instance, the NGO International Council on Monuments and Sites established a committee in 1976 to encourage research and conservation of vernacular architecture. There was also the occasional obscure anthropological paper, often using some aspect of local architecture as a vehicle to advance current theories, such as mid-century structuralism looking at stylistic features for dualistic symbolic and metaphoric allusions (Bourdieu 1970; Cunningham 1973; Littlejohn 1967).

Towards the end of the century, vernacular architecture started to establish itself as a recognizable field (Oliver 1969; Rapoport 1969), with conferences, such as those of

21 Bourdier and Al Sayyad's (1989) volume is partly an exception, although the contributing architects and anthropologists supply separate chapters.

the International Association for the Study of Traditional Environments begun in 1988 (Bourdier and Al Sayyad 1989), focusing not only on aesthetic issues but also technical and social ones. A turning point for many was the mid-sixties' exhibition *Architecture Without Architects* at the New York Museum of Modern Art, with its copiously illustrated handbook (Rudofsky 1987). It sought to promote vernacular buildings of the world to the status of classical architecture, and although it gives a rose-tinted view of life and buildings elsewhere, it was influential. The arrival of vernacular architecture as a serious subject was beyond doubt by the close of the century with the publication of the monumental *Encyclopedia of Vernacular Architecture of the World* (Oliver 1997). Research today encompasses a range of issues beyond straightforward documentation of vernacular building styles, such as their role in the reinvention of tradition, in identity construction, including ethnicity, heritage and tourism, and in environmental and social sustainability.

Architects and vernacular architecture

Parallel trends are evident in the recent history of the architectural profession (Glassie 2000). While awareness of vernacular architecture dates back two hundred or more years (as evident in the work of architects such as George Devey and Edwin Lutyens, whose styles owe much to vernacular buildings, and Laugier's 1753 book which, echoing Rousseau's noble primitive fantasy, recommended that architects look to the 'beginning of human history' for guidance; see also Fergusson 1865), interest remained largely latent until recently (Marchand 2009:59–63). The profession considered vernacular architecture rude and primitive for decades (Rykwert 1981), compared to modern construction featuring the latest manufactured materials, regardless of praise for local traditions by such prominent architects as Frank Lloyd Wright, Adolf Loos and Le Corbusier – for instance, Wright thought that vernacular traditions offered more promise than 'self-conscious academic attempts' at architecture (Oliver 2003:9).

In 1946, the Egyptian architect Hassan Fathy, having studied Nubian mud-brick architecture, planned the New Gourna township at Luxor drawing on local building designs, albeit the project failed for various reasons (Fathy 1986). Others who pioneered consideration of vernacular architecture included South Asian architects Bashirul Haq of Bangladesh, Charles Correa of India and Geoffrey Bawa of Sri Lanka, who were strongly influenced by their regions' traditional architectures. The work of Alexander and collaborators (Alexander 1979; Alexander *et al.* 1977) marked growing, albeit unorthodox, interest in vernacular styles, seeking to identify and adapt architectural features cross-culturally and empowering building occupants to influence design. According to him, 'There is one timeless way of building. It is thousands of years old, and the same today as it has always been. The great traditional buildings of the past, the villages and tents and temples in which man feels at home, have always been made by people who were very close to the center of this way.' (Alexander 1979:7).

It is since the new millennium that vernacular architecture has become firmly established within mainstream architecture, and considerable numbers of papers and books published on the topic. Increasingly, architects study and seek ideas in the vernacular, many acknowledging that there is something to be learnt from the design solutions, use of materials, spatial organization, heating and cooling arrangements, and so on, of the many building traditions worldwide. They vary widely, however, in their opinions as to its merits, with repeated warnings from some quarters not to romanticize such building customs and overlook their inconveniences (Lawrence 1983). The majority consensus currently is to draw on both the traditional and modern for inspiration, combining the tried and tested with the new, exploring the scope that new technologies and materials offer for innovation.

This new school of architects frequently talks about buildings that are in harmony with their natural environment and the human spirit, connecting occupants with the wider world. Increasingly, they draw inspiration from local building traditions as exemplars of sustainable and non-alienating architecture. They seek a renewed accommodation between building and natural habitat, using local materials to design structures that not only fit the surrounding landscape but are also ecologically appropriate, with growing concern for planet-friendly building activity. Some architects draw eclectically on vernacular architecture, while others found their designs on the local traditions of their regions. Architects whose work illustrates modern views on vernacular architecture include Samuel Mockbee, with his contemporary Alabama traditions, and Paolo Soleri, with his 'arcology' idea. They combine modern architectural practice with vernacular inspired designs to come up with striking innovations, sometimes described as 'organic' and 'earthy'.

Vernacular architecture and sustainability

A growing awareness from the 1990s of worsening environmental circumstances globally, epitomized by warnings about climate change and unsustainable use of resources, partly accounts for this renewed interest in the vernacular (Bay and Ong 2006; Smith 2005). Increasingly, some professionals ask what vernacular architectural traditions might teach us about sustainability (Heath 2009; Ryan 2011). When we are in unfamiliar surroundings such issues often strike us forcefully. The impact of contemporary construction activities became starkly evident to me when living in the Persian Gulf, where urban developments are patently unsustainable in the long term, featuring high-rise plate-glass buildings that require expensive energy-demanding air conditioning to keep them occupiable. They pay scant attention to Arab architecture evolved over millennia to cope with the region's climate (Andrews 1999; Hawker 2008; Ragette 2003). Realization that the construction industry erects many energy inefficient buildings and uses a significant proportion of the world's resources non-renewably (in addition to consuming much energy and contributing significantly to global greenhouse-gas emissions during building activities) has prompted many architects, who wish their profession to be seen as responsible, to think seriously about replacing environmentally damaging conventional architectural designs and practices with alternative 'green' ones. In their search for inspiration and viable options, they have turned to vernacular architecture for insights into environmental adaptation, realizing that local builders have built sustainably for generations using traditional techniques to erect apt structures, mostly utilizing locally available natural materials without threatening to deplete supply, and judiciously employing largely renewable sources of energy, such that their activities are generally in harmony with the surrounding environment.

The wind tower of traditional Arab architecture illustrates the potential of turning to vernacular structures for ideas (Coles and Jackson 2007; King 1998). It has proved its worth, extending back over 5,000 years, directing local wind flow to cool houses. The structure ensures fairly constant and comfortable interior temperatures, particularly when combined with locally made mud bricks, the thermal properties of which keep interiors relatively cool in the day by absorbing heat and then relatively warm at night by radiating out the stored heat (Marchand 2009:42). And the tower uses zero additional energy to cool and ventilate buildings, so attracting interest in the search for energy-efficient designs that reduce carbon footprints, in response to global warming predictions. Current research seeks to optimize the design of this ancient construction, using wind-tunnel and computer-flow modelling to study aerodynamics and thermodynamics and devise control algorithms for high-tech versions. The 45 metre high wind tower in the courtyard of the Institute of Science and Technology at Masdar City in the United Arab Emirates (which aims to be the world's first carbon-neutral and zero-waste city) is a prototype, featuring an automatic monitoring system that operates louvres according to wind direction and a misting arrangement within the tower to keep temperatures comfortable in the courtyard below (Conroy 2011).

It is encouraging for many to see architects talking 'green design' in response to concerns about resource depletion, environmental degradation, growing populations, energy worries and economic meltdown, but it is necessary to beware of idealizing the environmental credentials of local building styles in the search for ways forwards, particularly

Plate 1.9 Wind tower feature, Al Fahdi, Bur Dubai, United Arab Emirates.

where considerable change occurs (Lawrence 1983). Such change is another reason for interest in vernacular traditions, many of which are changing rapidly (e.g. Rensel and Rodman 1997) along with other cultural practices, a feature of what today we call globalization. We have lost, and are losing, much valuable knowledge and many skills, which may have important sustainability lessons. Such loss prompted early anthropologists to invent the maligned practice of salvage ethnography. Resource depletion, often following the export of local materials such as hardwood timber, in addition to people opting to use various imported manufactured materials such as plastics, laminates and concrete, sheet metal and glass, have led to large changes in local building styles. Changes in house building in north China illustrate the trend, the arrival of concrete and steel, together with electricity, rendering obsolete the traditional *kang* 'adobe cooking stove and under-floor heating system' that was a central feature of houses. The heated brick platform required deep knowledge of materials and design, together with lost skills and experience, to construct and operate correctly relative to house size and heating requirements (Flitsch 2008).

Plate 1.10 Wind tower adjacent to Souq Waqif, Doha, Qatar.

Economic, political and social transformations, often in response to globalizing pressures, also impact heavily on traditional building practices worldwide. Changing lifestyles can render traditional structures functionally obsolete. New designs featuring novel materials, frequently inspired by outside influences, may better meet the needs of contemporary communities; for instance, with a shift from extended to nuclear family dwellings. The relentless growth in urban populations, frequently crowded into slums, often further renders traditional structures difficult and impractical to build, and the resulting overcrowded shanties diversely built of miscellaneous, often scavenged materials, may pose serious environmental problems with heavy demands on local services and pollution.

Vernacular architecture and identity

With these sweeping changes in circumstances, many communities no longer construct some of the buildings depicted in the vernacular architecture literature. Yet ironically, while globalization and the technological change facilitating it are eroding vernacular building, they are also promoting interest in keeping it alive. Global communication technologies, such as the television and telephone, and now the internet and mobile, together with ubiquitous air travel, pass information rapidly in all directions. Communities in the remotest rainforest and desert regions hear about new designs and ways to construct buildings featuring exotic new manufactured materials. Meanwhile information about traditional ways of construction reaches a wider audience and is more readily available to interested parties, such as architects seeking sustainable inspiration (Heath 2009). Interest in vernacular ways may not be confined to such outsiders, but may also concern insider builders as they become aware that global consumer culture threatens to engulf their cultural identity. In this regard, the use of local building materials not only promotes sustainable use of resources, but may also contribute to maintaining identity: 'Building materials are ... the key to retaining local distinctiveness. Developed over many centuries, craft construction techniques were entirely geared to the most abundant materials available in any one region... Skills and craft techniques were handed down through many generations ... so that craftsmen evolved with intimate understanding of how materials behaved.' (Ryan 2011:56).

Buildings can serve as potent symbols of regional and cultural identity (Abel 2000; Powell 1983). Such a sense of belonging is not only emotionally and intellectually important, anchoring individuals historically and geographically to a place and society with its moral compass and behavioural values, but also politically and economically significant, in battles for recognition and rights in the wider world. Many communities today, even urban ones living in 'modern' houses, may keep alive some aspects of traditional architecture to demonstrate their cultural inheritance. These may serve as weapons in the battle for their rights to land, resources and education, namely the independence to determine their way of being in the world. A good example is the *wharenui* 'big house' of the Maori, which is the community meeting hall built on the *marae* 'ceremonial clearing' and decorated with carvings in distinctive style (Brown 2009). Dotted around Aotearoa (New Zealand), such structures are a constant reminder that the Maori settled there first, mutely indicating European colonial usurpation.

Another strand in the current revival of interest in vernacular architecture concerns the empowerment of marginalized communities. There is growing awareness in some development circles, notably those informed by a participatory ethos, that local architecture can serve the interests of those living in poverty. It has the potential to supply people with low-cost housing, a key factor for those who lack financial capital. Involving local users from the design stage onwards, such an approach to building is likely to exploit locally available resources more effectively. The resulting structures are also more likely to be acceptable to inhabitants, as they will have a certain familiarity. This approach may not only prove effective in addressing long-term housing problems, for instance in slum areas, but it may also prove useful in times of disaster. An appreciation of vernacular architecture may profitably inform the construction of both emergency and transitional shelters, as it involves skills people possess and again often features locally available materials. Furthermore, accommodating displaced persons in shelters of familiar design may reassure them and help them recover from traumatic experiences, as locally appropriate and acceptable (Broadbent and Broadbent 2004; Corsellis and Vitale 2005; Davidson *et al.* 2006).

Although it has taken some decades for folk architecture to gain acceptance (as Hassan Fathy's attempts in Egypt over half a century ago witness), several national and international bodies now subscribe to 'culturally appropriate building'. In using familiar architectural styles, materials and technologies to build private dwellings and public infrastructure, this approach addresses the specific problems and needs of the poor within an ethnic context. While the buildings may feature certain innovations that exploit new manufactured materials and technology, these neo-vernacular structures are more socially and environmentally acceptable than conventional modern buildings with their ubiquitous rectangular design and concrete-block and tin-roof construction. Furthermore, such a culturally sensitive approach to urban and rural development may revive pride in local building traditions.

While cultural identity is not a notable issue for New Guinea Highlanders currently, who are more concerned to 'catch up' with the industrial world's modern technology and material goods, it is probable that it will become a concern at some point in the future, if events elsewhere are any guide. This further relates to the 'why bother' question. In addition to lessons noted earlier, for instance about sustainable resource use, it is possible that records such as this one of vernacular architecture and local building practices – part of a larger holistic study of cultural ways of being in the world that are changing dramatically – may prove a useful resource if generations to come have cultural identity concerns, particularly if these focus on architectural traditions. Already many younger persons who have grown up away from the Was valley, in urban centres for example, do not know how to construct houses such as documented here. They are like many others from around Papua New Guinea. Concerns about the loss of local architectural traditions, for instance, prompted the University of Technology's 'Village Studies Project' that focused on building construction (Ruff and Ruff 1990). While this book inevitably seeks to understand aspects of Highland life from my own cultural-intellectual viewpoint, the heirs to this heritage may nonetheless find it of some value in defining their cultural identity in a globalized world, if they seek to rediscover aspects of their ancestors' ways, maybe even 'reinventing' tradition sometime in the future.

2
A dwelling perspective

Vernacular architecture has achieved a place in architectural theory and practice today beyond that of a mere intellectual curiosity. Through its association with environmental, technological and political-economic issues, it serves as a source of design inspiration, possible exemplar of sustainable resource use, and an identity anchor and political tool in a world characterized by rapid and sometimes bewildering change. But the literature on it is limited in some regards, notably with respect to consideration of practical construction matters, with its focus instead on cultural tradition and custom.

The study of vernacular architecture often features in the heritage domain, possibly involving reinvention of tradition, and as a result is frequently associated with the preservation of historical buildings. Consequently it focuses largely on finished structures, often ancient ones, giving descriptive accounts of buildings and their architectural points – external and internal features, ground plans and landscape relations (as illustrated by the quotation from Ruskin in the previous chapter) – often lavishly illustrated with drawings and photographs together with some history. The previous chapter's brief account of Was valley houses is in this vein. The publications of various heritage bodies illustrate the genre; such as the Royal Commission on the Ancient and Historical Monuments of Wales, where an old anthropological college friend serves as a Senior Historic Buildings Investigator and whose work on the houses of Radnorshire (Suggett 2005) and recent retrospective on Smith's (1988) classic *Houses of the Welsh Countryside* (Suggett and Stevenson 2010) are exemplars of this vernacular approach.

Few studies of local architecture pay much attention to the actual work involved in constructing the buildings. This reflects the contemporary division of labour between architects and builders in our society. Regarding Western architectural practice, 'Drawing ultimately remains a separate enterprise from the making on the building site' (Marchand 2001:180). Architects produce plans and illustrations for new buildings, leaving the building contractors to deal with the technical aspects of construction, assuming a skilled labour force that has the experience and know-how to undertake the practical work without need of how-to-do-it instructions, and the ability to solve local problems as they arise.

A similar bias is evident in anthropology. Discussions of material issues infrequently consider the technical aspects of the making of things; as some archaeological critiques note (discussed later). It is a longstanding attitude. In his account of the construction of the Trobriand *bwayma* 'yam storehouse', for instance, Malinowski comments (1935:240): 'However one may scorn the purely technological interest ... details of construction and of manual procedure remain cultural facts which must not be neglected ... [and] the following account may irritate the sociologist because it is over-technical.' He goes on, reflecting the holistic perspective central to his functionalist approach, that the account may also 'upset the technologist because it is too much larded with social, economic and magical digressions. Yet both would be stultifying their own interests were they to keep these two aspects of the subject in watertight compartments.'

How to do things

While interest in material culture has increased noticeably since *Made in Niugini* was published, it has largely focused on finished objects, as noted in the previous chapter. It considers these either as things in museum collections, tracing their 'biographies' and contemporary significance, or as commodities, interpreting their ownership and use with respect to social identity and status. There is consequently little on practical know-how, such that after reading these studies you are scarcely more able to engage with the technology than when you started, not having learned much about the tasks involved in, for instance, making an artefact or building a house. It reflects attitudes in some academic circles, where it is unfashionable to take an interest in doing practical things; for some it is a matter of pride not to be able to do even simple tasks such as mend a fuse or paint a door. A colleague made me sharply aware of this when he told me that he was decorating his lounge and burst out laughing when I visited expecting to see him wielding a paintbrush; he was taken aback that not everyone had decorators in to do such work! It arguably indicates unspoken class attitudes, even snobbery (Leach 1984); technical matters, such as making and repairing things, concern blue-collar workers not white-collar intellectuals who engage in scholarly debate.

Not all academics are so elitist. Some archaeologists take an interest in the making of things, albeit their discipline's focus on dead subjects inevitably limits their scope (see Chapters 4 and 5). The *chaîne opératoire* ('operational sequence') approach is currently popular in this respect (Lemmonier 1992:26). It considers the sequence of technical processes and associated social arrangements involved in the production of things, from the acquisition of raw materials through their step-by-step manufacture into objects, to their use and eventual discarding. Some dispute its claim to originality, pointing out that the 'reduction sequence' idea in lithic studies precedes it by a century (Bradley 1975; Shott 2003), and sparking debate in some archaeological circles (Tostevin 2011). Indeed, it is arguable that the central point of the approach, namely that making things comprises a series of stages, goes back to the first craft manuals many centuries ago – see below. It may even extend back to the proverbial 'living in cave' times, as it seems such an obvious idea (you cannot thatch a house before constructing the roof, nor construct the roof before building the walls etc.). In fact, I used such a sequential approach to structure the *Made in Niugini* accounts of artefact manufacture, after the step-configured activities of my Wola friends, who have terms for the various stages that comprise the making of many artefacts (albeit the informal hands-on way of learning means that the stages are often implicit, the making of things proceeding as a continuous flow), without any awareness on our part at the time of the formal 'operational sequence' approach.[1]

The term *chaîne opératoire*, if not the idea, is attributed to Leroi-Gourhan (1993:114, 323), who extended on Mauss' (1935) argument that embodied techniques afford a unique perspective on human sociality (Beaune 2013; Bidet 2007). Initially associated with the reconstruction of stone-tool manufacture, by recreating the steps and techniques thought necessary to make such tools, its use has spread in archaeology to the production of other things such as pots, with attempts even to introduce it into anthropology (Leroi-Gourhan 1968; Lemmonier 1976, 1983, 1986, 1992). Proponents claim that the approach relates holistically to wider society, for any production process concerns not only technical but also social organization. The presumed actions needed to make something may tell us more than the final object itself, as a socially situated product, about social arrangements. Some even think that the *chaîne opératoire* approach may take us beyond speculated social context to cognitive issues, involving ancient patterns of thought, symbolic interaction and embodied knowledge. The sequential engagement of long-ago individuals with materials and things affords an opportunity, they argue, to reconstruct past persona. The making, using and disposing of things that are today archaeological remains undoubtedly represent decisions and choices made by persons long ago. The result of meaningfully determined action, they encapsulate the agency of those who engaged in their manufacture and use, they 'retain in their constituent parts the memory of their predecessors' (Audouze 1999:172).

While it is undeniable that we cannot make sense of how archaeological finds came into being without inferring the makers' intentions and consequently thoughts, it is unclear how reconstructing the hypothesized stages in making such things affords access to wider social issues and human cognition; the *chaîne opératoire* concept being in these respects, 'Almost mystical in its ramifications' (Odell 2001:80). Indeed it is arguable that to assume stages everywhere may be ethnocentric, as it is possible to think of manufacturing processes as continua (see Chapter 7; Ingold 2011b:53; Shott 2003:101–2). Nonetheless, such processes imply stepwise procedures (it is not possible to produce a polished stone axe head without first quarrying a suitably sized piece of stone and then roughly knapping into an axe blade shape), even if actors do not consciously think this way or, as often with my Wola friends, it is implicit in their actions. In other words, procedures may depend on tacitly held awareness, which gets us into even deeper epistemological water, and not only in archaeological contexts. Nonetheless, the opening up of cognitive issues in prehistoric settings is suggestive, as discussed later.

We have to turn to craft handbooks and house-building manuals if we seek detailed practical information about making things and constructing buildings.[2] These comprise a long literary tradition dealing with the manufacture and use of things; for instance, some of the more prominently cited books in the history of technology include the works of Theophilus (1979 [*c.*1125]), Biringuccio (1942 [1540]),

1 Some of the, otherwise puzzling, intellectual excitement in archaeological circles with ideas such as reduction sequence and *chaîne opératoire* was that they contributed to the break away from previous generations' concentration on chronological sequences (Shott 2003:101).

2 Many academics find the detail in such books boring (e.g. Tilley 2007), and while the difference with intellectual discourses is considerable, there is surely no call for dismissive attitudes, as Ingold (2007:36) points out, commenting that if craftspersons 'were required, as part of their training, to read works in the tradition of material culture studies, they would find them just as tedious and irrelevant… Because the interpretations they offer, of the social meanings of things, are divorced from any practically grounded understanding of materials and their properties. It is the same with a cookbook that – with its endless lists of ingredients and quantities – would look dull and boring on the scholar's shelf, but in the kitchen could be a source of inspiration and insight'.

Agricola (1950 [1556]), Piccolpasso (1980 [1556–9]) and Besson (1999 [1578]), which go into, then state-of-the-art, metal working, pottery throwing and diverse engineering activities. The handbook genre remains popular, not only with professional artisans, but also with those who enjoy amateur handicrafts, these texts being precursors of today's do-it-yourself books – witness the best-selling sections in many bookshops with volumes on decorating, knitting, carpentry, upholstery, horology, lapidary and so on.

The classic building text in this vein is Marcus Vitruvius Pollio's multi-volume *De Architectura*, known today as *The Ten Books on Architecture* (1999). A Roman architect and engineer, he argued that built structures should exhibit three qualities, known as the 'Vitruvian virtues': *firmitas*, 'solidity', *utilitas*, 'usefulness' and *venustas*, 'beauty'. He thought that humans should imitate nature in building, using natural materials like other creatures. The book, codifying existing architectural practice rather than innovating with novel ideas, was of considerably wider scope than modern architectural texts, encompassing engineering, art, landscaping and craftwork. It is the only major architectural work surviving from classical antiquity, and was the authoritative text on building until the Renaissance, when Leon Alberti (1988) published a reworking of it in 1452 as *De Re Aedificatoria*, which influenced many artists, including Da Vinci and Michelangelo, who were designing buildings. It was subsequently followed by Philibert de l'Orme's *Le premier tome de l'Architecture* in 1567 (1981) and Andrea Palladio's important work *I Quattro Libri dell'Architettura* in 1570 (2002),[3] whose design principles informed the work of many prominent architects such as Jones, Bernini, Vanbrugh, Wren and Adam among others.

It is not only European civilizations that adhere to Vitruvius' three principles. The buildings of the Was valley unselfconsciously reflect them too, being solidly functional and elegant structures that comply with the sustainable use of materials suggested in the injunction to emulate nature. In subsequent eras many books appeared, including in the nineteenth century Pugin's (1836) controversial *Contrasts* and Ruskin's (1849) *Seven Lamps of Architecture*, the conservative views of which contrast markedly with those of the twentieth century, such as Le Corbusier's (1923) *Vers une architecture* and Gropius' (1935) *The New Architecture*, which promote modernism. A plethora of popular texts are also available today, such that books on house design and building, from initial blueprints to site preparation and from bricklaying to plumbing, feature prominently on the afore-mentioned do-it-yourself shelves in bookshops. They range from standard texts for builders (Browne 1974; Chudley and Greeno 2010), to manuals for self-builders (DeCristoforo and DeCristoforo 2007; Rock 2010) and, increasingly popular, guides to 'green construction' that parallel the Was valley building tradition (Evans, Smiley and Smith 2002; Haun 2008).

These craft studies and house-building manuals presume that readers intend to make or build things, that is, engage practically with the activities they describe. The assumption is that you have or will acquire the necessary skills, which a book cannot teach, for they require physical practice. This brings us to tacit or experiential knowledge, an increasingly topical issue. While on one level this book is a contribution to vernacular architecture studies, on another it is an exploration of the implications of tacit knowledge for our understanding, paying close attention to the doing of construction activities, namely how to build things. Tacit decision making features prominently in Was valley house building; for instance in choice of raw materials, which features little deliberation, individuals just getting on with it. If, as is increasingly evident, much of what people know exists tacitly in what they do, then it is necessary to focus on associated practical matters if we wish to further our understanding of their interactions in, and manipulation of, the world. What they know implicitly concerns not only the material world but also social behaviour in general, which features a considerable tacit element. The study of things, I argue, gives a firm anchor 'out there' to test propositions, advancing on sociological supposition.

In the interval that it has taken me to get around to writing this book (see Chapter 1) interest in the experiential aspects of human behaviour has grown noticeably. Some recent work deals with tacit issues in the context of material culture studies and comparative architecture (Ingold 2009, 2013; Marchand 2009, 2010; O'Connor 2005, 2006; Warnier 1999). This literature focuses on a range of issues, from subjective descriptions in the postmodern reflexive vein, focusing on the experience of deploying skills in undertaking certain tasks, to philosophical discourses that draw largely on the phenomenological tradition of 'being in and acting on the world', to more conventional investigations of the sociological aspects of practical activities such as the transmission of skills between generations. After reading these studies, you will not necessarily know how to make or build things. There is, again, seldom much, if any, detail on what you need to do to acquire the skills that constitute the tacit knowledge; that is, the activities that you need to engage in and the physical actions that you need to master. In short, there is scant guidance to the doing that you need to practise to acquire the know-how.

The tacit aspects of knowledge and behaviour have long intrigued me. In *Made in Niugini*, for instance, I refer to *The Wheelwright's Shop*, Sturt's elegiac account of Surrey wheelwrights, in discussing the experiential aspects of skilled knowledge and the challenges these pose for academics who trade in words. It was in *Give and Take* (1979a:8) that I first inadvertently stumbled on the topic

3 Ingold 2013:47–59 discusses the emergence of architecture in Medieval Europe in an anthropological context.

of tacitness while contrasting the normatively conditioned 'social robot' approach to human behaviour with the consciously calculating 'self-interested' individual. It is a theme that I follow up in *From Land to Mouth* (2010a:49–64) in some detail, in exploring agricultural activities as skilled knowledge deployed without discussion. In *A Place Against Time* (1996:297–9), I delve into the related experiential aspects of environmental knowledge, such as judging soil fertility through practical cultivation, not discussion of soil properties. And in *Managing Animals in New Guinea* (2003:257–60), I investigate the tacit aspects of animal knowledge that persons acquire through hunting game and herding pigs, using my attempts to access such skills, such as a bungled participant-observation episode trying to keep my own pigs, to illustrate the challenges we face.

In this book I address the experiential quality of knowledge and behaviour, and related philosophical – notably phenomenological – issues, more closely than heretofore, having previously prepared the way, initially unwittingly, to engage with this subject of growing topicality that was distinctly marginal when I started out. In doing so, I tackle the question: how can we deal with knowledge of house building when much of it is manifest in constructing a house? Even though the Wola are explicitly aware of aspects of the house-construction process – as the vocabulary for various components and aspects of house building shows (see the appendix and various tables) – builders engage tacitly in many associated activities. The tacitness is of two varieties: persons engage in tacit actions in constructing houses and they unquestioningly accept customary architectural arrangements. The importance of paying close attention to ethnographic evidence in exploring these skilled and unspoken aspects of life is increasingly evident, which returns us to ethno-determinism. While it is crucial to pay close attention to the ethnography in furthering our understanding of tacit issues, it is challenging to produce a reliable record. In order to capture some of the unverbalized aspects of these activities it is necessary to employ other media, such as drawing and film, as in the movie *House at Honael*, made as part of this study to document house construction activities;[4] albeit diagrams and photographs do not pass on tacit knowledge or sensory experiences, so much as record visual aspects of associated behaviour that we cannot catch in words (Pink 2009:97–114). The retrospective realization that tacit knowledge was featuring centrally in my work shows the implications of ethnography leading enquiries, with issues initially unclear and emerging through my struggles to engage with it. Nonetheless, while the ethnography leads investigations, I patently and somewhat paradoxically, deploy it to address experiential questions that I think relevant.

Tacit knowledge

While the subject of tacit knowledge may be of growing interest in today's academia, there is nothing new about pointing to the significance of such oblivious know-how in human affairs. Indeed, as with many ideas, we can trace its roots back to ancient Greek philosophical distinctions between practical and social wisdom (Baumard 1999:53). The epigram- and riddle-writing pre-Socratic philosopher Heraclitus, for instance, refers to tacit awareness in his opening '*Fragment*' stating that:

> of this account, which holds forever, people prove forever uncomprehending, both before they have heard it and when once they have heard it. For, although all things happen in accordance with this account, they are like people *without* experience when they experience words and deeds such as I set forth, distinguishing (as I do) each thing according to (its) real constitution, i.e. pointing out how it *is*. The rest of mankind, however, fail to be aware of what they do after they awake up just as they forget what they do while asleep.
> (Robinson 1991:11, Fragment 1 – ital. in original)[5]

While awareness of skilled knowledge is ancient, Michael Polanyi's work (1958, 1966) is widely regarded in academic circles as a landmark in its study.[6] There are some earlier works that mention experiential knowledge, such as Beckmann (1777), who refers to technology involving naturalistic methods, and Mauss (1935), who talks of bodily techniques as the first and most natural of human instruments. According to Baumard (1999:8), Piaget (1954 [1937], 1950 [1947]) prefigured Polanyi in

4 See the film *House at Honael: House Building in the Papua New Guinea Highlands* at the website associated with this book (www.community.dur.ac.uk/paul.sillitoe/wola/houses). I am grateful to my colleague Steve Lyon for assistance with the website.

5 Arguably the maxim carved on the temple at Delphi, γνῶθι σεαυτόν 'know yourself', hints at a similar awareness of what we are unaware; in warning that we need to realize that we are not god-like and have limits.

6 Michael Polanyi was a physical chemist who turned social philosopher to challenge the prevailing view of scientific knowledge, arguing that 'complete objectivity as usually attributed to the exact sciences is a delusion' (1958). He came out of the laboratory to argue that the idea that science proceeds entirely by drawing value-free on propositional knowledge summarized in formulae overlooked the unspoken skills, personal attributes and creative intuition that feature centrally in such work – some time ahead of the now popular field of the sociology of science that is busy enquiring into how scientists' informal talents, including interactions with one another, influence their work (Collins 1974, 2001a; Knorr-Cetina 1999; Latour 1999; Latour and Woolgar 1986). Later Polanyi attempted to extend his discussion of tacit knowledge to encompass art, myth and religion (Prosch 1986).

identifying subliminal aspects of actions. And in a New Guinea context, Thurnwald (1916:390) concluded his Bánaro ethnography a century ago with the following comment on tacit knowledge: 'The institutions of the social organization dominate the individual to such a degree that his actions become almost automatic and are generally no more considered than is his gait.' But these pre-Polanyi commentators do not explore tacit issues in detail, their discussions not distinguishing these clearly from other aspects of knowing. And since Polanyi, other writers have discussed tacit knowledge, introducing their own jargon besides, such as Searle's (1983:143) notion of 'background', which refers to unpremeditated states that inform 'capacities' and 'practices', comprising a 'set of skills, stances, preintentional assumptions and presuppositions, practices and habits' (Searle 1983:154); it is 'intention in action' as opposed to 'prior intention'. And Dreyfus' (2002) refers to G-intentionality (G standing for gestalt), which he describes as an 'experienced causal connection' between an individual and the world that features direct physical interactions rather than indirect representational (i.e. in the head only) transactions.

While different disciplines – from social sciences to business studies, psychology to philosophy – vary in their approaches to tacit knowledge, also called experiential and informal knowledge[7] – they share common broad themes. They agree that it is internalized knowledge derived from personal experience, intuition and observation (Collins 2001a; Polanyi 1958; Wagner and Sternberg, 1985, 1986). It is knowledge we routinely use, such as the ability to recognize a person's face, irreducible to explicit propositional understanding that we can articulate, for what we know comprises not only logical propositions declarable in sentences, but also 'personal knowledge' beyond explicit formulation. We cannot know it vicariously from what others tell us. For instance, we can only know the colour green from seeing it; we cannot explain it to someone born blind. Neither the physicists' objective light-wave explanation (that light with a wavelength of about 510 nm is green); nor the neurologists' retina-brain process (that such light stimulates a neuron pathway of optical nerve bundles) can get us near the actual experience (Lakoff and Johnson 1999:23–6). Tacit knowing is manifest through performance and is context specific; it is more lived than learned, not being amenable to reduction to rules. It comprises a range of sensory and image information, featuring in visual, aural and tactile perception, mastering physical skills, tool use, discovery and invention, language acquisition and use, and the apprehension of categorical concepts (such as what the category 'tree' comprises) (Polanyi 1969a:123–8).

7 These terms may carry different, and varying, shades of meaning; just as the word tacit may also, from 'the silent through the unspoken to the implicit' (Ingold 2013:109).

Such knowledge makes up a substantial part of our intelligence of the world and of our actions in it, which often deploy technology. Much technical knowledge is tacit, comprising informally acquired skills or a 'knack' for doing things. The expertise of craftsmen is built on years of experience and they often find it difficult to express what they know, such as the principles behind their skilfulness. Creative persons may derive inspiration and insights from such subjective bodily experience. Many human discoveries, including those of contemporary science, are the result of the integration of this unconscious with conscious awareness. Tacit knowledge transmission requires personal experience, maybe facilitated through hands-on training. Its acquisition involves 'learning by doing' (Wagner 1987:1236; Wagner and Sternberg 1985:454). When taken on board, such knowledge 'indwells', as Polanyi (1958) vividly put it. Another of his widely cited comments that catches the tenor of such knowledge is that 'We can know more that we can tell' (1966:4). In his assessment of the human predicament, Koestler (1967:207) affirms the essence of such knowing:

> There seem to be several factors which determine how much, if any, conscious attention is to be paid to an ongoing activity. First, the acquisition of a skill by learning requires a high degree of concentration, whereas with increasing mastery and practice it can be left 'to look after itself'; ... the canon of the skill function[s] unconsciously, [and] applies equally to manipulative, perceptual and cognitive skills.

Unaware awareness

We possess a range of such predispositions – including bodily acquired skills and things passed on through culturally mediated social experience – of which we have no clear understanding and cannot readily express using language. They involve learning to share a skill or habit that we may not even be aware of, as it is not always known explicitly to actors. How many cyclists, for instance, realize they practise counter-steering? The term Polyani (1966) coined for such unaware informal up-take is 'subception'. This implies that certain behaviours, such as skilful activities and cognitive processes such as judging distances, feature operations inaccessible to consciousness, or at least that conscious attention is not necessary in undertaking them. In this regard, Polanyi (1958) distinguishes between 'subsidiary' and 'focal awareness' with respect to tasks that feature the use of a device/tool. He refers to knocking in a nail with a hammer (Polanyi 1958:55; Polanyi and Prosch 1975:33), which is reminiscent of Heidegger's (1962:48, 98) discussion, who takes tacit knowledge to be practical ability gained through habituation. He distinguishes between 'present at hand' (*vorhanden*) – awareness of a physical thing (which usually occurs when it stops functioning as expected, such as when a tool breaks); and 'ready to hand' (*zuhandenheit*) – obliviousness to a thing (the tool functioning normally, where the user focuses on doing something with the tool).

A learner exhibits awareness of subsidiarity, paying attention to the tool (concerned not to hit a finger with the hammer), whereas for a competent individual it is an 'unseen' extension of the body and such a skilled person maintains focal awareness on the overall objective (the accomplishment of the task, say nailing on roofing battens). When we learn to rely on such sensory awareness while focusing on something else, it becomes tacit knowledge, assimilated into our unconscious. And as such it becomes somewhat opaque, beyond explicit awareness. We cannot explain exactly how the feel of holding a hammer informs our use of it. Such knowledge is personal and 'knowing is action' (Polanyi 1958), in that it takes a skilled knower to unite subsidiary elements with focal action to achieve some end. The operation of the subsidiary/particulars and focal/whole distinction draws on the part-whole premise of gestalt psychology regarding perception and cognition, where the whole is greater than the sum of its parts, namely that the individual elements of any action do not entirely make up the outcome of the total activity (Polanyi 1969b:138–9). The subsidiary parts are integrated into the perceived whole focal act, which exceeds their aggregation, as we have 'powers of perceiving coherence' and can see the 'varied and changing clues' of a shifting entity 'jointly as one single unchanging object' (Polanyi 1969b:139).

When using a hammer to knock in a nail, we are aware of both but in different ways. We know that we are holding the tool but our attention is focused not on it but on hitting the nail squarely. In a way, the hammer is an extension of our arm and we are aware of it only as we are aware of our arm, such that it is difficult to 'determine where the body begins and ends' (Marchand 2012:269), the tool 'becoming part' of the person. Recent neuro-scanning evidence correlates with subsidiary awareness, showing that the brain not only focuses visual perception on the functional part of a tool, such as a chisel's cutting edge, but also locates a sense of touch there (Takahashi *et al.* 2009). It is an aspect of proprioception, that is, our unconscious perception through our bodily senses of space and movement (such as where our limbs are). From an embodied knowledge perspective, it relates to our sense of self. The tool by implication becomes 'invisible' to the user, just as our arms and fingers are to us in everyday life (unless something happens to bring them to our attention, such as they do not function as normal, usually brought to our awareness through a pain sensation – missing the nail with the hammer and hitting the thumb). When we learn to use a tool or speak a language or play a musical instrument, we 'make ourselves aware of these things as we are our body, we *interiorize* these things and *make ourselves dwell in them*' (Polanyi 1969b:148 – original ital.). In this way we augment our body's capabilities (Polanyi, 1958:59; Polanyi and Prosch 1975:37).

A tool becomes something through which we act, it is interiorized to achieve a purpose, and expands the effectiveness of our interaction with, and experience of, the world. The sense of hammer-holding is of an instrument, not an object of our attention. 'I know the feelings in the palm of my hand *by relying on them for attending to the hammer hitting the nail*. I may say that I have a *subsidiary* awareness of the feelings in my hand which is merged into my *focal awareness* of my driving the nail.' (Polanyi and Prosch 1975:33 – original ital.). 'Dwelling in' a tool implies unconscious commitment. Focal and subsidiary awareness are interdependent (Polanyi 1958:56), in that if we pay attention to the tool we interfere with focal awareness and are likely to use the tool clumsily and risk hitting our fingers rather than the nail. We change the character of the activity. The 'centipede's dilemma' tale catches this aspect of tacit knowing, where a frog asks a centipede in what order she moved her many legs and she could no longer move for thinking about it, having previously left her legs to manage themselves; the frog then ate the hapless centipede.[8] The integration of subsidiary and focal awareness is irreversible, while in explicit propositional knowing we can logically move back and forth between hypotheses and data without losing anything (Polanyi and Prosch 1975:39–42).

An apt example of these issues from the Was valley concerns men's handling of axes, a skill which they acquire from a young age. The use of stone axes by some young men working with me on house construction illustrated the way in which focusing attention on the tool can unbalance skilled performance. They were entirely familiar with the subsidiary use of steel axes, which they wielded with confidence, but they were unfamiliar with stone axes, which they used hesitantly at first. These require a different technique, in which instead of putting your full body weight behind a blow, you have to master a sort of recoil routine, where you 'soften' the blow by absorbing some of the shock. Stone axes demand more use of the forearms, relying on a bending-from-the-elbow pecking action to deliver blows rather than swinging from the shoulder. If you use your full body weight, as in a shoulder swing, you are likely to damage the axe – as we discovered early in our work, with several

8 Psychologists refer to such disruption of normally unconscious competence by conscious thought as 'hyper-reflection' or 'Humphrey's effect', after George Humphrey who mentions the 'centipede's dilemma', pointing out (citing William James, and before Heidegger or Polanyi) that 'Habit is everybody's private secretary, doing all the routine actions of everyday life' such that 'No man skilled at a trade needs to put his constant attention on the routine work. If he does, the job is apt to be spoiled' (1923:108, 109).

shattered stone blades. The men had not interiorized or 'dwelled in' the relevant skilled knowledge. Consequently, under the tutelage of older experienced kinsmen, they were initially cautious and conscious of the axe – the focus of their attention – and they were unable to work smoothly or effectively. When, through practice, they assimilated the axe-wielding knowledge – namely, it became the subject of subsidiary awareness – their focal attention switched from imitating the axe swinging action of kin to applying their new skill to the tasks at hand. They had mastered the knack of using a stone axe and their work rhythm changed so that they were able to undertake tasks like their older kin.

The focal and subsidiary awareness distinction does not imply, in according tacit know-how a prominent role in practical activity, that the skilful handling of tools is automatic. It is undoubtedly an 'error to regard such know-how as subconscious, as though practitioners could "do it without thinking", when in fact their work involves the most intense concentration' (Ingold 2013:111). When cycling, for instance, I may not think about how I know how to ride a cycle, but I concentrate all the time on traffic around me. Similarly, craftspersons may not think about how to use tools, but they concentrate on the task at hand, sensitive for instance to the properties of the material they are working on, such as the twist of the grain in wood, and adjust their actions accordingly. 'Even when certain tool-wielding actions are mastered and the neuromuscular system performs them in a seemingly automatic way', they are not mindless, for a worker is 'acting not merely *upon*, but *with* materials and *with* tools whose inherent qualities "respond" to the work' (Marchand 2012:267, orig. ital.). While self-evident to practitioners, this point seems novel to some academics to judge from recent discussions of materiality, where it features centrally, as discussed later.

The embodied mind

Another approach to tacit issues refers to embodied knowledge, which encompasses the practices, intelligence, routines and responsibilities known to us bodily without conscious thought. It comprises a range of approaches, from the philosophical to neurological (Clark 1997; Noë 2004; Rowlands 2010). It originates philosophically with the phenomenological view of being, notably that of such philosophers as Husserl, Heidegger, Merleau-Ponty and Sartre, which raises some profound questions about what we presume to know. It concerns experience, notably how we encounter 'phenomena': how they appear to us and what they mean, what we make of things. Its focus is subjective, as captured in Husserl's idea of the individual's 'life-world'; that is, the world as we personally experience it. The argument that we can only apprehend phenomena subjectively led to existentialism, the version of phenomenology that focuses on the part the human body plays in experience. (Though Sartre (1962) argued that the Ego is an object of consciousness and not a subject in it; and Merleau-Ponty (1962) took the absolute experience of the embodied subject to be in-the-world.)

According to this philosophy, knowledge is embodied as a consequence of a unity or correspondence between subjects and objects.[9] This embodied knowledge is quite different to discursive knowledge that we can converse over. Embodied actions and associated perceptions have a large habitual component, responding to both natural and cultural environment, many of which are acquired through imitation. Merleau-Ponty, who refers to such skilled bodily attributes as '*praktognosia*', uses a sporting activity to illustrate how we refine associated skills for coping with activities, drawing on past experiences and responding to events accordingly, without necessarily reflecting on what we are doing. Footballers rarely have time to think thorough how to play according to the rules of the game, team strategies and physical actions, rather they embody the game's principles, expressing their import in their performance. A mental representation of the objective is not necessary in this pre-reflective mode, which may be purposive without the actor consciously conceiving it – it involves being in the 'zone', in current parlance. Practical mastery of skills characterizes embodied intelligence and allows behaviour without reflective thought. When you motion to a friend to approach, no preceding thought motivates your action; you do not discern a signal in your mind – wave to John – and then act on it (Merleau-Ponty 1962:111). According to this pre-reflective cogito view, you do not perceive an interior state that precedes an incident's expression. Many human propensities would not be learnable without such a pre-reflective foundation. It is not by conscious thought that we learn much behaviour, such as dance sequences or language games, but through repeated efforts to embody them, which we modify until we achieve the 'right' actions.

The habit to which Merleau-Ponty refers is not a mechanical propensity to behave in a certain way. It is akin to a flexible skill or competence, one's habitual mode being constantly subject to changes, often small, mobilized in response to different situations with varying goals (1962:143). In the act of learning, we embody knowing unconsciously and the resulting intelligence of the body – as evident in its capacity to innovate – challenges philosophical emphasis on cerebral introspection. Habitude is neither formal knowledge nor involuntary action but 'knowledge in the hands, which is forthcoming only

9 A copy-typist's familiarity with a keyboard illustrates the idea of unity between subject and object. S/he will not consciously think about where the keys are – and in a sense s/he may not know (for instance, if asked, s/he may not be able to draw the keyboard layout). The relationship is between her/his trained hands and the keyboard, not her/his conscious mind and the keyboard.

when bodily effort is made, and cannot be formulated in detachment from that effort' (Merleau-Ponty 1962:144), a formulation that recalls the definition of tacit knowledge. According to the embodied view, 'the body just knows' how to engage in activities and with things. In his later work, Merleau-Ponty acknowledges that his idea of pre-reflective cogito is problematic (1968:179) in that it assumes a self anterior to conscious awareness, thought and language. The problem is accommodating this aspect of being. In seeking to access this primordial experience, he inevitably has recourse to language, such that his tacit cogito is a dialogical product, which inevitably smuggles in such concepts as consciousness and propositional logic. This is exactly the problem that this study faces in dealing with construction activities that depend on, by definition unspoken, embodied knowledge.

If we can only think within the limits of our embodied brains, this sets certain constraints on what we can know. Indeed it is arguable that as physico-neural beings, the body-brain relationship informs our thoughts. For instance, Lakoff and Johnson (1999:77) maintain that 'The mind is not merely embodied, but embodied in such a way that our conceptual systems draw largely upon the commonalities of our bodies and of the environments we live in.' (Lakoff and Johnson 1999:6). They extend on the investigations of neuroscience into the question of how processes of sensory perception inform the ways we understand certain concepts, such as space and colour. They argue that physical experiences – spatial awareness, bodily movement, manipulation of objects – relate directly to the way we understand the world. We draw imaginatively on 'preconceptual' bodily experiences to connect perceptions of the senses with our understanding of related concepts. We learn to keep our balance, for instance, through physical acts of trial and error without any need of concepts. We may then extend on such experiences by 'metaphorical projection' to formulate ideas: for instance, by using our bodily awareness of balance to conceive of things in equilibrium more generally, such as balancing between long and short objects; or to think about more abstract social ideas and values, such as weighing the evidence on two sides of a dispute (Johnson 1987:74–5).

We extend on these experiences metaphorically when reasoning about things, not only in language, but also in gestures, diagrams, activities etc. We see this, for example, in our association of the concept 'more' with the experience of 'up' and 'less' with 'down.' It is evident in expressions such as 'price rise' (a £25 theatre ticket not being higher than a £20 one physically) and 'bonds falling' (a 10 per cent decrease on the stock exchange likewise involves nothing physical). Similarly, if things are going well in life, we are on the up, but in sickness we are down.[10] Early life experiences of the world may reinforce such associations, such as lying down when ill or seeing liquid rising up when poured into and filling a container. In this way, metaphors project 'inference patterns from one conceptual domain onto another ... [which] allows us to reason about the target domain in a way that we otherwise would not' (1999:82), and Lakoff and Johnson go on to maintain that such figurative language cognition, deploying complex and versatile conceptual metaphors and categories, suggests that reasoning about abstract topics such as society and religion may originate in commonplace empirical domains, such as thinking about spatial relationships.

Complementing this philosophy, scientific investigations into neural activity seek to understand the biology of the embodied mind. The neurological approach concerns the anatomy, function and development of the nervous system, using a range of techniques from the molecular level, studying individual nerve cells and synapses, to the organ level, imaging motor and sensory activity within the brain (e.g. fMRI, PET, SPECT scans) to map neural circuitry. These data afford new ways to address complex questions about human cognition, allowing neuroscientists to see how thoughts, actions and emotions map to specific neural networks and thereby furthering understanding of interactions between the brain and the environment by tracking activity inside the head with respect to the outside world (Gazzaniga, Ivry and Mangun 2009; Gibbs 2006; Varela, Thompson and Rosch 1991). Rather than answering the conundrums of phenomenology, this research poses another series of questions in exploring the inputs to our brains from what our bodies' sense, as discussed later.

Never mind the mind

In arguing that we understand the world through sensory-motor-system mediated experiences and emotions, the 'embodied mind' line of reasoning challenges previous philosophical orthodoxy, which assumed that body and mind are separate. As Descartes asserted 'it is certain that I (that is, my mind, by which I am what I am), is entirely and truly distinct from my body, and may exist without it' (2008:115). This is a view that appears somewhat to parallel my Was valley friends beliefs, both previously – with humans having a *tokor* 'body' and a *wezow* 'life-force' that resides in the chest and informs the *konay* 'thoughts' (and departs at death to become a *towmow* 'ancestor-spirit') – and currently, having adapted these beliefs to accommodate Christian ideas about the everlasting soul. This dualistic mind-and-matter idea of being implies that knowing is something that occurs in the mind, whether located in the head or the chest, which is independent of the body, giving

10 Other metaphors include the association of affection with sense of temperature, as in 'warm greeting' or 'cold shoulder,' and comprehension with vision, as in 'see what it means' or 'blind to the fact' (Lakoff and Johnson 1999).

us a disembodied or reasoning 'knowing subject' and an embodied or sensing 'known object'.[11]

According to the phenomenological embodied view, we are patently our somatic selves and our lived bodily experience denies that the mind – our capacity to think and know – is separate from the corporeal body. It seeks to transcend the detachment of subject/mind from object/body by considering the existential body (Merleau-Ponty 1962:xii),[12] a radical alternative that focuses on perception as embodied experience. The intellectual and the material are intimately connected in such an embodied state of being, such that we cannot view the body simply as a material object or worldly entity, for to do so would be to ignore the contingent nature of thought. The incarnate mind is inseparable from the physical body of which it is a part, which in unison perceives and thinks, such that thinking is not a product of a disembodied mind, but the result of interactions between embodied-minded person and the world, which classical philosophy underestimates in viewing the body as an object occupied by a transcendent mind overseeing its functions. Only by avoiding rationalism's tendency to construe the meanings that we make of the world as the immanent property of the reflecting mind can we 'rediscover, as anterior to the ideas of subject and object, the fact of my subjectivity and the nascent object, that primordial layer at which both things and ideas come into being' (Merleau-Ponty 1962:219). The implication is that this 'knowledge in the hands' is primeval; when we separate ourselves from it, we engage in philosophical undertakings that reify the unhelpful mind-body distinction. The challenge is accessing the embodied mind convincingly.

The dualistic mind/body view rests on a ubiquitous religious belief that asserts that for the mind to exist beyond the body presupposes a soul (or *wezow*) that departs at death as a spirit force (or *towmow*) to exist in some afterlife. While phenomenology covers a range of views that make its crisp definition difficult, it represents a philosophical approbation of disbelief in the mind as a sort of divine spark. It continues the Enlightenment trend for bold philosophers to challenge church dogma; such as David Hume, who in his (in)famous sceptical 'Of miracles' essay argues that the 'truth of the Christian religion is less than the evidence for the truth of our senses' (1902:86).[13] It is philosophy moving with the times, as increasing numbers in recent generations doubt metaphysical theological doctrines about human existence and the transcendental meaning of life. I recall the day that I admitted to myself that there is nothing in it. Although I attended Sunday school regularly from a young age and subsequently church, even becoming a Sunday school teacher in my late teens, I was often puzzled by the Bible's teachings and increasingly had my doubts. It was at a church youth-club[14] retreat one weekend, where a young curate was on hand to discuss theological issues with us, probably on the grounds that we teenagers would take more notice of someone in his late twenties than a superannuated rector (who was giving some of us opaque confirmation classes at the time). I found the nerve to ask him how he could be sure there was a God, and he replied that he had faith and that I needed faith, and that was it. When I pressed him about evidence and cited things in the Bible that were contradictory, he became tetchy and I realized, like large numbers since the Enlightenment, that there is no evidence. Belief demands blind faith that an imaginary Being exists; the creation of metaphysical musings of a nomadic community living in the Levant thousands of years previously. Phenomenology, of which I was then blissfully unaware, is the search for an ontology that does not require a deity to explain things. It is a philosophical viewpoint that for those with a numinous bent is too bleak to countenance, coming down to there being no point to life beyond 'being' for a brief period on planet earth – each of us an astonishing soup of molecules in a skin membrane, 'in the world'. And it explores the implications for what we can aspire to know, as self-aware organisms living on the congealed-custard-skin-like crust of a cosmically relatively small molten ball of rock hurtling through space-time.

Empiricism and its discontents

The advance of science has contributed significantly to the undermining of theological doctrines, with rigorous scientific enquiries invalidating scriptural teaching, revealing their mythical and inconsistent quality. These enquiries endorse the empirical philosophical approach to our understanding of the world, displacing metaphysical debates that assume the existence of a divine power. While empiricism embraces a range of views about what constitutes knowledge, all empiricist philosophers assert that knowledge comes only, or primarily, a posteriori from sensory impressions, emphasizing the role of experience and evidence. It contrasts with rationalism that asserts we may derive knowledge by a priori reasoning, intuition or revelation. For instance, Locke argued that the mind is a blank tablet that acquires knowledge through life via sensations and reflection. Hume sceptically argued

11 See A.J. Strathern (1996:1–6) for a helpful commentary on the dualistic Cartesian mind-body view.

12 What he called the 'body-subject' or more generally, later, the 'flesh' (Merleau-Ponty 1962, 1968).

13 In a nutshell, his argument is that we can only believe in a miracle if the evidence for it is so strong that it will demand another miracle to prove it false; that is, if the alternative is belief in a greater miracle.

14 Frankly, the club's attraction, looking back, was that it afforded an opportunity for a lad – who attended a single-sex school and (in those days) Boy Scouts – to meet girls!

that all ideas derive from 'impressions' or accumulated sensory experience, of which ideas are faint copies. Empiricism was subsequently central to the advance of the scientific approach, which requires one to test hypotheses against, preferably experimental, observation, with the pragmatists, such as Peirce and James, clarifying these methods. The positivists such as Frege and Russell sought to use mathematical logic to boil all knowledge statements, notably those of science, down to a verifiable code free of the ambiguities of language, and so demonstrate that woolly metaphysics concerns pseudo-problems. And Dewey, whose work, such as *Art as Experience*, directly informs current embodied arguments, subsequently formulated instrumentalism, which argues that interrelated past experiences constitute reality.

The ethno-determinist approach avowedly subscribes to this empiricist philosophical tradition with its insistence on the primacy of ethnographic evidence and attempt at scientific-like rigour – to give some content to the 'science' claim of 'social sciences' such as anthropology. But that is not all, as this philosophical excursus seeks to make clear in response to those who criticize the approach as naively positivistic and put down its focus on ethnography as promoting mere information gathering, rather than furthering academic anthropological enquiry that contributes intellectually to furthering understanding of humankind. I dispute this 'essentialism' and argue that imaginative constructions-cum-deconstructions that are not 'determined' by ethnography may contribute to social metaphysics but not to social science. These differences relate to Merleau-Ponty's (1962) point that philosophical arguments oscillate unhelpfully between two opposed points of view: empiricism and intellectualism. This is a venerable distinction that reflects a deeply rooted division in the Western intellectual tradition from Aristotle and Plato onwards.[15] On the one hand there are empirical arguments, which require concrete evidence from the world 'out there', that can be perceived with our senses and back up contentions (such as analysis of experimental data). On the other there are metaphysical arguments that demand belief in unverifiable innate ideas and the power of pure reason, which give us conjectures that depend on sheer imagination (such as discussions about divine beings). The 'embodied mind' thesis has roots in the empiricist tradition. Its argument that the mind is not something separate from the body cuts the ground from under much idealistic metaphysical philosophy.

Yet, while phenomenology endorses empiricism – in responding to the success of scientific enquiry – it simultaneously affords scope, paradoxically, for its censure. The dense philosophical argument admits contradictions, perhaps inevitably in view of the contrariness of human 'being in the world', which religious texts have played on brilliantly for millennia and the accommodation of which is central to structuralist interpretations. Phenomenology focuses on subjective 'lived experience' – encompassing perceptions, emotions and discriminations – and seeks to determine the properties of consciousness through reflective attentiveness. It concerns physical-mental appearances as opposed to concrete reality, maintaining that reality comprises things and events as perceived by the embodied mind, and not existing for us independently of human consciousness. The intention is to reveal the basic structures of human consciousness by focusing on 'phenomena', understood as things apprehended through concrete experience by the consciousness and not outside it. Philosophers vary in their interpretations of the approach. In confronting the mind-body dualism, for instance, Merleau-Ponty (1962:72) challenged the assumption that we can unambiguously distinguish the outside world from the thinking subject. The 'I' that knows is mixed-up with the 'it' I know. We cannot, after all, treat our body, object-like, as something that may or may not be present; it is always there for us, as we cannot be without it, its absence being inconceivable (Merleau-Ponty 1962:91). In a typical paradoxical play, he suggests that 'I am not in front of my body, I am in it or rather I am it ... If we can still speak of interpretation in relation to the perception of one's own body, we shall have to say that it interprets itself.' (Merleau-Ponty 1962:150).

In this way, phenomenology prefigures some aspects of subsequent 'postmodern' thought, particularly existentialist versions with their focus on subjective experience. It lends support to the argument that objectivity is not possible because subjectivity inevitably intervenes in understanding anything in the outside world. Radical representational versions imply that things out there have no independent existence – there is no objective world – it exists only in the mind. We cannot achieve objective understanding, for we can only ever have our own individual subjective understanding of 'out there'. It prompts some to criticize any approach that has a whiff of empiricism about it, such as arguing for the primacy of ethnographic evidence and striving for some scientific thoroughness, it being deeply unfashionable to refer to facts among many social researchers currently caught up in the postmodern debate, where subjectivity is the only true reality. Anthropologists who subscribe to this postmodernist view resort to autobiographical 'reflexive' accounts of their fieldwork experiences, as the only reality accessible to them. They also engage in the 'deconstruction' of anthropological texts, seeking to show how the author's subjectivity informs or distorts accounts.

In conceding the unavoidability of a certain subjectivity, ethno-determinism attempts to accommodate to

15 In discussing the schism, Gellner (1985:7) humorously refers to it as positivist Babbits versus Hegelian windbags.

postmodern criticism, not by adopting an egocentric focus, but by exploring the limits to thoroughly evidencing and understanding others' experiences, practices and ideas, which I understand to be a central aim of anthropology. The ambiguity of the position is captured by Nagel (1974:448) who, in wondering if it makes sense to 'ask what my experiences are really like, as opposed to how they appear to me?', notes that 'We cannot genuinely understand the hypothesis that their nature is captured in a physical description unless we understand the more fundamental idea that they have an objective nature (or that objective processes can have a subjective nature)'. In short, this book uses an account of observable construction activities to challenge contemporary interest in reflexively hyper-aware deconstruction. After all, if some of us do not construct, there is nothing for others to deconstruct. The implication is not that imagination has no part to play; even the most positivistic experimental science relies on leaps of intuition to further understanding. But we need an imagination tied to real world facts – things out there that we can sense bodily – not pure reason, free-floating intellect.

Individuals again

While the focus of phenomenology, particularly existentialism, on individual experience has stimulated postmodern-fuelled concerns for subjectivity, it relates to considerably wider issues than the contemporary reflexive turn. The individual figures prominently in Was valley social life, as mentioned in the previous chapter, and argued at length elsewhere (Sillitoe 1979a, 2010a), albeit doing so has met with considerable criticism from those wedded to the collectivist sociological perspective. According to a prominent colleague, for instance, 'the individual is not an analytical category'. But opposition is expectable in arguing for more attention to the individual because it challenges the widely accepted 'the social group is where it is all at' view that has long dominated social anthropology, which assumes that for those living in small-scale societies community interests subsume individual ones, and the collective-interested moral person has long eclipsed the self-interested manoeuvring individual (Ingold 2011b:221; Just 2004).

We see the extent of anthropology's focus on the social group to the exclusion of the individual throughout the twentieth century, for instance in its appropriation of such terms as idiographic and nomothetic, which relate to the distinction made earlier (see Chapter 1) between enquiries in specific geo-cultural locations and general cross-cultural investigations (Ingold 2008:70–2). The neo-Kantian philosopher Wilhelm Windelband first used these terms to denote two approaches to understanding, to distinguish clearly between the humanities and sciences (Windelband and Oakes 1980:167). Idiographic refers to particularization, and concerns things as unique phenomena, often subjectively, such as the historical description of events and individual acts, and involves ideological judgements. Nomothetic refers to generalization, and endeavours objectively to derive laws, such as in natural science, to account for classes of phenomena universally, and involves empirical observation.[16] When introduced into psychology by Gordon Allport (1937), idiographic came to refer to the individual, studied as a singular agent with her/his own unique experiences and specific life history. Nomothetic, on the other hand, came to refer to a group or class of persons, whose values and behaviours are taken as generally typical of the population.

When Radcliffe-Brown (1952:2–3) subsequently introduced the terms into anthropology, the individual disappeared. Idiographic refers to normatively structured interaction within a particular social group, and comprises ethnography. Nomothetic, in contrast, refers to cross-cultural comparison that seeks to formulate general laws or theoretical propositions about human behaviour universally, and characterizes anthropology. It is a distinction that we introduce to students in 'Anthropology 1' courses, more as something that they need to know historically than for its usefulness today (Ingold 2008:70). The distinction between ethnography and anthropology – one investigating particular socio-cultural arrangements and the other advancing understanding of humankind respectively – sits uncomfortably with ethno-determinism, which advocates that intellectual problems of universal relevance emerge from engagement with specific socio-cultural contexts, which as postmodernism points out, inescapably implies working in a cross-cultural comparative way (Sahlins 2002:12), unless you are sociologist working in your own community.

In retrospect, the idiographic and nomothetic discrimination has proved more confusing than enlightening in the social sciences, where both perspectives blend together, with individuals behaving not as free-wheeling beings but culturally habituated persons, who behave according to generally shared values and norms, albeit not programmed like automatons but behaving as calculating self-interested agents. The existential tension is caught by Koestler (1967:56, 246) who observes that:

> On the level of the individual, a certain amount of self-assertiveness – ambition, initiative, competition – is indispensable in a dynamic society. At the same time, of course, he is dependent on, and must be integrated into, his tribe or social group. If he is a well-adjusted

16 The distinction was subsequently critiqued and refined; for instance, natural science involves both the idiographic collection of particular facts and their nomothetic explication using general scientific principles, which parallels the ethno-determinist approach.

> person, the self-assertive tendency and its opposite, the integrative tendency, are more of less equally balanced ... the individual retains his character as ... a part-whole, who *qua* whole, enjoys autonomy with the limits of the restraints imposed by the interests of the community.

Consequently, a person is a

> Janus-faced entity who, looking inward, sees himself as a self-contained unique whole, looking outward as a dependant part. His self-assertive tendency is the dynamic manifestation of his unique wholeness, his autonomy and independence ... Its equally universal antagonist, the integrative tendency, expresses his dependence on the larger whole to which he belongs: his 'part-ness'.

It is noteworthy that Koestler (1967:67) was among the first to draw on von Bertalanffy's idea of the 'dividual', which has subsequently featured prominently in recent reassertions of the collectivist sociological perspective in New Guinea (Anderson 2011; A.J. Strathern and Stewart 1998; M. Strathern 1988) via the so-called 'new Melanesian ethnography', a viewpoint that differs markedly from my own.

During the period that I have sought to further understanding of life in the Was valley, social philosophy has also increasingly given attention to the individual, moving beyond a static view of society comprising solidary normatively structured groups to a dynamic one that sees human behaviour unfolding as an emergent network process. The tension between the idea of the collective or 'social structure' and associated normatively ruled behaviour, and the individual or 'social agent' engaged in negotiating, even recreating, the socio-cultural rules to which they are subject in the strategic pursuit of personal aims, had emerged by the 1990s as a prominent issue (Just 2004, 2005:2). The intellectual roots of some commentators – such as Bourdieu, Deleuze and Latour – go back to Tarde, marginalized for much of the twentieth century as the individualistic rival of the collectivistic Durkheim, who won the debate at the time; Radcliffe-Brown subsequently introducing his superorganic society wholesale into British anthropology. According to Tarde, sociology should concern observable interpersonal processes because it is the interactions of individuals that comprise any social aggregate (Candea 2010; Latour 2002; Tarde 1899). This clashed with the collectivist view that asserts we cannot understand society by investigating the actions of individuals, on the grounds that there is a superordinate intangible entity that imposes norms that guide the behaviour of persons; whereas Tarde argued that interaction between tangible individuals generates accepted practice, with innovation and imitation fundamental forces.

While focusing on the individual may disquiet social anthropology, which acquiesces to the socio-determinism of sociology, attention to the individual complies with the approach taken by economics to human behaviour. Not that seeking to switch disciplines and styling this book as an economic study might lessen criticism. Indeed, while it seeks to further understanding of the Was valley 'economy', in so far as it considers material aspects of life (Chapter 1), it actually challenges the assumptions of formal economics, which are, paradoxically enough, of doubtful relevance to local arrangements (Sillitoe 2010a). The Highlands socio-cultural context transforms presumed universal economic principles, starkly revealing for instance the inadequacy of taking only a narrow range of individual emotions – desire, selfishness, envy – as underlying human behaviour; whereas the Was valley obliges us to allow, after Hume, for a wider range of human emotions – respect, altruism, generosity. This gives us a more rounded understanding of human behaviour and allows for its contradictory character, as some economists are now realizing with calls for economics of well-being and happiness in addition to income and expenditure.

Considering emotions returns us to tacit issues informing our behavioural responses and social interaction, which are implicit in large part. While the focus of economics on the individual squares with tacit concerns, tacitness further challenges economic assumptions, such as the one that actors make rational choices to maximize their self-interested satisfactions when faced with limited means (see Chapter 3 on choice in the context of house location; Sillitoe 2010a:126–68). Acknowledging that economic behaviour, like any other social behaviour, features tacit awareness has significant implications for the rational decision-maker model. When shopping, for instance, we often make habitual choices without actually thinking consciously about them, sometimes with emotional undertones (such as I buy this brand because my mother bought it). We see it in the behaviour of persons in supermarkets (those cornucopia of choice), many simply putting in their trolleys what they usually buy, without shopping around for the best deal. Few shoppers stand for several minutes, for instance, surveying the selection of every imaginable (and some unimaginable) breakfast cereal, deliberating over what packet to choose. One of the aims of advertising is to get consumers to break with this tacit behaviour and choose something new – the advertised product.

While emotions underlie human behaviour, anthropology neglects them largely – in overlooking the individual 'as not an analytical category'. Instead a reified notion of society's interests, such as solidarity, stands in. The imposition of the Western abstraction of 'society' – thought of as a moral code that informs collective social behaviour with shared expectations, responsibilities, obligations and so on, inculcated during socialization and passed down the generations to give us historically founded arrangements – on those who have no such idea, as in the Was valley, demands care (Sillitoe 2010a:65–72). And to assume that the society concept accounts for human behaviour is to give at best a partial interpretation, if not a distorted one, in failing to address the individual emotional roots of social interaction and its self-interested aspects, in which actors seek to manipulate social expectations and so on to personal advantage. Furthermore, tacit knowledge, which is of interest here, is, by definition, of the individual person, a locus that demands attention to individual action, albeit conditioned and constrained in some measure by communal expectations and commitments.

Mindfulness

While phenomenological philosophy has a subjective aspect, picked up by existentialism and postmodernism, it also has an objective side. For instance, Merleau-Ponty refers to the body as an 'object', suggesting the existence of an objective world 'out there'. In an ethno-determinist response to subjectivity, this chapter draws attention to the less widely acknowledged implications of some features of this more positivistic aspect of the knowing individual. For phenomenology seeks to transcend the idealist view, which reduces the world to the impression that we have of it in our heads (or phenomena as hidden 'things-in-themselves'), and the opposing realist view, which has the external world existing out there independently of what our minds make of it (or phenomena as revealed 'things-in-themselves'). The notion of 'intentionality', which Husserl owed to Brentano, is central to this endeavour. It has consciousness as always about something, the 'intentional object', and 'intentionality' is often referred to as 'aboutness'. While the consciousness may instantiate an object in various ways – through sensory perception, recollection, symbolic representation – the object does not change. It recalls Kant's notion of the 'thing-in-itself' that he posited to explain how humans make sense of the world, which is the reality that stands beyond the efforts of human understanding but towards which our understanding orients itself. These intentionalities or ways 'about' the object may differ, but it remains the same object out there whether perceived directly or recalled later for consciousness to be 'about'. The Houses of Parliament are the Houses of Parliament whether I stand in Westminster looking up at Barry and Pugin's edifice, hear its bell's hourly chimes on the radio or remember seeing it as I crossed the river Thames!

The intentional perspective challenges the representational view of consciousness, which argues that we can only apprehend reality indirectly as mental images and not directly. Husserl (1982) argued that consciousness concerns something other than itself. This is the intentional object, which the phenomenological method seeks to apprehend and describe in its 'immediacy', as it presents itself to our consciousness through intuitive prehension of sensory experiences before we conceptually elaborate on them. The goal is to explicate the structuring of acts of consciousness through 'eidetic reduction', vividly experienced imagery. In this way, 'intentionality' features in our individual conscious structuring of our experience of self and things, encompassing the analysis of enabling conditions such as bodily skills and cultural habitus, including language. The relationship between consciousness and things means that phenomena are as they seem to us, and with being and appearing thus coinciding, the method investigates the constitution of meaning. The idea of 'intentionality', having us always directing the consciousness towards things in acts of perception, analysis and imagination, connects with the empiricist tradition. It assumes an objective world out there for consciousness to be 'about', which the ethno-determinist position accords with, in its emphasis on ethnographic evidence. The parallels do not stop there. The 'intentional object' attempts to engage openly with the reality of experienced phenomena to identify their actual qualities, unlike closed philosophical approaches that rely on abstract theories to comprehend their features. The assertion that we need to focus on experience of things, accords again with ethno-determinism, namely allowing ethnographic actuality to lead enquiries with an open mind.

The phenomenological approach consists of realizing the presence of something and immediately grasping its meaning intuitively. It seeks to comprehend the 'essential structures of the directly given'. The apprehension of phenomena has a somewhat transcendental quality, as something that appears, comprising sudden perception of intelligible content. It excludes anything that does not immediately present itself to the mind. The immediacy of the experience is central to exposing the elemental structure by pre-empting any prior assumptions about existence and causality. According to Husserl's (1982) *epoché* method,[17] it is necessary to 'bracket' out the existence of the rest of the world to achieve this understanding and focus attention on awareness of experience, by intentionally directing consciousness towards something. This version of phenomenology seeks to return us to a primordial experience of the thing, to perceive it without prejudicing perception with what we already think we know, take for granted or infer, as this diminishes understanding of the reality. It is necessary to rely on an intuitive grasp, allowing for 'naïve' experience of the object to inform knowledge of it, suspending or 'bracketing' presuppositions, beliefs and prior judgements.

17 After the Greek sceptics' idea of abstaining from belief, further signalling how this philosophy was a response to the death of the idea of deity.

It seems that 'bracketing' somehow involves accessing the subconscious directly, inviting parallels with tacit understanding, as picked up in subsequent embodiment discourse; for instance Merleau-Ponty approaches embodiment as the body's practical capacity to act, which equates with tacit knowing. But the approach presents problems in seeking to apprehend phenomena supposedly without any presuppositions, restricting attention to the authentic facts before us, untainted by everyday prejudices, scientific theories or philosophical schema. While ethno-determinism agrees with the injunction about focusing on the experience of things, namely the ethnography leading enquiries, it does not subscribe to the proposition of 'bracketing out' or suspending prior beliefs, which suggests focusing on a non-existent unsocialized individual (Sillitoe 2010a:63, 110). The idealistic presuppositionless state of 'phenomenological reduction', which is reminiscent of the philosophers' hypothetical 'state of nature', assumes the suspension of our knowledge and beliefs about the world and presumptions about the factual existence of phenomena, to allow the 'transcendental ego' to be the true source of knowledge. It may be imagined in a metaphysical sense, but is impossible to conceive in reality, for all observers are situated in the world and cannot empty their minds, such that it is not possible in reality to act on, as opposed to positing, some imaginary 'pure' or prior uninformed consciousness.

It is here that ethno-determinism draws on postmodern concerns about subjectivity, in contrast to aforementioned autobiographic responses. We always have our culturally programmed understanding. There is no such human being as an unsocialized individual. We clearly cannot escape from our own socio-cultural backgrounds and personal histories. We may or may not be born with a blank hard-drive, to up-date Locke's imagery, but we do not invent our own software through our sensory experiences in the world. Our understanding is programmed for us in large part by our socio-cultural heritage, which is the accumulation over millennia of our ancestors' experiences and reasoning (arguably extending back to the pre-hominid era). We review this inheritance recurrently against our experience of the world and may modify it in the light of that experience (as scientific enquiry, for instance, has modified beliefs about being, making religion redundant for many), and we hand on what we think we know to the next generation, and so on into the future. This perspective complies with later phenomenology, of Heidegger and subsequently Merleau-Ponty, who argue that we have to think of humans' 'being-in-the-world'. Knowledge not only exists 'in the body' but also 'in the world', in the sense that humans pass it on culturally across generations and inscribe some aspects in their surrounding environment, posing the question as to where the knowing body ends and the world begins, taken up in later chapters.

Dwelling in philosophy

According to Heidegger (1962), who went on to explore the ontological meaning of Being itself through the temporal dimensions of experience, our encounters are 'in the world', which we do not bracket out in interpreting experience, rather we consider these encounters and their meaning 'being-in-the-world', according to context and our relations to things. This returns us to the holistic and hermeneutic issues of the previous chapter, which here situate understanding as an aspect of *Dasein*, 'existence',[18] a manner of being informed by our socio-cultural conditioning and experiences. Appositely for a study of house building, the notion of 'dwelling' features in this phenomenological discourse (Heidegger 1993). The act of dwelling, as a social activity, captures how our immersion in a socio-cultural tradition informs our experience of the phenomenal world and understanding of it. According to Heidegger (1993:350) we are dwellers, for it is in the nature of being to dwell. He uses the image of 'dwelling' (*wohnen*),[19] which he imagines as a house in the forest with paths leading to it, to evoke the character of our 'being-in-the-world' (*Dasein*) and to give meaning to life; a trope subsequently drawn on by some anthropologists (Anderson 2011; Ingold 2000; Weiner 1991).

Dwelling here relates to pre-reflective experience rather than conscious deliberation, concerning tacit matters in the first instance not theoretical reflection, and revealed to us by the way we are in the world, which is intuitively familiar. Using the dwelling metaphor, Heidegger (1993:350) argues that 'we do not dwell because we have built, but we build and have built because we dwell'. In other words, we do not build to dwell but because we dwell. And thinking relates to dwelling in the way that building relates to dwelling. Our experience of dwelling- or being-in-the-world fundamentally informs our thinking, just as dwelling motivates our building. We do not see and interpret the world as if set apart from it, but we experience it – it is revealed to us – as 'dwellers' in it, via our interactions with our environment (1962:95). Embodied inherence in the world is more fundamental than cognitive perception. Our basic way of relating to things is through practical activities, such as the hammering in a nail discussed earlier, which Heidegger distinguishes from representational intentionality, such as thinking about the hammer. The connection of tacit familiarity with conscious reflection involves 'interpretation', not necessarily propositional, and concerns hermeneutics, in that a part of the implicitly

18 'Being-in-itself' is the conventional translation of Heidegger's (1962) obscure *Dasein* concept, which lingers uneasily between an abstract term for 'being human' and a concrete one for actual human beings.

19 Meaning, among other things, 'to stay in a place' or 'to dwell at peace' (Heidegger 1993:350).

experienced whole comes explicitly to our attention, such as in the famous breaking hammer that suddenly makes us consciously aware of the tool. Phenomenology here concerns the practice of 'letting things show themselves'. Interpretation depends on the object appearing consciously as something, 'disclosing' it within the surroundings of the world, part of the totality of subjective experiences and routines. It is given linguistic signification by 'assertion', that is association with a particular sound sequence (words), which is affirmed, as is its truth value, by the world disclosing integration of interpretation and understanding. This disclosure of knowing is prior to that of verbal propositions such as 'a hammer is heavy'.

There are parallels with the idea of 'prehension' in process philosophy, which refers to interaction with something that involves perception but not necessarily cognition.[20] According to Whitehead, who argued that apart from experience there is nothingness, it comprises 'uncognitive apprehension' (1925:69). He introduced the term prehension[21] in his empirically focused investigations into the relations between people and their proximate environment to account for those relations that depend not on conscious knowledge but are more akin to 'grasping' aspects of the world. This 'pre-epistemic' grasp is where we connect with the world and we subsequently glean from it and accumulate our knowledge of things through a process of abstraction. According to Whitehead, there are two kinds of physical prehension: that registered by the senses as pure appearance: 'presentational immediacy' (such as what our optic nerves convey visually to us); and that comprised of diffuse feelings and environmental influences: 'causal efficacy' (such as feelings of joy, fear, anger – Hume's passions). Both of these combine via an automatic unconscious process, which comprises the 'symbolic reference' that connects sense perceptions with causal factors. This understanding again challenges the mind and matter duality, for according to Whitehead 'the sharp division between nature and life' established by Descartes 'poisoned all subsequent philosophy' (1938:204).

The phenomenological approach, focusing on unconsciously embodied knowledge, reveals, as Heidegger points out, 'That which is ontically *closest and well known*, is *ontologically* the farthest and not known at all; and its *ontological* signification is constantly overlooked' (1962:69, orig. emph.). Featuring intuitive perception, knowing depends on 'being-in-itself', conditioned by living within a particular cultural and historical context of practice and understanding. And he goes on to argue that ' not only must this entity not be missed … but it must be made accessible by a positive characterisation' (1962:69). This suggests going from the unconscious, where context is often implicit, through the semi-conscious to conscious awareness of experience. This book explores whether we can achieve such a 'positive characterisation' regarding tacit knowledge. While we can attempt to describe our experience of something (Husserl's and Merleau-Ponty's 'pure description of lived experience') and then interpret it by relating to context (Heidegger's social and linguistic 'hermeneutics'), it is arguable that any process rendering such experience explicit will destroy it, which is maybe what Heidegger (1962:69) is hinting when he refers to 'the mode of average everydayness, even if this is only the mode of fleeing *in the face of it* [*Dasein* 'existence'] and forgetfulness *thereof*' (orig. emph.), which is otherwise a mystifying injunction.

We know that when we seek to gain intellectual control of the phenomenon of embodiment in human action, of embodied 'knowledge', we hamper our ability to do the things we do as corporeal beings. There is no clear distinction between individuals and the objects of their concern; self-awareness melds with wider discernment. Accordingly, we do not apprehend reality by collecting together and arranging experience of it to derive universal propositions that correspond, to varying extents, with the world. If such knowing is essentially irreproducible in the explicit propositional terms that are the focus of Western academic endeavours, this may explain why philosophical endeavours to access it result in such convoluted and inaccessible arguments. They are wrestling with the impossible. We are left with the problem central to this work of how we can at least achieve a somewhat accurate view (if not true understanding) of aspects of life where tacit knowledge is central.

Ethno-determinism offers one way to tackle these phenomenological conundrums and assess the limits of our understanding of tacit-knowledge issues. It seeks to capture something about the unconscious and uncommunicative aspects of experiential knowledge and associated behaviour and their consequences, and further understanding of how humans are in the world and how they order their lives in it. While the term ethno-determinism may have been coined as a provocative nose-thumbing gesture to certain ideological-cum-theoretical views (structuralism, Marxism, post-theories of various kinds), it is proving unexpectedly useful in framing attention on some current issues, even achieving some intellectual leverage in respect of certain strands of philosophical thought and potentially contributing to various related intellectual debates. It draws, as mentioned previously, on a long-established anthropological tradition dating from when our forerunners first stayed in communities for extended periods to further understanding of peoples' lifeways. This is not to imply any criticism of them for not seeing parallels

20 Roy Ellen (2006:28–9) uses the idea of prehension in discussing the cognitive process of classification, albeit in a somewhat different way. He uses it to underline the significance of ethnographic situation in understanding classificatory behaviour.

21 From the Latin *prehendere*, 'to grasp or seize'.

with contemporary phenomenological debates, being too busy trying to make sense of Trobriand, Inuit, Australian Aboriginal and American Indian ways to have time to delve into arguments then emerging in philosophy, sometimes in obscure controversies, comprising subtle and confusing blends of subjectivity and objectivity.

The phenomenology of ethno-determinism concerns our ability to access, or not, tacit knowledge in furthering understanding of other culturally framed ways of being in the world. It takes its cue, regarding whatever limited understanding we can achieve, from the actors who possess the tacit know-how. Anything else would amount to the imposition of our own subjective rationalizations and speculations on others. We see this in social-science writing that advances hypotheses to account for social value codes and associated patterns of observed behaviour in terms of its theories, sometimes featuring teleological explanations of surmised sociological outcomes. There is nothing wrong in such endeavours, so long as we are clear about what we are doing and do not suggest that they represent local understandings – I have engaged in them myself, for instance in accounting for exchange institutions in reference to social order in the acephalous Was valley (Sillitoe 1979a).[22] The conundrum is that the actors themselves are apparently unaware of the inferred sociological implications – they just drop out of their tacit culturally informed social interaction – such that the outside observer appears to 'know' more about their behaviour than the actors themselves who are responsible for it; a conundrum explored further in the next chapter. Understanding such experiential aspects of being is surely central to furthering knowledge of human behaviour and associated socio-cultural arrangements.

We need ethnographic evidence, as comprehensive as possible, to guide and support, and subsequently assess, whatever inevitably partial insights and interpretation we achieve. The problem with social behaviour is that it is gone as soon as it appears, leaving no physical trace to further assess the already invisible tacit awareness that informs it. While an account of house building may seem an unpromising, if not jejune topic initially, it may contribute to the 'construction' of a sounder understanding of tacit issues, focusing on knowledge with respect to 'things out there', namely houses and other structures, which afford something tangible to relate to, and against which to test ideas. It is the concreteness of houses after all that gives them their evocative power, as seen in Heidegger's (1993) use of the dwelling metaphor to evoke the way we are in and experience the phenomenal world. *Built in Niugini* is consequently not just about constructing houses, it is also about constructing understanding. In focusing on physical things, it accords with the phenomenological approach that maintains that the way we embodied beings relate to phenomena is fundamentally through concrete interactions, where interpretation depends on objects 'disclosing' themselves to consciousness, as aspects of lived experience, and where such embodied 'knowledge' blurs the distinction between cognitive agents and interpreted objects.

22 Alternatively, those who now fight shy of attempting any such explanations of social interaction – following criticisms of notions such as structure and norm, or more recently structuration and habitus – may take refuge in reflexive accounts, feeling intellectually secure cowering in their own postmodern navels, where their existential reality is as valid as anyone else's.

3

Locating the house, locating the social

The first task in building a house is deciding where to build it. In the Was valley, this involves consideration of site factors that we can take as falling into two categories: physical characteristics and social aspects. These distinctions are used as a heuristic device; the Wola do not customarily make the distinctions, although we may assume that they are intrinsic in some measure to their decision making. Indeed, they do not ordinarily think consciously about the issues as presented here, but, in keeping with this book's tacit theme, they draw implicitly on their knowledge and the ramifications of siting a house in one location compared to another. Often they have no reason to consider these issues, either explicitly or implicitly, as they frequently decide to rebuild a house on its current site, where the present inhabitants or their forebears decided to build previously, and the location has proved sound in the experience of those who have lived there subsequently. This discussion of issues is not entirely as alien as these comments suggest, for although people may not consciously juggle these issues in deciding where to build a house, they can discuss some of them, at least when asked questions about why they locate houses where they do.

Generally speaking, the choice of house location involves consideration of some issues similar to those that feature in the choice of garden sites (Sillitoe 2010a:126–68). We are not considering verbally inexpressible skill in these instances, but rather the grey area between formal and tacit knowing; that is implicit, what is normally acted upon without being brought into conscious focus, but which actors can discuss if prompted to do so by some incident, such as having an outsider – for instance a curious anthropologist – asking questions about their actions. We can consequently document and discuss some of this knowledge, although it is more doubtful if we can appreciate its integrated consideration in the performance of deciding the location of a house. This is because some of the knowledge, notably that relating to social issues, is tacit, and also because making a decision assumes an intimate knowledge of any local territory, its topography, history of occupation and so on.

Physical factors

Regarding our first category, a site's physical characteristics, the topography of the land is an obvious issue. There is little naturally level land in the Was valley that readily lends itself to being a house site – what there is, is usually poorly drained, even waterlogged – and so men have to find places where they can excavate a flat site of sufficient size. They look for locations where they can level the ground with relatively little effort. These are places on gentle slopes, where a house-yard sized platform can be readily excavated. Another consideration regarding slope is stability. The terrain of the Was valley is precipitous with many steep slopes, and occasional earth movements, sometimes catastrophic landslides, are an ever present threat, especially following periods of intense rainfall (Sillitoe 1996:134–6). It would be reckless to level a house platform on anything more than a gentle slope. The threat of landslides is something that people have at the back of their minds, with folk memories of families engulfed and buried alive. An indication of a safe site, people point out, is where rock outcrops at the surface. They say that such places are unlikely to experience serious earth movements. Outcrops of limestone are, in fact, a common feature of ridges and knolls in the Was valley.

People favour ridges and knolls for another reason, which is that the 'sun sees' such locations. When I pointed out a level site at a place called Tetaenda that was adjacent to a friend's garden, for instance, and suggested that it would make a good location for him to construct a house – albeit the garden was only about a quarter of an hour's walk away from his current house – he and others disagreed vocally, saying '*natom suw na henday*', the 'sun would not see the place'. The location was situated in the bottom of a steep-sided fold and was consequently in the shade for much of the day. This was bad because a house in such a place would be perennially damp. After rain, which falls almost daily, it is necessary for a house to be in the sun when it next returns, so that it dries out. A house that is constantly damp rots quickly. Such a house is also unhealthy, being clammy and cold, and its occupants will fall sick. The location of houses higher on valley sides may also put them above the

	<1 min.	1–5 min.	5–10 min.	10–15 min.	15–20 min.	20–25 min.	25–30 min.	30–45 min.	45 min.–1 hr	1–2 hr	>2 hr	Totals
					Mixed vegetable gardens (em gəmb)							
no. gdns	60	6	1	2	0	6	1	0	1	0	0	77
					Taro gardens (ma em)							
no. gdns	10	12	4	4	2	5	4	2	10	3	4	60
					Sweet potato gardens (hokay em)							
no. gdns	54	79	64	64	31	40	31	54	39	5	0	461
					All types of gardens combined							
totals	124	97	69	70	33	51	36	56	50	8	4	598
per cent	20.7	16.2	11.5	11.7	5.5	8.5	6.0	9.4	8.4	1.3	0.7	100

Table 3.1 Distance from houses to gardens.

chilly and damp morning mist that often fills valleys; Clarke (1971:100) points this out for the Maring too, and goes on to comment that some of the idiosyncratic answers that he received when he asked 'why they placed their houses on lofty sites' indicate that individuals had not previously given the question any thought, but tacitly accepted it as customary practice.

Interestingly, regarding the location of many houses on ridges and knolls, no one in the Was valley has ever mentioned that this is because these are good defensive positions (unlike some New Guinea people).[1] The fighting that occurs in the region is not the sort where men defend fortified positions. Skirmishes are fluid and fast-moving encounters in which the front line can move considerable distances in short periods of time (Sillitoe and Sillitoe 2009:49–65). In the event of a rout, any houses in the area overrun, whatever their location, are likely to be set on fire. It is surprise raids on homesteads, particularly at night, which men have to guard against, especially if their houses are near the frontiers of their *sem* territories and more likely to suffer raids than those located in the heartland, where raiders are unlikely to penetrate without serious risk to their lives. Regarding defence against such raids, the Wola point to thick brakes of *gaimb* sword grass[2] around homesteads, which are impenetrable without cutting a path (the Enga similarly deploy cane grass for defence, Meggitt 1977:48). These make a stealthy approach impossible. Men may supplement weak points, such as paths, with high stake fences that have narrow gateways, sometimes with stiles, which they can block off with heavy timber in times of hostilities. These barriers may also serve to keep foraging pigs out of house yards.

Another consideration with respect to the location of houses is convenience of access to at least some of the residents' gardens (Sillitoe 2010a:128–39). For instance, one-fifth of gardens are less than one minute's walk away from the nearest house of the family cultivating the area, increasing to one-third of gardens within five minutes' walk (Table 3.1).[3] People visit gardens regularly, and woman almost daily, to cultivate and harvest food, and so distance is a significant consideration. Regarding convenience, ready access to building materials, on the other hand, is not much of an issue. Men do not construct houses where they necessarily have easy access to the raw materials they need. They may transport these over considerable distances, even from remote forest locales, to a house site.

Social factors

The convenience issue brings us to our second category, social considerations informing house location. House-builders tend to look for locations adjacent to kin living in the same vicinity because it is convenient to reside near those on whom one can call for assistance with everyday tasks if necessary. It also ensures the mutual security necessary in an acephalous political environment, kin looking to one another for support in disputes, especially if it appears they may spark violence. At root we might relate this tendency to locate near kin to the human urge for sociability, which has been central to our species' success. The sociability consideration is evident in the Was valley with the siting of many houses around the *howma* 'grass clearings' mentioned in Chapter 1, where people periodically engage in collaborative activities such as pig-kill festivals. It is usual to find *enza* 'men's houses' adjacent to such *howma* clearings with some associated *tenda* 'women's houses' nearby.

Sometimes the reverse occurs, and a breakdown in social relations influences the location of houses. If relatives are embroiled in an acrimonious dispute, one party may move away, building a house elsewhere, either in the same neighbourhood, or if the split is particularly bad, on the territory of another *sem* 'family' where it has kin (Sillitoe 2010a:117–18). For instance, after a furious dispute in which a man who had served as local government councillor for Haelalinja accused his younger brother of having intimate relations with his second wife, he moved with his family from the Was valley to a place called Sezinda in the

1 Such as the Chimbu, Goilala, Etoro, Enga, Daribi and Tairora, according to Brown 1973:29; Hallpike 1977:55; Kelly 1977:21; Meggitt 1977:48–9; Wagner 1967:18–19; Watson 1983:65.

2 *Miscanthus floridulus.*

3 For an account of the three garden types, see Sillitoe 2010a:16.

neighbouring Nembi valley, never to return – although his first wife, who was unhappy at the move, did return on occasion to live in the *tenda* 'woman's house' of her husband's father's sister's son.

The behaviour of the one-time councillor's wife relates to the issue of how families decide where to locate their homes. The decision-making process is similar to that described for determining where to clear new gardens (Sillitoe 2010a:135–6). It is more by implicit agreement than through open dialogue that most families decide on the location of houses. But in some families a dominant man may decide where to build a new house with little or no apparent consideration of others, even in respect of a *tenda* 'woman's house' for a wife and children; albeit he has to exercise some care not to infringe a wife's rights to expect her home to be reasonably located with respect to the factors discussed here, or she may desert him, returning to her natal kin, who will support her if they agree that he has infringed her rightful expectations (in which event it is likely to cost the husband, as he will be expected to pay his wife and her kin some compensation for his poor behaviour to induce her to return and live in his household). The behaviour of the councillor's first wife is of this sort, albeit she returned to be with some of her husband's paternal kin with whom she had warm relations and not her blood kin.

In many cases families have no decision to make because, as pointed out, men frequently rebuild houses on their previous sites. The data with the interactive maps at the Wolaland website (Wolaland.org) show that 32 per cent of houses were located at the same place for at least 15 years, dropping to 21 per cent on the same site for 20 or more years[4] (some of the newly sited houses surveyed during this period were constructed adjacent to old house sites and required little, if any, site work, as they occupied the environs of a previously levelled house-yard area. Many of these houses also featured the reuse of some building materials, which can be remarkably durable (Saini 1966).

Overt decisions about where to situate a house revolve less around the specific considerations discussed here, which largely inform decisions in a tacit way, and more on the general issue of the *sem* neighbourhood in which the family is going to live; for instance, will it be among the husband's kin or the wife's kin. A family will potentially have residential access to several *sem* territories – wherever its members have extant exchange-validated social relations with some of those living at these places – such that a man's father's mother's or mother's father's or his wife's father's or her mother's relatives will likely acknowledge that he and his family can reside in the same *sem* neighbourhood as themselves. A family may settle any internal differences as to place of residence, and which kin to live near, by maintaining two homesteads at locations on two different *sem* territories. The Wola refer to such families as *suw pa kab* 'two places' (Sillitoe 2010a:218–19). They usually maintain gardens in both places and migrate back and forth between them, spending some months at each in turn.

Land rights factors

The determining issue regarding house location is whether current residents recognize that a family has rights to cultivate land at the place – which implies extant transactionally substantiated social relations with them – for a family is unlikely to build a house on a territory where it has no gardens, as signalled by the fact that one of the factors informing house location is proximity to at least some of a family's gardens. Nearness to gardens usually means having land-tenure rights at a place, although such a connection is not obligatory to build a house there. But it is usual. While people maintain that whether a family has rights of tenure or not to a site suitable for a house is not necessarily an issue because the amount of land involved is too small to quibble over (unlike land for gardens, where people do watch tenure rights closely – Sillitoe 2010a:85–125), in reality a family is unlikely to build a house on land where it does not have at least *semonda* connections. Depending on relations, a family may build a house on land claimed by another *semgənk*, if it is particularly convenient, after making sure that relatives with land rights there do not object, which is unlikely as the area involved will be trivial. But even so, to ensure security of occupancy, persons may decide to build on land to which their associated *semgənk* has rights of ownership.

An episode when building the house documented in this book (see Chapter 7) illustrates how tenure can be an issue, regardless of widely held opinion that the area required for a house is inconsequential. It turned out that tenure rights to the land we cleared and levelled at Honael for the house platform were in contention. The man who disputed our right to build a house on the site intervened physically several times, and obstructed the work, until we were able to mollify him with an offer of a payment. The site was located within an abandoned garden area that a relative of the disputant, who was helping with the construction of the house (a close friend who often helped me in my work), had cultivated previously and he was happy for us to build there. But it transpired that one of his deceased relatives (his FMBS), who was the disputant's father, had cleared and gardened the area before him, meaning that the disputant and his descendants arguably had prior rights to the land – that my friend only had temporary cultivation access and did not have inheritable land rights (Sillitoe 2010a:95–6). On one occasion violence almost broke out, when the disputant pulled up some of the *dol* 'stakes' that we had erected as one side wall of the house – having painstakingly sighted them in a straight line – and tore off the top row of *gamonk* 'beading', following which my friend and he

4 There are 150 houses in the sample.

grabbed fence stakes and tried to hit one another, but others intervened and disarmed them. After the near fisticuffs another relative of the pair (a FBS of the person disputing our rights to build on the Honael site), who was also helping with the house construction and who subsequently took possession of the house for his family home, agreed to pay the disputant two pearl shells and K8 in money[5] if he backed off. He took the K8 on offer there and then, and stormed off, but that evening returned the money, saying that he did not want payment after all.[6] Such disputes over the location of houses are, however, unusual.

While families tend to build houses close to those of other kinsfolk, resulting in residential clusters that people recognize as named *sem* 'family' groups – both *semgənk* 'small families' and aggregations of these into *semonda* 'large families' (as mentioned in Chapter 1) – people do not talk about recruiting persons to these groups *per se* as political entities. There are no recruitment rules to these collectivities beyond the rules of land tenure that control access to the land and other resources on the territory associated with the *sem* corporation. It is their association with geographical territories that gives them permanence, with land featuring significantly in their constitution, as we have seen. They reflect the haphazard outcomes of everyday life and social interaction, with networked individual relations radiating out from them across and beyond the Was valley. These groups have a patrifilial bias, as noted, both in recruitment (men prefer to reside where their fathers did and they probably grew up, as they 'belong' here more than elsewhere) and in ideology (persons with female ancestors in their pedigrees may be distinguished from those with all male connections to *sem* founders[7]). It is a matter of informal cluster filiation as opposed to formal group affiliation. While many men may prefer to reside at their fathers' places – labelled as a process of so-called 'cumulative patrifiliation' (Barnes 1962)[8] – this is because having grown up there they know the locales intimately and have long-established relations with others, rather than because of any socio-politically informed agnatic-descent ideology or residence rule, as I have argued at length previously (Sillitoe 1979a, 2010a). Although an agnatic preference is evident, there are no physical differences between the house sites of those who exercise paternal claims and others who locate their houses on land claimed through their mothers or wives.

The agnatic bias in the composition of local groups suggests, as mentioned above, that men dominate decisions about where to locate the family home. When planning to build a new house, they are unlikely to discuss their plans at length with others, although they may mention them in passing during conversation. This accords with implicit judgment informing house site selection: men proceeding to build houses at particular locations without needing to engage in discussions to assess the various criteria that may inform their decisions. But women are not excluded from contributing to decisions about where to build houses, although the extent of their influence varies. It would be unusual for a man to start on the construction of a house without the issue coming up during everyday family conversation with his wife (or wives) and any other female members of his household, and for them to talk it over. It would be unwise to ignore their opinions, as pointed out, particularly if he intends to construct a house on land claimed through connections to their kin.

The temper of such informal household deliberations varies as much as the personalities of those involved, family relations featuring a wide range of interaction.[9] The men in some households may seek, as described, to impose their wishes on female relatives, but they have to beware of pushing them beyond a certain point, such that they return to their natal kin pleading ill-treatment. The opposite can be found in households where the woman is a forceful character and the man is laid-back, more likely to agree to her demands, to have a peaceful life, although she would be unwise to push him to make decisions that his kin think unacceptable and find herself facing his annoyed relatives. Many relationships lie between these two: some families comprise a relaxed pair who reach decisions harmoniously, with occasional differences, and others comprise two strong characters where quarrels are common and they sometimes come to blows; albeit, it is not always the man who commits the assault, although it often is, sometimes prompting the woman to leave for her natal place. If there is a strong desire on the part of both a man and a woman that they live in different places, such as both of their natal locales, they may decide to be a *suw pa kab* household, maintaining houses at both places, migrating periodically between them.

5 K = Kina, the national currency of Papua New Guinea.

6 I never discovered whether he experienced some shame at extorting the payment, given the questionable status of his land-tenure claim, or whether he felt satisfied that he had made his point over the land-rights position. I was told that he was disgruntled at me giving three pearl shells each to two of the men helping to build the house, and he may have been trying to force me to pay him something, and was discomfited when the strategy backfired.

7 My friend who had last cultivated on the aforementioned disputed Honael site had a female ancestor in his *sem* affiliation that the disputant who had an all-male ancestoral connection mentioned during their argument, implying that he had stronger rights.

8 These points and the following pertain to the descent-group controversy that was once prominent in debates about the nature of New Guinea highland society (Barnes 1962; Langness 1964; Sillitoe 2010a:45–6; Strathern 1972).

9 See also Sillitoe 2003:298–9 and 2010a:135–6 on family discussions about pigs and garden location.

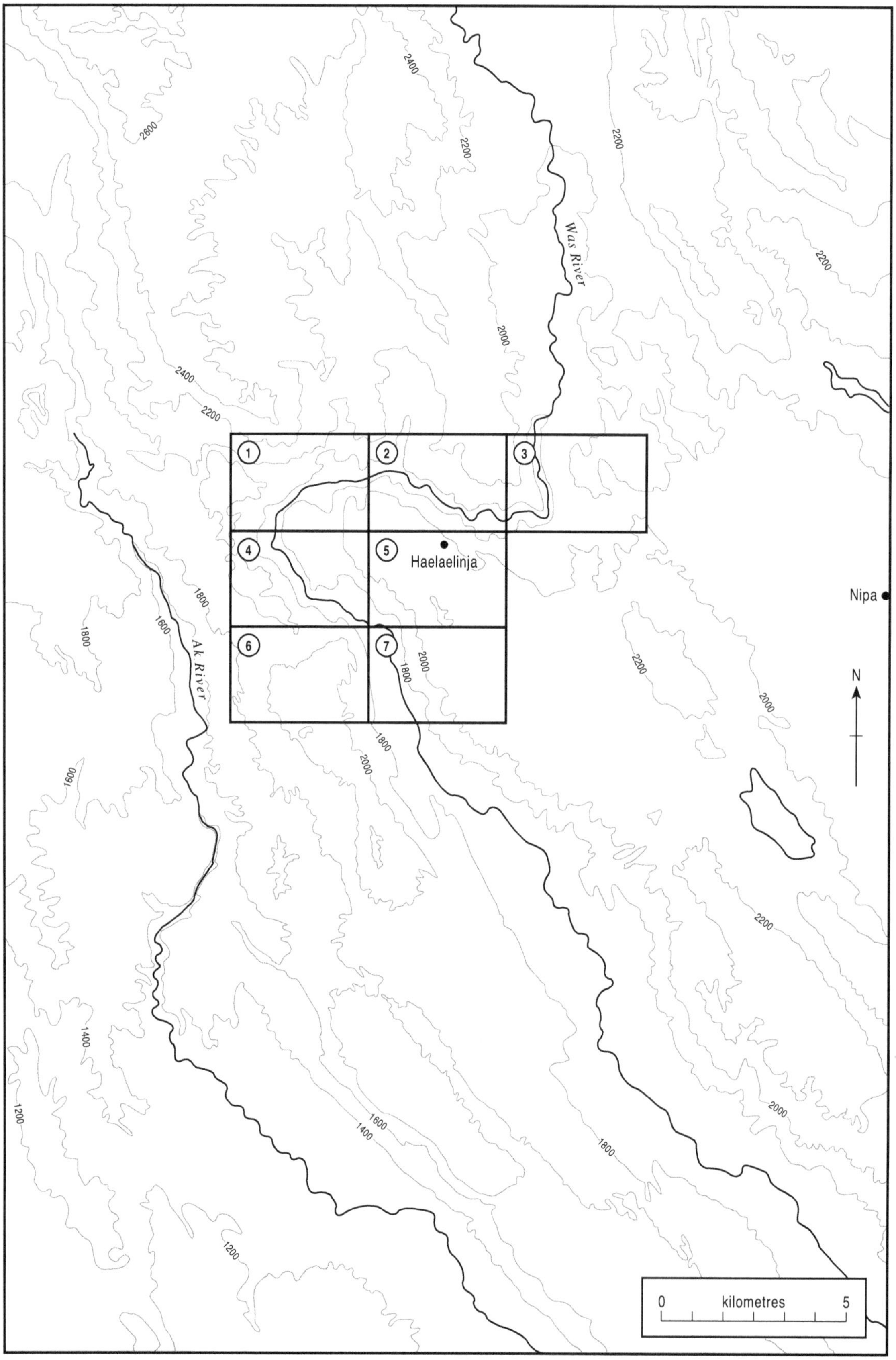

Map 3.1.1 Series of maps showing location of houses and residence pattern in part of Was valley.

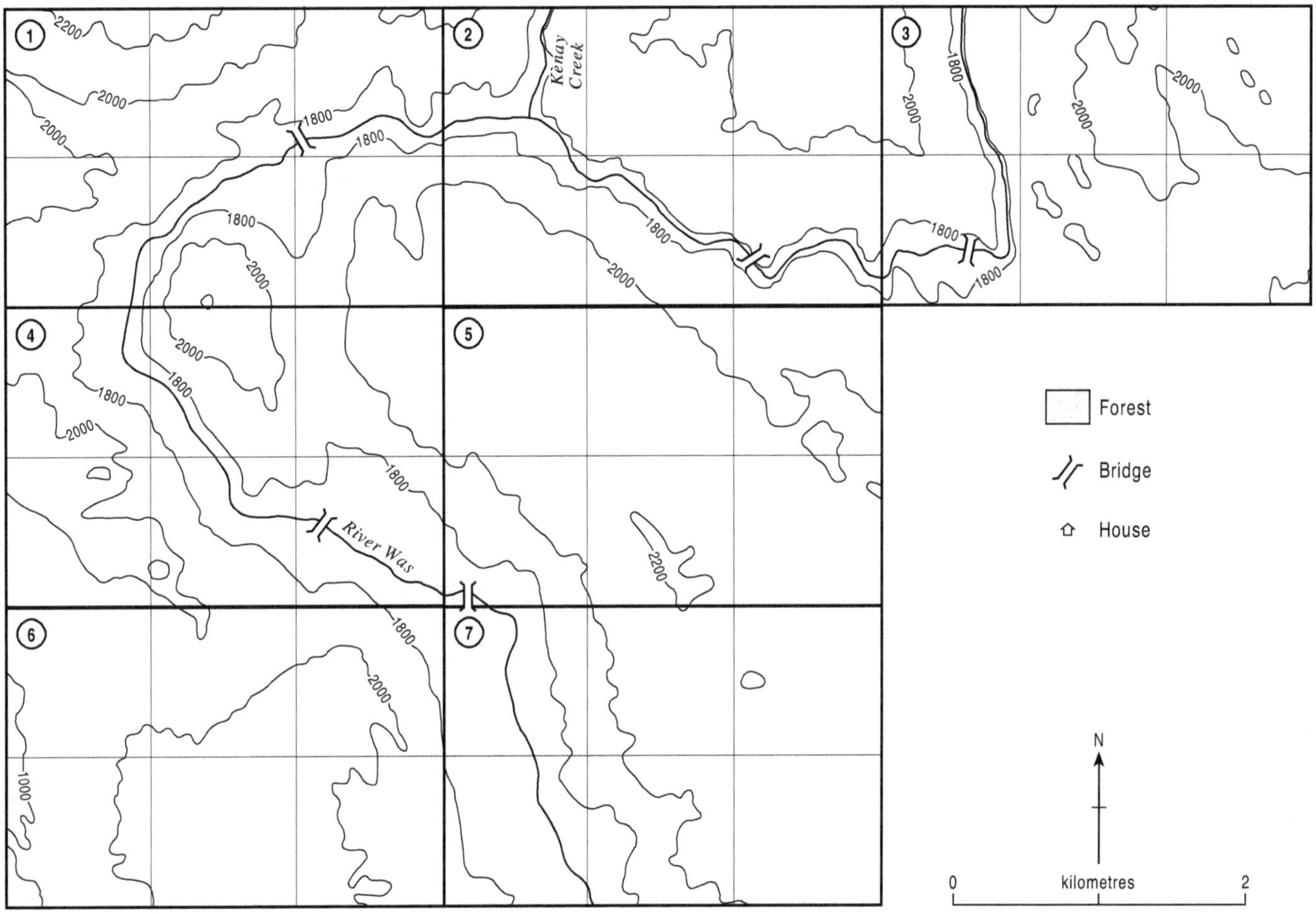

Map 3.1.2 Series of maps showing location of houses and residence pattern in part of Was valley (numbers by houses link to those on following genealogies).

The series of maps shows the residence pattern at one period of time on the territories of two neighbouring *semonda* –Aenda and Ebay – in the Was valley (Map 3.1), and the associated genealogies (Figure 3.1) give the persons resident in each numbered house (see Clarke 1971:101–2 for a map showing house distribution in the Ndwimba valley and Pataki-Schweizer 1980 for settlement patterns in various Eastern Highlands communities). They show the kin-clustered nature of Wola homesteads and the patrifilial-cum-bilateral constitution of associated *sem* groups. Several families maintain two homesteads at different locations on the same *sem* territory, usually in the vicinity of gardens that are considerable distances apart. Some families also had houses on other *sem* territories (off the map); these are the *suw pa kab* families. The interactive maps available at the Wolaland Interactive website (Wolaland.org) show the residence pattern over a quarter of a century and detail how the residential membership of the *sem* corporations changed during that time (see Brown and Brookfield 1967 for comparative data from Chimbu). These data show the variable kin composition of these residential groupings of relatives.

Tacit social knowledge

The residence pattern, with houses in any vicinity comprising variably composed clusters of kin, is the result of informal social decisions. Indeed these decisions are less made than happen, and again concern tacit knowledge, this time social, not practical. This follows the distinction between 'cognitive tacit knowledge' that informs behaviour through taken-for-granted 'know-that' normative values, and 'technical tacit knowledge' that involves 'know-how' and practical skills (Baumard 1999:59; Nonaka 1994:16–17; Nonaka and Takeuchi 1995:8–9; Ryle 1949:25–61).[10] Tacit knowing is prominent in daily social life; we engage in many acts of which we are focally unaware: 'ninety percent of life is just being there' (attributed to Woody Allen).[11]

10 It is arguable that Ryle's (1949) distinction between 'knowing that' and 'knowing how', which seems obvious when pointed out, was only necessary as the English language uses the same verb for engaging in an ideological informed behaviour and undertaking a practical task, thus obscuring the discrimination – rendering it tacit, in fact.

11 Cited in Clark 1997:9, epigraph to Chapter 1.

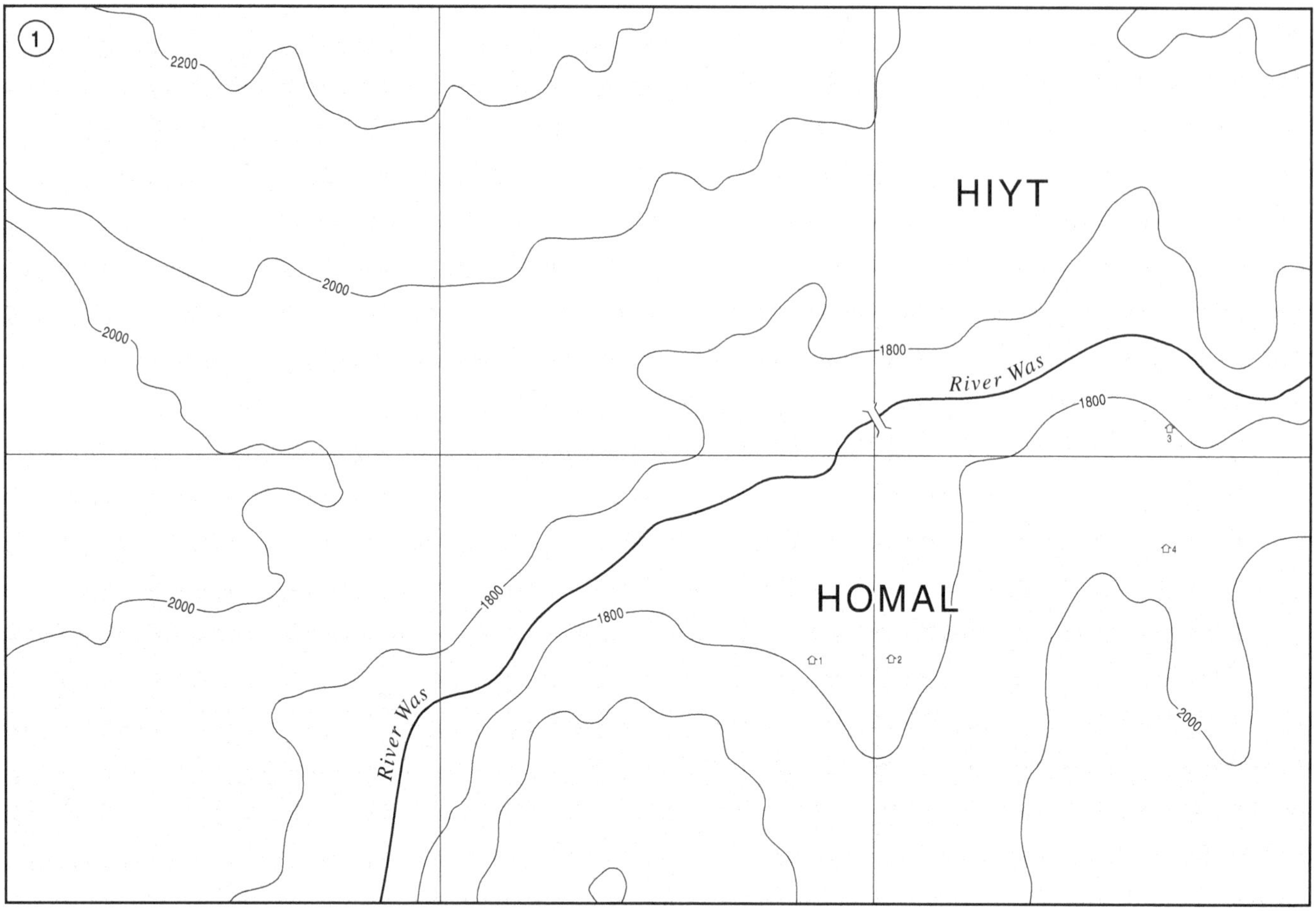

Map 3.1.3 Series of maps showing location of houses and residence pattern in part of Was valley.

It encompasses the everyday and accepted presumptions that inform our behaviour, such as assuming that our mothers will nurture us, unconditionally without question, and a sibling help us, with a dash perhaps of rivalry. Language competence is another manifestation of tacit cultural knowledge: speakers not pondering grammar and vocabulary when they communicate but drawing on this knowledge subconsciously, speaking without deliberately thinking about syntax.

The tacit character of much social behaviour complies with the argument that our emotions are a prime mover of behaviour, as propounded by David Hume, whose ideas are a precursor of today's cognitive science. He argued that it is our 'passions,' as he called them, that impel us to act and motivate behaviour, not our 'reason.' In other words, the primary motivations behind our actions are emotional desires such as our needs for affection, respect, happiness, security, fulfilment etc., which we more feel than think and often cannot verbalize, acting on impulse and empathy. We perceive the passions first-hand as vivid, strong and impulsive 'impressions' or sensations, not second-hand as ideas or mental representations of other things.[12] In his relentless drive against metaphysical flights of imagination, Hume argues that the 'causes' of these motivating emotional perceptions are unknown, as the 'original qualities of human nature, which I pretend not to explain' (Hume 1896:13). Unaware at the time of Hume's philosophy, I argued for a similar position in seeking to understand aspects of Wola social interaction, taking the emotional drives of respect and revenge, for instance, as basic issues regarding behaviour (Sillitoe 1979a:76, 120). It led to fierce criticism from the mainstream of social anthropology that thinks society not the individual should be the focus of enquiry.[13] It was with some relief that I discovered Hume, who lived nearby in

12 This summary glosses over the finer points of Hume's argument, such as his distinction between 'direct passions' (such as joy, fear, hope and despair), where he includes some instincts, and 'indirect passions' (such as pride, shame, love and hate), all of which we experience (e.g. as pleasure or pain) and that 'cause' intentional actions.

13 For instance, a reviewer for a press that rejected the *Give and Take* manuscript told me off because 'This is not how anthropologists think or write in the late twentieth century.' I

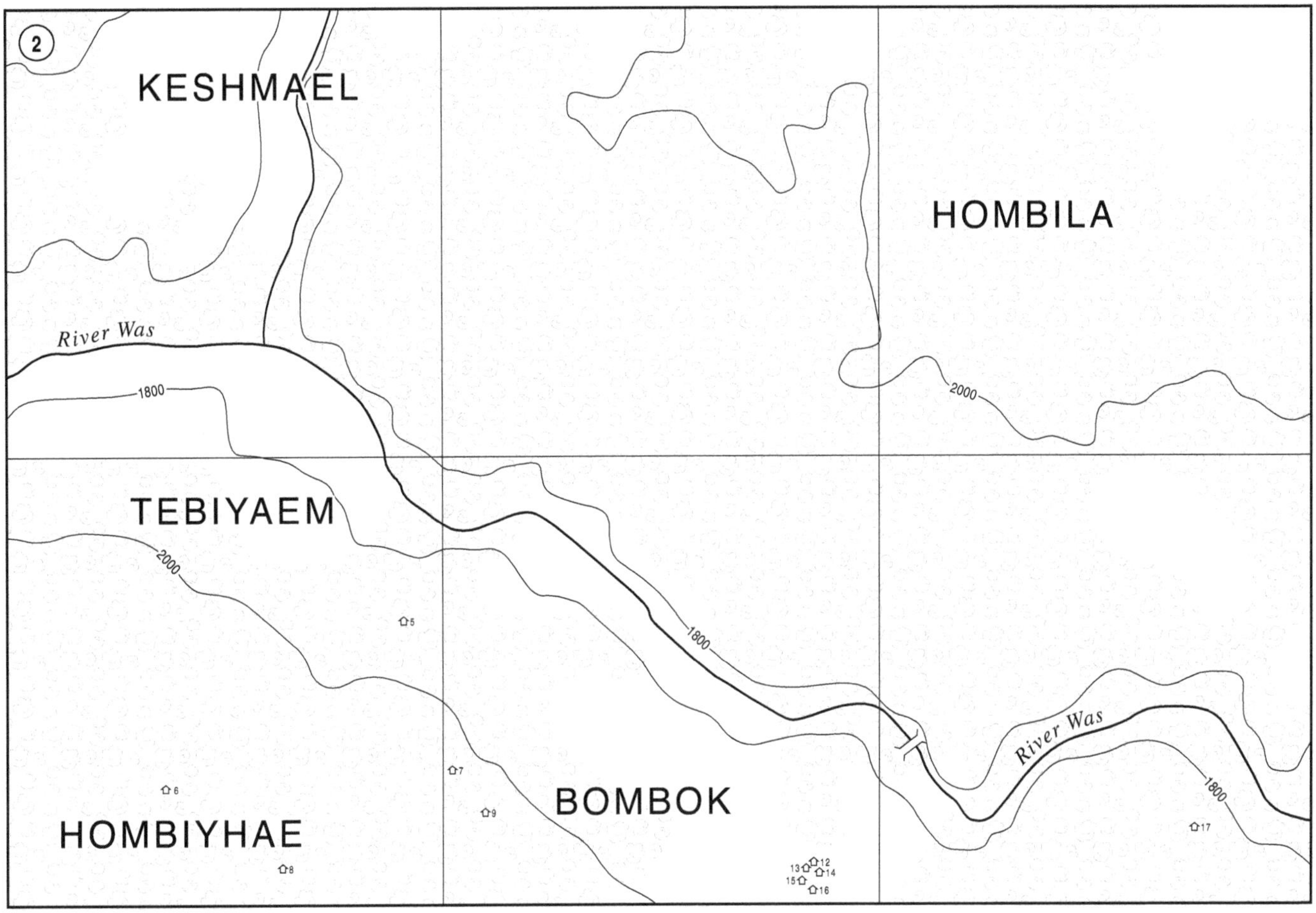

Map 3.1.4 Series of maps showing location of houses and residence pattern in part of Was valley.

the Borders, for it renewed my confidence to learn that I was unwittingly rehearsing his philosophical position, as 'the greatest philosopher of the eighteenth century, itself a century of greats.' (Stangroom and Garvey 2007:68) – albeit in, for him, the strange location of New Guinea.

The passions, or emotional sensations, are the 'original qualities of human nature' that we can locate empirically through experience and observation, whereas the ideas that we formulate to further understanding of them and the behavioural sequences to which they give rise are abstractions open to varying interpretations. Ideas represent something beyond themselves, namely the impressions that they copy, and we can only give cognitive coherence to ideas by tracing them back to the impressions from which they originate. In this view, the emotions necessarily circumscribe reason, which involves the manipulation of ideas. As Hume put it, in a widely cited sentence in the *Treatise*: 'Reason is, and ought only to be, the slave of the passions, and can never pretend to any other office than to serve and obey them.' (1896:217), which was controversially the reverse of the traditional view that the passions should be subject to reason. The argument is not that reason plays no part in the initiation of action. It may provide information, notably about means for attaining given ends that are designated by our passions, so influencing the direction of the will. But reason by itself cannot provide any impulse to action, which demands passion. Our reason orders ideas and pursues abstract knowledge about social relations in order to understand passions' ends. The concept of society itself, for instance, is such an abstract, culturally informed, product of reason, which I have wrestled with in the Was valley context, where people have no such verbalized equivalent and its imposition on their social interaction threatens misrepresentation and confusion (Sillitoe 2010a:66–71).

If reason alone cannot dictate action, it cannot be responsible for morality. As Hume explained: 'Morals excite passions, and produce or prevent actions. Reason itself is utterly impotent in this particular. The rules of morality, therefore, are not conclusions of our reason.' (1896:239). Moral judgements depend on 'sentiments', which in turn involve 'sympathy'. These sentiments of sympathy arise from our common 'feeling for humanity' and serve as a

have been attempting to respond to such criticisms and make my position clearer ever since.

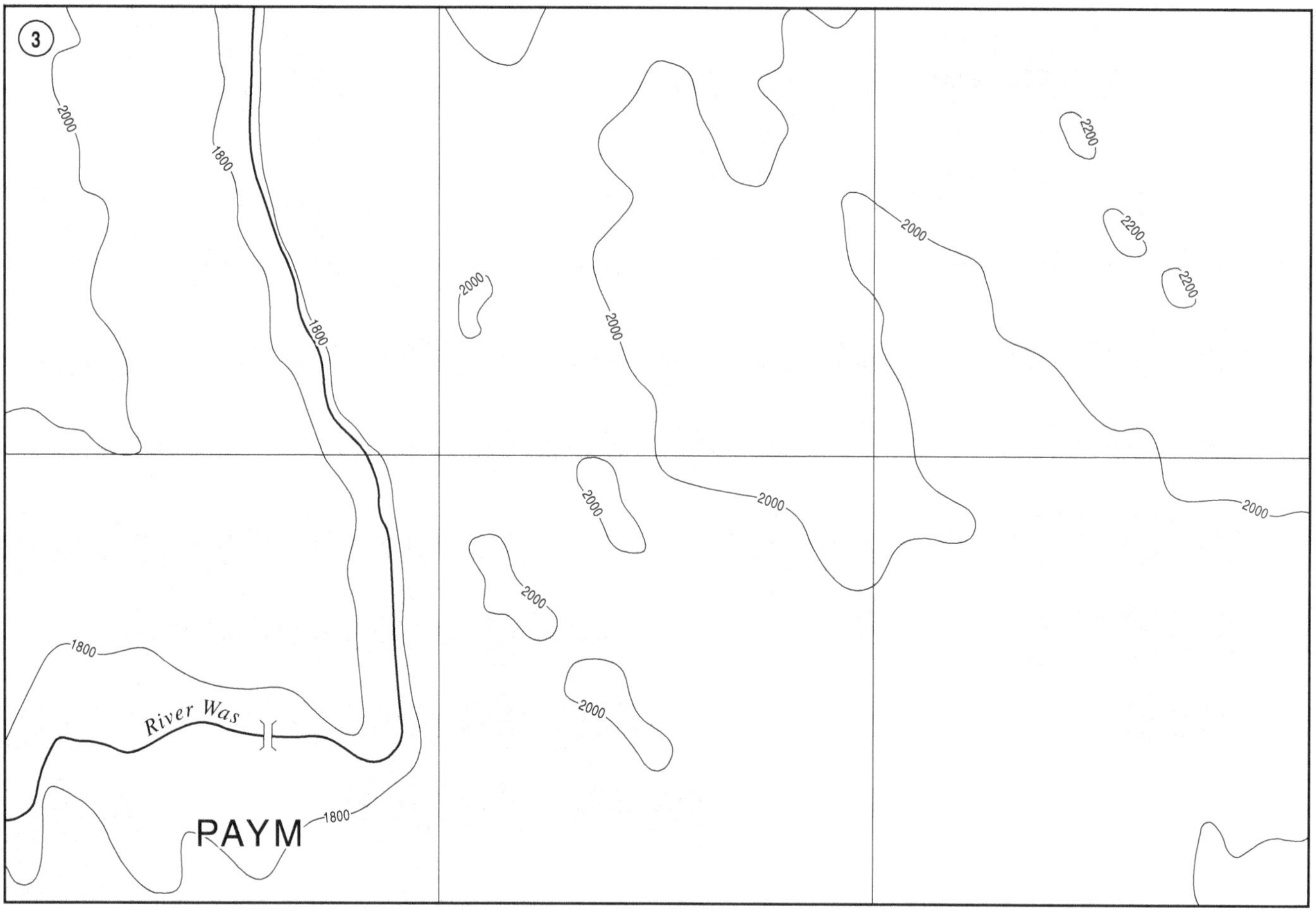

Map 3.1.5 Series of maps showing location of houses and residence pattern in part of Was valley.

source of approval and disapproval for actions that promote social interaction that increase prosperity and happiness for those around us, prompting us to perceive the advantages of sociability. Moral sentiments promote a preference for socially favourable actions; while they do not preclude self-interest (as central to Hume's friend, Adam Smith's thinking). They favour prosocial, in the community-interest, actions. We approve of dispositions that serve the interest of the social collective (such as obeying laws, justice, sound government and 'good-manners'), for 'the laws of nations, of modesty, and of good-manners ... are mere human contrivances for the interest of society' (1896:294).

There is further interaction between individually experienced emotions and socially derived expectations, in that values and beliefs are socio-culturally conditioned, passed on from one generation to another, even if actors are largely unaware of this consciously, absorbing them as implicit knowledge (Collins 2001a:110–11). They experience moral approbation of their behaviour emotionally as a pleasure and disapprobation as a pain, and they receive these judgements as sentiments others communicate to them, which they heed as social beings inculcated according to the same moral code. In informing behaviour, such emotive tacit understandings involve intangible issues. They are socially moulded ideals and norms that are so ingrained that actors assume them without question; albeit individual experience may modify their views of these moral issues. So in the Was valley, a family may move its place of residence following some experience that prompts it to question the faith it can have in certain neighbours behaving according to some widely held expectations. If sufficient persons have similar dissonant experiences, it may lead to changes in the canon of values and beliefs passed on to the next generation with no debate or need of spoken consensus.

The challenge faced by social studies is that while communities inculcate persons with shared values that inform them about the proper way to behave, they are apparently unaware of the resulting regularities in their individual and collective behaviour, and the social implications. Nearly a century ago, Malinowski (1922:22) made the same point in a Melanesian context, observing that while 'people usually do feel or think or experience certain psychological states in association with the performance of customary acts, the majority of them surely are not able to formulate these states, to put them into words'. The knowledge is tacit: actors do not apparently

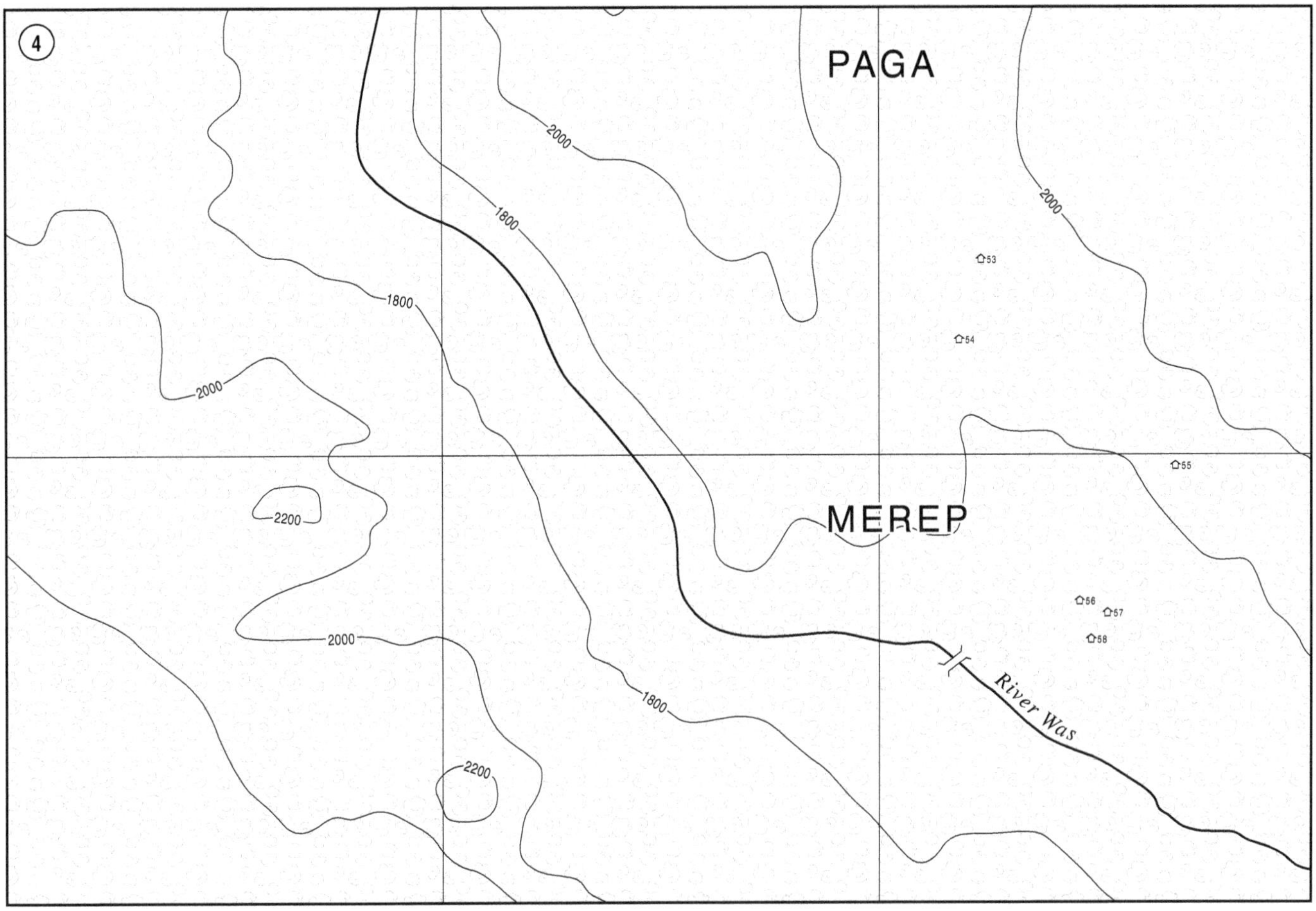

Map 3.1.6 Series of maps showing location of houses and residence pattern in part of Was valley.

know what they know. Consequently, those who possess such knowledge may not be aware that they are passing it on to others. It comprises a repertoire of intelligence, skills and ways of doing things so internalized that it is beyond thought; actors simply 'do', they do not have to 'think' about their social actions, which youngsters take on by imitation. When we say that tacit knowledge informs activity, we are saying that the actors have little or no conscious awareness of what guides their actions. The ascription of such a state is theoretical, to explain behaviour. The comparative poverty, cognitively speaking, of the subjective, often mechanical, experience is striking compared to the richness of the content we may analytically ascribe to the behaviour. And the connection between the two is difficult to address with the relation between conscious and unconscious states so complex and poorly understood.

According to one commentator, such tacit understanding 'exists in a practical state in an agent's practice and not in their consciousness or rather in their discourse' (Bourdieu 1977: 27). The resulting socially imbued propensities dispose them to act in certain ways according to circumstances. This approach is phenomenological, seeing social interaction as informed by direct experience, and actors as drawing on embodied dispositions in appropriate contexts. The manner in which such knowing informs social interaction Wittgenstein alludes to as 'following a rule' and Bourdieu, after him, more accurately according to today's agency viewpoint, as 'tacitly knowing a rule' (Gerrans 2005:55). These behavioural habits, inculcated via social relationships, likely involve relations of power, in turn signalled by embodied behaviours, which Bourdieu and Foucault call 'techniques of the body', as seen in the way persons of a particular station comport themselves and speak. Furthermore, this ineffable knowledge, which affects actors' 'agency' in regulating their social activities, does not, as noted, so constrain behaviour as to rule out manoeuvring by them for perceived advantage, personally, or for associates (e.g. kinsfolk). It results in variation as well as adherence to tacitly learned established social arrangements, something that is a potential source of social change.

Accessing tacit social 'know-that'

While the actors are unaware, those trained in social research can apparently convert such collectively shared implicit knowledge and associated individual manoeuvring

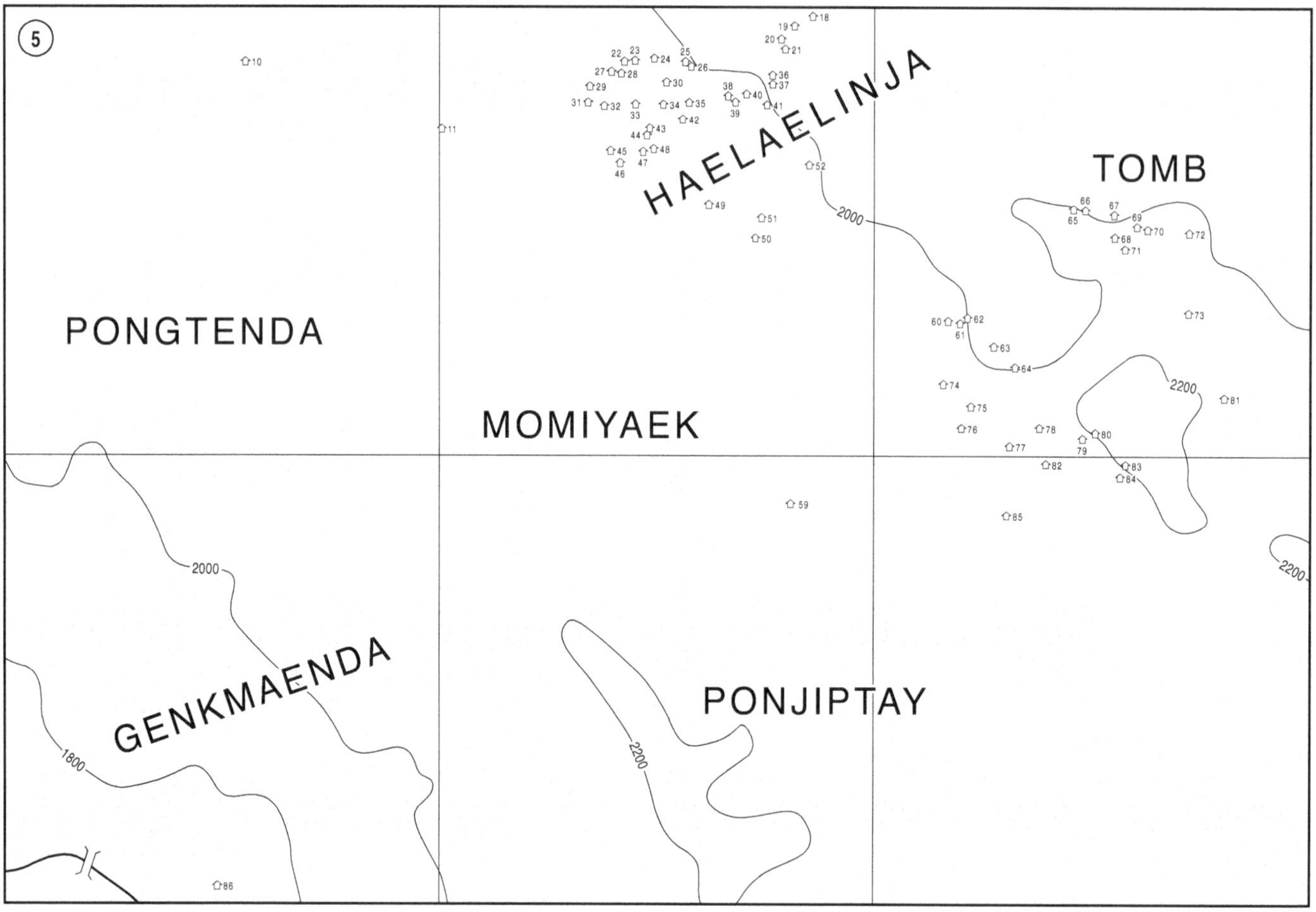

Map 3.1.7 Series of maps showing location of houses and residence pattern in part of Was valley.

– which is of the actors and manifest in their conduct and what they say – into explicit knowledge. Researchers purportedly have ways to make such 'know that' intelligence that is implicit in behaviour explicit, by deploying concepts such as function or process, structure or system, norm or habitus. They use data on behavioural regularities and variations, and actors' comments during such sequences, as evidence to support their theoretical insights into, and interpretations of, social life, though these are inaccessible to the actors themselves. Regarding descent, for instance, I recall Meyer Fortes once explaining to me that it was a 'heuristic device' that helped make sense of people's behaviour in stateless contexts. The problem is that such stratagems, while they may assist the outsider sociological imagination gain some traction on others' behaviour, are open to challenge, lacking the authority of any indigenous voice – predictably, if we accept that related behaviour features implicit knowledge. So while the descent model may have leverage for some in certain African contexts, others have questioned its relevance (Holy 1979; Verdon 1982), and it has certainly proved to be a distraction I think in New Guinea Highlands' contexts (Sillitoe 1979a:24–46, 2010a:112–16).

In contrast to this behavioural approach that seeks to reveal the implicit knowledge embodied in everyday conduct, there is an opposed intellectual one that accords actors a capacity to represent their social behaviour in terms of propositional knowledge and value systems (Holy and Stuchlik 1981, 1983). Such commentators question the pervasiveness of tacit awareness and posit indigenous models, attributing to people some sort of unverbalized explicit knowledge of their cultural systems. They detect a faculty for abstract cogitation on social forms, such intellectual codification often featuring manipulation of unfamiliar symbols – as in structuralism – for instance of mask designs, ritual acts, dance sequences or mythical narratives (Lévi-Strauss 1968:281–2; Schneider 1965:28). While people may not discourse reflectively on the implications of their socio-cultural arrangements, they project their thoughts symbolically on to what they build and make, extending intellectual meaning to these beyond their practical use (Campbell 2002; Scoditti 2012).[14] And

14 Some interpretations of indigenous ideas about socio-cultural arrangements in the Massim region illustrate the genre, which focus on canoes and include the anthropologically renowned

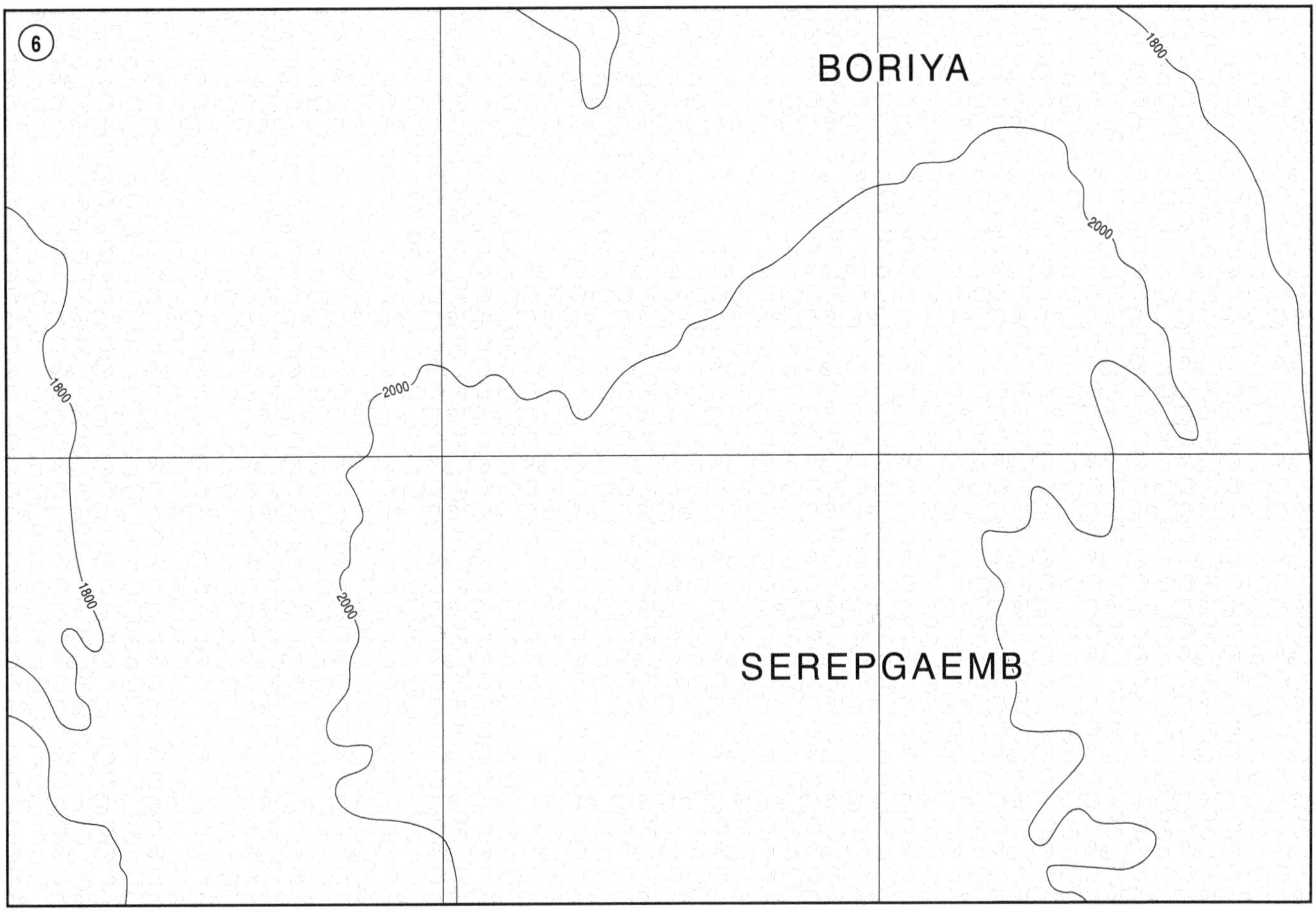

Map 3.1.8 Series of maps showing location of houses and residence pattern in part of Was valley.

they repeat this symbolic representation from one domain to another, from house design to settlement pattern and rituals to myths, the same patterns apparent to those able to discern them, culture bearers imbuing all realms of their social life with significance and so establishing a meaningful culturally mediated existence. Such interpretations are tendentious, claiming that those who have the necessary insight can reveal these alien intellectual 'discourses' that somehow – albeit often opaquely – parallel Western propositional knowledge. They are arguably double intellectual impositions; firstly, ascertaining social arrangements and values in idealistic sociological terms and secondly, somehow identifying these in exotic indigenous categories. They are also arguably patronizing, in assuming that people must have models of the world that parallel researchers' theoretical concerns.[15]

While the disagreement between these behavioural and intellectual approaches is profound, it is tangential to our tacit concerns. What is of interest is the implication that

kula. The forerunner is Munn's (1986:124–57) discussion of canoe symbolism in constituting values, notably in exchange contexts, arguing that it represents men's individual and collective pursuit of identity, the associated tension between equalitarian and hierarchical relations, and featuring various space-time transformations. According to Campbell (2002), the region's carved canoe prow boards encode meanings and values about everyday and spiritual life, notably again with respect to male ideology, such as men's wish to achieve immortality, and overlap with messages conveyed in rituals and myths. In more recent work, Scoditti (2012) argues that the spiral elements in prow boards, which he thinks parallel those in nautilus shells, are central to an aesthetic schema inspired by Monikiniki, a mythical hero, the spiral metaphor serving as a 'visual text' on various cultural issues such as the concept of time. In contrast, Malinowski was apparently unable to penetrate any such deep intellectual symbolic discourses, responding to a query from Haddon that 'there is no sense, no meaning in the whole of a design' (Campbell 2002:4, who frankly acknowledges the dubiety of such structural interpretations – 2002:6).

15 This is not to imply a challenge to the basic structuralist point that all humans have the same intellectual capacity regardless of culture (Lévi-Strauss 1972), only the assumption that our concerns and views feature the same deep unconscious cultural structures, downplaying the profound influence of social and historical context on our being in the world.

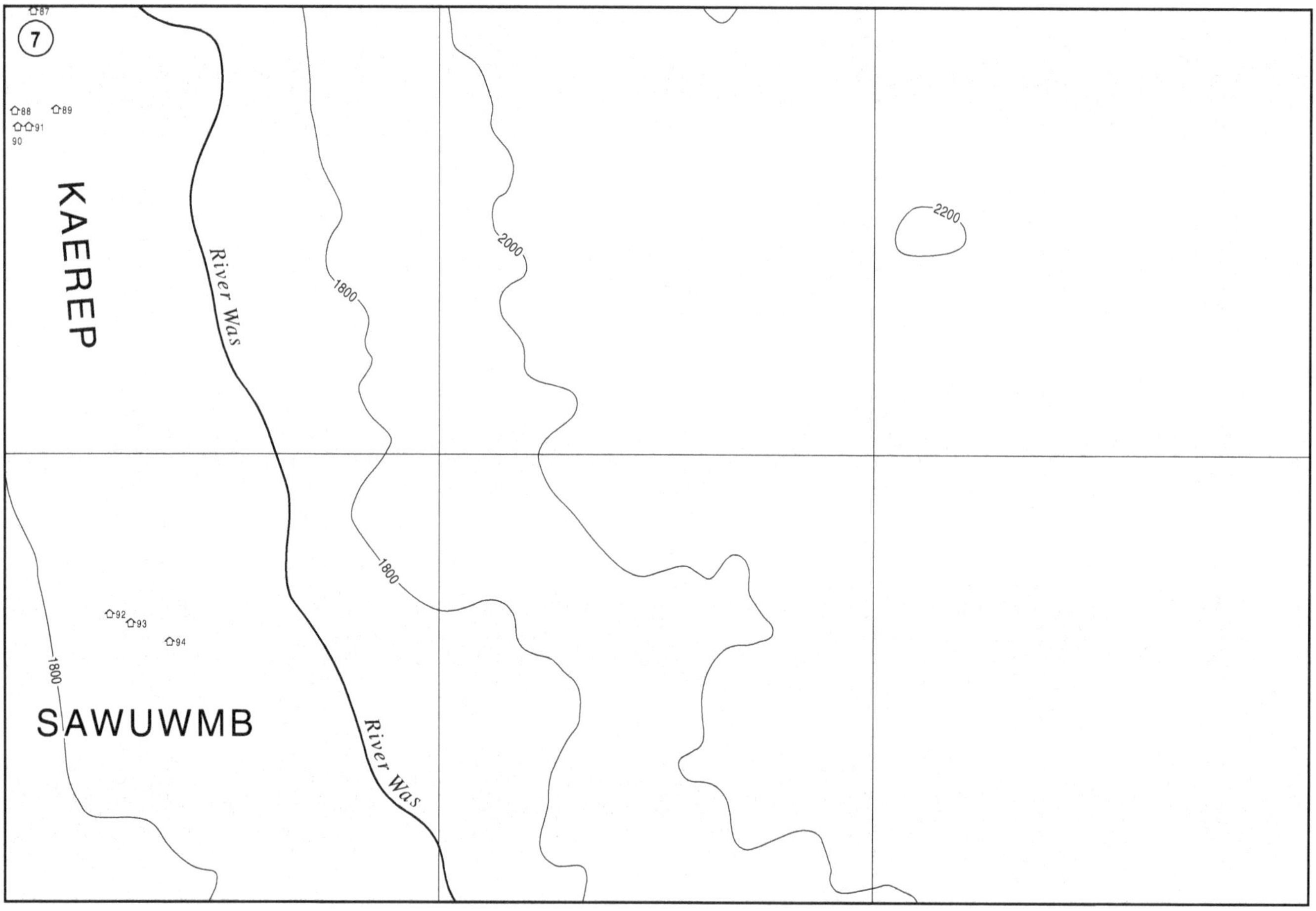

Map 3.1.9 Series of maps showing location of houses and residence pattern in part of Was valley.

social research methods offer a way to access tacit knowledge. Indeed Bourdieu (1977) claims to have formulated a 'theory of practice' in this respect. But a review of his influential 'theory' suggests that it is little more than a restatement of the central conundrums that have concerned French social philosophy from the time of Comte and Durkheim onwards, disguised under some new jargon such as 'habitus' (≈ norms),[16] 'logic of practice' (≈ function) and 'structuring structures' (≈ social structure) (Farnell 2000; Gerrans 2005; Just 2005). While acknowledging that there are observable regularities in human social behaviour – statistically demonstrable through survey data etc. – Bourdieu (1977) sets out to challenge the view (shadowing the earlier argument between Fortes and Leach – Sillitoe 2010a:5) that they are directed by normative rules. He argues that the concept of habitus accounts for regularities in behaviour without need of such consciously held rules. It comprises 'strongly interiorized principles' that are collectively generated and passed on such that persons acquire a feel for the proper way to behave socially without necessarily comprehending the sociological implications, such as the ensuing structure detectable in statistical regularities of behaviour. These result from individual social agents unconsciously following strategies that are generated by and accord with their habitus. The shift in focus from Durkheim's norms as 'social facts' to Bourdieu's predispositions as 'interiorized principles' is not particularly radical, both see persons compelled in a socially determined way to reproduce in their behavioural strategies what in the first instance went into creating the predispositions that inform their strategies (Farnell 2000;

16 The idea of habitus originates with Mauss (1935 – A.J. Strathern 1996:11-13; Farnell 2000:399-401; Just 2005:3) who in his discussion of embodied knowledge maintains: *'J'ai donc eu pendant de nombreuses années cette notion de la nature sociale de l' «habitus». Je vous prie de remarquer que je dis en bon latin, compris en France, «habitus». Le mot traduit, infiniment mieux qu' «habitude », l' «exis», l' «acquis» et la «faculté» d'Aristote (qui était un psychologue). Il ne désigne pas ces habitudes métaphysiques, cette «mémoire» mystérieuse, sujets de volumes ou de courtes et fameuses thèses. Ces «habitudes» varient non pas simplement avec les individus et leurs imitations, elles varient surtout avec les sociétés, les éducations, les convenances et les modes, les prestiges. Il faut y voir des techniques et l'ouvrage de la raison pratique collective et individuelle, là où on ne voit d'ordinaire que l'âme et ses facultés de répétition.'*

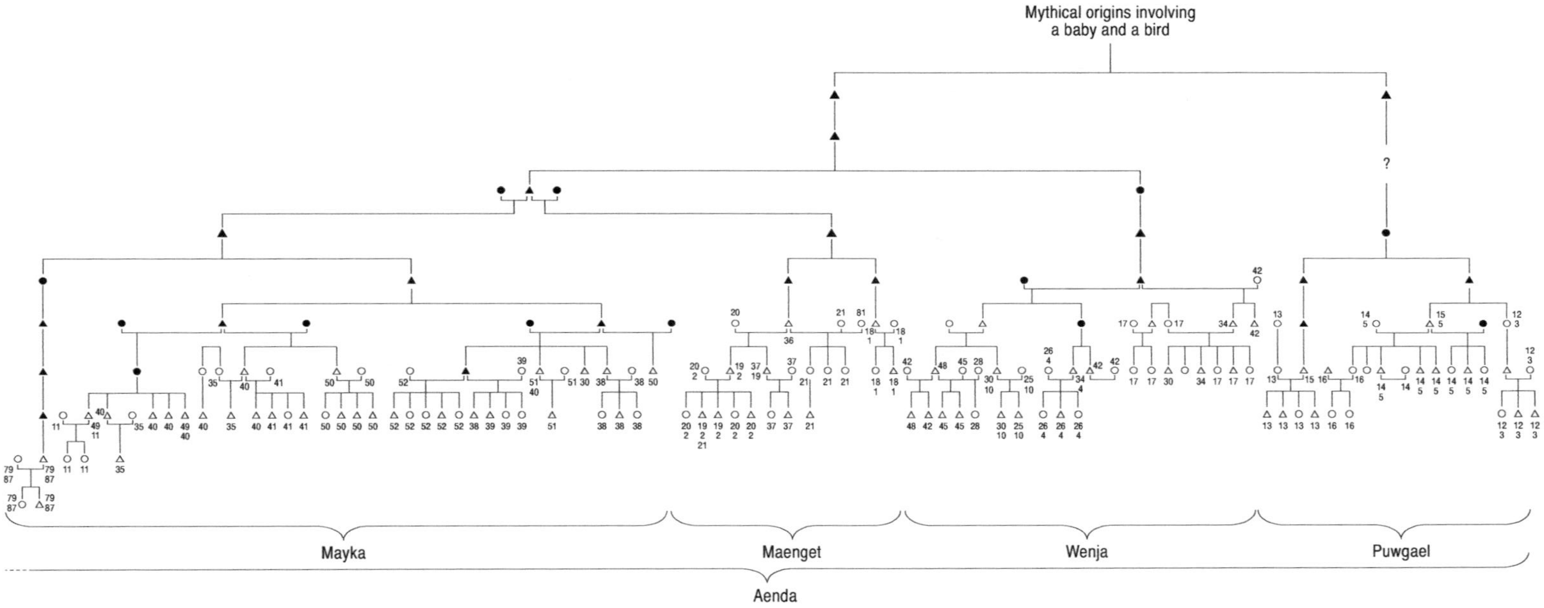

Figure 3.1.1 *Series of genealogies linked to residence maps.*

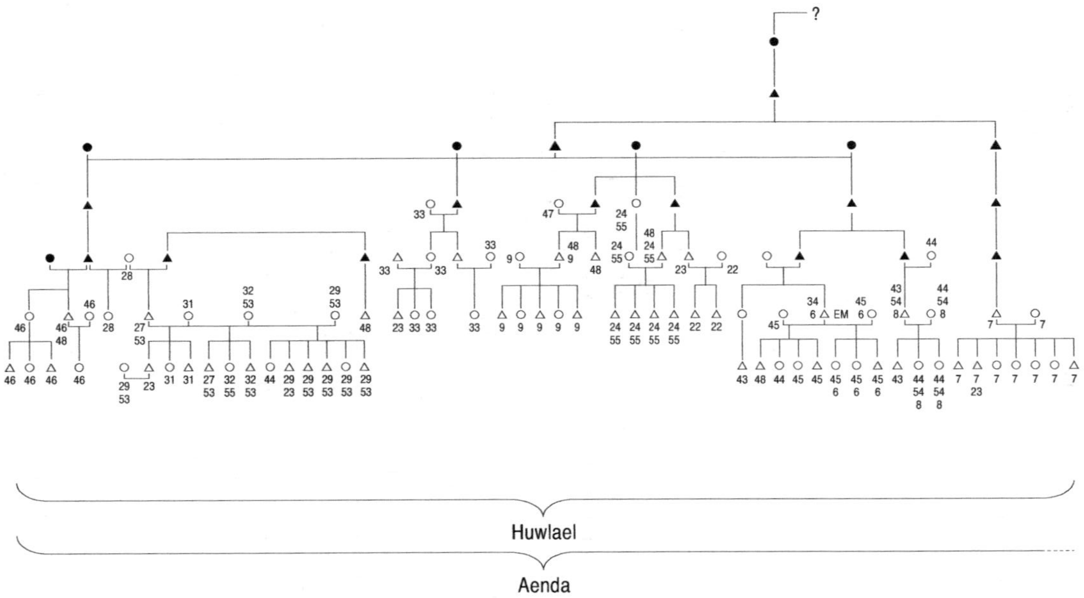

Figure 3.1.2 *Series of genealogies linked to residence maps.*

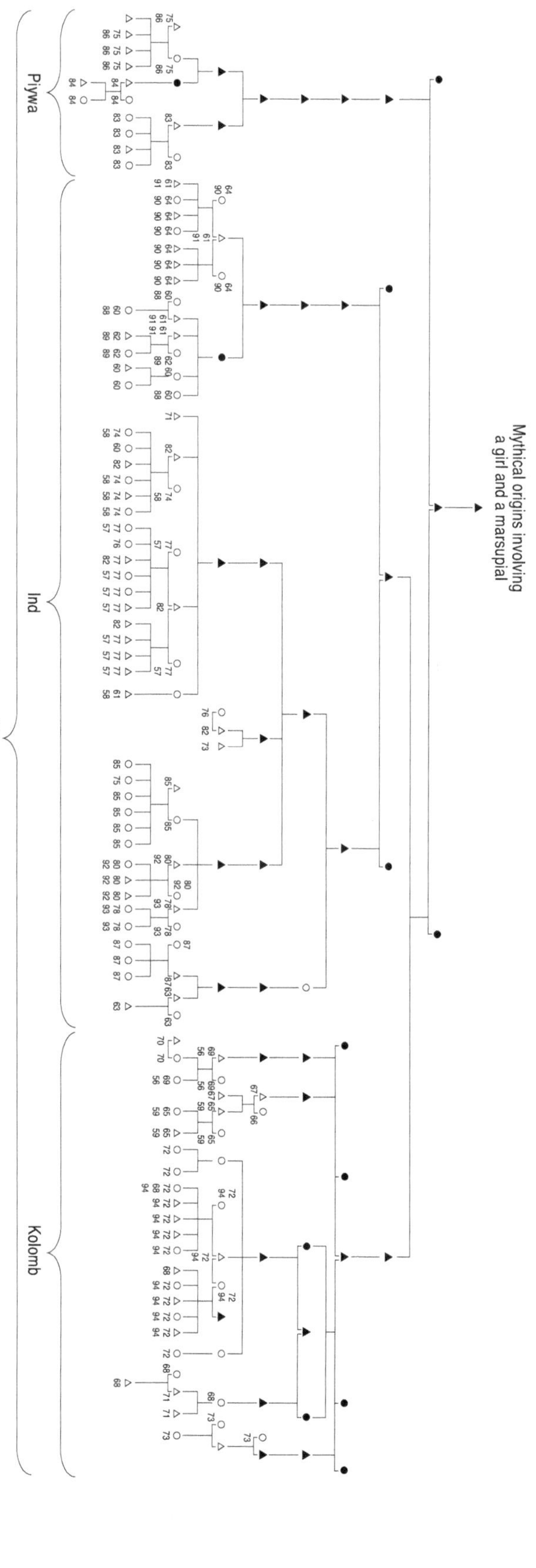

Figure 3.1.3 Series of genealogies linked to residence maps.

Gerrans 2005).[17] In short 'invocation of the *habitus* does not advance matters much', other than affirming what is widely accepted that 'most of social life does not consist in people consciously following rules, [it] leaves everything just as it was before' (Just 2005:16).

The socio-cultural and historical environment into which we are born comprises our 'particular conditions of existence' and informs the predispositions of habitus, which account for how individual human agents become social beings, and in so doing perpetuate the regularities that comprise collective interaction through time. The circularity of the argument illustrates the hermeneutic contention of parts (individuals) situated within complex wholes (society): predispositions channelling the individual into social ways and the consequent actions of the individual feeding back to reconstitute those ways. It is noteworthy that individual agents do this subconsciously, or rather their 'thinking' is tacit, being simultaneously embodied in their 'practice' (behavioural actions), and this 'logic of practice' (thought incorporated in action) feeds back to reaffirm the collective constitution of the 'interiorized principles' that govern their tacitly informed behaviour, all the while 'the principles embodied .... are placed beyond the grasp of consciousness, and hence cannot be touched by voluntary, deliberate transformation, [and] cannot even be made explicit' (Bourdieu 1977:94). The question of how individuals imbue and propagate the regularities of collective social life, which involves both social and psychological issues, demands an account of the process by which the habitus achieves these ends, but social practice theory fails to specify the socio-psychological mechanism by which these culturally and historically conditioned internalized predispositions acquit themselves. The habitus is 'a "black box" ... an intellectual no-go area whose realization within the individual cannot by definition be articulated without betraying the non-articulate nature of the predispositions and embodied forms of knowledge to which it gives rise ... [it is] a problem disguised as a solution' (Just 2005:16, 20).

The claim that social research methods can access tacit knowledge is apparently beyond substantiation. The 'theory of practice' proposition (Bourdieu 1977) is unconvincing, contradictory even, in implying conversion of tacit knowledge ('practice') to explicit knowledge ('theory') without telling us how to do it (see Chapter 11), as tacit knowing is currently by definition incommunicable with words. It seems that writers do no more than infer the existence of such tacit practice – or more boldly explicit knowledge represented by exotic symbols – from

17 I am grateful to Roger Just for catalysing my thoughts on the habitus concept, about which I long had suspicions, provoking insightful debate during a seminar presentation at Durham, subsequently published (Just 2005).

subjects' behaviour and comments, which is sociology as normal. The phenomenological approach challenges such explicit formulations that impose a rational post-event analysis, of which the actors predictably are largely, or entirely, unaware, when the mainspring of much implicit-knowledge-informed social behaviour is emotional not rational. Whatever way we look at it, we have the paradox of social commentators apparently knowing more about the social lives of those they study than the tacit rule-following actors know themselves. This cautionary point is a recurrent theme of Bourdieu's work, that 'the way we know and understand the world of others does not reduplicate the way they know it and the way they are able to act within it and reproduce it' (Just 2005:16). Practice theory's criticism of normative sociology takes us in a psychological direction, but has scarcely anything to say about the mental processes that allow individuals to reproduce the regularities – or transposable dispositions – observed in human social behaviour, and consequently it has little to contribute to the tacit knowledge-cum-skills problem addressed here. The practice theory approach of Bourdieu is resolutely socio-determinist.

It is difficult to see how we can progress sociologically beyond inference to proof in view of the intangible nature of the evidence, which, comprising behaviour and words, disappears as soon as it appears 'out there', and has significance and meaning; even specifying actors' intentions, subject to various interpretations, even disagreements after the event. Furthermore, to the extent that we may agree that social researchers are successful in explicating such human behaviour, it is implicit not tacit knowledge that they reveal, according to the definitions used here (i.e. it is knowledge that is somewhat consciously accessible). It may also be discerned to some extent by actors, if some experience prompts them to reflect on their behaviour and bring it into conscious focus, albeit they infrequently do so and cannot readily articulate what they know. Such knowledge comprises a cognitive disposition to assume that things are the way they have been until now, an implicit 'knowing-that' somewhat expressible in words if brought to our conscious attention. In short, these are issues that we do not consciously consider, unless events challenge our beliefs. Although infrequently, and perhaps not readily, verbalized, these implicit-cum-tacit social assumptions shape our perceptions of the world and inform our behavioural interactions. Operationalizing them can depend in considerable measure on empathetic insight and intuitive inspiration.

Whatever the accessibility of social phenomena cognitively to actors, social data comprise notoriously slippery 'facts' such that their evidential value and interpretation is constantly open to challenge, which puts them in a somewhat different class to 'know how' comprehension of concrete material things. This view is clearly at odds with the one that thinks we can 'consider social facts are things' (Durkheim 1964:14) that we can know unambiguously and analyse, and to which many social commentators subscribe (Murphy 1972:37–45);[18] which is not to deny that the socially inculcated, value-laden rights and responsibilities to which individual persons subscribe, often subconsciously, indubitably inform their behaviour, and in that sense their acts are subject to them. Yet it is widely acknowledged that our society concept refers not to a thing but a culturally relative abstraction derived from the behaviour and values of interacting individual persons. It is the abstract quality of the idea that arguably prompts many to write indirectly, avoiding the subject of sentences, for fear of criticism for reifying an abstraction; such that sociological writing can be a baffling labyrinth of passive expression compounded by the use of impenetrable jargon. Albeit, if you seek the subject of many sentences it turns out to be society!

Was valley social arrangements

While we may discuss the idea of society according to an agreed social philosophical logic, there is no way of resolving different understandings of it by reference to a concrete benchmark 'out there', particularly considered as a culturally informed abstraction subject to varying interpretations.[19] As an idea, the only proof of its existence is the word we use to express it. There is no such word in the Wola language that translates as society or phraseology that suggests any analogous concept, which indicates that it is a foreign notion (Sillitoe 2010a:65–72). The implication is not that the Wola cultural imagination is devoid of abstractions – it supports a rich panoply of supernatural beliefs involving ancestor spirits and forest demons, and with the arrival of missionaries people have readily embraced Christian beliefs involving a trinitarian God and malevolent Devil (Sillitoe and Sillitoe 2009). The absence of any abstract society-like idea and the presence of ethereal supernatural beliefs contrast with conceptualizations that relate to concrete things, for which there are words, such

18 For instance, Radcliffe-Brown (who during the creation of social anthropology played a significant part in importing the views of Durkheim's *L'Année Sociologique* into the discipline, views that continue to hold sway in many guises) declared that 'I regard the social structure as a reality' (Radcliffe-Brown 1953) – ironically enough, in response to another school of French social philosophy.

19 The use of the term society for a collection of individuals brings to mind the set of all sets that is not a member of itself paradox (Sillitoe 2010a:68–9), which results in gathering things that have some resemblance to one another into a generic category or set that differs from its constituents. The constitution of the set of all sets recalls Hume's distinction between 'simple ideas', related to the 'impressions' or sensations, and 'complex ideas', built up from them (which also foreshadows the nomothetic view – see Chapter 2).

as *aend* 'house' or *iybtor* 'bridge' and *showmay* 'pig' or *ma* 'taro', that our senses can detect whatever our cultural background, assuming that we all see, hear, smell, taste or touch the same thing 'out there'. This is not to suggest that persons inculcated into different cultural traditions perceive of these phenomena the same way (as vernacular architecture studies show regarding various ideas about dwellings and ethnobiology demonstrates with respect to different ways of classifying animals and plants), only that we share analogous concepts.

The apparent absence of any equivalent concept to our idea of society does not imply that those who live in the Was valley fail to recognize social rights and responsibilities, nor that social order somehow emerges spontaneously from everyday interactions (Lederman 1986:64). The word *maen* equates to 'customs' or 'norms' that inform behaviour in various contexts, which persons can explain, such as the complex exchange procedures that formalize a marriage. But whenever I try to draw my friends into a discussion of particular *maen* 'customs', asking them why they do things the way they do and expect others to behave in certain ways according to these norms – for instance why they engage in specified exchange sequences at a marriage – they invariably tell me that they do so because it is *naun shumbaen maen* 'our ancestors' custom [or way]' or *naun shumbaen bismiyuwp* 'our ancestors did so'.[20] They are reluctant to go any further and engage in discussions about the social implications of *maen* informed behaviour. In other words, they tacitly accept these customary stipulations and are not in the habit of interrogating them further.

Another response to the why do you follow *maen* 'customs' question is that persons do so because it is *mbinyom konem* 'their idea' or *mbinyom turiymon* 'their wish', hinting again not only at the tacit nature of such behaviour but also signalling the importance placed on allowing room for individual action. While there is a notion of customary ways to proceed, for instance an elaborate vocabulary for exchange sequences and an expectation that transactions will follow a certain routine, there is considerable scope for variation depending on the situation and actors involved at any event. If one is looking for some expression of the Was valley equivalent of the notion of social organization (Firth 1951), as opposed to society, it exists perhaps to some extent in the complex conventions that govern exchange transactions. They represent the society issue the other way round, less abstractly, as they delineate expected ways to interact in sequences where persons associate. They comprise a complex vocabulary for which we in turn have no equivalent words in the English language. This transactional lexicon might serve as a New Guinea Highlands' equivalent of the much cited (albeit possibly apocryphal) case of Inuit snow categories that are often cited to illustrate how the words people employ reflect in some measure their knowledge and understanding. The use of inappropriate terms may hinder comprehension, as seen in the way the gloss 'gift', for exchanges, obscures the nature of associated transactions (Sillitoe 2010a:412–13). Our 'gift' vocabulary is insignificant compared to the plethora of Wola terms for socio-political exchange transactions (Sillitoe 1979a), and the bland gift and commodity dichotomy widely imposed on such lexicons (Gregory 1982; Strathern 1988; Weiner 1992; Godelier 1999) scarcely does justice to associated transactional ways of relating to fellow human beings. It is necessary to master the complex exchange vocabulary to come anywhere near understanding what people are doing, for while our mercantile tongue can subdivide the commodity side of transaction into several sorts of deals (such as cash sale, lease, hire purchase, cheque, credit card etc.), it does little more than talk of the gift, which is the reverse emphasis to that of the Highlands' languages.

It is arguable, furthermore, that the very idea of the gift[21] is inapposite, as something freely given with no expectation necessarily of a return (in anthropological terminology, an act of 'generalized reciprocity'). It is difficult to see how one can square the generosity of the gift with the notion of equality because, as Aristotle observed in his *Ethics* (2012), the gratitude generated implies a sort of inequality, akin to the debtor to creditor relationship. He thought that a 'noble person' would find the position of beneficiary unacceptable and would need to even the score by making a return, or reverse it by making a more generous return. It is a point that Kant (2001:123) picks up on his lectures on ethics, arguing that it is a:

> breach of duty to oneself if one accepts favours; for he who accepts favours creates debts that he cannot repay; he can never even the score with his benefactor, since the latter first did him the kindness of his own accord; if he returns the favour, he does it only insofar as the other preceded him in this, and thus remains forever owing thanks to him … But he who pays promptly for everything can act freely, and nobody will hamper him in doing so.
>
> (Kant 2001:123)

20 These phrases, and those following, are equivalent to such Pidgin terms as *pasin bilong tombuna* 'customs of the ancestors' and *tombuna mekim olosem* 'ancestors did the same', and *tingting long en* 'their thoughts' and *laik bilong ol* 'wishes of everyone'.

21 The notion of the gift that features in anthropology also comes, like that of habitus, from Mauss, who in his essay *The Gift* (1970) debates what obliges gift recipients to repay. The essay is the target of Derrida's (1992) critique of the idea of the gift, which, along the lines of Aristotle's and Kant's arguments previously (Sillitoe 2010a:412–13), further supports the contention that the term gift is inappropriate for Melanesian transactional activities.

This is the logic that informs the socio-political exchange arrangements (in anthropological terms, 'balanced reciprocity') that feature prominently in the acephalous Was valley polity, which places a premium on equality and individual liberty. The constant giving and taking that institutionalized exchange obligations promote results in no party coming out on top, all are equally creditors or debtors; albeit some, the *ol howma* 'men of renown', more so than others There are no benefactors who give gifts as favours, only transactors who engage in exchanges that entail obligations to give back.

Aspects of Wola ideas about social organization are also evident in the construction of houses (for instance, with respect to relations between women and men) and the distribution of houses across the landscape (for instance, with respect to the constitution of social groups). The Wola idea of *sem* 'family' relates to both houses and social groups, and throws further light on conceptions of society-like issues. The *sem* term is polysemous, as mentioned previously, and the English gloss 'family' only partially covers it. On one level it refers to a group of persons roughly equivalent to our idea of a family, who occupy the same homestead, which may comprise a single house or two or more houses, depending on the family group's adult composition. Such a *sem* comprises a man and his wife or wives and their offspring, and possibly other close kin such as unmarried siblings and widowed parents. People may identify these as so-and-so's *sem*, naming a member of the group, often a father figure. So we may have *Ongolon sem* 'Ongol's family', where Ongol is the senior male who is father to some members; such identification is not invariable though, for instance a child may refer to the group by naming a peer member of it rather than that individual's father.

Several closely related nuclear-cum-extended *sem* families that live in close proximity to one another, such as the *sem* of adult siblings and first and second cousins (both male and female, although given the patrifilial bias, more often male) comprise the next level of *sem* grouping, which, as noted above, speakers may refer to as *semgənk* 'small families' (somewhat counter intuitively for English speakers, as they are larger than the so-and-so's *sem* 'families'). These groups bear a name, often of some ancestor, such as Mayka and Kolomb, and as corporations in which land rights are vested they persist through several generations regardless of the comings and goings of any individual so-and-so's *sem* families. Several such *semgənk* 'small family' groups that occupy the same territory comprise the final level of *sem* grouping, which, again as noted previously, speakers may refer to as *semonda* 'large families'. These groups also bear a name, such as Aenda and Ebay, and continue through several generations as corporations occupying the same territory across which are interdigitated parcels of land claimed by constituent *semgənk* 'small families', title to which may shift with the possible disappearance of some of these smaller groups or the expansion and fissioning of others.

The foregoing map and genealogies illustrate the composition of such groups and their residential dispositions. It is noteworthy that these identifiable social groupings refer to actual collections of related persons, albeit the groups' compositions may fluctuate over time; they are not pure abstractions. When people talk about them, they have in mind groupings of particular individuals known to them. If asked about *sem* generally, interlocutors will invariably ask what one you are referring to. This relates to the above point about people not engaging in abstract discussions about social issues. When it comes to social matters, my Was valley friends are empiricists.

They identify no larger groups; for instance, they recognize no grouping equivalent to the Was valley populace nor do they entertain any idea that all Wola speakers comprise a named population or 'society' in conventional anthropological parlance. Indeed, before Australian intrusion into their valley, residents of the Was region did not know the full geographical extent of the population that spoke their language, let alone have any idea that they comprised some sort of identifiable linguistic and cultural grouping. The absence of such a concept has proved something of a problem, my adoption of the term Wola to identify such an anthropologically recognizable linguistic and cultural entity leading to some confusion (Sillitoe 2010a:67–72). While accepted anthropological practice, I can see in hindsight that if I had come up with, and adhered to, the tenets of ethno-determinism from the start, I should have developed my understandings around the above locally identified social groups. Although this may have avoided some of the aforementioned conundrums encountered when struggling to avoid criticism for reifying the idea of society, in bucking accepted practice, it may have engendered even deeper confusion and would also imply accepting without question the tacit nature of much social behaviour without further comment. This dilemma catches the problem addressed in this book.

The phenomenological approach to social life accommodates the individual versus collective antinomy of social philosophy that I have used in an attempt to further understanding of the implications of acephalous socio-political arrangements in the Was valley, where the mediation of these two mutually exclusive views of behaviour involves ramifying socio-political transactions, featuring what I have termed 'the competitive sociability of institutionalised exchange individualism' (Sillitoe 1979a; 2010a). On the one hand, there is the view that humans follow social procedures as moral persons, so-called 'legalism' associated with sociology, though we know that they constantly seek to modify and evade these for personal advantage, and furthermore that social life is too complex and open-ended to comprise such a consistent system. On the other hand, there is the view that humans behave as self-focused individuals intentionally out to satisfy their own ends, so-called 'voluntarism', associated with economics as pointed out, though we know that the existence of socially ingrained regularities inherited from forebears necessarily limits the horizon of agents' choices, and furthermore that asocial chaos is more likely than sociable order in such unbounded circumstances.

The intentional motivation in any interactional sequence may emanate from the individual, but it is subject to cultural conditioning, as argued, with persons acting to maximize emotional approbation and 'cultural capital' – comprising respect, influence, alliances and authority etc. – resulting in a positive feedback loop, such that changes to tacit individual drives remain within certain socially tolerable limits and do not undermine them, threatening disorder. An intriguing philosophical conundrum is the extent to which the will features in tacitly informed social acts. Actions that we have limited awareness of appear less willed than those where we are fully conscious of what we are doing. The extent to which we voluntarily behave according to a 'set of rules of which we are largely unaware' is consequently debatable, as opposed to the 'set of rules' obliging us to act in a certain way. In short, it is uncertain how much the actor directs the act and the act directs the actor. This conundrum interestingly underscores the above discrimination between the collective persons who habitually follow the social norms that their community has inculcated during socialization and the selfish individuals who willedly scheme to circumvent or change these to advance personal interests. While all societies have to accommodate this riddle, the political challenge is stark in stateless ones, and its confrontation in the Was valley has featured centrally in my attempts to understand the social order (Sillitoe 2010a).

Inevitably, such tacitly inculcated knowledge about the way to behave circumscribes individual freedom of action. Total freedom for human beings is not feasible, as imagined by some social philosophers when they talk about the time humanity 'lived in a state of nature', which amounts to a purely imaginary condition (the palaeo-anthropological evidence shows that humans have always 'lived in a state of culture'). There can be no such thing as socially unfettered liberty, as humans are inevitably members of, and conditioned by, socio-cultural orders. But there can be greater or lesser degrees of political liberty, in the sense of subjugation to or freedom from domination and rule by others, with acephalous orders at the less subjugated pole, where all are equally fettered by social arrangements and none privileged above their fellows by social institutions and arrangements, and no individual is able to subject other persons to their will. It is also noteworthy that the acephalous ideology of equality does not imply that all persons are the same or exactly equal in all regards (Sillitoe 2010a:49), which is inconceivable, as Hume (1902:155) pointed out: 'however specious... ideas of *perfect* equality may seem, they are really, at bottom, impracticable; and were they not so, would be extremely pernicious to human society'. Some individuals are more prominent on some occasions and may seek to manipulate others to their viewpoint, but they cannot coerce them. An upshot, as briefly outlined in the previous chapters, is that the *sem* socio-local groups do not comprise political entities led by dominant characters – the fabled 'big-men' of New Guinea. It is the genius of the Wola socio-political order to achieve such an egalitarian association, and juggle the implications, which I seek to figure out, unlike the actors who tacitly behave it.

Turning to 'know-how'

If we wish to go further and get onto a more secure intellectual footing, we might return to Hume's argument, which likewise focuses on tangible individuals and not intangible society in asserting that the 'passions' motivate behaviour. Furthermore, if we accept that the emotions feature significantly in behaviour, this accounts in some measure for the contrary nature of behaviour and the unavoidably inconsistent character of ideas advanced to explain/interpret it, for our emotions are inherently contradictory. For instance, we feel both love and hate, happiness and sadness, fulfilment and frustration, selfishness and generosity and so on. Indeed, in order to experience any such emotion, we need to be familiar with its opposite: to feel happy implies familiarity with the notion of sad, for without sadness we should not know when we were happy, as we should only ever presumably feel one or other emotion and have nothing to compare it with – we should just be. Some of the ancient philosophical traditions of the East build on the contrast of opposites, such as for instance the widely known Chinese notions of *ying* and *yang*. A parallel dialectical tradition features prominently in some Western philosophical schools, such as those that explore the logic of binary thought.

The contradictions prompt me to question reference to 'theory' in social contexts, which I understand in the technical sense to relate to an established explanation that consistently uses a body of agreed principles systematically to account for known facts. It is more appropriate in social contexts, I argue, to talk of 'ideology', in view of the rapid fashion changes in principles-cum-values used and the turnover of interpretations (Chapter 1).[22] The contrariness of emotionally motivated behaviour suggests that any second-order sociological thinking about behaviour will

22 The tacit aspects of socio-cultural life also facilitate social science's theoretical fashions, affording the opportunity for one unfalsifiable interpretation after another.

inevitably have to accommodate contradictory ideologies (theories) because human behaviour by definition features a paradoxical amalgam of opposites. There can never be one correct explanation/ interpretation using an agreed code of principles, as opposed to the moment's fashionable one using the hottest ideas. We see this in the differing ideologies of the twentieth century. These range from pragmatic explanations of cultural arrangements as ensuring survival and existence (functionalism) to ontological interpretations of culture concerning being and identity (symbolic-interpretivism). They encompass social solidarity interpretations (structural-functionalism) versus social conflict views (Marxism), even a dialectical view (structuralism) that seeks to understand how humans accommodate contradictions intellectually, which contrasts with the phenomenological-cum-existentialist turn (postmodernism) that throws its hands up at the prospect of any objective stance.

So how can we get on to more secure intellectual ground? In respect of the emotions underlying intentional actions, Hume clearly sets the hermeneutic limits for his times by declaring that to enquire into the causes of the 'passions' is to engage in baseless metaphysical speculation, which he criticized fiercely. If we wish to go beyond these limits, we have to engage in empirical investigation of the emotions and associated tacit awareness, which points us today in the direction of psychology. In other words, it is necessary to access the cerebral 'black-box', delve into actors' cognitive activity, to secure a firmer hold on social behaviour and to further understanding of the tacit aspects of social learning and perception, as discussed in later chapters. It implies the integration of sociological enquiries with cognitive psychology and neuroscience (Bloch 1998, 2006; Gerrans 2005) – albeit we shall presumably encounter other limits to our understanding, such as the variable interaction of molecules along neurological pathways or the interminable philosophical conundrums of determinism that argue predestination undermines free will and moral responsibility.

While the 'know that' tacit knowledge that informs social interaction appears to differ from the 'know how' skilled variety that involves 'automatic' physical sequences, another way forward is to work in the latter domain, focusing on tangible phenomena 'out there' in the material world. It is an approach that I broached with colleagues in a seminar presentation that comprises the next chapter that addresses archaeology, as the discipline that focuses on the study of things. Some of the *chaîne opératoire* literature, for instance, promises, as mentioned earlier, to afford access to socio-cognitive issues, albeit opaquely, which is perhaps inevitable in archaeological contexts given that access to 'know-that' social intelligence is proving so challenging in contemporary settings. Its reduction to a psychological black box, currently undermining much social theory, compounds the problems, rendering the opacity almost pitch dark. A focus on tacit knowledge of the 'know-how' sort may afford archaeology an opportunity to make a unique contribution to social theory by helping to lever ajar the black box, which may further open up an avenue to developing more securely founded socio-cultural insights into the archaeological record.

4

A seminar

What can archaeology and anthropology do together?

The following edited version of a seminar opens a line of argument that runs through the book, focusing on buildings to reappraise what constructions of the material world can tell us about being human, notably about socio-cultural arrangements. It floats the idea of possibly accessing tacit knowledge by initially focusing on material things rather than immaterial interactions. It argues that the 'know-how' skills that feature in the construction of such things as houses 'out there' in the world may further understanding of tacit knowledge of the 'know-that' variety, 'in here' in our heads, which informs social behaviour. Archaeology features here, for unlike anthropology, it pays close attention to material issues. It also affords an opportunity to address the wider issue of the relation between the disciplines that recurs in this book, to contribute to recent efforts to address the 'trade deficit' in ideas between archaeology and anthropology (Garrow and Yarrow 2010:4) and to explore opportunities to build further connections between the two subjects. The question is not only the extent to which archaeology might advance a theory of behaviour founded on tangible material things to further understanding of the intangible aspects of social interaction in extant communities, but also the possibility of accessing the social aspects of extinct communities more securely with no actors to talk to.

To the seminar: thank you for letting me come and talk to you today, as I have some problems that I am hoping you can help me sort out.[1] A senior colleague once told me some time ago that a law of academic life is never to start a seminar paper with an apology. So this is not by way of an apology but as an explanation of what follows. I have no polished paper to present to you, but rather some data, ideas and a number of unanswered questions to pose. In this sense, I think that what I am going to present is more like a postgraduate student's attempt at a seminar than a hoary professor's paper. This is nothing to apologize for, come to think of it. Some of the best seminars that I have ever attended have been given by postgraduate students who, groping their way to some understanding of their research material, offer their audience a real opportunity to contribute briefly to their enquiries and participate in their intellectual quest.[2] Some of the worst seminars that I have ever attended, on the other hand, have been given by veteran faculty who, embroidering on ideas that they have built their careers on, present papers that are so polished, even already published, that the audience is left wondering what it can add or if its comments will make any difference.

The 'do' in the seminar's title denotes my problems, initially in the reciprocal sense of what archaeology and anthropology can do for each other. The issue has puzzled me for some years; for instance, following some fruitful collaborations with archaeological colleagues that have not seemed to me to yield fully the anticipated intellectual returns (Hardy and Sillitoe 2003; Shott and Sillitoe 2005). On my part, I have admittedly been somewhat hazy about relating materially focused ethnographic research to archaeology, by situating things in wider socio-cultural contexts, being largely motivated by a gut feeling that such ethnography should help further understanding of long-gone communities that depends on material evidence, mostly comprising remains of buildings and things. Rather than talk in general terms, I wish to root discussion, at least initially, in a specific context and consider structures found in the New Guinea highlands.

The Tebiyaem excavation

The first problem that I have will be familiar to you as archaeologists. It concerns the interpretation of an excavated site. It is located at a place called Tebiyaem in the Was valley of the Southern Highlands Province of Papua New Guinea (Map 3.1.2). It came to my attention while surveying some gardens there, when I noticed a flat area

1 I am grateful to archaeological colleagues at Bristol, Durham, Newcastle, UCL and York universities for allowing me to present this paper to them and agreeing to me recording our discussions to summarize here. You will find the PowerPoint presentation that I prepared to accompany the seminar at the website for this book (see www.community.dur.ac.uk/paul.sillitoe/wola/houses).

2 It was more faithful too to the origin of the term seminar – from Latin seminarium 'nursery' – in asking participants to join me in playing with a problem.

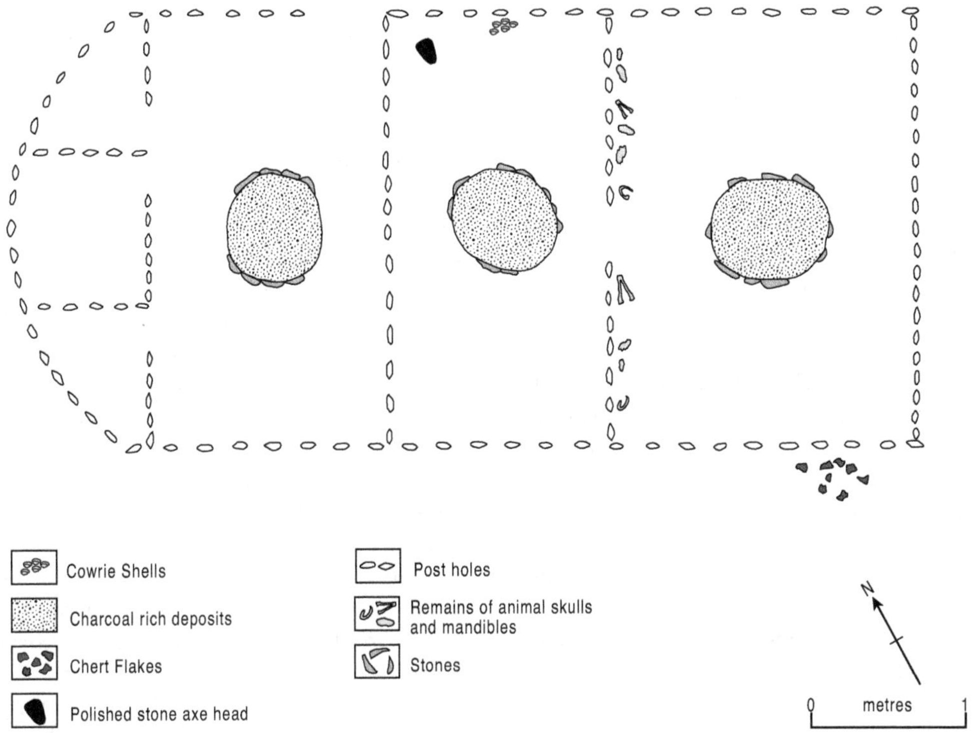

Figure 4.1 The excavation at Tebiyaem.

of land under thick secondary regrowth, which could only logically have been the result of some human activity, in view of the rugged landscape and the distance above the river (some 300 or more metres away downslope) that precluded any natural terrace erosion. My friends agreed that it looked like the sort of platform that people level to build some structure on. But no one, not even the oldest residents in the area, could recall seeing, or being told about, any building there. The only structure that they mentioned was a *seblaenda* shelter that previously stood on a nearby prominent knoll, surrounded by several majestic palm trees (*Gulubia* sp.) still standing at the place. It had housed some prehistoric stone artefacts (such as mortars, pestles and club heads), which people believed served as the abodes of the ancestor spirits of Puwgael *semgənk* 'family', and was where members sometimes offered pigs when someone was sick, rubbing oil and ochre on the stones (see Sillitoe and Sillitoe 2009:98–9). People show the same disinterest in the history of these ancient objects as they do old residential sites, saying that persons occasionally found them, often after they had been exposed by some earth movement, such as a landslide (the objects have long puzzled archaeologists).[3]

While those living in the Was valley, like others living across Melanesia, evidence strong connections to place, their interest in the history of their place, beyond recall of events in the recent past (that is those within living memory, which largely come up for discussion when disputes occur over rights to cultivate land), is strangely nonchalant. There is scant concern for prehistory. The lack of interest, even arguably the absence of any idea akin to prehistory, poses the question: what is the relevance of archaeology to such people's worldview; a tricky aspect of the relationship between archaeology and anthropology? The implications for archaeologists of local people having no interest in evidence of their predecessors' activities are considerable, especially in these times of increased local collaboration, which such views presumably compromise – for instance, such communities are likely to support the destruction of archaeological sites in the name of development. What would the value be of the excavation of the Tebiyaem site; how might findings of interest to archaeologists be related to the local population, and by extension to anthropology?

An archaeology colleague at the University of Papua New Guinea in Port Moresby (who spent some time in Durham as a Commonwealth Visiting Scholar) was looking for somewhere to undertake a dig so that students could gain some experience of excavation, and when I told him about the Tebiyaem site, he thought that it would serve his

3 Archaeologists have excavated mortars from sites in the highlands of Papua New Guinea dating back 3,000 to 8,000 years BP and conjecture on their use; for instance, found in regions suitable for taro cultivation, some speculate that 'mortars and pestles were used to process taro and nuts into a paste-like pudding on ceremonial occasions' (Howarth and Swadling 2009:29).

purposes well. A dig also served my purposes, in wishing to know more about the location, after the manner of Edward Schieffelin (1982), who arranged for archaeological colleagues to excavate the site of a cult house used by Kaluli speakers on the nearby Papuan Plateau, so that he could further his understanding of the rituals conducted there and check what people told him about them (the rituals had occurred in living memory).

The dig revealed the remains of some structure, comprising a rectangular pattern of post-holes together with some charcoal-rich deposits and other remains resulting from human activity (Figure 4.1). The charcoal gave a ^{14}C date of 510 ± 30 years BP. Some charred pandan nut shells (probably *Pandanus julianetti* or *Pandanus brosimos*) were found in the right- and left-hand (but not central) charcoal deposits (Sillitoe 1983:103–11). The stones occurring with the charcoal deposits were pieces of local limestone (Sillitoe 1996:127–8). The flakes of chert came from the same source, being broken pieces of nodules that erode from the region's highly folded Miocene limestone; a few of the flakes had bulbs of percussion, others were shattered nodule fragments (Hardy and Sillitoe 2003). The polished stone axe blade comprised metamorphosed basalt that outcrops in the Western Highlands (sourced to a quarry there, close to my archaeology colleague's home region – see Chappell 1966). The five cowrie shells (*Cypraea moneta*) found on the site are further evidence of the movement of things over large distances, probably originating in the Gulf of Papua to the south. The animal remains comprised the skulls and mandible fragments of mountain cuscus (*Phalanger vestitus*), long-beaked echidna (*Zaglossus bruijni*) and ringtail possum (*Pseudochirops corinnae*), together with the mandibles and canine tusks of pigs (*Sus scrofa papuensis*) (Sillitoe 2003).

This is all the evidence from our modest excavation. Now it is your turn to participate in this seminar and help me solve my first problem, which is, as you have probably guessed, to figure out what to make of this evidence. I am hoping that as persons experienced in the interpretation of such fragmentary material evidence, you can come up with some ideas (as I am used to observing and talking to the human actors responsible for such things). And maybe the reader would like to ponder what it was that the excavation uncovered before reading on.

Summary of discussion

Some participants boldly speculated about the excavated remains. Several suggested a house, one proposing that different clans occupied the various sections and another colleague drawing parallels with hierarchical Pacific island communities where persons of different rank have separate rooms and fireplaces. Further suggestions were a menstrual seclusion house or a workshop, with different stages of production carried out in the demarcated areas and waste deposited outside.

Other colleagues were more cautious and initially discussed the excavation evidence. Several commented on the small size of the structure unearthed, and the particularly small spaces at the semicircular end. The other feature that attracted several comments was the arrangement of gaps or entrances, notably the apparent absence of access points into two of the spaces: 'How did they get in, there is no entrance?' This observation prompted some to question if the structure was roofed, particularly in the absence of any large central post-holes to carry a roof beam. Someone pointed out further that while the single narrow entrance would allow some light in, it would otherwise be a dark environment if it was roofed. The uniform size of the relatively small post-holes further suggests that it may not have been a load-bearing roofed structure at all, but an enclosure. In this event the staked walls may have been low enough to step over to enter; even if roofed some of the walls may have been no more than waist height, not extending all the way to the roof. But some persons pointed out that it is strange that there are openings into some parts and not others, and speculated that the posts, particularly given their small size, may have been removed to allow access and then have been put back. It was even suggested that users could have adjusted the size of the space depending on how many persons were involved in whatever went on there. Other participants thought that the apparent access conundrum suggested the remains represent more than one phase of activity, erected over a period of time such that the enclosed areas were not contemporary but built and used sequentially at different times. Another possibility was that the area was enclosed after some activity that involved the fires.

The number of large hearths in such a small structure attracted further comment. Some persons suggested that it may have been a sweat lodge. The absence of many stones in the fire pits ruled out earth ovens. Someone suggested a smoking house or butchery place. The animal bones attracted many comments, notably their neat alignment along one wall, which together with the remains comprising skulls/heads was suggestive of some ritual act. But the absence of bones from meaty parts of animals, such as long bones, suggests that whatever the rite was, feasting did not feature. The neat pattern suggested a collection of *objets trouvés* to someone, who pointed out that 'people collect things they want on display for some reason'.

The discussion of excavated features such as post-holes and animal-skull arrangements led to further interpretation. Many colleagues saw a ritual structure, with barely space for more than one person in each room, maybe a corpse laid out or where someone slept. As one colleague quipped, the 'fallback position is always ritual', although arguably my introductory comments mentioning the *seblaenda* ritual structure led the audience in this direction to some extent. But some participants were cautious about engaging in interpretation, particularly of an extravagant kind, with archaeology's current return to focusing on material evidence, in part a response to chastening previous experiences – as caricatured in Macaulay's (1979, 1980) send up of an archaeological discovery in the future, of an American motel buried in 1985 and interpreted as a Pharaonic-like burial chamber with ritual paraphernalia. Nonetheless, the clear division of features looked ritualistic to some colleagues, with animal bones neatly patterned in one area, exotic items (polished stone axe and cowrie shells) in the middle section, pandanus nut shells occurring in the two end hearths only, and the heap of chert on the outside. The arrangement of areas suggested functional differentiation, possibly featuring a sequence culminating with an animal sacrifice in the final section. Some colleagues asked if there had only been a single fire in the hearths or no fires at all, and whether the pits contained accumulated ash deposited from fires for offerings made elsewhere. The apparently clean-swept clay floor intrigued others, who saw the structure prepared for abandonment with things laid out in a certain way. (One likened it to an ephemeral structure excavated in Amazonia that had 'a strange archaeological profile' with things laid out and kept clean, thought to have had something to do with ancestral animals.)

Some experienced colleagues asked how the ground plan compared with contemporary houses. It is a question that any archaeologist is likely to ask, for having familiarity with a region inevitably informs one's understanding of excavated remains, particularly if fairly recent (albeit many archaeologists work with ancient remains that bear no resemblance to today's structures, which is when some of them turn to ethnographic analogy for inspiration). Many seminar attendees were at a disadvantage, having no such regional knowledge, as was shown by a participant who did have regional experience, who asked if there was any evidence that the small cells housed pigs, because if so it was a woman's house. The question about parallels with any of today's structures serves as an entree to the next part of the seminar.

The Wola house

What we have just discussed addresses, in part, one aspect of my paper title's question, namely: 'what has archaeology to offer anthropology?' The next problem that I have concerns the other aspect: 'what has anthropology to offer archaeology?' Now, I do need to apologize to you, whatever orthodox academic wisdom. According to the postmodern point of view, all academic discourse is a construction contrived to support a particular argument and subjective worldview, and the aim of deconstruction is to reveal these biases, which extreme interpretations consider fictions. Such deconstruction is straightforward in this case, as I have presented you with a fictional dig and in doing so, I have been deliberately disingenuous, in withholding some evidence from you, hopefully to further (not score) a point.

I know exactly what the structure depicted in the 'dig' diagram is: it is what the local people call an *enzatenda*, a 'men's-women's house' (Chapter 1). The right-hand and central rooms are occupied by men and the left-hand room with the bell-end is occupied by women (usually the male occupant's[s'] wife or wives and children, sometimes with other female relatives such as unmarried sisters and widowed mothers). The right-hand room, which is entered by stepping over the low fence that comprises the front wall, is the men's day room where they sit and talk and cook their food, and on occasion arrange displays of wealth during socio-political transactions. They also display the skulls of animals on the wall there as hunting trophies. The middle room is where the men sleep and store their possessions, often wrapped in bundles. Women sit and cook in the left-hand room and sleep there too. The enclosures in that room are pig stalls (as one colleague surmised), animals spending the night there. The charcoal deposits mark the central fireplaces in each room.

What I have told you so far is trivial and many archaeologists (as noted) will ask if there is any such ethnographic evidence that may help in the interpretation of an excavation. But knowing that the structure is a house does not get us far in understanding the socio-cultural world of those responsible for constructing it; other than the living arrangements informing us that male-female relations involve a certain physical segregation, suggesting perhaps some constraints on interaction between the sexes. And the remains found around the structure indicate a stone-tool-using culture that has dealings with pigs and engages in hunting forest animals. What further ethnography would an archaeologist find useful? There are two levels of evidence: that relating directly to the material remains and that relating more widely to the culture responsible for these.

Regarding the first level – ethnography relating directly to the material remains –I can tell you about the raw materials used in building such a house and I can give you details about its construction, having closely documented the building of such a structure. I can specify the building materials (give their vernacular and botanical identifications) and document the uses to which they are put. I can also tell you the amounts of materials required to build such a house (for instance, the length of rattan vine used and the number and dimensions of sapling poles used as rafters) and the time taken to collect and prepare these things (see Chapter 6). Going on, I can detail the construction of the house, from laying out the plan and driving in the wall stakes (the post-holes revealed in the 'excavation'), to erecting the rafter and lathe frame of the

Plate 4.1 A reconstruction of an Iron Age house at Flag Fen, Cambridgeshire.

roof and thatching it. A virtual blueprint, if you like, such that you could build a similar house, if you had a mind to do so. Again, I can tell you the time that it takes to complete the work involved in each stage of the construction process. Indeed, I can go even further and document the technical differences in working with stone and steel tools, and give a comparison of their relative time efficiencies (see Chapter 7).

What use is this sort of ethnographic data to archaeologists? I ask this question because, while there is a long tradition of drawing on ethnography in archaeology, the discipline does not seem particularly interested in such data, which strikes me as odd because they relate directly to the sort of material evidence that comprises the greater part of the information available to archaeologists (particularly prehistorians). The remains of buildings, after all, constitute a significant part of archaeology's evidence, as any quick flick through reports and journals confirms with numerous drawings of, sometimes enigmatic, patterns of post-holes and stones marking wall foundations together with remains of fireplaces and assorted residue of domestic objects, often largely refuse. Presumably experimental archaeologists who engage in the reconstruction of prehistoric structures, for instance houses in locations such as Butser Farm (Reynolds 1979:29–45, 93–103),[4] may find such comparative ethnographic data useful, which pay attention to material issues and relates these to society more widely, whatever the attitudes of other colleagues who confine themselves to the interpretation of excavated remains.

I have had the privilege of living in a region where people had until recently depended on stone tools (both ground-stone axes and chert blades, as unearthed in our 'dig', the latter still knapped and used – Hardy and Sillitoe 2003) and I have expended considerable effort in documenting the use and comparative efficiencies of these tools in comparison to steel ones – arguably in one of the last opportunities to study habitual stone-tool users (Sillitoe 1979c, 1988a, 2010a). I did not expect postmodern anthropological colleagues to show much interest in these data but the lack of interest shown by archaeological ones has, I confess, surprised me. Now it is your turn again to participate in this seminar and answer my second question,

4 Butser Ancient Farm is an experimental archaeology site on the English South Downs in Hampshire, featuring an Iron Age settlement (see www.butserancientfarm.co.uk). Reynolds (1979:32-35) gives some information on raw materials and structural details of a reconstructed Iron Age house.

Plate 4.2 Roof construction of Iron Age house, with parallels to New Guinea Highland house (see Reynolds 1979:38–9, 98–9 for details).

which is: what sort of material ethnography is of interest to archaeologists and why?

Summary of discussion

Many colleagues were quick to reassure me that archaeologists find the above sort of ethnographic data useful, particularly those relating to things that occur in the excavated record. Prehistoric archaeologists often 'use contemporary data to look backwards', in interpreting what they find and creating theoretical models. But several participants disagreed, warning of the dangers of drawing analogies between communities that are 'too long ago, far away', that is historically and geographically distant. Others argued that the aptness of any analogy is the issue, and yet others that whereas most archaeologists would be wary of drawing on ethnography in such detail they might use it in formulating general principles.[5]

Plate 4.3 Wall construction of Iron Age house, unlike New Guinea Highland house.

Another issue is that few anthropologists bother with material ethnography, such as making and using artefacts, or if they do, they concern themselves with issues that are remote from archaeologists' material interests, such as the symbolic import of things (for instance persons' expressions of identity in what they consume). 'Archaeologists are interested in forensic issues not of interest to anthropologists.' Anthropologists do not, for example, consider processes associated with site abandonment, what prompts it and the things left behind, which are central to furthering archaeological interpretation of excavated finds.[6] These are not issues that concern many communities, as illustrated by my friends not recalling the Tebiyaem house site. A seminar participant pointed out that while the largest body of archaeological data from New Guinea concerns ditches (such as the six phases of drainage ditch at Kuk going back 9,000 years – e.g. Bayliss-Smith and Golson 1999), there is scant ethnographic information on ditches (such as dimensions, who digs them and how dug, and relation to crops). 'No ethnographer wants to touch this

---

5 Ethnographic accounts of buildings, and the relationship between architecture and the use of space, may offer useful analogical sources to archaeologists in their search for principles to inform the interpretation of excavations, many of which feature building remains, supplying information on 'what influences domestic architectural design and the use of space and how each is related to the other' which features 'complex interaction between domestic structures and spatial organization … as expressed in technology, socio-political complexity, symbolism, and economics' (Kent 1990:1–2).

6 There are a few exceptions, see for instance Moore 1982 (for models of patterns of discards related to gender, types of rubbish and notions of cleanliness, among Marakwet of East Africa); Shott and Sillitoe 2004 (for calculations of use-life distribution of Wola artefacts to further understanding of archaeological assemblage formation).

stuff because too boring.'[7] In response, a colleague pointed out that prehistorians are likewise picky about what they research; for instance, 'in Melanesia we are most interested in the Lapita era and stone monuments'. Consequently, archaeologists, not expecting to find relevant information do not scour the ethnographic record for it, but have instead embarked on their own ethnographic fieldwork in the ethnoarchaeological tradition (e.g. Binford 1978, 1983; David and Kramer 2001; Gould 1980; Hodder 1982a, 1982b; Kramer 1979; Longacre and Skibo 1994).

In addition to interpretation of excavated sites and finds, archaeologists are increasingly interested in the socio-cultural processes that lead to the material they unearth. They are not only concerned about 'the how but also the why', such as why people use the materials they do and why they do things the way they do. A palaeobotanist commented that he found ethnography useful to fill gaps regarding organic materials not found in the archaeological record. 'If we know people use certain plant material, we can look for phytoliths in the pollen record.' Availability of resources and the time it takes to do things will have influenced choices and decisions. The technological choices persons make, what materials and methods they decide to use, relate to variability, for actors may achieve the same end result using different materials and processes. Archaeologists have scant hold on the variability issues, although it critically influences the excavated record. The selection of materials also relates to the movement of some things from their source, how they travel and distances they cover, which may further understanding of archaeological distributions (such as Langdale axes across England – e.g. Bradley and Edmonds 1993:59–154).

Colleagues who were relaxed about ethnographic analogies spoke about seeking patterns of behaviour in relation to things: what does the production and use of objects tell us about social organization, to what extent are they 'manifestations of social relations'? It is a view that I held but now question – see later. Even more questionable were suggestions that 'connections between the real world your data refer to and the cosmological world' might inform archaeological interpretations. What people think of associated activities is currently of interest to some archaeologists, particularly any symbolic ideas. The 'relationship between symbolic thought and material properties of things would interest us'. For instance, 'whatever they were thinking about in thatching a roof'. When I pointed out that large, arguably intractable, epistemological problems attend accessing such thoughts (as I go on to discuss in the seminar), and that Was valley houses have few symbolic associations (which actors can consciously express anyway, as opposed to foreign intellectuals somehow divining and imposing), I was told that it is important archaeologists know these issues. Some were of the opinion that people saying they do things because their ancestors did them that way is a significant 'symbolic' point, indicating that present values rest on ancestral authority, the attitude being that 'the ancestors figured out the best way to do it'. Several commented on the value of knowing how the relationship between men and women links to the architecture of houses, which could further understanding of gender-based issues regarding resources and behaviour, such as the possibility of food taboos leading to men and women cooking food in separate places.

Some participants, who were chary of seeking direct ethnographic parallels, thought that ethnography could usefully serve as a check on experimental archaeology. Someone commented about being 'conscious of the difference between a skilled person who produced a thing excavated and our self-taught attempts to make it.' Another person mentioned the 1950s Danish stone-axe experiment (see Steensberg 1980:18–33) in which many axes were damaged. Hence data about the time people invest in making things, such as knapping stone tools, are consequently useful for critical appraisal, giving an idea of what habitual skilled workers might have done in the past, which may not match what we do today – a comment that relates directly to tacit know-how. And construction data are useful (for instance looking at contemporary dry-stone walling in England has yielded useful information about prehistoric stone-walled structures). But others saw problems. Archaeologists may 'not even know for example if a structure was thatched', and it is difficult to know 'what to make of such ethnographic detail'.

Another issue is geographical scale. Archaeologists are interested in more than objects; for instance site-landscape relations are of interest to them, albeit they are little explored in the Pacific as there are relatively few sites excavated across the region. While fieldworking anthropologists largely focus on single communities, just as excavating archaeologists focus on single sites, anthropologists infrequently consider their place in the regional landscape, and if they do, they concern one locale, whereas archaeologists consider many sites in interpreting landscape relations – a colleague mentioned the alignment of sites and astronomical correlations. Nonetheless anthropology can help to understand landscape issues, such as how people use different land zones for farming, settlement, rituals and so on. Furthermore archaeologists may find helpful information about how actors make land-use-cycle decisions, such as where to site a house, when to abandon it, what then happens to the land, all of which relate again to choice. Also, while excavation may reveal

7 Although again there are a few exceptions, see for instance Lemonnier 2012 and Sillitoe 2010a on Baruya and Wola garden barriers respectively.

how many houses were on a site, how many were occupied simultaneously is unknown (see Chapter 3).

There is also the time-scale issue. The house-building ethnography details construction for one period in time, whereas archaeology considers much larger time spans drawing on comparatively scanty evidence. One commentator asked how comparable the two kinds of evidence are, using an evocative movie analogy, where anthropology watches only the last frame in detail, while archaeology aspires to reconstruct the entire movie, albeit only able to see an isolated frame or two occasionally, maybe not even whole frames, with large gaps. Several colleagues thought that the *longue durée* perspective of archaeology is something that should complement anthropology's *courte durée* view of humanity. While anthropology gives rich detail now, archaeology gives time depth, furthering understanding of processes of change in the deep past that ethnographic fieldwork largely overlooks, even with today's heightened historical awareness. Archaeology is consequently much taken up with how cultures transform over long time periods, notably targeting issues of technological change and hybridization; even 'culture clash' (such as occurred with the Roman Empire's arrival in England). There is the issue too of the meaning of time in the past, which ethnography reminds us is relative; our view of it as a scarce resource is not universal (it may even be odd in the context of global socio-cultural arrangements) and time-consuming activities may not have the significance we think.

Differences in the time persons spend on activities is only one issue regarding the change that results with access to new materials and technologies, albeit it is 'useful to quantify how communities spend their time, what they put energy into'. But new things 'change the culture' too, have knock-on effects on domains such as kinship and gender relations. In this respect, many participants, perhaps predictably, showed considerable interest in the stone- and steel-tools data: how the transition affected the way houses are built, any shift in materials used, and changes in understanding of the construction process, in addition to the extent of wider cultural transformations. They also expressed an interest in technical issues such as how long it took to make a stone axe and how they were hafted and used; whether all axes are made of the same stone; did different tasks require specialized stone tools; how long axes lasted in use; and the marks left on blades by various materials. These are a questions rarely asked by anthropologists, another example of them infrequently collecting data of interest to archaeologists. Those happy with more speculative analogies asked if there was competition over access to quarries, did making and using stone axes feature rivalry between men, how many axes did households possess, and did they acquire a biography as passed around and between communities, and did this add to their cultural value.

Time is also a disciplinary issue. The questions that interest archaeologists change and intellectual fashions influence the evidence they use, such as what they take from the ethnographic record. 'Fashion-driven writing is common in archaeology, not always starting out from the excavated evidence.' The need to have a hypothesis to initiate an archaeological research project exacerbates this tendency, pushing enquiries in a certain direction. Whereas in the 1970s–80s heyday of ethnoarchaeology, data such as those on Wola house construction would have attracted some interest, it is less today with the advent of post-processual archaeology, or interpretative archaeology that is archaeology's rendering of postmodernism, which stresses the subjectivity of interpretation. It is a response to processual archaeology, or the one time 'new archaeology' of the mid-twentieth century (which argued for a scientific approach to achieve objective interpretations of archaeological material), and comprises a disparate range of thought encompassing such diverse approaches as structuralism, neo-Marxism and phenomenology. There is some overlap with anthropology here, albeit borrowing abstract ideology. Archaeology needs to push back, using the assets of its material perspective.

The search for 'theory'

Now let us turn to the second level of evidence – that relating more widely to the culture responsible for material remains (which some colleagues raised in discussion). This links to what I think is archaeology's intellectual equivalent of the Holy Grail search, namely to develop a theory[8] that will allow archaeologists to hypothesize more soundly on the socio-cultural arrangements of those responsible for the remains they excavate. The implication is that some archaeologists think that it is possible to develop a body of theory that starts with a collection of material objects – such as some stone tools, patterns of post-holes, charred remains of food and animal bones – and securely further understanding of the socio-cultural arrangements of those who manufactured and used the things beyond current speculation. It may also allow them to make original contributions to social theory. The goal is no longer to play second fiddle to social philosophers, sociologists and anthropologists, who currently dominate social theorizing, while archaeologists hover on the outside

8 Although I prefer 'ideology' (Chapter 1) to 'theory', I continue to use the word 'theory' here, as the term that features in discussions about the feasibility of a materially focused approach to understanding socio-cultural life generally. The distinction reflects the intellectual tension in archaeology between attempts at idealistic social-science theory building to interpret cultural context and the use of natural-science theory to explain empirical evidence.

of this social-science ensemble trying to seize on ideas that might illuminate their finds, but rather to make unique contributions to the advancement of social theory.

When I briefly discussed something along these lines, in the context of my study of the manufacture and use of artefacts in the Was valley (Sillitoe 1988a), critics accused me of naivety. In an otherwise generous review of *Made in Niugini*, Peter White (1990) for instance, an authority on Pacific archaeology, comments that my suggestion that the study of artefacts will take on increasing importance in this way 'presents more problems than this review can encompass, especially since Sillitoe does not (?cannot) elaborate on how this development might be achieved, or even what a general theory of culture based on artefactual knowledge might look like. Regretfully I believe, however, that Sillitoe is profoundly wrong.'[9] It is such comments that in part prompt me to return in this seminar to the possibility of, and implications of, material-culture grounded theory and pose the questions that I am putting to you. It will presumably require anchoring in actual human behaviour with respect to things, which implies referring to ethnography, albeit opportunities to collaborate with anthropologists may be limited, who are busy with 'glocalizing' issues, and if they consider material issues at all, tend to focus on the use of introduced things in novel contexts, such as cotton textiles replacing barkcloth, rather than locally produced goods that might serve to advance a materially focused theory.

There is nonetheless a long history of drawing on ethnography to advance understanding of the people and cultures responsible for excavated remains. While many associate ethnographic analogy with 1960s processual archaeology and its 'mid-range theory', we can trace the practice back to the Enlightenment when early antiquarians turned to reports from the new colonies to fire their imaginations about life in ancient times (Orme 1981:2–13; Piggott 1989). The practice was well established by the Victorian era, before archaeology and anthropology emerged as separate disciplines, with anthro-archaeologists drawing on ethnographic accounts, corresponding with colonial contacts and sometimes visiting 'primitive' people, whose ways of life they assumed reflected that in prehistoric Europe, before 'civilization'.[10]

Archaeologists continue today to draw on the ethnographic record when interpreting excavated remains, albeit circumspectly (Coudart 1992). While ethnographic analogies open the range of possibilities considerably, we can never be sure of their appropriateness (as pointed out in seminar discussions); the convincingness and force of any analogy varying with differences in time, place and culture compared. Nonetheless, in considering an extinct way of being in the world, reviewing evidence from extant orders with similar material assemblages is likely to throw up useful pointers to interpretation, beyond relying on our own imaginations, conditioned by an alien high-tech lifestyle. An example is the use of near-contemporary Polynesian ethnography from the other side of the globe to illuminate Celtic cultural arrangements two millennia ago, with the striking similarity between the hilltop structures of the European Iron Age and citadels in parts of the Pacific region and associated material evidence (Firth 1927; Sillitoe 2012a).[11] We are unavoidably biased culturally and analogies help us confront our assumptions by presenting us with alternatives, lessening our projection of our own limited, and possibly inappropriate, cultural experience. The past is another culture too. In contemporary postmodern spirit, many 'post-processual' archaeologists now present the range of possibilities and leave the reader to judge their aptness in furthering understanding of excavated remains (Fowler 2004).

Regardless of this long and ongoing tradition, the integration of ethnography in archaeological projects, to advance understanding of excavated remains, continues to be limited. Why?

The problem of evolution

The relative paucity of ethnographic evidence relating to the material aspects of life is one reason for the limited disciplinary integration (as noted in the seminar discussion). Anthropologists fought shy of material issues

9 As an aside, I might note for the record, that I have addressed some of the shortcomings that White (1990:73) identifies regarding knapped tools (Hardy and Sillitoe 2003) and I have also made a collection of such tools for future study (Shott and Sillitoe 2005).

10 An intriguing example of the use of ethnographic analogy to address issues of archaeological interpretation from this period concerns the parallels drawn between Australian Aboriginal flaked tools and supposedly 'prehistoric' European eoliths – roughly chipped pieces of flint originally thought to be artefacts made by early humans but subsequently, and embarrassingly, judged 'geoliths', flakes of natural origin (Ellen 2011:298) – an early lesson in the pitfalls awaiting the unwary in drawing on ethnographic parallels.

11 The Polynesian analogy illustrates the challenges of bridging between the material aspects of existence and wider socio-cultural arrangements and value-loaded human behaviour, and the way values inform social-science theory. In speculating on British Iron Age society (Sillitoe 2012a), I take equality and liberty as key values of tribal orders and review the archaeological evidence focusing on features common to such small-scale socio-political arrangements – such as subsistence independence of households, scant development of specialized labour or markets, dispersal of power among many and the importance of kinship in social life – and put forward an interpretation that supports recent challenges to the orthodox one that has hierarchical orders lead by nobility and is reminiscent of later medieval through to contemporary European socio-political arrangements.

following the excesses of our Victorian ancestors – who used ethnography crudely to advance their evolutionary views, seeing European society as the civilized epitome of human development, and the poor benighted 'savages' somehow stuck in the prehistoric past. When the tradition of fieldwork got underway and showed that the supposed 'primitives' were not primitive after all, anthropologists preferred to focus on sociological topics. The Victorian assumption of superiority rested largely on the technological capability of Europe, which measured in material terms was certainly more advanced than elsewhere– that is, what it could do to intervene in nature, to modify and manage the environment to improve material well-being and comfort for (at least some) human beings.

The technological capacity justified colonialism (indeed facilitated it, firearms outgunning bows and arrows) and continues largely to underpin its successor, international development, which seeks to share with others the benefits of industrial technology (albeit growing numbers of people are questioning these purely materialistic assumptions as the associated environmental and social costs become increasingly evident). Anthropologists perhaps subconsciously feared that by focusing on the material conditions of life elsewhere they would feed evolutionist, and subsequent diffusionist, conjectures. After all, the material assemblage of a hunter-gatherer band (whose nomadic lifestyle prohibits the possession of many material goods) does indeed look paltry compared to that of the average European household.

Anyway, the emergence of the social sciences and the challenges they posed appeared more intellectually interesting than enquiries into material issues, which, while some of us may now think a narrow-minded attitude, has left archaeology out on a limb to this day. The ideologies of the social sciences however, are strongly fashion driven and change with the decades (as pointed out), or rather rotate carousel-like, and it is currently the turn of evolution again. While it is enjoying something of a revival, the theory of evolution has scarcely contributed to the renewal of interest in material culture in anthropology, possibly for fear of repeating Victorian errors about equating evolutionary progress with technological capacity. But eschewing the equation of evolutionary success with technological standing seems to deprive social theorists of any yardstick equivalent to that of the reproductive success of biologists (see Chapter 11), who measure evolutionary success according to the rate genes pass to the next generation (Hodder 2011:165–6).[12]

The problem of ethnography

Attempts to advance a materially focused social theory on which archaeology might draw to further understanding of social arrangements in bygone ages have largely focused on identifying possible correlations between material existence and social life – this is certainly what I had in mind in *Made in Niugini* when suggesting that research on extant technologies might furnish comparative data for such theory building. It seems a reasonable approach. A defining feature of humanity, after all, is our use of technology to intervene in the world and manipulate nature to our ends. Another is our propensity to cooperate socially, and linked to it our capacity for complex symbolic communication, notably through language. These features of humanity are intimately interconnected; the rub is that they are interconnected in a multitude of culturally mediated different ways. But the use of ethnographic evidence to further such a theory has to date proved relatively unsuccessful.

The paucity of ethnographic studies of technology notwithstanding, the variable quality of the ethnographic record generally hinders its use in formulating a body of theory to advance understanding of the cultures responsible for the material remains excavated by archaeologists. It varies from thoroughly reliable (allowing for the unavoidable strictures of postmodernism about the inescapability of a certain distorting subjectivity) to downright dubious (the sort of easy-target ethnography favoured by postmodernists in their damning critiques). And it is not always easy to assess quality from the written record – for instance, it is often unclear how well the researcher spoke the local language, which is a key issue regarding reliability (regardless of the dated arguments of some anthropologists who maintain the contrary, after Mead 1939). The confusion that surrounds the terms that anthropologists use compounds the variable quality issue further, their discipline never having formulated an agreed vocabulary of clearly defined terms; consequently words such as clan, lineage, chief, sorcerer and so on have a range of different meanings, which makes accurate comparison and argument difficult. (This problem prompts some of us to use vernacular terms, where they exist, for key concepts – such as *sem* for social groups in the Was valley – in

12 The driving force behind Darwinian evolution, namely the survival of the fittest, seems irrelevant, for with increased technological capability we have families shrinking in size, which is hardly conducive to passing on genes (some demographers pin their hopes on this trend bringing down world population growth, particularly in materially poorer nations). And we have attempts to keep alive infirm individuals, such as premature babies and the chronically ill, and including those with serious genetic defects and persons in vegetative comas, and the infirm elderly, who would perish if 'red in tooth and claw' nature prevailed. Furthermore, we use in vitro fertilization to allow otherwise childless couples to have offspring, thus subverting nature's course, which is for such persons not to reproduce and pass on their genes. In short, we have culture interfering with biology, which has led to the idea of gene-culture co-evolution. But these issues, while interesting and discussed further in Chapter 11, are taking us away from our theme.

an effort to avoid unnecessary confusion and arguments at cross purposes such as litter the discipline, leaving translation up to those working on cross-cultural issues, who presumably use a consistent terminology).

It is arguable that regardless of these shortcomings with the ethnographic record, it should be possible to winnow out sufficient reliable data to attempt a correlation between the material aspects of peoples' lives and other features of their socio-cultural orders. While anthropology prides itself on advancing understanding of the human condition through cross-cultural comparison, such large-scale comparative studies are unpopular. They demand considerable effort to compile databases of reliable and comparable ethnographic information (anything less, such as use of off-the-shelf databanks, e.g. the Human Relations Area Files, will result in pointless garbage in and garbage out). Also, until recently, rigorous analysis was challenging, if not impossible, and the search for correlations between socio-cultural features across cultures has on the whole yielded poor or dubious results for the effort put in. But ready access to powerful computers makes analysis of large bodies of data feasible today – although my early attempt at such computer-aided comparative analysis of New Guinea ethnographic data (Sillitoe 1978) suggests that some of the above problems will persist, gremlin-like, as it yielded ambivalent returns, partly due to the dubious nature of some of the data and the problem of terminological confusions.

Even if we undertook such a large-scale comparative analysis, sifting the ethnographic data for correlations between material culture and other aspects of peoples' lifeways, it is possible (on previous evidence) that we might only come up with relatively trite conclusions given the complexity of connections between material existence and human socio-cultural arrangements, as the observations of crude evolutionism and functionalism underline (Darlington 1969; Malinowski 1960). Correlations such as, for instance, people with the wheel can potentially engage in direct trade transactions with geographically more distant communities as they are able to transport larger volumes of goods over longer distances and are likely subject to the authority of governments that organize and maintain roadways; or those who depend on a stone-tool technology lack factories featuring a specialized division of labour to speed up production and organization according to owners-managers-workers with associated exploitative labour arrangements, so ruling out Marxism, a popular ideology with social scientists until recently.

The proposal that it may be possible to devise a body of theory that will allow archaeology to extrapolate more reliably from material evidence to peoples' socio-cultural arrangements is looking rather forlorn. And the recent thoughts of an authority on this issue, who talks of an 'integrated archaeological perspective' and attempting 'a synthesis of archaeological theory' (Hodder 2011:154), makes rather pessimistic reading too, although he seeks to remain optimistic, concluding that while having 'failed in my attempt at integration' nonetheless there is 'some value in striving for a synthetic and integrative account, however partial' for a 'focus on things allows one to bring together aspects of much social theory related to materiality, embodiment, memory construction – with behavioural, ecological, biochemistry, and materials science approaches' (Hodder 2011:173–4). But I am left wondering what the point is of bringing together thing-focused aspects of social theory, beyond the end of bringing them together.

Now it is your turn again to participate in this seminar and answer my third question (which the reader might want to consider too, before reading on), which is: does this brief review of the problems facing archaeologists in drawing on ethnography to formulate a theory that could help further understanding of the socio-cultural arrangements of those responsible for excavated material parallel your understanding, or am I off message?

Summary of discussion

Some colleagues responded positively to the idea of 'archaeological socio-cultural theory', of 'one body of theory for human culture accommodating data from ethnography and excavations', but pointed out that it will not be easy to formulate. Others were sceptical. Some thought my comments supported an anthropological view of archaeologists as unable to cope with social theory and understandably responded robustly. They maintained that materially based theory is progressing far more than I suggest with much interesting social theory straight out of material culture studies, which may be so, but these studies involve living actors. The problem remains of devising a theory that rests on things only. No one could tell me how we can go from a collection of objects used long ago, with no access to those responsible for making and using them, to social theory. Others acknowledged that archaeologists more often adapt socio-cultural theory from other disciplines than formulate their own to interpret things in the past. 'We pull in stuff from many different disciplines' and consequently are 'always inevitably coming along behind'. Nonetheless, archaeology has made considerable progress. The advance, for instance, from the regional time charts of the 1950s that sought to document migrations, to today's sophisticated interpretation of scientific site evidence, set in landscape context and extrapolated ancient natural environment, is impressive, drawing on a range of theories and encompassing ethnographic analogy to accommodate cultural variation through time.

Several participants pointed out that it is necessary to talk about socio-cultural theories and not theory, and shifting the perspective from thinking generally about humanity to thinking relatively about particular cultures. Archaeology, like anthropology, confronts the complexity of humanity, covering a wide a spectrum of cultures (e.g. from palaeolithic to Roman to Anglo-Saxon), and to think one theory will apply is inappropriate. Instead archaeologists use, magpie-like, whatever theoretical concepts are best in any particular circumstances, 'we use whatever explanatory mode seems appropriate'. One colleague described how he uses creolization techniques to illuminate the Roman to Anglo-Saxon transition. But some colleagues thought that this theoretical omnivorousness presents problems, because 'we keep all the theory, we hoard it – traditional, processual, post-processual'. The absence of any periodic theoretical review to focus ideas results in a complicated state of affairs. Theory is not a parcel of overall concepts and integrated ideas but 'a set of discombobulated theoretical tools, where we plug in whatever we think useful' with no theoretical overview. And for some in archaeology, where science and humanities backgrounds meet, 'social theory is a dirty word', being cover for speculative interpretation as opposed to factually based explanation, and they give short shrift to talk of 'materiality theory' or any other theory. Someone pointed out how, looking back, the use of Marxist theory when in vogue a couple of decades previously looks distinctly unconvincing today. Whatever their camp, all agree with the caveat that any explanation or interpretation is provisional, for we can never say definitively what happened in the past using excavated artefacts and other archaeological traces of human activity long ago.

Asking why archaeology is 'not out there with other social sciences' irritated some colleagues. They did not think that ethnography might help overcome the considerable barriers that confront the advance of archaeological socio-cultural theory. Many others thought that it can help, as the intertwined history of anthropology and archaeology shows from the early antiquarians to recent ethnoarchaeologists. But ethnographic parallels are currently less fashionable intellectually, with critics questioning the validity of cross regional-temporal-cultural analogy (Wylie 1985). One colleague cited Arizona University's Kalinga Ethnoarchaeological Project (Longacre and Skibo 1994), questioning the appropriateness of drawing parallels between changes in contemporary Filipino material culture, notably ceramics, and the archaeology of the south-west United States. Another commented forthrightly 'I fail to see how you can draw parallels between New Guinea ethnography and mesolithic European stone-age society', which was not my intention, directly anyway.

Others could see more subtle comparisons, pointing out how ethnography may correct our biases and further our understanding of actions that resulted in the material found in the archaeological record. For instance, those responsible for such things may have made more use of certain senses than we do. We rely heavily on sight, but hunters in forests may rely more on hearing; while temperate European forests are different to tropical New Guinea ones, it is possible that hunters relied similarly on auditory senses, which may influence how they used things that we find today. While ethnography may help us to ask what things 'meant to people responsible for remains and not what they mean to us', such meaning is often beyond our reach. Inevitably it comes down to what they mean to us, as the post-processual approach argues. The scientific explanations of evidence afforded to us are likely radically different to the causes attributed by ancient users. For example, if excavation reveals that people placed dung around food-bearing plants, we are likely to see the act as fertilizing the soil and increasing yields because that is what scientific theory tells us would happen; but those responsible long ago, while observing the same end result of increased yields, may have thought that they occurred for quite different reasons, such as because the practice kept at bay evil spirits that reduced crop yields.

The point that I was trying to make is that analogies refer to ethnographic specifics; that is, they relate to what the 'new archaeology' of the 1970s processual era appositely called mid-range theory. They do not seek to advance a meta theory of humanity's situation, after evolutionary theory for example. This is the larger question that I am asking: is it feasible that archaeology could come up with a material-cultural focused theory with sound predicative powers? My doubtful comments about currently resurgent evolutionary theory drew some heavy criticism. 'Evolution is more sophisticated now than you allow.' New technology offers new ways for anthropology and archaeology to work together, regarding how cultural arrangements and technology change over time, with previously unheard of computing power available to search for correlations. We no longer 'compare one type of descent system in New Guinea with one in Africa' but draw on the enormous data sets now available, and 'using all such evidence we do end up with a general socio-cultural evolutionary theory about how certain cultures emerged or why they are as they are'. Evolutionary theorists no longer see evolution occurring in stages but as a complex interlinked process through time. In respect to these shifts, some colleagues lamented that many anthropologists seem to have little interest in cross-cultural generalities, which might reveal relevant patterns in the ethnographic data.

Others thought, as I do, that the study of changes in human cultural arrangements over time amounts to the study of history not evolution; that is, they are unique series of culturally and temporally specific events rather than general shifts in cultural arrangements through time. They associated archaeology more with the humanities than the social sciences, and several times spoke of 'telling stories of daily life', overtly positioning themselves as mid-range theorists. (Someone pointed out that interest in the unearthing of Richard III's remains under a Leicester car park focused on what they might tell us about his life and events in medieval England at that time.) In this view, archaeologists rely on the scientific collection and analysis of excavated data for sound evidence, without which their accounts verge on fiction. And it turns out that the problem of reliability is not only an anthropological one regarding ethnographic accounts, but also an archaeological one regarding excavation reports. Many assemblages are not compatible, authors having coded data in different ways, vitiating attempts at detailed comparisons, such that 'an article on for example Middle Bronze Age pottery tends to be a description of key sites put together with some general points, few people go through the entire run of ceramic assemblages'. The structure of archaeological site reports is another issue, with some sections (for instance, on stratigraphy, pottery, flint tools and so on) delegated to sub-disciplinary specialists, often analysing and writing to different standards. The result is a patchy and fragmented record, each chapter more addressing fellow specialists (the flint chapter written for others interested in lithics and so on) than contributing to an integrated account. On top of this, 'we know that what we excavate is only a very partial record of life' and its representativeness is unknown. It is unavoidably a more limited record than that found in most ethnographic accounts of what anthropologists observe, let alone what they do not see. And archaeologists go on to use this scanty evidence to talk about regional cultures extending over epochs (such as the mesolithic culture of Britain), which is even more questionable than anthropologists working with one small community and then extrapolating to an entire language group (such as referring to all Wola speakers).

Advancing understanding of the tacit

My comments so far question any move onto more rarefied levels of theorizing, namely the proposal that archaeologists not only develop 'theory' that allows them to further understanding of the socio-cultural arrangements of the makers and users of assemblages of excavated objects, but that they also strive for perspectives that will allow them to contribute to the advancement of social theory generally. This seems a truly forlorn hope in view of the disconnect with wider social issues, as seen in the apparently limited value to archaeology of ethnography on the manufacture and use of things, and the inadequacies of ethnographic analogy in helping to deduce from material objects the wider socio-cultural context of the makers of the things. But it is here that I can see a way that archaeology may break away from its limited use of ethnography to date, to contribute to current social theory debates. So here is the up-beat and radical part of my seminar, although what I have to say initially may suggest otherwise, intimating a deepening of the divide between archaeology and social science.

Social research is increasingly grappling with the topic of tacit understanding that underpins much behaviour. By tacit understanding, I refer to internalized knowing that comes from personal experience and intuition (see Chapters 2 and 3). We all act on such knowledge – learned as bodily acquired skills and through culturally mediated social experience – of which we have no clear awareness. It is difficult, arguably impossible, to communicate this knowledge verbally to another person, as the linguistic root of the term tacit implies (from Latin *tacitus* 'silent') (Polanyi 1958; Wagner and Sternberg 1985:438–9). As Isadora Duncan, one of the founders of modern dance, put it: 'If I could tell you what it meant, there would be no point in dancing it.' (Bateson 1973:137). It relates to things that we know how to do without consciously being aware of what we know.

Tacit knowing informs much of our understanding of and actions in society. We engage daily in many social acts that we are not focally aware of, which are informed by taken-for-granted presuppositions – how many of us pondered our behaviour in entering this seminar room and sitting down around the table? The tacit character of such social behaviour complies with David Hume's (1739) argument that our emotions, or our 'passions' as he phrased it (desires such as our need for affection, respect, happiness etc.), motivate our actions, not our 'reason' (as mentioned in previous chapters). We experience the 'passions' first-hand as lively sensations, not second-hand as ideas. We more feel than think about such motivations of our actions, often unable to verbalize them, acting on impulse and empathy. This phenomenological approach stresses that direct experience informs social interaction, that behavioural responses are inculcated via social relationships and that actors draw on embodied dispositions appropriate to the context. The associated tacit awareness is predicated on socially formed expectations that are so embedded in our psyches that we assume them unquestioningly.

Evolutionary theory again

We can imagine, in the manner of Enlightenment philosophers, primeval emotions similarly motivating our distant ancestors who, as they increasingly developed social awareness and realized the benefits of resulting cooperation, may have accorded higher value to actions predicated on feelings that promoted sociability – we can imagine the 'hidden hand of sociality' at work, to paraphrase Adam Smith (1993), in whose economic and moral ideas we can see his friend Hume's influence. The resulting society identified today by social commentators came into being without conscious planning – there was no gathering of the sage and grey-haired to debate and delineate the shape of any social order.[13] It was, of course, not one society that emerged but a large number of different social formations built on varying moral codes that all promoted sociability in various ways. The culturally informed interaction of individual persons is the product of many generations of tacit behaviour informed by commonly held values, which may shift over time. The proverbial mists of time shroud the actual course of events, as Hume (1985:467) observed nearly three centuries ago, referring to 'the original contract' by which the 'most ancient rude combinations of mankind were formed', noting that it is futile to ask 'in what records ... this charter ... is registered. It was not written on parchment, nor yet on leaves or barks of trees', and those interested in social evolution today can likewise only speculate on their possible course.

We pass on these socio-culturally informed patterns of interaction largely unconsciously to our descendants (social reproduction – as mentioned in previous chapter), inculcating the next generation with the values that underpin them and inclining them to behave in ways predictable to other actors (Collins 2001a:110–11). While we collectively pressure one another to conform to these largely tacitly held value-laden rights and responsibilities, this does not so constrain behaviour as to rule out manoeuvring by actors for perceived advantage, personally and for associates. And the variation in actors' emotional responses adds to the fickleness of human behaviour; for instance, a shy personality is likely to respond differently to a gregarious one with respect to feelings of affection. The scope for variation on established tacitly imbued social arrangements is a potential source of social change, with each generation perhaps modifying to some extent on attendant expectations and obligations. While persons may be disposed to assume that things are the way they have been until now, they are not immutable, as archaeology dramatically shows, and this adds to the complexity of social interaction, with behaviour changing over time. Individual experience may modify an agent's understandings; for instance, events may challenge their beliefs. If sufficient persons have experiences that challenge their way of doing things, they may lead to changes in the canon of values and beliefs passed on to the next generation (as pointed out in the previous chapter), without any debate or need for spoken consensus.

Archaeology furthers our awareness of such change over long periods of time. The revived interest in human evolution, focusing on such change, which some think is the theoretical way forwards for the discipline (Shennan 2008), offers only a limited opportunity to participate in current social-science debates. The evidence, as I have pointed out, suggests that we have not evolved in the biological sense since modern humans emerged in the Middle Palaeolithic some 100,000 years ago with, in evolutionary time terms, an astonishing increase in size of the brain, which it is widely speculated correlates with growing social awareness and co-operation that demands considerable cognitive capability, and leads to culturally informed behaviours that undermine biologically framed evolutionary assumptions, which some seek to circumvent with the idea of gene-culture co-evolution. Furthermore, to argue for social evolution with modern humans is to enter into a culturally relative minefield, where deciding what is evolutionary progress involves value judgements that can only be justified on chauvinistic cultural grounds, particularly if we discount the Victorians' equation of progress with technological change, which is what shows up largely in the archaeological record. It seems unlikely that engagement with current evolutionary debates will lead to a theoretical breakthrough for archaeology, which has been wrestling with long ago socio-cultural change for decades, in the humanitarian tradition of history rather than the scientific one of biology, which these comments suggest is the appropriate course regarding cultural change.

Accessing tacit social 'know-that' or not?

The challenge for social studies (as we have seen) is that while communities imbue individuals with shared values about appropriate ways to behave, the actors themselves are apparently unaware of what they are doing with respect to picking up and passing on behavioural regularities, for instance the sociological implications. They inculcate a way of doing things beyond conscious thought, such that persons act without having to think about their actions. They draw on this implicit knowledge in considerable measure through empathy and intuition, not reason and rational thought. In a sense this doubles archaeology's double bind, not only having to cope with no actors to talk to in relating the material aspects of life to wider socio-cultural arrangements, but also attempting to relate these to what was in large part tacitly known and not even overt to the persons involved long ago. But while actors may be unaware of these issues, social researchers it seems can transform such collective tacit knowledge and attendant social stratagems into propositional knowledge. Through observation of social interaction and actors' comments on their behaviour, researchers claim to uncover the tacit

13 Politicians and policymakers in states with central governments may now try to think through explicitly what they believe are the best social arrangements for those subject to their suzerainty, but the contradictions that characterize human emotions – particularly in a collective – constantly thwart their plans. It may be possible to model, as agent-based modelling becomes more sophisticated with computing power, the social arrangements that emerge when different individuals interact who have varying emotional drives and implicit knowledge (as when our palaeolithic ancestors' brains suddenly expanded in size), but I should be less sanguine about so modelling social change in complex contemporary society, where chaos rather than probability theory seems more relevant.

knowledge embodied in their everyday actions, using these data to support their theoretical arguments about the character of social arrangements (see Chapter 3). The fullness of researchers' insights contrasts with actors' unawareness of the innate aspects of their behaviour.

The contradiction at the heart of the social sciences is that researchers seemingly know more about the behaviour of actors than they do themselves, the implications of their tacit social convention-following apparently being inaccessible to them (as pointed out in the previous chapter). Whatever strategies researchers use to uncover the sociological implications of others' behaviour, they are questionable without the actors' mandate, which is inevitably unforthcoming with implicit knowing informing social interaction. The argument that actors may express abstract views on these social issues using unfamiliar idioms need not detain us – e.g. as in structuralism, featuring exotic symbolic codes such as designs on things, songs and myths (again, mentioned in Chapter 3). The way those with the necessary insight maintain they expose such cognitive structures is obscure, depending on unverifiable metaphysical leaps of imagination of the sort challenged by Hume and subsequent empiricist philosophers.

The assumption that current social research methods can get at tacit knowledge seems untenable. Scrutiny of Bourdieu's (1977) influential 'theory of practice', for instance (see previous chapter), suggests that it goes over the arguments rehearsed by social philosophers for generations, masquerading under new jargon. It appears that sociologists do no more than surmise access to what is implicit in people's behaviour and comments – or more brazenly, explicit, communicated via strange symbols. How they can get beyond theoretical supposition to firmer intellectual ground is a challenge given the intangibility of tacit social evidence (as we have seen), which, like actions and speech, has a brief existence 'out there', and its import, and even actors' purposes, being open to differing interpretations. In order to secure a firmer hold on social behaviour and further understanding of the tacit aspects of social learning and perception, it will be necessary to get 'inside actors' heads', which implies the integration of social research with psychology and neuroscience. There are currently increasing moves in this direction (as pointed out), with the complex relation between conscious and unconscious dimensions of social behaviour poorly understood and ripe for investigation. But these moves would appear to shut archaeology even more firmly out from contributing to further developments in our understanding of social behaviour, because we obviously need living human beings to conduct necessary socio-psycho-neurological investigations.

Using 'know-how' to access the 'know-that'

But there is potentially a place for archaeology in these moves to further our understanding of tacit knowledge with its firm anchorage in the material world. Whatever conscious access actors may have to social phenomena, the data that comprise evidence of these are, as we have seen, particularly difficult to pin down convincingly, such that their significance and interpretation are often open to challenge. It is increasingly questionable to treat these data as 'facts' at all rather than impressions, particularly with the postmodern turn. In this event, how are we to progress our understanding of tacitly underpinned social behaviour? One way forwards, as this book argues, is to work initially in domains where unspoken knowledge relates to tangible phenomena 'out there' in the material world, where the outcomes of the exercise of tacit knowledge are evident, involving practical engagement with real things such as raw materials, tools and end-products that our senses can readily detect, more often than not with little ambiguity. Though pointing this out may invite opprobrium where postmodernism holds sway and talk of facts is unpopular, and subjectivity is seen to frame our understanding, such that things only exist for us in our minds and each of us may perceive of these differently, which is the existentialist extreme.

There may be a route towards social-theory building that involves material existence, focusing on tacit knowledge of the 'know-how' variety, which underpins the skills that feature in any technological regime, to further understanding of tacit knowledge of the 'know-that' variety, which underpins social interaction (distinctions introduced in Chapter 3). While these 'know-how' skilled kinds of tacit knowledge, which involve instilled physical sequences and comprehension of concrete material things, appear to comprise a different class from the 'know-that' variety that informs social behaviour, there are significant parallels that suggest that furthering understanding of one will further understanding of the other. This returns us to ethnography that details the manufacture and use of material objects – such as the building of a house – not now as such activities may relate to the material remains of the kind that archaeologists work with, but with a view to researching such activities to further our understanding of tacit-knowledge issues that are increasingly exercising social scientists. Technology may remain as unpopular a topic as it was when I wrote *Made in Niugini* (as opposed to the artefacts that result from the application of technology, particularly those comprising museum collections, in which considerable interest is now shown), yet its study may well help social research out of the philosophical maze that has kept it going around in dizzying circular arguments for generations, or at least more securely establish the limitations of such research.

It may be naive, or at least premature, to anticipate the emergence of a materially based theory of social issues that will allow archaeologists to reconstitute the social arrangements of past communities with confidence, as opposed to participating in current largely fanciful speculation. Another way to engage significantly in current sociological debates is, I argue, by commenting on the shortcomings of current interpretations of 'know-that' tacit social knowledge from the perspective of research focusing on 'know-how' human cognitive capacities, which may in turn lead onto the prize of intellectually more securely founded materially focused social theories. There is an emerging strand of social research that is enquiring into the tacit aspects of knowledge to which archaeologists might potentially contribute. In view of the discussion here, the material focus of archaeology should afford some confidence in so entering the sociological fray, material things existing 'out there', affording objective benchmarks against which to test the veracity of arguments. One way is through the tradition of ethnoarchaeology, conducting further fieldwork on the making and use of things to further understanding of similar extinct technological regimes and possibly their socio-cultural dimensions (Binford 1978; David and Kramer 2001; Gould 1980; Kramer 1979; Longacre and Skibo 1994). Here the proposal is to extend on such work to advance understanding of tacit skills. Another way is through the tradition of experimental archaeology where persons seek to recreate the skills necessary to manufacture objects found in the archaeological record. Here the experimenters themselves, that is the tacit skills they develop, might be the focus of further research, drawing for instance on neuroimaging (Malafouris 2013; Malafouris and Renfrew 2010; Stout *et al.* 2000; Stout and Chaminade 2007, 2009 – see next chapter).

It is possible that archaeologists who are accustomed to thinking about invisible human behaviour and who depend heavily on objects, the production of which features drawing on tacit knowledge and skills, may be in a strong position intellectually – or at least occupy a different position, one accustomed to dealing with such issues – to advance our understanding of tacit issues as they relate to social behaviour. By definition, without the interacting human beings there is no society, which is the nub of the archaeologists' problem – all they have by and large are some material remains, often largely garbage of bygone communities. And they have been struggling for decades to achieve a secure social-theoretical foothold, which some see as necessary for intellectual respectability. A materially focused approach may afford them some needed traction on sociological debates, drawing on their strengths enquiring into the material conditions of human existence – prompted by the physical remains that comprise much of their evidence.

Summary of discussion

Some seminar participants were at pains to point out that 'archaeologists are well aware that the idea of society is problematic' (I had not intended to suggest otherwise). With the post-processual turn they are now interested in the relation between people and things, placing excavated evidence in social context. And practice theory (after Bourdieu 1977) informs a considerable body of archaeological work, being well suited to 'theorizing about the rhythms of daily life', such as 'looking at the lifetime of a house and persons moving through the spaces and what is left behind'. But my observations about tacit knowledge seem too left of field for some colleagues, judging by their comments; such as 'Gordon Childe wrote about people having rites to help in life, which had a tacit component'. Some were up front about it, 'not being clear about the role of cognitive psychology'. But they were happy to discuss the acquisition of tacit skills through informal learning, with one pondering how lumberjacks learn to run on logs coursing down rivers and another colleague, foreshadowing discussion in Chapter 11, asking if we learn about tacit knowledge as participating observers or observing participants. Others argued that post-processual archaeology engages with embodied issues, being 'all about the body and placing persons in the landscape', albeit they are imaginary persons. Furthermore, it would be an error to focus on things alone and not include abstract ideas, such as society, as ideas inform how people behave. But 'how can we access society when nothing is left but a few things?' The problem for archaeology is how to recover and deploy such culturally driven immaterial ideas convincingly with only material things as evidence, other than as researcher impositions. Some might argue that their ideas today have universal relevance in making sense of human behaviour, even if there are no longer any actors to have such ideas. There are parallels with ethnographic contexts, where living persons may not apparently hold such ideas either, consciously anyway (e.g. of society, as discussed in Chapter 3).

Some commentators thought that promoting interdisciplinarity is a central issue. They suggested that we might seek to get back to an earlier time when anthropology and archaeology comprised a single integrated subject, although how we might dismantle existing disciplinary divides with associated academic power structures is difficult to see. It is questionable whether doing away with the disciplines would be helpful or not, for how are we otherwise going to organize (discipline) what we know and make sense of the world? We have developed separate specialist bodies of knowledge and skills; for instance, archaeologists' sophisticated techniques of scientific excavation. We do, however, need to collaborate to further understanding that we cannot achieve alone, and to devise ways to navigate the 'complicated plate tectonics of interdisciplinary relations'. One suggestion was that anthropologists and archaeologists could do fieldwork together, integrating research on a local level rather than a broad disciplinary one. Another way to foster interdisciplinary dialogue is to supply ethnographic data on objects that archaeologists find in the excavated record, such as stone axes (Pétrequin and Pétrequin 2011). This confirms my approach, which has been singularly unsuccessful so far! Disciplinary identity, ironically enough, is likely a contributory factor here. Regardless of the intellectual omnivorousness of post-processualism and calls for cross-disciplinarity, subject allegiance remains an issue. Some pointed out that as I am not seen as an archaeologist my work goes unnoticed, as I fail to attend the right conferences and so on. There is perhaps a regional aspect here too, with long-standing institutional links informing the movement of academics and circulation of ideas, for instance between Africa and Europe, such that East African ethnoarchaeology informs European prehistory; although one colleague countered that 'New Guinea ethnography appears everywhere in archaeology, as if equally relevant to everybody' and such uncritical analogizing is certainly problematic.

Some colleagues thought that my argument was unsound because my understanding of the relationship between anthropology and archaeology was dated. Why was I not paying attention to cognitive archaeology, they asked, which was already pursuing enquiries along the lines that I advocated? My reply, in brief, was that this emerging approach to archaeology only tangentially concerns itself with the tacit knowing that is the seminar paper's focus. The full answer is more complex and comprises much of the next chapter, for cognitive archaeology broaches other issues of interest to us.

5

Seminar postscript

Accessing minds, past and present

The seminar discussion of the long-standing relation between anthropology and archaeology (Binford 1962; Childe 1946; Gosden 1999; Spriggs 1977) accords with the observations of several recent conferences and publications (Coudart and Cleuziou 2005; Fahlander 2004; Garrow and Yarrow 2010; Hodder 2012; Rowlands 2004; Shankland 2012; Spriggs 2008; Thomas 2012). The growing frequency of discussions about relations between archaeology and anthropology possibly indicates that archaeologists are gradually gaining confidence to engage with the large body of ethnographic data on pre-global communities[1] that anthropology has accumulated over the previous century or so, and to take forward such research, using it to further our understanding of humankind's cultural repertoire and potential, as intimated in assertions that today 'archaeology is anthropology' (Gillespie, Joyce and Nichols 2003:155; Thomas 2012:219).[2] It is a trend that I predicted in *Made in Niugini* (1988:12), with anthropology increasingly abandoning the study of such communities, which were once its defining subject matter (arguably contributing to its current identity crisis), for the study of the globalizing forces that have swept them up, finding it embarrassing to associate itself with what was previously called 'primitive', 'exotic' or, worse, 'savage' culture and the associated domineering colonial heritage.

Regardless of these recurring discussions, any sense of progress is limited. They rehearse the same themes, such as the limitations and pitfalls of ethnographic analogy (the extent to which arrangements in extant communities might inform research on extinct ones presumed similar in some crucial respects, as mentioned in previous chapter), which suggests that there are persistent and intractable problems. These revolve around the same grand question that features in my seminar paper: how can archaeology further understanding of the socio-cultural orders responsible for what today comprises the archaeological record without living persons to discuss their way of being in the world? Several of the contributors to a recent conference volume seeking to explore and foster connections between archaeology and anthropology (Shankland 2012) comment on this question. In the introduction, the editor observes that anthropologists 'have the advantage of being able to talk to subjects ... [and] though the introduction of a sense of social complexity in the study of remains of the past has made the practice of archaeology much more immediate, and even much more human, in the end the crucial point is whether the microanalysis on any particular artefact does what it claims to do.' (Shankland 2012:9–10). And in the conclusion, we read that:

> Lacking direct access to the people whom it studied, confronted with evidence that was fragmentary and incomplete ... archaeology sometimes looked on anthropology with envy ... while anthropologists observe social practice, archaeologists see only its residues... Archaeologists lament their inability to 'recover' human beings in the way that they recover artifacts, while anthropologists apparently encounter human beings ... [through] face-to-face dialogue ... ancient artifacts, and the configuration of spaces and architecture, explicitly refer to the former existence of people who have now vanished.
>
> (Thomas 2012:219–23)

Another contributor vividly captures the implications, pointing out – with reference to pedological evidence of a long-ago wooden post – that archaeology is

> founded fundamentally on the proposition that we can talk about the lives people led in the past through the things that have persisted from that past to our present.... [which] we index through representational strategies that have become so much taken for granted that we may literally fail to see the lack of identity

1 Albeit not entirely pre-global by the time any foreign ethnographer arrives, whose presence by definition indicates the start of global connections.

2 According to Thomas (2012:219), for instance, with 'anthropologists seeking to redefine their objectives in the postcolonial era', archaeologists are 'dismayed at what they understood to be a retreat from issues of diversity'.

> between our use of *posthole* for soil texture and color variability, and what anyone digging posts today or in the past would have imagined by this term.... [and] this undoubtedly obscures ... our understanding that the trace we recognize is a sign of history, not a thing recaptured from a past lived experience and revived in our present circumstances.
>
> (Joyce 2012:119–22)

Asymmetrical relations

The result is an uneven relationship between the disciplines, as my seminar points out, with archaeology apparently limited by comparison to anthropology in what it can do to further our understanding of humankind's socio-cultural arrangements. But pointing out the 'inability of archaeological inanimate material traces to stand up to the requirements of a "proper" understanding of socio-spatial dynamics, compared to the testimony of animate social beings available in anthropology' is irksome to some, particularly as the 'postcolonial critique of anthropology and the reflexive anthropological discussions on fieldwork, ethnographic practice, thick descriptions, and deep understanding have indicated that accounts of social and cultural reality can be as incomplete and absenting of people in anthropology as they may be in archaeology' (Souvatzi 2012:187). In short, 'while archaeology may still see itself as confronted with the problems of missing human beings and a static record of action, there is growing acceptance that anthropologists also deal with evidence that is fragmentary and imperfect' (Thomas 2012:221). Apparently, the 'challenge lies not in the kinds of archaeological or anthropological evidence but in the kinds of questions we ask of it', for while 'neither discipline may be able to capture the finer points of the complexity of social reality, past and present, social reality has spatial and material components' (Souvatzi 2012:188). And it is consequently 'imperative to recognize that the distinction is not between an anthropology that deals with human beings who are physically there and an archaeology that is concerned with absence. Rather, we are simply addressing different modes of human presence and its dispersal.' (Thomas 2012:226).

However, neither of these injunctions seem to take us anywhere. It is disingenuous to seek to turn aside archaeology's problems with evidence – not being able to observe and talk to those responsible for excavated remains – by pointing to anthropology's problems with evidence, over which the discipline has agonized for decades. A blind and deaf person who points out to others that they do not see and hear everything that goes on around them does not reduce his own disability. It is undeniable that aspiring in anthropology to a holistic view of a community is precisely that, an aspiration (see Chapter 1). It is not possible to encompass all aspects of community life – we cannot escape the generalist versus reductionist conundrum, for in seeking to research into some aspect of life in depth we unavoidably lose sight of other domains (Sillitoe 1983:xiv–xv). There is also the endless postmodern wrestle of minimizing so far as possible our subjective views distorting what we see and hear. But there are few archaeologists who would pass on the opportunity of a Tardis[3] trip to Stonehenge or somewhere else in the distant past when ancient people were living there, to see what they were doing and, learning their language, talk to them about it. Undertaking such a trip, they would swap their current problems for those of anthropology, being unlikely to pay much attention to shards of pottery on a rubbish heap with potters there actually making pots and cooks using them (though hopefully they would retain their rigorous approach to data, sorely needed currently with the postmodern turn prompting some social commentators to scoff at the significance of facts).

Some colleagues think that the recent postmodern questioning of what anthropology has achieved with living subjects, including the growing interest in tacit issues, may arguably equalize relations between the disciplines (Garrow and Yarrow 2010). The gap between the disciplines is less than we may think, if for both what we know is not only subjective but much of it is also experiential and beyond words. The absence of archaeological data may even be an advantage: 'the substantive asymmetry that is often imagined between the discursivity of anthropological subjects and archaeology's lack thereof is not quite the problem that it is often imagined to be. Indeed lack of discourse could ... be imagined as a resource, in the sense that it acts as an imaginative stimulus.' (Garrow and Yarrow 2010:6). But how, exactly, and are both now equally incapacitated? The claim about symmetrical relations between the disciplines chimes with the current 'humanistic' approach to materiality, discussed in the next chapter, which features speculation on the symbolic import of things, sometimes almost metaphysical in its recourse to the imagination, in arguing that materials and objects have 'agency', impacting on human behaviour. This contrasts with an 'ecological' approach to materiality that is firmly connected to empirical evidence and plays to archaeology's strengths in reconstructing past environments.

The symmetrical relations argument seems to be at odds with archaeology's empirical strengths (and to clash with the ethno-determinist premise about the centrality of ethnographic evidence), in suggesting that the less data available to archaeologists (largely excavated remains, even midden rubbish[4] of long-ago communities – unlike

3 The police-box time machine in the Doctor Who science-fiction TV series.

4 As a student, I recall using the word midden with great respect as a term encountered frequently in archaeological reports, until I

anthropology with its dense fog of thick description and interpretation of contemporary communities) is an asset because it allows freer rein to the imagination, in fitting interpretations to the facts. From the ethno-determinist perspective, one of archaeology's strengths is the way it usually cleaves to the fewer facts available to it. The discipline uses archaeometric techniques that

> give a level of precision that may even become the envy of social anthropologists … the integration of huge amounts of shared data in archaeological projects is far in advance of anything that I know anthropologically… To say that social anthropological fieldwork does not need this level of contextualization of data because of its access to living communities would be greatly debatable … it is highly unlikely that social anthropological fieldwork routinely comes up to the standards of contemporary archaeological recording.
>
> (Shankland 2012:9–11)

We see this, for instance, in painstaking work on pottery shards, not only reconstructing pots but also ancient diets with analysis of food traces to determine what people cooked in them.

But none of this debate seems to get around the problem of having no actors to talk to. Nor, come to that, the challenges of having actors to talk to. The results of archaeology's painstaking empirical research and forensic scientific methods are unquestionably astonishing in their reconstruction of aspects of past life from artefact remains, plant pollen and animal bones together with data on ancient environments such as soils and topography. The careful collection and analysis of such evidence, often rubbish of previous ages as noted (from bits of broken pottery to metal objects corroded beyond recognition) relates in part to its scarcity. The care taken with evidence is an object lesson from an ethno-determinist perspective, and might profitably inform supposedly 'holistic' anthropological method and help curb the cavalier approach of those who, to the envy of archaeologists, have an embarrassment of evidence working in extant communities, although it is almost impossible to conceive of conducting ethnographic work at such a level of detail. You can arguably only achieve such thoroughness where you have so little evidence.

Cognitive archaeology and the extended mind

The cognitive turn in archaeology promises to overcome this grand problem. It represents the discipline's take on some phenomenological issues. It seeks to replace the 'post-processual' approach, which in its turn previously replaced the 'processual' one. The processual view assumed that material culture is evidence of, largely practical, aspects of previous socio-cultural arrangements and environmental relations,[5] and it dominated when I wrote *Made in Niugini*, with ethnoarchaeology popular. It was to this, then current approach, that I thought the New Guinea ethnography could contribute, setting the account of artefacts in wider cultural context – following the holistic ethno-determinist (then unchristened) approach (with theoretical changes afoot and structuralism on the wane in the social sciences) – believing that an in-depth study might further material culture theory, which was largely ignored at the time. The post-processual view criticized processualism for not paying attention to the meaning of material things and promoted, drawing on semiotics and structuralism, their study as symbols from which we might deduce something deeper about the ideology of previous socio-cultural arrangements and relationships, almost 'reading material culture as text'. 'Objects shorn of their ideological component in the processual model have now become objects abstracted as ideas in people's heads in the post-processual one.' (Boivin 2004:63). Disillusion set in with realization of the shortcomings of the interpretive turn across the social sciences. 'While artifacts might reflect cognitive ordering, the relationship between ideas and objects was not sufficiently straightforward for the one to be "read off" from the other.' (Thomas 2012:220). This yo-yoing reflects the materialist versus idealist dichotomy, and the empiricist and intellectualist one more generally (see Chapter 2), which improbable sounding cognitive archaeology purportedly seeks to overcome. An advocate of this approach is scathing of post-processualism: 'the prevalent archaeological tendency to treat as symbolic any aspect of the archaeological record that has no obvious pragmatic or utilitarian function is obviously wrong and should be abandoned' (Malafouris 2013:189–90).

Until recently, archaeology's 'in-the-head' view of the mind assumed that the cognitive states and processes of the long dead were inaccessible. It did not consequently enquire directly into the cognition of those responsible for excavated things, beyond acknowledging that past minds were behind their manufacture and use, both as goods and symbols. However, if material culture not only represents but also contributed to the shape of past thought, archaeologists may study ancient minds more directly. The argument here is that 'cognition and materiality intersect, mutually catalyzing and constituting each other … [featuring] the constitutive intertwining of cognition and material culture' (Malafouris 2004:53). The upshot is the initially implausible-seeming claim that the 'things studied by cognitive archaeology are literally parts of (no longer functioning) minds' (Wheeler

learned that it means 'rubbish heap'.

5 For instance, see David Clarke's *Analytical Archaeology*, then a star volume that seems subsequently to have black-holed (Clarke 1968), and some of the ethnoarchaeology literature.

2010:29). Furthermore, this perspective is apparently 'able to collapse the conventional mind/ brain tautology and mind/body dichotomy' (Malafouris 2004:57–8). Regarding my foregoing comments on symmetrical relations between disciplines, the approach is giving archaeologists renewed confidence in contributing to contemporary debates about humanity.

> We should not be intimidated by what is often perceived as a poverty of data. In actual fact we are uniquely placed to explore issues that elude other disciplines ... the complex interplay of mind and matter.... ours is a discipline that by definition brings together mind, matter and body, and thus offers unique opportunities from breaking down entrenched dualisms.
>
> (Boivin 2004:69–70)

Another commentator observes optimistically 'with our leitmotif of "thinking through material culture" the archaeologist need not be reduced to the unhappy state of being a frustrated mentalist condemned to materialism. If we accept that mind and matter achieve co-dependency through the medium of bodily action, then it follows that ideas and attitudes ... find themselves inscribed 'in' the object.' (Knappett 2005:34).

You may be wondering, as I did when I first encountered this literature, how archaeologists have resurrected prehistoric minds (Malafouris 2009). They do so by drawing on the 'extended mind' thesis (Clark 1997, 2003; Clark and Chalmers 1998; Rupert 2009), an increasingly influential perspective in cognitive science that views the mind not as brain-bound but as embodied, situated and distributed. It has emerged with the coming together of various disciplines, ranging from psychology to robotics, around the question of 'where is the mind'? According to an early proponent of the view, 'there are lots of message pathways outside the skin, and these and the messages which they carry must be included as part of the mental system whenever they are relevant' (Bateson 1972:458). It explores the extent to which mental processes extend beyond the head into the world beyond the body via the use of things that serve to facilitate thought, such as the laptop computer on which I am typing these words, which is allowing me to organize my ideas and keep a record of them outside my memory. In other words, the device is serving as an extension of my mind. Artefacts have featured in human lives and influenced our cognitive capacities since before the emergence of our species: 'Tools-R-Us, and always have been.' (Clark 2003:7). Note that while the extended-mind thesis is a version of the embodiment argument (Chapter 2), which focuses on the use of some material thing (for instance, the implications of using a laptop to record thoughts electronically), but takes for granted the associated tacit knowing (typing on a keyboard being a subsidiary act not consciously reflected on).

The extended-mind thesis is itself an extension, of the long-running 'internalism' versus 'externalism' debate in the philosophy of the mind, where internalism assumes that subjects' brain activity alone determines mental states (the previous archaeological perspective), while externalism maintains that when cognition implicates things exterior to a person it involves more than in-the-head representations. The target is the orthodox view that has mental processes – perceiving, remembering, reasoning – going on in brains only; as exemplified in the mind is to the brain what software is to a computer analogy (Malafouris 2004:54, 2013:31–2). It is an error to reduce human cognition to an analogue that can encode-store-retrieve data, because it involves more than an internal-memory organ that can represent and process information (see Chapter 11).[6]

The distributed-mind perspective goes further, giving us 'active externalism', in arguing that not only may mental content be external to the subject but so may some associated cognitive processes too (Malafouris 2009; Malafouris and Renfrew 2010:5). It proposes that we use objects in such a way that they serve as extensions of our minds, acting as a 'coupled system'. According to a landmark paper in the field (Clark and Chalmers 1998:7, 8), distributed cognition occurs when an:

> organism is linked with an external entity in a two-way interaction, creating a *coupled system* that can be seen as a cognitive system in its own right. All the components in the system play an active causal role, and they jointly govern behaviour in the same sort of way that cognition usually does. If we remove the external component the system's behavioural competence will drop, just as it would if we removed part of the brain. Our thesis is that this sort of coupled process counts equally well as a cognitive process, whether or not it is wholly in the head.

And the article goes on, 'If, as we confront some task, a part of the world functions as a process which, were it to go on in the head, we would have no hesitation in accepting as part of the cognitive process, then that part of the world is (for that time) part of the cognitive process.' These

6 For this reason, the use of computers to further understanding of cognition is arguably limited: 'Using computational simulations as a method for gaining information about the human mind, you might learn a few things about the representational structures that support inferential logic and problem solving, but you will certainly also end up with a distorted picture of how those structures relate to the environment, and probably with no picture at all of how those structures are enacted in real-life situations and in different cultural settings.' (Malafouris 2013:29).

views equate broadly with Rowlands' (2010) 'amalgamated mind', which seeks to bring together various strands of extended thought. He proposes that mental processes are not only encephalic ('in the head'), but also 'embodied' (involving other parts of the anatomy), 'enacted' (depend partly on action), 'embedded' (occur in the environment) and 'extended' (operate with things). He points out that phenomenological philosophy and the cognitive sciences have long considered mental processes as drawing on such a combination of neural, bodily and environmental resources. It is difficult to conceive of things any other way, where 'embodied engagements, not the isolated brain, create mechanisms for reasoning' (Malafouris 2013:105) or else we might as well have the brain throbbing away in a jar.

Material-engagement theory extends the mind

After alighting on the extended-mind idea, some archaeologists have used it to propose a new theory to further understanding of the remains of past objects and excavated sites: 'material-engagement theory' (MET). According to one authority, the theory focuses on 'a cognized or conceptualized materiality in which the physical and conceptual aspects emerge together. The notion of *engagement* thus holds what might be termed a unifying and hypostatic relationship towards the dualistic concepts of mind or thought on the one hand and brute matter on the other.' (Renfrew 2004:23). In a subsequent publication we are told:

> Material engagement theory considers the processes by which human individuals and communities engage with the material world through actions that have simultaneously a material reality and a cognitive or intelligent component.... The engagement that we are considering implies the development of new interrelationships between humans and the material world.
>
> (Renfrew 2007:122–3)

Another advocate gives the following crisp definition: 'MET seeks to provide an integrated archaeological perspective concerned with the interactions through time between cognition and material culture and with the consequences of these interactions for understanding the making of present and past ways of thinking' (Malafouris 2013:35). It is reassuring that someone as eminent as Lord Renfrew of Kaimsthorn thinks that a theory of material culture is feasible; maybe *Made in Niugini* is not so wide of the mark after all. While my views have changed since writing that book, it has always been evident that we are some way from a theory that might allow archaeologists to speculate more securely on the socio-cultural features of the ancient communities responsible for the objects they excavate (regardless of MET's claims) and, as Renfrew (2004:30) comments, need to avoid 'the hermeneutic fallacy that ... former meanings can themselves be recovered or reconstructed in their totality'.

The distance yet to travel in the quest for such a theory is evident in the lack of agreement over the meaning of terms, which makes it difficult to penetrate contemporary cognitive archaeology. The muddle of terms suggests a certain confusion of thought, different writers, sometimes even the same one, using terms in different ways, mixing mind up with thought, brain with intellect, senses with understanding, perception with cognition, and so on. Some are carried away by catchy images. A favourite is: 'Mind is a leaky organ, forever escaping its "natural" confines and mingling shamelessly with body and with world.' (Clark 1997:53). While for most of us a 'leaky brain' implies a fatal head injury, for cognitive theorists it indicates that 'the sort of stuff that makes up the mind [facilitates cognition?] can be equally found located within and outside the skin' and consequently 'from the perspective of material engagement, cognition has no location. The active mind [exercise of understanding?] cannot be contained.' (Malafouris 2013:37, 85) and human 'affective and emotional states and/or processes, literally, comprise elements in their surrounding environment ... Mind and matter are one.' (Malafouris and Renfrew 2010:8, 9). The brain is even redundant – perhaps just as well if it were leaking out – for 'the only representations that count and that may have any true significance in the study of mind are those to be found, beyond skin and skull, in the outside world' (Malafouris 2013:31), which even for an ardent empiricist is extreme, and may account for astonishing claims, such as, when digging soil 'it makes as much sense to say that the brain is part of an extended spade as the opposite' (Gosden 2010:40), whereas in my gardening experience my brain remains firmly in my head.

The reluctance of material-engagement theorists to define terms does not give confidence (to this reader anyway). 'One could say that things exist in a constant state of *ontological deprivation*. They remain fluid, formless and plastic; waiting to take the shape of our cognitive projections which inevitably vary in different times and places. Thus, no precise and closed definition of things should be expected.' (Malafouris and Renfrew 2010:4). It is difficult to know what to make of 'ontological deprivation' as it suggests things lack any 'nature of being', have no reality or existence. The reason for this reluctance is that 'Too much clarity ... could be misleading in a contest where transgressing the common wisdom about minds and things is often a precondition for success. Material engagement ... does not have to commit *a priori* to a certain ontological description about minds and things.' (Malafouris 2013:9). The upshot is some odd assertions, concerning for instance agency, with inanimate things apparently able to act, and 'a person may serve as an "extension of the material agent ...

[for] Agency as an emergent property cannot be reduced to any of the human or the nonhuman components of action" and one can consequently talk of "the active nature of material culture.' (Malafouris 2013:147–8).

The unwillingness to define terms further compounds confusion by partly facilitating claims to overcome dualistic thinking, in allying material culture studies with '"Situated cognition" ... that seeks to counter the Cartesian representationalist view of cognition' (Knappett 2004:45) after phenomenology, embodiment and existentialism. In criticizing those who continue to approach 'mind and matter ... through the Cartesian lenses of symbolic representation' material-engagement theorists 'seek to shift the focus away from the isolated internal mind and the demarcated external material world towards their mutual constitution as an inseparable analytic unit' (Malafouris 2004:53). Nonetheless, cognitive theorists continue to refer to internal and external, mind and body, persons and things, thinking and doing, seemingly unaware that if they argue using the English language and concepts, they are not going to escape such dualisms. The original dichotomy of Descartes, (see Chapter 2), which had the physical body and spirit mind interacting via the pineal gland, was challenged long ago – for instance, by phenomenological philosophy (notably existential and embodied views) and the natural sciences (such as psychology and neuroscience) – but some contemporary writers apparently challenge all dualistic ideas. Yet, 'For all that archaeologists and anthropologists might themselves call an absolute ontological separation between 'people' and 'things' into theoretical doubt ... their own methodological and epistemological practices do ultimately *start* from this difference.' (Garrow and Yarrow 2010:10). On the plus side, the aim is 'not to break things apart into a series of easily manipulated isolated entities, comprehensible in themselves yet incommensurable as a whole, but rather to understand how things are enmeshed and related, and to understand through what mechanisms those linkages are effected' (Malafouris 2013:53), which suggests a welcome commitment to the principle of holism.

The 'rejection' of dualism informs criticism of the post-processual approach because it relies on 'dualities [such] as symbol and referent, or signifier and signified.... [whereas] a deeper unitary relationship ... underlies the apparent duality' (Renfrew 2004:23). According to material engagement, the deeper relationship is, in part, that things stand for what they are, which is how we come to know and conceptually extend on their qualities. 'The material sign instantiates rather than symbolizes. It brings forth the concept as a concrete exemplar and substantiating instance ... [it] does not refer to something existing separately from it, but is a constitutive part of what it expresses and which otherwise cannot be known.' (Malafouris 2013:97). The argument uses the idea of 'cognitive projection', linking ideas to things (such as the idea of weight to our physical experience of the differing weights of objects – see Chapter 2), which allows humans to formulate abstract concepts by anchoring them to concrete things that our senses can perceive, and this process was 'the precondition for the emergence of symbolism' (Malafouris 2013:105). However, once the association is made and culturally established we presumably have scope for post-processual symbolism, and encounter all the usual conundrums of symbolic interpretation. While it may be that 'Things act most powerfully at the non-discursive level, incorporating qualities (such as color, texture, and smell) that affect human cognition in ways that are rarely explicitly conceptualized ... [and] bring about meaning in ways that language cannot.' (Malafouris 2013:95), archaeologists, like other academics, nonetheless use words to represent and interpret associated ideas and cannot escape the semiotic impasse criticized by material-engagement adherents.

While it is unexceptionable to suggest that perception of the qualities of things – such as heavy or light, rough or smooth, hard or soft – may inform associated concepts,[7] how are we to account for more abstract emotional ideas – such as revenge or forgiveness, honesty or deceit, hope or despair (note, again formulated at root as dualistic oppositions of the kind that structuralism stresses as fundamental features of human cognition, for how could we have one of these ideas without its antithesis: hope without despair)? It is such emotions or 'passions', following Hume (see Chapter 3), that underpin human behaviour. According to the extended-mind school the idea of 'surrogacy' indicates how real-world material objects can help us humans conceive such ideas, which may otherwise be beyond our ken. A 'surrogate situation' is 'any kind of real-world structure that is used to stand in for, or take the place of, some aspect of a certain "target situation"' (Clark 2010:24). It allows 'human reason to cope with *coupling-resistant* (absent, counterfactual, even impossible) targets[8] by means of local *coupling-friendly* surrogate material structures. Such surrogate structures allow human reason to reach out to that which is absent or distal or otherwise unavailable while at the same time providing a concrete arena in which to deploy perceptuo-motor routines of a fundamentally world-engaging kind.' (ibid.). Another perspective refers to physical things acting similarly as proxies for mental notions as 'material anchors for conceptual blends', arguing that 'it is possible to reason effectively about unfamiliar concepts, which would otherwise be unstable, if they can first be blended with stable familiar concepts' (Hutchins 2005:1560, 1574).

7 As Hutchins (2005:1555) points out, 'the association of conceptual structure with material structure' is a 'general and ancient human cognitive phenomenon.'

8 Note he discusses religious beliefs here.

These arguments are reminiscent of Lakoff and Johnson's (1999) embodied metaphorical-extension view about the imaginative use of 'preconceptual' bodily experiences connecting sense perceptions with conceptual understanding; they maintain, for instance, that 'Our brains are structured so as to project activation patterns from sensorimotor areas to higher cortical areas ... [which] allows us to conceptualize abstract concepts on the basis of inferential patterns used in sensorimotor processes that are directly tied to the body' (Lakoff and Johnson 1999:77 – see Chapter 2). And while the suggestion is that the association of ideas with material culture may allow us to 'come to believe things that [we] would never otherwise come to believe', there is no proof that linkage to physical things necessarily promotes belief in metaphysical abstractions. Indeed, pondering the prospect prompts a pioneer of the extended-mind approach to question a key assumption, musing that while 'material means may enable me to use basic biological skills of perception and manipulation to penetrate cognitive domains that I, at least, would otherwise find impassable. Is this enough to warrant the claim that the material objects are parts of the machinery of thought and reason? I am, quite frankly, not sure.' (Clark 2010:26).

Others question the extended-mind hypothesis further, although the philosophical twists and turns of their arguments need not long detain us, as they take us away from our central concerns (Rowlands 2010; Rupert 2004, 2009). For instance, some challenge the 'parity principle' that assumes that internal and external factors contribute equally to behaviour, arguing that the workings of brain-located memory differ entirely from environment-located things that may act as 'memory' prompts, as external devices that stimulate internal remembering (e.g. the internal memory is sensitive to what psychologists call the 'generation effect', which relates to use of mnemonic devices). The material-engagement theorists respond that 'the parity of outer and inner cognitive elements does not mean to imply an identity but instead a synergy; a constitutive intertwining of brains, bodies and things' (Malafouris and Renfrew 2010:7), which still implies parity. The use of the term 'enactivism' causes further confusion, some writers equating it with to 'enact' or 'do', in embodied-cum-tacit ways, when it actually contradicts the extended-cognition thesis: 'enacted minds are not extended minds' (Wheeler 2010:35). Enactivism concerns independent, bounded, self-organizing ('autopoietic') entities (Di Paolo 2005; Varela *et al.* 1991) and is 'committed to identifying the cognitive system with the living system' (Wheeler 2010:35), where the boundaries of both systems coincide, which is inconsistent with the extended-mind hypothesis, where the living system remains within the organism while the cognitive system extends out to include abiotic external things. Another target of critics are 'coupling arguments' (Adams and Aizawa 2009; Aizawa 2010) that observe the influence of bodily and environmental factors on cognitive processes and fallaciously jump to the conclusion that these are themselves cognitive processes. They conflate 'constitution' with 'causation', in confusing 'interaction' with things for their 'participation' in thought, and fail to distinguish cognitive from non-cognitive processes. In short, they push the boundary of cognition too far.

Implementational versus vital materiality

In response to these different views, Wheeler (2010) distinguishes between 'implementational materiality' and 'vital materiality'. In the former, persons depend on things to assist them with cognitive tasks, such as a pencil and paper to record information or compute complex sums. It recalls Gell's (1998:36) distinction between 'primary agents', persons able to engage in intentional acts, and 'secondary agents', entities that assist persons in pursuing their actions. It adopts 'a relational approach to the interface of mind and matter, such that information is deemed to be distributed between them' (Knappett 2004:44) and follows the extended-mind thesis, where 'The key insights are practical and straight-forward, highlighting the many ways body, action and material surround can simplify and transform the problems that face the biological brain.' (Clark 2010:23). In this form, it is an uncontroversial age-old view, arguably a statement of commonsense.[9]

The residents of the Was valley could recognize the 'implementational' view, that we use things to assist us in our thinking. They employ materials in this way when they use sapling wands, cane stems and lengths of vine to mark out the plan of a house and measure the required dimensions. The use of slivers of wood or twigs as *iysh hul* 'wooden bone' markers/counters when engaged in numerical calculations is another example; for instance, in working out the distribution of things among people. As is the use of *pong liy* 'knot hits', where individuals tie knots in a length of rope to mark units of something; for instance, denoting the passage of so many moons, serving as a time record of a period agreed for some purpose (Sillitoe 2007:265–6). Indeed, if asked about the use of such devices, persons are likely to respond along the lines of '*obuw naun konay bizuwt bay sa munubtuw tomb*', 'they help our thinking when we count'.

The straightforward 'implementational' view may encompass the associated notion of 'distributed cognition', which recognizes that 'cognition must be extended beyond

9 According to Wheeler (2010:32) the 'implementational' approach to the extended mind, where things play an information-processing or computational role, is an extension of the philosophy of the mind's version of functionalism, which maintains, after Putnam, that 'a mental state is constituted by the causal relations that it bears to sensory inputs, behavioural outputs and other mental states'.

the individual in order to accommodate broader cognitive events that include interactions among people, artifacts, space, and time' for a 'cognitive system that involves more than one individual has cognitive properties that differ from the cognitive properties of the individuals who participate in the system taken in isolation' (Malafouris 2013:67, 79). In other words, knowledge is distributed between the minds of persons who comprise a community, who use a range of shared 'extension devices' to manipulate and communicate intelligence among themselves. It seems that the distributed mind equates with the collective conscious or more broadly the idea of culture, where 'human intelligence "spreads out" beyond the skin into culture and the material world' (Malafouris 2013:3). Again, this is not an original observation. It is presumably what our proto-human ancestors groped towards as they evolved language.

The 'vital materiality' approach argues that when things are necessary for 'the enactment of, and partly constitutive of, certain cognitive systems or operations, they often do so in virtue of a kind of causal contribution that is, in some way to be determined, a product of those thing's essential materiality' (Wheeler 2010:29). It seeks to extend agency beyond the human actor to the things that feature in any cognitive process. In any activity, 'the information specifying where the situation can lead is not entirely within the agent's head, but is in some way also held within the object' (Knappett 2004:46). It is in this way that material engagement theory purports to access minds in the past. In this scheme 'recognizing and examining the agency of things essentially means penetrating the dialectic of resistance and accommodation that emerges through, and is constitutive of, the process of material engagement' (Malafouris 2013:148). It results in some apparently surreal images, with references to the 'cognitive life of things' (Malafouris and Renfrew 2010) that 'envelop our minds; they become us', a tool being 'an agent imbued with cognitive and social life' (Malafouris 2004:44, 60); so things have the 'ability to be, at the same time, mental and physical' and bid us to engage in 'understanding the ontology of the artefact' (Malafouris 2013:80, 148). For instance, a Greek Bronze Age clay tablet from Mycenae with signs on it (Linear B script, which is a precursor to writing as on this page), 'actually instantiates memory or participates in the cognitive processes responsible to how the Mycenaean remembers' (Malafouris 2013:80).

The panpsychic proposition that inanimate things can engage in animate acts is reminiscent of beliefs in animism. According to the animist worldview we have 'spirits residing in matter' (Pels 1998:94), people attributing a soul and agency to other creatures and inert objects. The 'vital materialists' make the animistic association, referring to material engagement as 'methodological fetishism'; fetishes being things that can affect the course of events by dint of their material presence and otherness (Ellen 1988:223–6; Pels 1998:94–5). It marks a 'return to the things themselves as socially alive and active in a primary sense' (Malafouris and Renfrew 2010:5) and takes seriously 'the state, well attested historically and ethnographically, in which humans intend things to be animate and to have intentions ... a scenario that is valid, real, meaningful, and certainly philosophically defensible ... that a non-biologically-alive entity possesses intentionality, far from being a naive figment of cultural imagination, is the natural extension of the intentional stance' (Malafouris 2013:146). Others are thinking along similar lines. For instance, the advocates of 'symmetrical anthropology' who argue that the study of behaviour should not prejudice the contribution of either things or humans by assuming one is inert and the other active (Latour 1993, 2005). Consequently, a 'More animistic notion of things should be encouraged so that we can look at what sorts of circumstances things will have a determinant effect on people.' (Gosden 2009:107). But there are problems, over and above whether or not one can subscribe to an animistic worldview, which require a large improbable ontological shift for most of us. The adoption of a fetishist approach seems, for instance, to contradict a key claim of material-engagement theory, for it suggests a reaffirmation of the Cartesian dualistic spirit-body perspective – even an expansion of it, rather than a challenge to it – in attributing animistic agency to things, thereby extending the notion of the soul (mind) from animate beings (which somehow exists within but independently and in some senses beyond them) to inanimate materials with which they share it.

It is perhaps surprising to find such arguments in the twenty-first century, albeit the emphasis arguably differs from that of our nineteenth-century forebears (e.g. Tylor 1871) by focusing on material agency in terms of the interface between humans and things, not 'naive' ways of thinking about reality (although see Holbraad 2011). 'We do not animate things by putting a "spirit" in matter; instead we seek to discover the "spirit" of matter.' (Malafouris and Renfrew 2010:5). Nonetheless, the 'vital materiality' approach appears to vindicate the argument that the extended-mind hypothesis confuses facilitation with causation. There is a significant difference between using something, such as a pencil and paper, to aid thought, and doing the thinking. The pencil is only 'active', of course, when used by a human being to facilitate and communicate thoughts (I have never had a pencil left on my desk write a word while I have been out of my study, or come to that, while present in it). Indeed Clark makes this point in proposing the extended-mind thesis: 'In rejecting the vision of human cognitive processing as *organism bound*, we should not feel forced to deny that it is (in most, perhaps all, real-world cases) *organism centred*.' (Clark 1997:123 – note that he is referring to thought, not the mind). While objects, from clay tablets to notebooks, may assist our cognitive processing, they do not themselves engage in cognition; they help us remember, for instance, but do not have memories of their own. Similarly, things may be the object of human thoughts and actions, but they do not intentionally shape or actively participate in such behaviour independently. According to one critic, even if we accept the proposition that embodied engagement with material things – interaction between artisan, tools and materials – structures the 'fundamental phenomenological space' and admit even the proposition of 'vital materiality': 'Without further argument, it is at least uncertain that this phenomenological point *alone* could ever mandate the claim that things-beyond-the-skin *must* be counted parts of the cognitive system, as opposed to mind-external elements that play a causal role alongside mind-internal elements in structuring experience.' (Wheeler 2010:36).

The idea of an animating numen, apparently resurrecting Descartes' dualistic ghost, suggests that the mind is an entity, albeit ethereal. The slippage from the mind's activity of producing intangible intellectual phenomena (that may prompt acts on tangible things), to it resulting in something tangible, intimates further confusion. Appropriately enough, in a novel featuring the creator of Sherlock Holmes – that arch solver of mysteries – Julian Barnes (2005:38) catches the enigma in attributing to Conan-Doyle the following spiritualist musings: 'The mind was an emanation of the brain ... something purely physical in character; while the soul, as far as such a term could be admitted, was the total effect of all the hereditary and personal functionings of the mind.' The mind is not synonymous with the brain, which is something, but is the emergent property of largely cerebral, neural bio-chemical activity. The cognitive outputs of the mind are akin to the acoustic ones of a symphony, which depend on the co-ordinated orchestration of the instruments of many musicians (who draw extensively on tacit knowing) to give a certain ineffable sound sequence that briefly exists; not as a thing, but as a fleeting acoustic effect that stimulates the aural senses and brain activity. The representation of music in a score does not give us the music itself; anymore than the written word captures events and experiences; the symbols can only stand weakly for associated thoughts and memories.

The material-engagement theorists are aware of the apparent wackiness of some of their claims – such as 'Things ... impose their own dynamics, consciousness, and temporality.' (Malafouris 2013:246) – and are anxious that their metaphorical flourishes should not appear literally to sweep them off into an animistic world. For instance, Malafouris admits that the material-agency approach 'appears to succumb to the sin of anthropomorphism. Speaking about things as agents seems to imply a personification of the inanimate and thus an illegitimate ascription of human form and attributes to the non-human.' (2013:130) and he does not wish to be associated with such a 'gray zone ... where supernatural powers dwell and acquire material form and substance' (ibid.). Rather his concern is with 'the relational ontology of material engagement' (2013:146). A comment on Mycenaean clay tablets illustrates the veering away from the 'vital' view back towards the 'implementational', making the argument difficult to follow as it dodges first one way and then the other: 'The Linear B tablets, by "being there" in the "outside" world, enable the Mycenaean scribe to substitute visual recognition for recall, thereby transforming a difficult "internal" memory problem into an easier "external" perceptual one ... [such] information ... becomes available "out there" for other people to use, comment on, transform, or incorporate.' (Malafouris 2013:82), which assumes of course that you know the code, such as our alphabet, necessary to decipher the communication, otherwise – without a Rosetta stone – it just comprises unintelligible marks; evidence of a mind at work long ago but with no clue as to its cultural-historically situated thoughts.

Things that constitute the mind

So what do the cognitive materialists have in mind with their more outrageous sounding comments about the mind? They want to establish that cognition occurs during engagement with the material world; featuring a combination of external and internal aspects, such that the mind 'emerges from contextualized processes that take place "between" brains, bodies, and things' (Malafouris 2013:78). Again culture, both material and immaterial, plays a part, as we inherit much of our knowledge from forebears and contemporaries, 'the human brain ... is an artifice of culture as much as it is a product of human biology. The "inner" mental domain ... shaped by and inseparably linked with the "external' domain' (Malafouris 2013:236). Consequently, 'We do not only create the world; it also creates us, and the history of human social and cultural developments cannot be somehow divorced from the history of human involvement with the material realm.' (Boivin 2004:69). Thus far, nothing is new. This is the basic premise of economic history, seen for instance in accounts of the 'industrial revolution.' Similarly, recognition that cognitive processes may feature interaction between embodied brains and exterior things is long standing, embellished recently by the extended-mind thesis.

The radical point, it seems, is the contention that minds and artefacts are interdependent and constitute one another, such that neither can exist without the other, and so, given 'the constitutive intertwining of mind with the material world ... our ways of thinking are not merely causally dependent upon but constituted by extracranial bodily processes and material artifacts' (Malafouris 2013:227). Human cognition needs both brains and things to function. Artefacts 'attention-sculpt', target and secure cognition, such that we can manage to think some of the thoughts we do. It is not simply that our use of pencil and paper or pocket calculator facilitates complex calculations, but that our interaction with these things actually makes such cognitive processes possible, without them we could not conceive such intellectual operations at all. Consequently, 'agency should not be perceived as a fixed property of humans but as the emergent product of our engagement with the world' (Malafouris 2013:148). Furthermore, things may transform our understanding and act as catalysts for new thoughts. In other words, 'artefacts enable us ... not just to act and perceive differently but to think about the world in brand new ways ... provide the material leverage for new kinds of tool-enhanced thinking to emerge and to be maintained ... participating in the *construction of new modes of thinking*' (Clark 2010:27).

These ideas excite archaeologists because they suggest that it may be somewhat possible to access ancient minds. For, if 'past ways of thought need not just be *expressed in* material culture, but now can also be seen as partly *constituted by* material culture ... past ways of thinking can be now studied in a more direct fashion' (Malafouris and Renfrew 2010:9). On these grounds, it is arguable, for instance, that the Mycenaean clay tablets prompted an 'extended reorganization of the cognitive system ... a Linear B tablet, more than simply amplifying memory, brings about some radical changes in the nature of the cognitive operations involved and in the functional structure of the system as a whole ... a different set of skills and affordances is introduced, and those skills and affordances radically reconfigure the cognitive ecology and the dynamics' (Malafouris 2013:81). But how exactly do such material developments and interactions change the functional structure of cognitive systems when the fossil evidence shows no change in the size and structure of the human brain since the Pleistocene emergence of our species?

While it is plausible that pre-human hominins experimenting with the manufacture and use of material artefacts, such as the Olduvai axe makers, may have contributed to the biological evolution of the human brain, it is dubious to imply an ongoing process, that 'the hallmark of human cognitive evolution is metaplasticity ... that allows for environmentally and culturally derived changes in the structure and functional architecture of our brain' (Malafouris 2013:241). There is a danger of almost racist stereotyping here, in suggesting that those who developed industrial technology, with its complex gadgets, have more evolved brains with superior 'structure and functional architecture'.[10] It makes many wary of cultural evolution and potentially ethnoarchaeology too – adding to the criticisms mentioned in the previous chapter (Fewster 2001). For instance, suggesting that those who until recently used polished stone tools – such as my Was valley friends – might teach us something about the way people in the distant past who used such tools may have lived runs the danger of implying that such people are less intellectually endowed, even throwbacks, which is emphatically not, I think, the intention.

So how do things change the brain? A key concept here from neuroscience is plasticity. It refers to changes in neural pathways in the brain caused by environmental stimuli and personal experiences, which may include natural (e.g. forest loss), material (e.g. new artefacts), behavioural (e.g. social status) and individual (e.g. bodily injury) changes. Neuroscience shows that the adult brain cortex, previously thought physiologically static, can create new synaptic links, reorganize neural pathways and even sometimes create new neurons throughout life (Pascual-Leone *et al.* 2005).[11] According to evolutionary psychology, the brain comprises many domain-specific, genetically blueprinted modules or specialized circuits evolved as adaptations by our Pleistocene ancestors, overlain by social enculturation and life experiences that exploit the brain's developmental plasticity to shape its cognitive-processing configuration.[12] It refers to the plasticity of neural connections, that is, the programming of the cortex. It involves changes in the software, not our Pleistocene-evolved brain hardware, so all humans are equally capable of it – as my Was valley friends demonstrate, happily adapting their behaviour and lives not only from polished-stone axes to steel tools, but also to motor vehicles, aeroplanes, mobile phones

10 Malafouris is aware of this, albeit in denying any such implication he seems to contradict his material-engagement argument: 'I am not assuming that the possession (or lack thereof) of a certain technology or material innovation correlates directly or causally with possession (or lack thereof) of a certain capacity or ability.... Material culture is not a cause in the linear mechanistic sense; it is cause in a dynamic ontological sense.' (2013:241).

11 While the environment plays a central role, genetics can also influence the process, which includes changes in glial and vascular brain cells in addition to neurons, and varies with age.

12 Epigenetics is a related and rapidly growing field that studies heritable changes to gene expression, including neuronal structure, that do not involve alterations to the DNA nucleotide sequence. Changes in the external environment – life experiences – inform these modified expressions of otherwise inflexible DNA code (Bird 2007; Dias and Ressler 2013; Riggs *et al.* 1996; Spector 2012). The result is 'transgenerational epigenetic inheritance', the mechanisms of which are subject to intense research. (A well-researched example of epigenetic modification is DNA methylation, where methyl groups attach to CpG sites on the genome and affect the expression of nearby genes, usually damping them down – CpG sites are regions of DNA where a guanine nucleotide follows a cytosine one.) The implications are not only considerable for the way the brain is configured, as discussed here, but also for the nature versus nurture debate, with 'nurture' apparently altering what we thought of as 'nature'.

and so on. It is argued that modular networks reduce the energetic costs of connections between nodes (a significant consideration with the human brain using one-fifth of our energy intake, at resting metabolic rate). While there is some debate over the extent to which module structures are genetically specified evolutionary adaptations, there is agreement that neuroplasticity is the human brain's key psychological adaptation, affording responsiveness to experiences of local environmental and social conditions, and that cognitive architecture is the outcome of a mutually causal-reciprocal relationship between gene expression and socio-environmental interaction (Buller and Hardcastle 2000, 2005; Clune *et al.* 2013; Ward 2012).

The cognitive archaeologists focus on the role that artefacts play in moulding the brain's synaptic configuration, stressing the dynamic nature of relations between people and things (Gosden 2009:109; Malafouris 2009). They ask 'how do things shape the mind?', picking up on the implications of the apparently unique predisposition of the human brain to reconfigure itself, partly in response to us making and using material culture, where the activities that we engage in help differentially to shape parts of the cortex neurologically. The following description from a material-engagement proponent, for instance, could come directly from an introductory text to neuroscience: 'things change the brain. They effect extensive structural rewiring by fine tuning existing brain pathways, by generating new connections with brain regions, or by transforming what was a useful brain function in one context into another function that is more useful in another context. More than that, things extend the functional architecture of the human cognitive system, either by adding new processing nodes to the system or by changing the connections among existing nodes.' But then it goes further, reverting to gobbledegook, asserting that the 'process of extended reorganization does not simply refer to an activity-dependent change in the neural architecture … It refers to an outward expansion of the cognitive system in order to forge extra-neural connections objectified through material culture, bodily action, and learning.' (Malafouris 2013:247).

In summary, the argument is that in using a thing, such as a pen, I can knowingly/ consciously cause an effect 'out there' in the world that may assist me in thinking about something (e.g. using a pen I can cause ink to run onto paper, making signs to store and manipulate information). But at the same time I can unknowingly/ unconsciously cause an effect 'in here' in my head that may affect my thinking about something (e.g. by promoting some neuronal changes in the brain, featuring a reformulation of certain synaptic connections). Note that it is not the pen that is causing these effects, but me using it.

The question is to what extent is it possible to access these cognitive processes, that is, past minds, from the remains of objects, which returns us to the central problem of archaeology of having no living mindful subjects. There is the empirical response of processual theory that artefacts represent the product of blueprints that persons long ago had in their mind; they had the idea of a pot, for instance, and made and used such a thing. And there is the speculative post-processual response that artefacts may be symbols that communicate some intelligence; the whole pot may have stood for something, for instance, such as the womb or decoration on it may have conveyed some information or represented some emotion. And there is the tentative cognitive response that artefacts are aspects of extended minds and so afford access to those minds; from the 'implementational' view a pot may have acted as a measure, for instance, and assisted people in calculating exchange rates, and from a 'vital' view its production and use served to configure brain architecture to some unknown extent. According to this last cognitive perspective, 'Material signs do not represent; they enact. They do not stand for reality; they bring forth reality.' (Malafouris 2013:118), in which case things do not act in post-processual fashion as symbols representing something but as catalysts for new thoughts.

While the extended-mind thesis prompts some intriguing discussion in cognitive archaeology, it assumes tacit know-how – like the social sciences – leaving us with the problem of how we can understand and represent human behaviour in which such knowledge features prominently. Use of the thesis in archaeological contexts rests on conjecture, with no live minds to test them against, which brings us to the seminar paper's mention of neuroscience contributing to understanding of tacit knowing, for experiments involving neuroimaging with contemporary subjects may give cognitive archaeological propositions some much-needed firm foundations (Malafouris 2013:164; Malafouris and Renfrew 2010:9; Stout and Chaminade 2007, 2009; Stout *et al.* 2000).

Neuroarchaeology and the tacit again

Some neuroimaging work in experimental archaeology, involving the making of stone tools, indicates the direction of enquiries (Stout and Chaminade 2007; Stout *et al.* 2006; Stout *et al.* 2009). Positron-emission-tomography investigation of brain activity involving both experienced and inexperienced experimenters knapping replica Early Stone Age (Oldowan and Acheulean) tools shows increases in 'effective visuomotor coordination and hierarchical action organization' (Stout *et al.* 2009:15) as toolmaking becomes more advanced. The imaging patterns for knapping Oldowan tools show reliance on the parietofrontal perceptual-motor and grasp areas, together with bilateral activations in the dorsal intraparietal sulcus associated with vision, suggesting that enhanced sense perception of motor activity characterized early hominin technology rather than conceptualization of actions. The patterns with Late Acheulean knapping show increased activity in both hemispheres in the ventral premotor and inferior parietal frontal lobe circuits, and in the right-hemisphere ventrolateral prefrontal cortex, reflecting a transition to the planning of more complex action, with flake removal demanding higher levels of intentional organization. The overlap of the latter activation patterns with language circuits suggests that toolmaking and speaking share something in common regarding capacity for goal-directed complex actions (Stout *et al.* 2009; Stout and Chaminade 2012).

While neuroimaging promises to advance understanding of the cerebral aspects of cognitive processing, it clearly has limitations in archaeological contexts (Malafouris 2013:168–9), again deriving from the absence of living minds. It is difficult to know how to assess inferences about cerebral activity in long-ago skulls from imaging data of experiments with modern subjects, especially when the skulls were those of different pre-human hominin species. Furthermore, the brain activity observed under laboratory conditions likely differs from that which occurs in the complex real world, where situational factors such as culture, history, personal experience and social interaction influence behaviour. In short, neuroimaging techniques overlook the dynamic materially informed aspects of thought and behaviour, and consequently the 'role of brain-imaging, however convincing and enchanting as a statement about the inner workings of the human brain it might be, should under no circumstances be confused … with that of a device about to delimit the realm of the "truly cognitive"' (Malafouris and Renfrew 2010:9). It is an error to assume from a neuroscience perspective that the hands and tools of artisans do no more than execute the orders of their brains. The way neural activation occurs within given cerebral regions when making and using things confirms, after material-engagement theory, that these external actions complement, indeed are continuous with, the brain.

Archaeological interest in neuroimaging concerns its contribution to our understanding of how changes in the human brain possibly correlate causally with the formation of material and wider culture; it does not aim overtly to further our understanding of tacit know-how per se, as I suggest may be a way forwards. But the two endeavours are compatible. For instance, 'The activation of ventrolateral, but not dorsolateral, PFC [prefrontal cortex] indicates that Acheulean toolmaking is distinguished by cognitive demands for the coordination of ongoing, hierarchically organized action sequences rather than the internal rehearsal and evaluation of action plans.' (Stout *et al.* 2009:14). In other words, it involved tacit knowing. And such 'higher levels of intentional organization and technical competences in stone knapping can emerge only through deliberate practice and skill acquisition' (Malafouris 2013:167). In short, the experiments indicate that the evolution of the human brain, with respect to tool making and use, featured experiential embodied knowledge. If artefacts show that using tacit know-how was significant long ago, it potentially deepens archaeology's central problem of no living subjects and the feasibility of accessing past minds from the remains of objects, in view of the problems we face accessing such unconsciously held knowledge with living subjects. On the upside, our cerebral hardware – the size and structure of the human brain – has remained the same since modern humans emerged, so any work on contemporary subjects may well reflect long-ago cognitive processes, but not those from before the arrival of *Homo sapiens*; surmising what pre-human hominin brains were up to being far more speculative.

Moreover, neuroimaging affords an opportunity to investigate the extent to which bodily actions comprising the use of things not only involve tacit conduct that individuals may not consciously think about, but also the extent to which these actions may actually contribute to the configuration of the brain, that is constitute human cognition, how persons and things 'respond to and participate in each other's coming into being' (Malafouris 2013:8). It is intriguing to ask, for instance, if the skills and knowledge involved in building and occupying houses in the Was valley result in different neuron configurations to those of populations that construct buildings using mud bricks or erect concrete and reinforced steel structures. The neurological evidence of plasticity suggests that some differences are likely, which deepens the problem addressed in this study, namely the extent to which it is possible to understand Was valley house construction and use, as the Wola know and experience it. It is arguable that lacking the same synaptic connections in our heads as those we conduct fieldwork among will somewhat compromise the accuracy and soundness of any ethnographic account, which adds further to the conundrums surrounding the ethno-determinist approach.

In summary, neuroimaging may serve both the purposes of furthering understanding of long-ago minds and of contemporary tacit knowing and cerebral plasticity. Some cognitive archaeologists recognize the connection. For instance, as Knappett (2004:44) comments, while 'some information may simply be stored within the agent's mind … a great deal of information is not immediately accessible and "on-line" in this way; rather it is situated and embodied, emergent in particular contexts of action' and consequently there are 'artefacts in given situations that serve to prompt sequences of actions almost "unthinkingly"'. Advocates of the extended-mind thesis see similar connections with tacit knowing, suggesting for instance, with respect to 'the big question' of the 'cognitive processes, perceptual mechanisms, and moods' that 'extend beyond the borders of consciousness' that it is 'plausible that it is precisely the nonconscious part of them that is extended' (Chalmers 2011:xiv). Another material-engagement theorist plunges directly into a brief ethnographically informed exposition of well-known issues surrounding tacit know-how and the problems of capturing skills in words, drawing on contemporary Greek pottery, arguing that 'we should replace our view of cognition as residing inside the potter's head, with that of cognition enacted *at the potter's wheel*' (Malafouris 2004:59). But cognitive archaeology only refers tangentially to experiential knowing that is of interest here, and excessive claims, such as the following, undermine the tacit point: 'the efficacy of material culture in the cognitive system lies primarily in the fact that it makes it possible for the mind to operate without having to construct a mental representation – that is, *to think through things, in action, without the need of mental representation*' (Malafouris 2013:237).[13] This arguably over-advocates the role of tacit knowledge, for surely persons have some idea, or a mental representation, of what they intend to make or we should not have large numbers of similar objects – such as Acheulean axes – or a culturally informed material culture to talk of, but a thoroughly random collection of stuff?[14]

Whatever way we look at it, drawing attention to the shortcomings of other disciplines, such as anthropology, does not diminish the challenges facing archaeology. Its central problem remains 'that its otherwise thing-oriented methodology suffers from a deficit of human conformation' (Holbraad 2011:23). The focus on long-ago populations inevitably prompts a concern for the material aspects of life, as objects and their environmental setting are its only evidence of past behaviour, conditioned by unknown socio-culturally relative ideas and values. Consequently, it seeks a sound material-culture-framed theory. The point that some of the knowledge associated with things is often unconscious among the living, either held tacitly as skill or communicated subliminally as symbols, compounds the challenges. There may be some way to go yet before anything appears approaching such a theory that will allow archaeologists to interpret more securely on the cultural arrangements of long-ago communities from their material remains, but attention to cognitive processes and tacit-knowledge issues promises, as I argued in my seminar, to allow them to make increasingly sound assessments of such ancient arrangements at a level approaching more closely the impressive work that archaeometrists do in dating and reconstructing past environments, and may also possibly allow them to contribute to contemporary social-science debates. While the extended-mind thesis that informs the cognitive turn in archaeology is an interesting perspective on embodied issues, it does not advance us far in addressing the question of the extent to which it is feasible to access tacit knowledge explicitly. But on another tack, cognitive archaeology relates to the materiality approach that has recently emerged in material culture studies, which takes us off in another direction regarding this question as discussed in the next chapter.

13 Elsewhere the same author maintains the reverse 'The point is not to deny altogether the existence of mental models, schemata and internal planning procedures as active in the course of any creative process, but rather to recognize them as the temporally emergent and dynamic projects of situated activity … plans and models are always too vague to accommodate in advance the manifold contingencies of real-world activity.' (Malafouris 2004:59–60); such contradictory claims characterize material-engagement theory and threaten to undermine its legitimacy.

14 But according to 'Material Engagement Theory this is the wrong way to approach the question of artefact metaphysics.… What this backward reading means, to use the example of tool making, is that we start with tools in the material world and then trace their origin in the knapper's intracranial mental world … instead of seeing in the shaping of the handaxe the execution of a preconceived "internal" mental plan, we should see an "act of embodying". In tool making, most of the thinking happens where the hand meets the stone. There is little deliberate planning, but there is a great deal of approximation, anticipation, and guessing about how the material will behave … Knapping, then, should be seen more as an active "exploration" than as a passive "externalization" or "imposition of form".' (Malafouris 2013:235–6).

6
Building materials and materiality

Before they start to construct a house in the Was valley, men collect together most of the materials they will require, such as stakes and bark sheets for walls, and ridge-poles and rafters for roof, together with lengths of rattan vine and strips of bark for lashing. The only materials not amassed are those ready to hand adjacent to the house site, such as cane-grass stems, some saplings and miscellaneous foliage; together with the grass thatching materials, whether kunai or cane grass, which would soon start to rot if left for any period stacked up in piles.

The reason people give for collecting the materials needed to erect the main framework of a house, before they start to build it, is that they have to complete the work, once started, within a few days or the resulting structure will be unstable. The framework of a house is self-reinforcing: the rafters trussed to the walls and lashed in turn to the ridge-poles, such that the components of the symmetrical structure mutually brace one another. If a man was to drive in the wall stakes of a house and leave these for many days while he collected the materials needed to continue, or went to attend to some other task, the stakes would start to become wobbly on exposure to the wet and windy weather that is usual in the region (Sillitoe 1996:57–63), or worse, if pigs used them as scratching posts.

Another reason given for completing the construction of houses promptly is that if a partially competed structure is left open to the elements, particularly the rain, for any length of time, the result would be a cold and damp house. Ensuring that a house is warm and dry is a major concern of occupants, as seen in such practices as the prohibition on individuals who have lately engaged in coitus approaching recently built houses because they are believed to have a chilling effect and the screening of new houses so that the gazes of such persons do not fall on them.

Access to resources

All residents on a *sem* 'family' group's territory have equal access to the materials they need to build houses and to land on which to site their dwellings. This accords with the egalitarian stateless ethos that characterizes social life in the Was valley, which is further evident, for instance, in neither the structure nor the location of houses signalling anything about the occupants' social status. No families live in larger, better equipped, more palatial houses than others, and differences in social standing do not threaten to compromise equality – even though some men and women are recognized as more able than others at certain esteemed activities, notably those related to socio-political exchange (see Chapters 1 and 3). The construction and siting of houses illustrates how an egalitarian order compromises any idea of established superiority or leadership. Those of higher standing do not have access to better locales or superior materials, something that exemplifies the lack of correlation between leadership (there are no leaders as such) and status in this stateless context.

While some materials are more plentiful in certain locales than others, the availability of resources is broadly speaking the same across Wolaland; people living in one region do not have access to resources that are unavailable to those elsewhere. No individual or group has exclusive monopolistic rights to any particular raw materials. All have access to the resources they need, which works out at around the same amounts, and their supply is sufficient to meet all demands for building materials. This does not mean that individuals have open access to construction materials that occur everywhere in their region. They cannot gather what they need to build a house from anywhere. There are precepts that determine who can use what. These rights reside with either social corporations or certain persons.

The Wola region is divided into a large number of broadly defined territories in which a specified, albeit quite widely defined, range of related persons collectively claim rights to land and resources. They comprise the corporations that bear the relativistic label *sem* 'family' (Lederman 1986; Sillitoe 1979a) – the patrifilially oriented bilateral kin corporations described in previous chapters. In addition to controlling access to land for house building and garden cultivation (Sillitoe 2010a), they also govern access to the forest for collecting raw materials. Rights to land and the natural resources occurring on it feature significantly in the disposition of *sem* 'family' groups, their existence depending on their members joint confirmation of rights to territorially defined resources. They are one of the ways the Wola convey and act on their kinship and identity relations: meeting kin-defined responsibilities regulates access and passage of rights between generations. People claim the right to collect and use raw materials through acknowledged exchange-validated connections to these kin-defined, locally constituted *sem* 'family' corporations.

The local arrangement of resource rights can baffle outsiders. The *sem* territories are not discrete, for instance, with clear boundaries, but comprise tracts of land interspersed with similar areas claimed by others, the territories of *semgənk* 'small families' interdigitated patchwork-like with one another. And the fluid approach to *sem* affiliation is further confusing on first acquaintance. A *sem* includes those who live on a territory and cultivate gardens there, who comprise residential groups of kin that vary in size and composition, and other persons who have the right to reside on the territory and cultivate gardens there, but living elsewhere have not acted upon it. Individuals can only legitimately gather building materials on the territory of the *sem* 'family' where they live, which is most probable as materials here are those nearest to hand, or on territories elsewhere where they are recognized as kin with such rights, which means they actively maintain meaningful exchange-confirmed social relations with those living there (e.g. with mother's or wife's fathers' kin). In other words, everyone can freely collect raw materials found on territories to which they can trace a kin connection, through either parent or spouse, which those living there acknowledge as extant (that is, the relationship is socially alive and not merely a dimly remembered distant connection).

The forest is a common resource, where all persons recognized to have rights of access are free to collect the raw materials they require to build a house. All those affiliated to the access-rights-holding *sem* corporation are collectively custodians and rightful users, handing on their inalienable rights to their descendants in perpetuity. Those who reside on the territory and know intimately what it has to offer largely exploit its resources. Access to some raw materials is freer and easier than this account of customary rights suggests, with those belonging to gatekeeper *sem* corporations not interpreting their sole rights strictly, particularly to abundant resources found on their land. While people should not gather materials on territories where they have no rights, the collection of some commonly occurring materials is unlikely to provoke any challenge. But to take other materials, such as certain species of uncommon palm tree, is likely to lead to a dispute (the same applies to hunting animals – people turning a blind eye to small common ones, while taking exception to strangers bagging large infrequently seen creatures on their land – Sillitoe 2003:125). There is little incentive anyway to seek house building materials on foreign territories because not only will you have to transport them a considerable distance to your house site elsewhere, but the availability and location of resources at such places are largely unknown to those unfamiliar with them.

A person can collect raw materials anywhere on the territories to which the *sem* he is related to control access rights, where no one else claims prior land rights. This *sem*-land is *hombuwnon suw* 'everybody's ground' and includes virgin forest and some secondary regrowth. But individuals do not enjoy unlimited access to all the resources occurring on the territories where they have the right to live and cultivate. In comparison to their laissez-faire attitude to the collection of materials in abundant supply, individuals can in principle only collect those in limited supply, such as certain palms, from locations where they have secure access rights, such as land claimed by the small circle of kin comprising a *semgənk* 'small family' to which they are affiliated. But such jealous guarding of resources is uncommon. People generally treat the territory of the *semonda* 'large family' as a single resource pool to which distant *semonda* as well as close *semgənk* relatives have equal access; the more restricted approach applying only to a few scarce raw materials and to claiming rights to garden land.

Furthermore, the raw materials found in some locations belong to individual persons, not to the land-right-holding *sem* corporation collectively, which any of those with recognized connections can use. These include plants that persons cultivate or tend, both in gardens and elsewhere, such as screw-pines and she-oaks, and also those that colonize abandoned garden sites such as robust straight-trunked saplings suitable for house roof rafters. These belong exclusively to those who have previously cleared and worked the land, who, together with their descendants, have prior rights to use it again. Entitlement to such resources passes to, and through, both males and females. A person may claim rights to materials found on land used previously by his father or paternal relatives (*abon suw*), maternal relatives (*amon suw*) or affinal relatives (*weren suw*), where he has acknowledged prior cultivation rights, perhaps sharing the resources with siblings or other close kin, depending on their relations. Claims by children of the last user take precedence, with other claims negotiated according to the closeness of the relationship to previous users. Both boys and girls learn what land their relatives have used within living memory, knowledge gained informally as they grow up, but parents do not divide resources up between their children before they pass away; they inherit shared rights and sort out who gets what materials as and when necessary.

Shortly put, rights to land determine rights to resources (Sillitoe 2010a:85–110); those who have previously gardened areas, who have prior entitlement to cultivate there again (or whose close relatives do, if deceased), consequently also have first call on raw materials occurring at these places. Individuals may claim similar prior rights of use to any raw materials they find before others on territories where the land-right-holding *sem* collective recognizes they have connections allowing commons access, usually by marking with a *showaip*, that is, attaching a bundle of leaves to them, often fern fronds, to publicize their acquisition.[1]

Someone wishing to collect any quantity of building materials from land to which others have priority use rights should ask permission of title-holders to avoid a possible dispute and compensation claims. They usually give permission freely if they have no plans to use the materials themselves, on the unspoken understanding that they may in turn request access to some particular materials on the asker's land in the future, unless they are on bad terms, when they may press for a payment. If they have no intention to use certain currently available materials (for instance they do not anticipate building a house in the foreseeable future and using the saplings growing on an abandoned garden site for roof poles), then it makes sense to let a relative collect and use them (for by the time they next build a house the saplings will have grown too large to serve as rafters anyway) and thereby establish an informal 'reciprocal bond' on their relative's resources, redeemable retrospectively.

The right to use materials cultivated in gardens and around house yards, such as *taembok* bamboo, predictably resides exclusively with the person who plants them. They belong to those who tend them, and to help yourself without asking is theft. Again there is some latitude regarding access, with relatives occasionally helping themselves to modest amounts of such cultivated materials without first seeking the owner's permission – anything more and they should seek consent or risk a dispute and compensation demands. The owner usually gives permission, subject to the above provisos about his plans and relations with the requester. These factors, together with the amount and type of the materials, may also determine whether the owner demands payment or not.

The value put on raw materials depends not only on current circumstances but also supply, with only those that are relatively scarce having a transactable value. These are largely cultivated materials that belong to individuals. Persons may also demand some payment if they go to the trouble of gathering raw materials on someone else's behalf, to recompense them for their efforts, even if the materials are common and freely available. Raw materials are the property of those who collect them (assuming they are collected rightfully), and they can demand a return from others wishing to use them, although with common and readily collected materials this is unlikely, individuals will gather these for themselves rather than pay others for having amassed the stuff. Again, a range of situational factors informs any payment demand, such as the kin connection between the parties, current state of relations and indebtedness to one another for previous assistance. When owners demand something for materials they may be expressing unhappiness about relations with the asker, not for payment as such.

Environment and vegetation communities

All of the raw materials used in construction are locally available. The composition of the vegetation communities supplying them are largely influenced by altitude and topography (that also influence settlement patterns and land use, and so human activities too), which give rise to many local micro-environments and -climates. The region is mountainous, rugged and sharp in relief, with fast-flowing rivers running along the valleys (CSIRO 1965). It was uplifted, folded and faulted in the recent geological past, and largely comprises sedimentary rocks, mostly limestone, with igneous rocks of recent volcanic origin on the western margin. Frequent earth tremors indicate continuing movement today. The soils are young (largely Andosols and Inceptisols) and, variably rejuvenated by volcanic ash, fairly fertile. The climate is generally equable 'Lower Montane Humid' (McAlpine *et al.* 1983:160), characterized by high rainfall (annual average 3011 mm) and cool temperatures (mean daily 18° C), with altitude having a moderating effect. The same weather largely prevails throughout the year with no seasonal effect on growth – although the Wola distinguish two seasons, *ebenjip* and *bulenjip*, that equate with the Southern Hemisphere's summer and winter (Sillitoe 1996:55–62) – albeit unpredictable fluctuations can occur, such as overly dry or wet conditions that can negatively impact on food supplies.

In the Was valley, they distinguish nine vegetation communities, which range from climax montane forest to secondary grassland, and include wetlands, rocky and alpine vegetation, herbaceous and grassy successions of recently abandoned gardens, and cane grass and secondary forest of longer term regrowth (Sillitoe 1996:179–200). People identify several hundred different plants. Cane grass predominates in the valleys, between 1,600 m and 2,000 m, where most gardens occur, with some soft-wooded secondary trees and clearings of shorter grass. Patches of undisturbed forest occur on steep, uncultivable land and locales of unfavourable aspect. The wildlife is relatively meagre, consisting largely of small rodents and some birds. Lower montane rainforest predominates on high mountains and watershed ridges, over 2,000 m, and in unpopulated river valley areas, with a few pockets of regrowth and occasional gardens. It comprises many species of trees and other plants, and the wildlife is notably richer, with populations of marsupials, rodents and birds.

1 There is no taboo-like magic or sanctions associated with *showaip* claim-markers to frighten off thieves.

Only some of the vegetation communities distinguished by Was valley residents supply materials used in house building. These include the *iyshabuw* 'primary rainforest', which covers upwards of 50 per cent of the Mendi to Augu valleys, extending across unsettled valley and mountain areas. It is diverse, including trees, vines, palms and tree-ferns among other vegetation (Sillitoe 1995). The southern beech (*Nothofagus* spp.) dominates, with other trees occurring in considerable numbers in places – such as oaks (*Lithocarpus* and *Castonopsis*), figs (*Ficus*), elaeocarps (*Elaeocarpus*) and gamboges (*Garcinia*) – giving large tracts of mixed aspect. The forest is structurally uniform by and large (Edwards and Grubb 1977, 1982), with the canopy at about thirty metres. The sky is visible in places, while elsewhere the forest is closed with thick foliage. The trees support a range of climbing plants, including woody lianas and palms (rattan), an important source of lashing materials for house building. Mosses thrive in the damp environment, where low cloud is common, particularly on ridge tops, and rain falls almost daily, with thick mats covering trees. The ground cover may be dense with stands of saplings and shrubs, impenetrable tangles of climbing bamboo (*Racemobambos*), and masses of leafy herbs including gingerworts (Zingiberaceae) and various ferns. There are locales where the ecology diverges and different plant associations occur, such as wet or waterlogged places, those where landslips occur, areas with bare steep slopes and those subject to extensive human interference (felling trees, tending pandans etc.), notably near to settlements. These locations and their particular resources are well known to those familiar with any of the territory.

Another vegetation community that supplies considerable amounts of building materials is the aforementioned *gaimb*, 'cane grass', which frequently establishes itself following land clearance for cultivation or natural disturbance such as landslides. After montane forest it covers the largest area; growing prolifically, it predominates in some parts of valleys. It is a considerably less varied succession than either montane forest or secondary woodland, often comprising extensive stands of tall cane grass (*Miscanthus*). Other tall grasses that occasionally occur include brakes of Job's Tears (*Coix*) and wild sugar (*Saccharum*), particularly along the edges of streams. Dense thickets of fern (*Dicranopteris*) also occur in places. Low grasses and herbs may cover the ground between cane clumps away from homesteads, where pigs do not root over the soil heavily. Cane grassland may also support some scattered low-growing, soft-wooded trees – such as nettles (*Pipturus*), ochnas (*Schuurmansia*), parchment barks (*Pittosporum*) and dillenias (*Saurauia*) – together with cultivated screw-pines (*Pandanus*) and she-oaks (*Casuarina*), often evidence of previous cultivations, sometimes growing with palm lilies (*Cordyline*) along old fence lines. The tree fern (Cyathaceae) occurs too, sometimes growing in small stands.

When the trees exceed a certain indefinite number, the cane regrowth more resembles *obael*, 'secondary woodland'; there being no sharp distinction between this succession and cane grassland, the two vegetation communities gradually blending one into the other. This is the alternative regrowth succession that may establish itself following clearance of land for cultivation or a natural disturbance. Such woodland is the other principal vegetation community that is a source of building materials. It is markedly different in composition to montane forest, and largely comprises fast-growing, soft-wooded trees – such as pipers (Piperaceae), spurges (Euphorbiaceae), nettles (*Pipturus*) and umbrella trees (*Schefflera*), among others – with a considerably lower canopy at between ten and twenty metres. The ground cover is generally less dense than under primary forest, consisting of various grasses, shrubs and herbs, including ferns.

Other locally identified plant communities, which relate broadly to topography and altitude, particularly as these inform human settlement and land exploitation – such as wetlands, grassy regrowth, crop gardens and alpine vegetation – cover insignificant areas in comparison and supply lesser amounts of raw materials. Another source of building materials is *em*, 'gardens'. The Wola distinguish different kinds, depending largely on crops cultivated (see Sillitoe 1983, 2010a on crops and cultivation). But they depend little on gardens under crops for construction materials; although they supply raw materials for some artefacts (Sillitoe 1988a). Rather, people turn to abandoned gardens under cane or kunai grass or sapling regrowth, or a mixture of both, for materials. Further sources of raw materials are *aend bort*, 'house yards', and adjacent *howma*, 'communal clearings', where people tend a limited range of long-term cultivated plants, such as bamboo, screw-pines, casuarinas, spurges and bead ashes, often as ornamentals. House yards may also indirectly supply other raw materials, sometimes serving as dumping grounds for left-over and unwanted materials collected previously for some other purpose.

The two levels of access to resources – vested in the commons collective and the individual land user – broadly correspond with the different locales that supply raw materials. The *sem* corporation collectively holds the rights to materials occurring on its territory's uncultivated land (including land cultivated long ago, where priority rights have lapsed because forgotten), notably montane forest and cane grassland. Both resident and non-resident *sem* associates are free to gather any wild, untended materials they need that occur in these places. The individual land users have sole access to the resources occurring in their gardens and house yards, together with those on their recently abandoned garden sites and areas under longer-term woody regrowth where they have priority access rights.

Knowledge of raw materials

While individuals may occasionally gather some commonly occurring building materials on others' territories, they infrequently do so, rarely visiting such unfamiliar places. They collect construction materials largely on the territories where they have rights to claim garden land, notably those where they reside. Familiar with these places, they know where to go in search of particular materials, making a mental note of where resources occur when they see them during their everyday activities, such that they know where located when needed later. All persons have such cognitive maps of their territories, semi-consciously plotting the location of resources. It is only infrequently that they have to travel large distances to find the materials needed to construct a house, say beyond a radius of one or two kilometres from their homesteads.

Residents know the territories where they live intimately, often from childhood onwards, regularly traversing them on foot. These comprise numerous named locales (see maps in Sillitoe 1988a:24, 1996:111); the place names representing condensed stores of information about the locations for those who know them. (There are two levels of place names: those for regions extending to one or more square kilometres and names for smaller areas within these regions.) The in-depth knowledge that residents have of these places is an advantage when it comes to collecting raw materials for a new house, having committed to memory the location of resources that may be useful in the future, often in passing and without deliberate reflection, such as a place where straight-trunked trees occur suitable for house rafters and ridge-poles. They may also draw on the knowledge and experiences of relatives regarding resources at certain locales. This combined intelligence comprises a significant accumulation of tacit knowledge, which persons may act on without need of deliberation. They 'just know' where to go to find a certain sort of tree.

This tacit awareness of resource availability illustrates the problems we face in reproducing such knowledge that builds on experience. In order to achieve a faithful understanding of such intelligence it would be necessary to learn the place-name code for some region, such that the mention of a name would conjure up an awareness of the resources there. For instance, someone might mention Momiyaek, a forested watershed locale, to a relative who is looking for a straight-trunked tree of suitable dimensions for a *bez* 'house ridge-pole', maybe adding *nobow Peson obael*, 'downslope of the secondary woodland regrowth of Pes's garden', to specify more precisely the location known to both, where there is a *bort* 'silkwood' tree that would make a good ridge-pole. Even if you did manage to get on top of such a place name code, which would imply first-hand knowledge of many named locations, it is difficult to see how you could communicate such experiential knowledge, given the restrictions of tacit knowing, other than by using the place-name code. While it may be feasible to conceive of a computer-aided searchable database[2] that lists all the named places known to persons living in a region together with a regularly up-dated inventory of the resources available at each, its use would not mimic local place-name practice, albeit it might contain the relevant intelligence that informs related tacit knowledge.

The collection of raw materials involves a further degree of tacit knowledge. If asked 'why have you collected these saplings here?' a man is unlikely to grasp readily the import of the question, responding to the effect 'they are good for rafters, can you not see?' and 'I just knew they were here on this abandoned garden area'. If prompted further, he may indicate features that make the poles good for rafters, but he is unlikely to have deliberated over these points previously, just knowing intuitively that they are good rafter poles. Similar to other skilled persons, Wola builders are not familiar with verbally rationalizing their decisions, and instead respond practically to the demands of the occasion. It would be a mistake to think that their deliberations are studied, even necessarily conscious in practice. They make judgments largely subliminally, assessing materials and knowing their properties without having to weigh these up purposively. It is unusual to hear persons discussing the features that make materials good for certain construction purposes.

The knowledge of where to find and what constitutes suitable raw materials is acquired less verbally than tacitly in the act of collecting materials. Individuals intuitively sense and evaluate the availability and suitability of resources at various sites. They accumulate the knowledge, from childhood onwards, through life's experiences, coming to know territories intimately, readily and almost automatically where they have rights of access. They infrequently have need to search around locales for construction materials, knowing their region's resources well, having grown up there more often than not and continually walking across it in their everyday lives, building up and maintaining a mental image of what is available. This account of the collection of construction materials is a weak representation of their practices and unavoidably distorts their implicit knowledge and decision-making process. While the Wola can talk when prompted about some of the features that make materials suitable for certain construction purposes and where to find the materials, they cannot explain how they reach decisions because they do not consciously reflect on them like this in practice – they know them in the doing.

2 A quite different proposition to the formidable task of actually compiling one.

The knowledge local persons have of their territories' resources gives an inkling of the dilemmas we face accessing tacit intelligence. A performance-attuned investigation of such understanding throws up challenges as to what we may think we can know. If the knowing is in the doing, how can we learn to know too? We need to acquire a local-like knowledge of various places in the Was valley region that supply construction materials and develop an ability to judge their qualities, and appreciate what it is to collect and transport them to a house site. In reality such an exercise is impractical. Even if you could achieve such skilled knowing, could you inform others about it using words, or would it be like many other skills, which you can only understand in the doing? The epistemological challenges are considerable; decisions involving knowledge as practice, as opposed to discourse, reveal the inadequacy of the written record to convey them. But I only have words and have little choice but to resort to the artificial discursive account attempted here, which can only convey an inkling of the tacit native view. We encounter the limitations of anthropological method, which ethno-determinism seeks to confront – promoting ways of more precisely presenting and understanding such knowledge, and resisting the pitfalls of misplaced abstraction evident in the Western tradition of distinguishing between skill and intellect as discrete aspects of knowledge (Marglin 1996).[3] While not referring explicitly to tacit knowing, recent thinking in material culture studies, which goes under the banner of 'materiality' and has parallels with the cognitive archaeology discussed in the previous chapter, suggests another way to engage with such knowledge by reversing our focus from what humans do to raw materials, to what raw materials do to humans. It speaks of materials having 'agency' and the implication is that they may affect behaviour without the human actors being aware.

Properties afforded by raw materials

Materiality is another approach to emerge in material culture studies since *Made in Niugini*, indeed it seems for some to supplant them (DeMarrais, Gosden and Renfrew 2004; Fahlander and Oestigaard 2004; Meskell 2005; Miller 2005; Tilley 2004; Tilley *et al.* 2006). It criticizes previous approaches to material culture for 'neglecting the potential social roles of materialities in the creation of social relations ... the human and the non-human dichotomy is misleading' and argues for consideration of the 'social driving forces of the material world' (Fahlander 2008:128). In arguing that we should study humans from the perspective of what objects 'do' to them, instead of studying objects from the perspective of what humans do with them, it seeks to go beyond considering the way materials 'passively' constrain what is technically possible to how they 'actively' shape behaviour.

While it is an unorthodox, and to the uninitiated a confusing, use of the word materiality – which dictionaries define, referring to things that comprise physical matter, as 'the state or quality of being material' – the observation that the materials available to us influence what we can do is scarcely new. We have presumably known this since we became consciously aware of the limitations that the environment puts on us, most crudely argued as environmental determinism. At times the materiality argument comes close to such old-fashioned reasoning: 'the material world is not a blank slate upon which may be inscribed any old narrative, it is a physicality which resists and enables, shutting down some alternative plots, and opening up others' and consequently the 'material world itself is a long unrecognised agent' (Boivin 2004:64).

The approach is again somewhat opaque, as recent definitions confirm, overlapping, and even confused with, material culture. The introduction to a pioneering edited volume on materiality, for instance, tells us that it focuses on 'how the very material character of the world around us is appropriated by humanity' (Graves-Brown 2000:1). While the introduction to an archaeology conference's proceedings defines materiality as 'not only the study of the characteristics of objects, but also the more general notion that humans engage with the things of the world as conscious agents and are themselves shaped by those experiences' (DeMarrais, Gosden and Renfrew 2004:2), another contributor to the same volume tells us that it refers to 'how the material character of the world is comprehended, appropriated and involved in human projects' (Pollard 2004:48). According to a further publication the same year, materiality assumes that the 'material or physical components of the environment and the social practices enacted in that environment are mutually reinforcing' and concerns 'how those material properties are enrolled in the life projects of humans' (Jones 2004:330), and yet another in the subsequent year tells us that while materiality refers to artefacts, it includes 'the ephemeral, the imaginary, the biological, and the theoretical; all that which would have been external to the simple definition of an artefact' (Miller 2005:4). And the introduction to an exhaustive handbook on material culture scarcely lessens the obscurity, telling us that materiality refers to 'the fleshy, corporeal and physical, as opposed to spiritual, ideal and value-laden aspects of human existence. Materiality can also be taken to refer to individual things, or collections of things, rather than to persons or societies' (Tilley 2006:3). In summary, materiality emphasizes 'the physicality of the material

3 Marglin (1996:228, 229) uses the Greek terms *episteme* and *tekhne* in making the distinction; the former being 'analytic' and breaking 'down a body of knowledge into its components', whereas the latter rests on 'intuition' and is 'as varied as the authority of recognised masters'.

world' and 'offers possibilities for the human agent' (Boivin 2008:26).

When social scientists other than archaeologists engage with the social dimensions of the material world, as they have increasingly in the last decade or so, the definition of materiality becomes even more confused, with what is encompassed ranging from the technical and structural to symbolic and aesthetic, in attempts to incorporate things into social life (Barck 1994; Brown 2004; Dant 2005; Glørstad and Hedeager 2008; Harré 2002; Leonardi *et al.* 2012; Pels *et al.* 2002; Preda 1999; Riggins 1994). The introduction of further baffling jargon compounds things, as in the following exhortation to include material things in discussions of social processes: 'we have to turn away from an exclusive concern with social relations and weave them into a fabric that includes non-human actants [= objects], actants that offer the possibility of holding society together as a durable whole' (Latour 1991:103).[4] The unexceptionable point is made that social relations include human interactions that may involve objects, which the author postulates comprise 'chained networks', arguing for consideration of the things that feature in our behaviour as straightforward objects and not necessarily as symbols that function somehow like text.

Whatever the confusion around the notion of materiality, it is undeniable that before this phenomenological inspired approach, most writers largely ignored the materials that feature in making things, focusing on interpretations of form and design (Boivin 2008; Conkey 2006; Ingold 2000; Kingston 2003; Tilley 2004). Until recently, they were interested in things that feature in human endeavours as 'meaningfully constituted' (Hodder 1982a:75), functioning primarily as notion-bearing symbols. The focus was on actors impressing culturally programmed designs on materials, overlooking, it is argued, that the properties of materials play a part too. The form of things was attributed to culture rather than matter, paying scant attention to materials and their properties (Ingold 2000:340). It is as though materials come to hand predisposed to the manufacture of things. In contrast, the materiality approach argues for a collaborative view of material and maker such that 'form unfolds within a kind of force field' (Ingold 2000:342), objects emerging as part of a senses-directed form-generating process and, acknowledging the phenomenological stance, the artisan working within the world rather than on it. Instead of thinking of materials as substances that humans act on, treating them as distinct from actors, the approach argues for consideration of their technical and social aspects, advocating 'enquiry into materials, their transformations and affordances' (Ingold 2007:3). After all, materials are central to the design of objects (Ashby and Johnson 2010:4) and their properties ensure 'technical functionality' (Were 2013:583), or put the other way, technical requirements dictate materials used.

The idea of 'affordances' features prominently in the materiality approach. It was formulated in perceptual psychology to argue for direct as opposed to inferential perception, and has subsequently been deployed in product design, including architectural plans (Chemero 2003; Dotov *et al.* 2012; Gibson 1977, 1979; Jones 2003; Shove *et al.* 2007; Stoffregen 2004).[5] According to Gibson, affordances indicate possibilities for action. He argues that we perceive the world according to the opportunities (affordances) that objects allow for action, such perceptions inevitably prompting certain courses of action. 'An affordance ... is equally a fact of the environment and a fact of behavior. It is both physical and psychical, yet neither.' (Gibson 1979:129). Affordances are qualities of objects or environments that we may perceive in a direct tacit way and which allow us to carry out actions, such as buttons to press and levers to pull. They are the 'action possibilities' that exist in any environment, whether or not individuals can see them, for they depend on their capabilities as agents to realize their potential.

The idea of affordance as used in material culture studies (Knappett 2005; Ingold 2007; Were 2013) refers to what we more commonly call the property of materials. We see this in deployment of the term; for instance, Were (2013:584) talks about 'affordance as a means to understand how the identities and properties of materials help shape the products they produce', and how it relates to the way 'materials are socialized in design', that is, people use them to make things. It refers to the opportunities that the properties of materials afford to use them for certain

4 The associated 'Actor Network Theory' (Latour 2002) recalls the heuristic framework that I put forward to further understanding of the stateless Wola socio-polity (Sillitoe 1979a:86–104), which draws on social network theory, as advanced by the 1960s Manchester School, to foreground the individual actor over the corporate group. When combined with the centrality of the exchange of things (both animate, such as pigs, and inanimate, such as pearl-shells) to socio-political life, the 'Theory' and such statements as 'actants ... holding society together as a durable whole' could come directly from Highland ethnography. It illustrates how the social sciences go around and around, with new jargon disguising old ideas.

5 The affordance argument seeks to challenge the orthodox indirect view of how humans perceive meanings or functions of things as a two-step process involving external perception and internal representation. The potentialities of things may be evident directly and not depend entirely on internal mental representations. This allows actors to assess what uses objects might 'afford'. For instance, the possible use of a wooden box as a seat can be seen directly without first labelling it as such, as 'the potential of an object for sitting on ... could be directly observed without first categorizing the object as "a chair"' (Knappett 2004:44). It is another aspect of tacit knowing.

purposes; for instance, the leaves of some plants, such as grasses, are linear in shape, flexible and durable when dry, which fits them for weaving things. Some commentators extend the term further to refer to the opportunities that particular materials afford for symbolic representation, for 'material may itself enunciate meanings' (Henare *et al.* 2007:4). What actors select may signal something; for instance, they may think that entwining blades of grass recalls networks of kin relations, alluding to the fabric of social life. In this dual sense, some claim that affordance brings together the polar 'ecological' and 'humanistic' perspectives of the materiality approach discussed below, to achieve a more balanced representation of human-thing relations (Fahlander 2008; Knappett 2007; Rio 2008; Were 2013).

The raw materials

So what are the affordances or properties of the materials used in house construction in the Was valley? Builders select materials for particular qualities that fit them to certain purposes, which they can explain if asked, for instance, 'why are you using this timber here?" But first, what are the materials that are suitable for building a house? Table 6.1 lists the raw materials, identified by Wola, common English and scientific names, and arranged alphabetically according to the Wola names, with which the entries open. The final column specifies the parts of the plants used, whether the leaves, wood or bark of a tree, for instance, in the building of a house.[6]

While the logic behind the list's classification of raw materials somewhat reflects Wola conceptions, it also modifies them. The table arranges the materials into categories that somewhat parallel the local classification, showing how people group together the plants supplying the materials and conveying something about the qualities that suit them to construction. The 'tree' category is the only one that complies faithfully with the Wola botanical classification scheme, corresponding to *iysh* in their system (Sillitoe 1995). There is no named category equivalent to 'pandans and palms' (nor indeed pandans or palms separately); the Wola only have names for particular pandans and palms and no named higher-level category grouping them together – there is only such-and-such a pandan or palm. And while they group many 'grasses' (and some herbs) into the category *den*, only *senz* kunai grass falls into this named category; as with pandans and palms, there are only particular kinds of cane grass and bamboo. The residual 'other' category includes some plants that fall into higher-level Wola botanical classes such as *kwimb*, 'moss' and *ya*, 'vine' (e.g. rattan), while others such as *henk* tree-ferns are pandan-like 'unaffiliated generics' to use the ethno-scientific terminology (Berlin 1992:171).

Regarding the identification of the materials, the common English names are more likely to evoke an idea of the plants for most readers than botanical identifications and are more often used in the text. Sometimes they are combined with the Wola name to ensure an accurate specification.[7] While the use of closely defined genus and species categories of botanical science ensures precise identification, we need to approach the equation of Wola names and scientific identifications with caution. The Wola frequently disagree over the naming of plants, sometimes displaying a surprising level of discord, particularly where identification features finer distinctions (Sillitoe 2003:62–71), and they recognize no authority to arbitrate and declare who is right, unlike botanical science where experts carefully define identification criteria and apply them consistently. While the Was valley classification system has a variable aspect alien to Western science, the naming of any plant is nonetheless restricted to a few options (for instance, no one is going to say that a beech tree is an oak) and it is possible to determine which identification a majority opt for and (admittedly somewhat contrivedly and using only a few respondents) correlate this, as in Table 6.1's list, with the botanical identification of the plant.

The preparation of materials for a house, referred to as *aend maenmaen bay* 'house prepare do', usually starts with the fashioning of stakes, both for the *aend dol* 'exterior wall' and *waenuw* 'interior dividing and exterior (front) walls', and also the *pinj* 'ridge supports'. The four trees listed as supplying wood suitable for these stakes have *buriy* 'strong' hard timber that rots slowly even when driven into the moist soil, such that walls last for years before rotting. Their timber also has a uniform grain such that it splits *duwn* 'straight', that is, even and true to give relatively regular plank-like lengths of wood. House-builders look for a tree with a straight trunk without any branches low down because these produce knots that upset the straightness of

6 See Pospisil 1963:257 and Clarke 1971:104 for some comparative data on Kapauku and Maring house building materials, respectively.

7 Some of the plants will be new to many readers unfamiliar with tropical botany, and lack any common name, even as a category. In these cases, the list either gives the common name as a contraction of the scientific family label, for instance melastome for Melastomataceae (accepted practice responsible for several common English names), or else it extends the common name of a similar species – for instance, labelling a *Litsea* species as a 'mangeao', the name used in New Zealand for *Litsea calicaris* (originally the Maori name for the tree). Where the same common name applies to more than one plant identified by the Wola – for instance, there are several 'spurges' and 'screw-pines' in the list – I combine the common English and Wola names to give a binomial that precisely labels plants, and suggests something about them at the same time for the reader; for example *bobae* spurge and *keret* screw-pine identify specific plants and convey an impression.

Wola name	Common name	Botanical name	Materials supplied
TREES[1]			
bat	silkwood	*Cryptocarya* sp.	Timber for *aend bez* 'ridge-poles'.
bobae	spurge	*Macaranga* sp.	Saplings for *pongow* 'jamming sticks', *polpen* 'rafter poles',[2] *gamonk* 'wall battens', *kuwngis* 'ridge-pole' and *iyb hedla* 'eave battens'.
bort	silkwood	*Cryptocarya laevigata*	Timber for *aend bez* 'ridge-poles'; saplings for *polpen* 'rafter poles' and *kuwngis* 'ridge-pole'.
en	spurge	*Macaranga* sp.	Saplings for *polpen* 'rafter poles', *kuwngis* 'ridge-pole' and *gamonk* 'wall battens'.
gwai	switchsorrel	*Dodonaea viscosa*	Saplings for *gamonk* 'wall battens', *polpen* 'rafter poles', *kuwngis* 'ridge-pole', front eave *sal gamonk* 'fringe arch batten' and *iyb hedla* 'eave battens'; leaves for *maembow pes* 'eave-gap sealant'.
haezuwmb	white magnolia	*Galbulimima belgraveana*	Timber for *aend bez* 'ridge-poles' and *tiriyor haekomb* 'door slats'.
hombolem	gazelle cherry	*Prunus gazelle-peninsulae*	Bark for *iyb hedla humbiy* 'eave tiles'.
huwshiy	cunonia	*Caldcluvia papuana*	Timber for *aend bez* 'ridge-poles' and *dombep* 'drainage logs'.
inom	stinging tree	*Debregeasia* sp. / *Maoutia* sp.	Bark *ya* 'strips' for lashing.
iysh huwniy	saplings	Several species	Supply timber for various purposes such as fireguard frames, scaffolding for thatching gutters, jamming sticks, various pegs etc.
kaeriyl	oak	*Lithocarpus rufovillosus*	Timber for *aend bez* 'ridge-poles', external *aend dol* 'wall stakes', internal *waenuw pongow* 'wall stakes', *pinj* 'ridge-pole supports' and *tiriyor haekomb* 'door slats'; saplings for *polpen* 'rafter poles'.
kol	spurge	*Phyllanthus* sp.	Saplings for *polpen* 'rafter poles', *kuwngis* 'ridge-pole' and *gamonk* 'wall battens'.
morowa	cypress	*Papuacedrus papuanus*	Bark for *iyb hedla humbiy* 'eave tiles'.
naep	casuarina	*Casuarina oligodon*	Fronds for *aend sal* 'gable fringe'; bark for *iyb hedla humbiy* 'eave tiles'.
pay	chestnut oak	*Castanopsis acuminatissima*	Timber for external *aend dol* 'wall stakes', internal *waenuw pongow* 'wall stakes' and *pinj* 'ridge-pole supports'; saplings for *polpen* 'rafter poles' and *kuwngis* 'ridge-pole'.
pel	southern beech	*Nothofagus starkenborghii*	Timber for external *aend dol* 'wall stakes', internal *waenuw pongow* 'wall stakes', *pinj* 'ridge-pole supports' and *tiriyor haekomb* 'door slats'; saplings for *polpen* 'rafter poles'.
romiya	melastome	*Astronia* sp.	Timber for external *aend dol* 'wall stakes', internal *waenuw pongow* 'wall stakes' and *pinj* 'ridge-pole supports'; saplings for *polpen* 'rafter poles' and *kuwngis* 'ridge-pole'.
sabok	sabia	*Meliosma pinnata*	Timber for *tiriyor haekomb* 'door slats'.
saemow	mimosa	*Albizia fulva*	Bark for *iyb hedla humbiy* 'eave tiles'.
serep	balubug	*Platea excelsa*	Timber for *aend bez* 'ridge-poles' and *tiriyor haekomb* 'door slats'.
shongaen	spurge	*Macaranga pleioneura*	Saplings for *polpen* 'rafter poles' and *kuwngis* 'ridge-pole'.
shonon	chenille	*Acalypha* sp.	Saplings for *iybhedla* 'eave battens', pandan-leaf tile *gamonk* 'battens' and *tuwmainj* 'wall beading'.
shuguwl	mangeao sp.	*Litsea irianensis*	Bark for *iyb hedla humbiy* 'eave tiles'.
shwimb	Indian bead ash	*Elaeocarpus dolidrostylus*	Bark for *iyb hedla humbiy* 'eave tiles' and *wil waenuw* wall facing.
timbol	spurge	*Homalanthus novoguineensis*	Logs for *kat tor* 'pig-stall roof'
waen	woolly cedar	*Trema orientalis*	Timber for *aend bez* 'ridge-poles'; saplings for *polpen* 'rafter poles' and *kuwngis* 'ridge-pole'; bark *ya* 'strips' for lashing.
wenet	engel	*Engelhardia rigida*	Bark *ya* 'strips' for lashing; bark for *iyb hedla humbiy* 'eave tiles'.
wok	gamboge	*Garcinia* sp.	Timber for *aend bez* 'ridge-poles'.

Wola name	**Common name**	**Botanical name**	Materials supplied
PANDANS AND PALMS			
aenk	karuga screw-pine	*Pandanus julianetti*	Trunk for *aend buwk* 'bark wall sheets', for *aend kor* 'bark firescreen' and *aenk tor* 'thatching gutters'; leaves for *hwimb* 'wall lining'.
goiz	palm	*Gulubia* sp.	Saplings for *gamonk* 'wall battens'.
keret	screw-pine	*Pandanus adinobotrys*	Trunk for *dombep* 'logs' surrounding fireplace; *dombep* 'logs' of pig stalls and *dombep* 'logs' of drainage ways.
mbet	betel palm	*Areca* aff. *macrocalyx*	Saplings for *gamonk* 'wall battens'.
nort	screw-pine	*Pandanus concavus*	Leaves for *shor howay* 'roofing tiles'; aerial roots for *gamonk* 'battens'.
aendashor/ paym	wild karuga screw-pine	*Pandanus antaresensis*	Leaves for *hwimb* 'wall lining' and for *shor howay* 'roofing tiles'; aerial roots for *ka* 'roof lathes' and front eave *sal gamonk* 'fringe arch batten'.
pundin	screw-pine	*Pandanus archboldianus*	Trunk for *aend buwk* 'bark wall sheets' and *aend kor* 'bark firescreen'.
GRASSES			
gaimb	cane grass	*Miscanthus floridulus*	Cane grass for *aend gaimb* 'thatch'; cane stems for *sezel* 'wall canes'.
gaimb henj	cane grass	*Miscanthus floridulus*	Cane stems for *ka* 'roof lathes'.
senz	kunai grass	*Imperata conferta*	Grass for *aend senz* 'thatch'.
taembok	bamboo	*Nastus elatus*	Bamboo strips for *ka* 'roof lathes'.
OTHER			
haeluwp	rattan	*Calamus fuscus*	Lengths of split *ya* 'vine' for lashing.
hathaen	stones	Limestone	Stones for *tem momberal* 'fireplace lining'.
henk	tree fern	Cyatheaceae (various species)	Trunk for *dombep* 'logs' surrounding fireplace and drainage channels
homb	sphagnum moss	*Sphagnum nov-guineense*	Wadding for *maembow pes* 'eave gap sealant'.
kwimb	moss	Frullaniaceae (various species)	Wadding for *maembow pes* 'eave gap sealant'.
tedbel	gingerwort	*Alpinia* sp. (sect. Dieramalpinia)	Leaves for *shor howay* 'roofing tiles' and *aend sal* 'gable fringe'.

1 Note that only the 'trees' category corresponds with a Wola botanical class *(iysh)* – Sillitoe 1996:169–77.

2 A range of trees may supply *polpen* 'rafter poles', in addition to those listed that are the most widely used; those named as unsuitable include *inom* (*Debregeasia* sp.), *kongol* (*Piper* sp.), *pakpak* (*Ficus microdictya*) and *pil* (*Ficus mollior*).

Table 6.1 Raw materials used in house construction.

the grain, making it difficult to split the trunk up evenly, as well as being physically much harder work.

The two screw-pines listed, which supply *aend buwk* 'house wall-bark-sheets' suitable for the interior lining of walls, grow straight and even trunks. The *aenk* screw-pine is a cultivated nut-bearing species and men look to fell *el* 'old' trees that no longer yield nuts (Sillitoe 1983:107). These trees are owned by individuals and if a house-builder wishes to use one belonging to someone else, he should seek their permission first or risk a dispute (especially as it is usual practice to plant the truncated crowns of such senescent pandans). The *pundin* screw-pine is free of such restrictions, as a wild tree. House-builders look for trunks of uniform diameter, from just above the last aerial roots up to the crown, to supply bark sheets of uniform dimensions (whereas a trunk that tapers to the crown will produce a sheet that narrows to one end, and one with a waist will give an irregular shaped sheet). They look for a stout trunk with no natural holes in the bark. They also avoid certain varieties of *aenk* screw-pine that have many *paepuwliym* 'flower' protuberances on their bark, which indicate a dense and brittle biscuit-like pith inside that is difficult to remove.[8] Although screw-pines are common, it can take a while to find suitable trees (Table 6.4).

8 The varieties are: *bort, tabuwn, kat, maelam tazay, meren, dor, humbuwm, mabiyp, hap, senim, mbul, emonk, maeraeng, ombohonday, honal, piliyhongor* and *tiyt*.

The large tough leaves of the *paym* screw-pine supply other wall-lining material; the other name for this pandan, *aendashor*, means literally 'house-leaf'. Men collect the dry brown leaves that comprise the bottom layer around the tree's multiple crowns or those that have fallen to the ground below. If these pandan leaves are unavailable, house-builders may use the green leaves of the cultivated *aenk* screw-pine (although this is something of a last resort, as they are not so strong), running these through a fire to soften and make them pliable, flattening the stout keeled leaves by pulling them under a piece of wood and then placing under a weight to keep flat until required. The final walling material is moss, preferably long-stranded, used to plug eave-line gaps, which is soon gathered, as it grows abundantly in the forest, festooning trees.

Several kinds of tree sapling and some young palms supply the *gamonk* 'batten' materials used in wall construction. These trees and palms are suitable as they grow long and straight with few, if any, low branches, which readily split into two to give long batten lengths. The *shonon* chenille saplings are suitable for the tops of interior walls as they are wand-like and flexible. Builders use *haeluwp* 'rattan' for most of the lashing for battens and other features. They collect lengths of the climbing palm's cable-like stem by pulling them down from the trees up which it grows, carefully peeling off the outer skin with its sharp protective thorns and coiling them up to carry back to the house site. It supplies a particularly strong and flexible binding material, favoured in the manufacture of a range of artefacts (Sillitoe 1988a). Other *ya* 'rope' used as lashing material on walls and eaves comprises strips of bark. The bark of the listed trees is relatively smooth and without many weakening irregularities (such as protuberances or holes) and peels off intact down to the fibrous phloem in long lengths suitable for binding.

The trees that serve as *bez* 'ridge-poles' grow suitable straight upright trunks when young (although not all do and it can take some searching to find two straight-trunked trees of suitable dimensions) and they have strong timber even as juvenile trees, sufficiently robust to support the rafters. A wider range of saplings-cum-young trees supply *polpen* 'rafter poles', which are of smaller dimensions; the principal specification is that they are straight. Builders avoid timber attacked by *mbwiy* 'woodworm'[9] such as *bort* 'silkwood', which although often straight and of suitable dimensions is soon weakened if infested. The *ka* 'roof lathes' comprise lightweight, straight materials, which come from split bamboo culm, slender cane-grass stems or split screw-pine aerial roots. The four or five rows of lathes over the eaves are different, consisting of split saplings such as used for wall battens. The robust bark from several trees, which comes away in tile-like slabs, may serve as a water-seal layer on the eave lathes. The best bark comes from *naep* 'casuarina' and *wenet* 'engel' trees, but on small houses occupied only occasionally, such as in forest locations, men may also use *shwimb* 'bead ash', *hombolem* 'Gazelle cherry' and *saemow* 'mimosa' bark, which rot more quickly.

House-builders usually thatch roofs, using either bundles of *senz* kunai grass or the leafy stems of *gaimb* cane grass (other types of cane grass such as *mokombez*[10] are unsuitable because too soft and weak stemmed). Occasionally they 'tile' houses with large leaves from *paym* and *nort* screw-pines or *tedbel* gingerwort. They consider all three roofing materials equally good for weatherproofing a house. The local availability of roofing materials largely determines the type of roof. In the Honael area, where the house documented in the next chapter was built, there is a lot of *gaimb* cane grassland, so few houses there are roofed with *senz* kunai thatch. It can take weeks to amass sufficient *senz* grass to thatch a house there – a task which demands considerable extra work transporting bundles from grassland over a kilometre away – compared to a day using *gaimb* cane grass (the data from this study suggest that it takes over six times longer to collect kunai grass compared to cane thatch for a house of similar dimensions).[11] A small number of *enza* 'men's houses' have *senz* kunai thatch, the residents having taken house-construction as an opportunity to demonstrate, and subsequently advertise, their generosity through their preference for this thatching material, and consequently having to reward relatives for gathering and transporting sufficient grass considerable distances – perhaps, today, with a meal featuring rice and tinned fish together with a small cash payment.

Any straight-grained wood that splits cleanly to give flattish plank-like pieces of wood suffices for making door slats. Likewise, a wide range of materials may supply *pongow* 'jamming-sticks'. Dense hardwoods that resist rotting in damp conditions are used to line the drainage ways around the eaves. Similar timber that does not burn through quickly is preferred for the frames around fireplaces. A

9 Powder post beetle (*Lyctus* sp.) larvae. As an aside on the kinds of experiences that constitute the happenstance character of the participant-observation technique, I learned about such woodworm-vulnerable timber after occupying a house built for visiting *kiap* 'patrol officer' parties that partly collapsed, people having erected it using any old timber that came to hand, probably feeling put upon by the administration that forced them to build such structures for no reward.

10 *Saccharum robustum*.

11 It took 43.7 worker-hours to cut and transport the cane grass to thatch the Honael house (Table 6.7), which had a roof area of 35.2 square metres, compared to an estimated 164 worker-hours to pull up and transport sufficient kunai grass to thatch another house at Honael, which had a roof area of 20.5 square metres; this works out at 0.8 square metres of cane grass thatch per hour compared to 0.13 square metres of kunai grass thatch per hour.

sheet of pandan bark suffices as the roof-guard above fires, usually an offcut from the wall-lining. Many trees supply timber suitable for other internal fittings such as pig stalls, which are often made from pieces of wood, for example excess wall stakes, left over from other construction work.

'Ecological' versus 'humanistic' materiality

While it is undeniable that the biophysical properties of materials contribute to the conformation of things that humans make, these characteristics do not inform perceptions of affordances alone, socio-cultural inclinations feature too (Knappett 2004:46). Wola individuals do not discover materials anew for themselves. Someone unfamiliar with rainforest plants is unlikely to perceive the potential use of rattan vine – an uninviting, thorny stemmed, spindly climbing palm – unless told about it. Decisions about what materials to use, or exploitation of their affordances, depend on understanding their properties, notably those that determine possible uses and how these qualities suit the thing to be made, as informed by past knowledge and experience. We inevitably perceive affordances with the aid of cultural intelligence, such that we understand the properties of materials according to context of usage, experience of use, relation to other things and so on. It features what Norman (1988:219) calls 'mapping' (in the context of the design of finished products and not selection of raw materials),[12] relating affordances to constraints on the possible use of something, such as physical limitations, social understandings and intellectual awareness. Although we learn about materials by culturally mediated knowledge and life experiences, associations that inform their classification and use, much of this is tacit knowledge, in that tactile encounters with materials inform our appreciation of their materiality and potential in making things and feed into complex doing-cum-design decisions. In other words, materials and their affordances are situated within complex cultural and natural environments that feature conscious and unconscious awareness.

The focus on the physical affordances of materials understandably accords with much archaeology, where the properties of materials comprise the only sound evidence, there being no actors to quiz on possible symbolic associations that depend entirely on researchers' imaginations (the justification of which is, as pointed out, a perennial challenge for archaeologists). But some criticize this 'ecological' approach to materiality; rehearsing again, in another guise, the venerable empiricists versus idealists philosophical debate (see Chapter 2). They argue that the 'ecological' view of affordance, as a relation between an object's physical qualities and a person's culturally informed mind that come together in the skilled actions of artisans, focuses on the natural properties of 'brute materials' and loses sight of people and what they think (Tilley 2007), such that 'humans seem to take a back seat to the trajectories of materials' (Knappett 2007:23). The 'humanistic' perspective laments 'the virtual absence of discussion of people and the meaning and significance of materials made (artefacts) or encountered (unaltered materials)', whereas 'consideration of the recursive relationship between people and things' makes necessary the 'concept of materiality' that goes beyond 'simply considering materials' and seeks 'to engage with the manner in which the material properties of things profoundly affect human conduct, both enabling and empowering people's lives and constraining them. The concept of things providing affordances to people is indeed very useful here.' (Tilley 2007:19). The 'ecological' retort is that the socially centred materiality perspective overlooks actual materials (Ingold 2007).[13]

The 'humanistic' response advocates an interpretative approach, seeing materials and things as having symbolic aspects, which is fair enough, as some do have such associations (discussed further in Chapters 8 and 9). The pandan leaves used in mat making by the Nalik of New Ireland, for instance, illustrate how 'through an analysis of material thinking in Melanesian society, we can begin to see how vital it is to draw attention to plant materials and their informed use in design as a means to articulate an understanding of the concrete [*sic*] social world' (Were 2013:597). In this community 'The adding of leaves side by side, and on top of each other, creates an image of the expansive and co-dependent nature of relations in which Nalik persons are implicated', but as the 'individual leaves together start breaking apart or decaying' the mat serves to map 'an image of localized, traditional social networks, sustained over a shorter, intense duration through its production and usage at life-cycle events' (2013:591, 593 and 596). The author points out that we have to see materials and objects in 'relation to social, political, economic, and natural environments ... not constituted simply of bare materials ... [but] as combinations of different relations,

12 Norman's (1988:23) concept of 'mapping', as an aspect of affordance, refers to the appropriateness of design for users, that helps them to understand what is intended by any feature, such as a handle in a certain location 'maps' to the idea that it is to open a door.

13 A similar argument occurs in anthropology over the priority given to natural versus cultural factors in exploring the embodied dimensions of cognition – under a confusing array of banners that seek to stake out particular approaches (such as 'anthropology of the senses', 'sensory anthropology', 'sensorial anthropology' and 'sensual culture'), the common theme being, whatever the semantics of these fine distinctions, reference to the five senses. It is an argument that, ironically for those who claim to be challenging dualistic thinking, mirrors the hoary old nature versus nurture debate (see, for instance, the Debate Section in volumes 18 and 19 of *Social Anthropology*).

entangled in natural and cultural processes' (2013:596), which reads as a convoluted affirmation of holism (that is central to ethno-determinism); as Tilley (2007:18) spells out in referring to materiality, 'developing a holistic and conceptual theoretical and interpretative framework'.

It seems fair to assume that people can comment on the affordances of raw materials that they use in making things, as I report here on the properties of the materials that feature in Was valley house construction, even though they may not have consciously thought about them previously. For instance, the response of some of my Wola friends to my questions about why they used certain timber or vines in house building suggests that they were unanticipated. They had seen relatives use these materials from childhood onwards and so they used them too, or maybe they thought that my questions were too obvious to merit a response. However, after some reflection, they could comment on the properties of materials that suited them to the purpose for which they used them; for instance, they pointed out that the timber they used for *aend bez* 'ridge-poles', such as *bat* and *bort* 'silkwoods', grew as young trees with straight trunks of the desired pole-like dimensions that have few branches low down to cause skewed forked trunks, and furthermore, the wood does not become brittle as it dries out but retains a certain 'give' that is good for the load bearing of rafters.

But the proposition that the affordances of materials bear a symbolic load is a different matter. While some raw materials may have symbolic associations for some people, it is questionable to assume that this applies to all populations, even in Melanesia, where the argument that raw materials and object design tangibly express the social world conveniently agrees with the contemporary view that materials and substances feature significantly in Melanesian sociality. In the Was valley I have never heard anyone talk of such associations regarding construction materials, even with repeated questioning of many persons in various situations and at different times attempting to winkle them out. It seems improbable that all raw materials will carry such symbolic loads, even where people talk about such associations with some of them – such as the Nalik of New Ireland with certain pandans, who according to Were (2013:585) respond to questions about the socio-cultural interpretation of raw materials and design of objects 'with highly intellectualized responses'. While it is possible that others may, in the interpretative tradition, use this Was valley account of the 'affordances' or properties of various building materials to suggest symbolic associations, and are free to do so, from the ethno-determinist perspective such imaginative conjectures demand caution if unprompted by comments from actors that point to symbolic connections awaiting ethnographically founded exegesis. Any such interpretations also have to confront the paradox mentioned previously, of the foreign analyst knowing more than the native actors about their behaviour and comprehension and comprehension are, when their behaviour the subjects of interest.

Preparing the raw materials

Whether or not materials carry a symbolic load, they mostly require preparation for use in making things. This encompasses, in an ethnographic context, aspects of both 'ecological' and 'humanistic' approaches to materiality. The properties of raw materials influence the work undertaken in preparing them; there is no point in taking an axe to a blade of sedge grass. The work, culturally informed, may feature social interaction as persons co-operate together in undertaking it. Bodily acquired tacit skills are likely to feature prominently. A significant amount of the time and effort spent on house construction in the Was valley involves the collection and preparation of raw materials. Table 6.2 lists the materials collected and used to build the Honael house documented in the next chapter.

The making of the stakes used in wall construction is the most time consuming and physically demanding of tasks in preparing materials (Table 6.3). It involves the splitting and sharpening of four types of stakes: *dol* 'exterior wall stakes', *waenuw* 'interior wall stakes', *pinj* 'ridge-pole supports' and *pongow* 'jamming stakes'; the last made from pieces left over from cutting the other three kinds of stakes.

The first job is to find a tree of sufficient size to supply enough stakes, but preferably no larger, as the bigger the tree the harder it is to fell and split up. When building the Honael house, we spent about two hours finding a suitable tree. We went first to a tree felled and split up for stakes by Hiyp (Saendaep's FBSS) at Haelbiy but when we found it, insufficient remained for our uses (see Chapter 7, particularly the genealogies in Figures 7.5 and 7.6 for details of persons). Then Neleb told us that he had felled a *pel* 'beech' tree at Ebera, for a house that he had intended to build, which was about the right dimensions to supply the number of house-wall stakes that we should need for the Honael house. Three men – Neleb, Saendaep and Kuwliy – spent the rest of the day engaged in *gənk bay* 'splitting tree trunk in half lengthways'. They split it in two for 9.45 metres using steel axes and wooden wedges (it would have taken approximately one-and-a-half to two times longer to complete this work using stone axes and wedges – Sillitoe, Stewart and Strathern 2002:38–41). The following day, they *bokor lay* 'chopped through trunk sections transversely' to get two separate tree-trunk halves of equal length (one to be further split up to produce wall stakes using steel tools and the other using stone). Next they *jiy bay* 'split the two halves of tree trunk into quarter sections longitudinally', working on one half-tree-trunk section using steel tools and the other using stone ones together with wedges to force splitting along the grain of the timber. Then they *tongok bay* 'split the quarter-trunk sections into stake-sized pieces',

Raw material	No.	Stone or steel tools	Dimensions	Used for	Amounts used
iysh pel 'southern beech' tree trunk	1	steel	circum. 137 cm, length 945 cm	*aend dol* and *hiytpaip* 'external wall stakes', *waenuw pongow* 'internal wall stakes' *pinj* 'ridge-pole supports'	all
aenk 'screw-pine' trunks	4	steel	circum. 58.5 cm; 48.5 cm; 52 cm; 51 cm	*aend buwk* 'bark wall sheets'	4 bark sheets: 66 × 470 cm; 58.5 × 419 cm; 63.5 × 592 cm; 63.5 × 505 cm
	4	stone	circum. 56 cm; 43 cm; 45.5 cm; 56 cm		4 bark sheets: 56 × 356 cm; 53.5 × 356 cm; 56 × 409 cm; 56 × 353 cm
	1	steel	circum. 66 cm	*aend kor* 'fireguard'	1 bark sheet: 132 × 56 cm
	1	stone	circum. 71 cm		1 bark sheet: 134.5 × 68.5 cm
	5	readymade (except 1 steel made)	length of house eaves	*aenk tor* 'thatching gutters'	all [reusable]
waen 'woolly cedar' tree trunk	1	steel	circum. 40.5 cm, length 610 cm	*aend bez* 'ridge-pole'	all
bort 'silkwood' tree trunk	1	stone	circum. 37 cm, length 666 cm	*aend bez* 'ridge-pole'	all
sabok 'sabia' tree trunk	1	steel	circum. 79 cm, length 163 cm	*tiriyor haekomb* 'door slats'	all
timbol 'spurge' tree trunk	1	steel	circum. 141 cm	*kat tor iysh* 'timber for pig-stall roof'	all
	1	stone	circum. 106 cm		all
iysh 'tree saplings' (various species)	30	steel	av. circum. 15.2 cm, av. l. 335 cm	*polpen* 'rafter poles', *aend kor* 'fireguard'	all
	27	stone		*kuwngis* 'ridge-pole'	all
iysh huwniy 'small tree saplings' (various species)	4	steel	various	*tor* 'framework' for fireguard	
iysh huwniy 'small tree saplings' (various species)	9	stone	various	*tor* 'scaffolding' saplings for gutters	
	?	steel			
gwai 'switchsorrel' saplings	11	steel	av. circum. 7.6 cm, av. length 457 cm	*gamonk* 'wall battens'	all
	14	stone			
	1	steel		*sal gamonk* 'fringe arch batten'	all
shonon 'chenille' saplings	14	steel	av. circum. 4.4 cm, av. length 305 cm	*iybhedla* 'eave battens' *tuwmainj* 'wall/ roof beading'	all
	20	stone			all
bobae 'spurge' sapling	1	stone		*pongow* 'jamming sticks'	

Table 6.2 Raw materials collected to build the Honael house (continued over).

Raw material	No.	Stone or steel tools	Dimensions	Used for	Amounts used
waen 'woolly cedar' bark strips	off 3 trees	steel	total weight 5.78 kg	*ya* 'bark strips' for lashing	5 trees-worth
inom 'stinging tree' bark strips	off 3 trees				
wenet 'engel tree' bark strips	off 3 trees				
waen 'woolly cedar' bark strips	off 3 trees	stone	total weight 2.72 kg		all
inom 'stinging tree' bark strips	off 2 trees				
haeluwp 'rattan' vines	10 stems	steel	total of 8601 cm	*ya* 'vine' for lashing	all except for 3 lengths totalling 1067 cm
	10 stems	stone	total of 7171 cm		
taembok 'bamboo' culms	3 lengths	steel	av. circum. 16.5 cm, av. length 640 cm	*ka* 'roof lathes'	all
	2 lengths	stone			all
naep 'casuarina' fronds		by hand	several fronds	*aend sal* 'roof fringe'	all
paym 'screw-pine' leaves	19	steel	weight 5.57 kg	*hwimb* 'wall lining'	all
	15	stone	weight 4.99 kg		all
huwshiy 'cunonia' tree trunk	1	steel	circum. 30.5 cm, length 615 cm	*dombep* 'drainage way log'	all
henk 'tree fern' trunk	1	steel	circum. 28 cm, length 432 cm	*dombep* 'fireplace surround logs'	all
	1	stone	circum. 34.5 cm, length 412 cm	*dombep* 'drainage way log'	all
keret 'screw-pine' trunks	1	steel	circum. 21.5 cm, length 366 cm	*dombep* 'fireplace surround logs'	all
	2	stone	circum. 23 and 24 cm, length 269 and 229 cm		all
	1	steel	circum. 21.5 cm, length 361 cm	*dombep* 'logs' of pig stalls	all
kwimb 'moss' balls	several	by hand	0.68 kg (dry weight)	*maembow pes* 'eave gap sealant'	all
gaimb 'cane grass'	84 bundles[1]	steel	5712 stems	*aend gaimb* 'thatch'	83 bundles – approx 5644 stems
	61 bundles[1]	stone	4148 stems		all
	~250 stems[2]	steel	av. circum. 4–5 cm, av. length 106 cm	*sezel* 'wall canes'	~248 stems
	~230 stems[2]	stone	av. circum. 4–5 cm, av. length 106 cm		~224 stems

1 A random sample of 5 steel cut bundles contained 50, 79, 50, 80 and 79 stems and of 5 stone cut bundles contained 62, 74, 61, 86 and 56 stems, which is an average number of stems per bundle of 68 for both.

2 Collected in 10 trips.

Table 6.2 Raw materials collected to build the Honael house (continued from previous page).

Plate 6.1 Splitting wall stakes.

Plate 6.2 Shaping wall stake.

cutting off various wall-stake lengths, which they pared down further to the required size and shape, before finally *waendilay* 'sharpening' at one end for driving into the ground. When they had finished making the stakes, they carried them in bundles on their shoulders to the Honael house site.

While stone and steel axes were used on similar-sized half sections of the same tree trunk, the stone-made stakes were chunkier and more roughly finished off, and fewer stakes were made from the same materials. The stakes split and pared with steel axes were trimmer; as noted in previous work comparing the use of stone and steel tools (Sillitoe 1979c:160); it is possible to achieve a better finish using them. These data show again how steel tools allow people to exploit natural resources more efficiently, with implications for their environmental stewardship. But in achieving the better finish they spend longer than would be necessary to achieve a standard of work comparable with stone tools, which has implications for assessing the comparative efficiency of stone and steel tools by time spent working on tasks; which is the sort of information that may prove of use to archaeologists (see Chapter 4). Another compromising factor was that the younger men helping with the construction work, although under the guidance of older experienced men, did not have previous experience of using stone axes (as discussed in Chapter 2); as evident when one of them, who initially failed to heed relatives' advice about how to use these axes, damaged the cutting edge of three tools. In order to compare the total times spent working with stone and steel axes it is necessary to adjust the totals on Table 6.3 to allow for the tree trunk being split in half using steel axes alone. If we discount this work by ignoring the first 8 rows on the table (i.e. only compare for the same work undertaken, subtracting 7 h, 6 m from the steel total), we obtain a steel-to-stone time ratio of 1:1.54. Or if we add adjusted times to the stone column

Tasks	Wola term	Time spent[1] Stone-tool side	Steel-tool side
clearing undergrowth around southern beech trunk[2]	*pel tay bort payay*		17 m
collecting wood for *holol* clubs and *ndul* wedges	*iysh ez bay*		13 m, 30s [3]
making *holol* clubs	*holol waendilay*		30 m, 20 s[3]
shaping *ndul* wedges	*ndul doi*	4 m, 15 s	1 h, 49 m, 30 s[3]
collecting *holol* clubs and *ndul* wedges	*holol ndul kab muk pay*	5m	9 m, 30 s
splitting trunk in half longitudinally using axe and *ndul* wedges[4]	*iysh gənk bay*		3 h, 12 m, 5 s
making two levers	*gəmb waendilay*		2 m
levering trunk apart	*iysh tay ongay*		52 m, 5 s
chopping through trunk crossways (producing two longitudinally split halves)	*iysh bokor lay*	2 h, 4 m, 45 s	1 h, 26 m, 40 s
rolling trunk to break at *bokor* cut	*iysh bol menay ndoborlay*	16 m, 45 s	
moving length of trunk for splitting	*iysh men esay*		24 m, 50 s
splitting one half of trunk in half again	*jiy bay*	20 m, 15 s	42 m, 55 s
knocking in *ndul* wedges during the *jiy* splitting	*ndul tuwkay*		5 m, 30 s
holding split open with legs to facilitate chopping	*ongor mon iysk kombilay*	1 m, 30 s	
rolling pieces of trunk over during splitting	*iysh munuw perayay*	16 m, 10 s	
splitting *jiy* lengths into stake sized pieces	*tongok bay*	1 h, 41 m, 50 s	51 m, 10 s
measuring off *dol* stake lengths	*el tomb wiy*	1 m, 15 s	
cutting *tongok* pieces into stake lengths	*tongok bokor lay*	3 h, 10 m, 45 s	17 m, 40 s
further splitting and trimming stakes[5]	*dol waenday pep*	6 h, 50 m, 30 s	8 h, 9 m, 15 s
rolling stake lengths down slope	*dol pendet bay*	58 m, 15 s	27 m, 55 s
carrying stake lengths down slope	*dol tok pay*	1 h, 43m	1 h, 45 m, 15 s
piling up stakes[6]	*dol wesuway*	12 m, 30 s	13 m, 30 s
sharpening stakes[7]	*dol waendilay*	7 h, 57 m, 15 s	2 h, 10 m, 15 s
retying or adjusting binding and axe head[8]	*kwiy tongay*	3 h, 16m	
lighting a fire	*iysh haeray*	7 m, 45 s[9]	
making rough cane-grass sunshade	*gaimbom nat aend bay*	4 m, 30 s	
shouting advice	*kort bay*	13 m	
cutting new path to Honael house site (with bush knife)	*haeret waeniy kaeray*	44 m	
carrying stakes to Honael house site[10]	*dol tok pay aend tay bort*	12 h, 5 m, 30 s[11]	
total times		**35 h, 37 m, 23 s**	**30 h, 18 m, 17 s**

1 In addition to above times, men spent a combined total of 10 h, 18 m during the 4 days working on wall stakes travelling to and from forested site at Ebera; some of this time spent searching for other materials.

2 Southern beech tree felled previously. See Ganonkiyba garden data (Sillitoe, Stewart and Strathern 2002:24–5) for comparative tree-felling times.

3 These clubs and wedges were also used in splitting up the half of the tree trunk worked on using stone axes – see Sillitoe (1988:68–70) for comparative stone tool times preparing *holol* clubs and *ndul* wedges.

4 For details of procedure followed splitting a tree trunk, see Ganonkiyba garden account (Sillitoe, Stewart and Strathern 2002:38–41).

5 A few steel-made stakes were also sharpened simultaneously while splitting and trimming.

6 Total number of stakes produced using steel = 114 (61 *dol*, 13 *waenuw*, 36 *pongow*, 4 *pinj*); total number of stakes produced using stone = 91 (46 *dol*, 25 *waenuw* [8 subsequently cut down to *dol*], 17 *pongow*, 3 *pinj*).

7 Stakes sharpened using steel = 60; stakes sharpened using stone = 66.

8 Two stone axe heads were badly chipped and one was slightly chipped, during this work.

9 Last five entries divided equally between, and added to totals of, stone and steel columns.

10 They transported 72 of the steel- and 70 of the stone-made stakes to the house site (leaving behind all *pongow*-sized pieces).

11 Approximately 1 h of this time was spent collecting 2 pandanus nuts.

Table 6.3 Making stakes.

Plate 6.3 Sharpening wall stake.

for the work done initially in splitting the tree trunk in half (adding 8 h, 50 m, 50 s), we obtain a similar ratio of 1:1.47.[14] Whichever comparison we take, it is evident that preparing the wall stakes for a house is a time-demanding task, whether using stone or steel axes.

The other principal walling material is *aend buwk* 'screw-pine bark sheets' (Table 6.4). The preparation of these involves the felling of screw-pines above the aerial roots (the bark where these grow would have holes in it) and chopping off the trunk section below the crown(s). The

14 The adjustment involves using data on club and wedge making in Sillitoe 1988a:69 to calculate the time it would take to make these using stone tools (which is 47 m for the clubs and 1 h, 41 m, 40 s for the wedges – i.e. about the same time), and using an estimate of 1.5 times longer to split the trunk in half longitudinally using stone axe and wedges (if we increase this estimate to 2 times longer, adding 10 h, 27 m to the stone total, we obtain a ratio of 1:1.52).

Plate 6.4 Cutting nicks in pandan trunk.

Plate 6.5 Beating pandan trunk with clubs.

Task	Wola term	Time spent using stone tools[1] pandan 1 (circum. 56 cm)	pandan 2 (circum. 43 cm)	pandan 3 (circum. 46 cm)	pandan 4 (circum. 56 cm)	Time spent using steel tools[1] pandan 1 (circum. 58.5 cm)	pandan 2 (circum. 48 cm)	pandan 3 (circum. 52 cm)	pandan 4 (circum. 51 cm)
searching for pandanus[2]	*aenk ez biy*	1 h, 1 m, 30 s	1 h, 14 m, 30 s	26 m, 15 s	3 m, 45 s	1 h, 17 m, 30 s	1 h, 30m	1 h, 5 m, 30 s	1 h, 56m
clearing round tree trunk	*aenk tay bort payay*			1 m, 45 s	1 m, 15 s		15 m		
felling pandan	*aenk bway*	3 m, 45 s	11 m, 45 s	7 m, 15 s	7 m, 30 s	2 m, 45 s	2 m, 45 s	2 m, 30 s	felled previously
making clubs	*gəmb waendilay*	15 m, 15 s		2 m, 15 s	2 m, 30 s	4 m	5 m, 30 s	3 m	8 m
cutting off required trunk length	*aenk tay bokor liy*	6 m, 30 s	5 m, 30 s	13 m, 45 s	9 m, 10 s	4 m	4 m, 40 s	3 m	12 m, 50 s
carrying trunk to clearing	*aenk iyshor tok pay*							15 m	2 m, 15 s
carrying trunk to house site	*aenk iyshor aenda munk bort tok pay*			19 m, 30 s	2 m				
cutting incisions in trunk	*bor mor homay*	33 m, 45 s	45 m, 15 s	52 m	41 m, 40 s	24 m	31 m	25 m	28 m, 5 s
beating trunk to soften pith	*buk liy*	27 m, 45 s	32 m	23 m, 5 s	26 m, 30 s	33 m	27 m	12 m, 45 s	29 m, 45 s
sharpening stakes to open up trunk	*puw henday gəmb waendilay*					3 m			
fetching stakes to lever open trunk (e.g. old fence stake)	*puw henday gəmb muk obay*		2 m, 45 s	4 m, 30 s	4 m, 30 s			2 m, 30 s	1 m, 45 s
opening up trunk	*puw henday*	31 m, 30 s	19 m, 30 s	31 m, 45 s	32 m, 15 s	17 m	31 m, 45 s	29 m	16 m, 30 s
removing pith from trunk	*buk iy muw tuway*	42 m, 15 s	12 m, 30 s	17 m	20 m, 30 s	1 h, 2 m, 15 s	54 m, 15 s	29 m, 15 s	18 m, 15 s
turning over bark sheet and pressing flat	*buk bol menay ma bat say*	5 m	4 m, 45 s	17 m	20 m, 30 s	4 m	8 m	3 m	
fetching vine	*ya muk obay*							2 m	
rolling up bark sheet and tying	*buwk bol menay diy ya tongay diy*	4 m, 30 s	6 m			4 m, 30 s	4 m	13 m	5 m, 30 s
carrying bark sheet[3] to Honael house site	*aend buwk tok pay*	12 m, 30 s	12 m, 30 s			45 m	47 m	8 m	17 m
adjusting stone axe	*kwiy tongay*	1 m, 30 s							
total times		4 h, 5 m, 45 s	3 h, 47 m, 0 s	3 h, 36 m, 5 s	2 h, 52 m, 5 s	4 h, 41 m, 0 s	5 h, 20 m, 55 s	3 h, 33 m, 30 s	4 h, 15 m, 55 s
total times preparing 4 bark wall sheets		**14 h, 20 m, 55 s**				**17 h, 51 m, 20 s**			

1 The work was undertaken over 3- and 2-day periods, respectively.

2 Searching for steel-worked pandanus at Iybweb (first and second), Haelaelinja (third) and Tagiy (fourth); and for stone-worked pandanus at Tetaenda (first and second), Honael (third) and Haelaelinja (fourth). For the geographical locations of places, see Sillitoe 1988:24 (Map 2).

3 Dimensions of bark sheets: pandan 1 = 505 × 63.5 cm; pandan 2 = 419 × 58.5 cm; pandan 3 = 592 × 63.5 cm; & pandan 4 = 470 × 66 cm.

Table 6.4 Preparation of pandan bark wall sheets (aend buwk).

Task	Wola term	Time spent	
		Stone-tool side	Steel-tool side
collecting *gwai* saplings[1] for *gamonk*	*gwai huwniy kaeruw muk pay*	1 h, 9 m, 15 s	1 h, 15 m, 30 s
peeling bark off *gwai* saplings[2]	*gwai huwniy shongol iyziylay*	42 m, 45 s	55 m
collecting *haeluwp* rattan[3] from Pongtenda	*haeluwp munu lenjay miy*	4 h, 14 m	4 h, 14 m
collecting bark strips[4] at Tetaenda	*iysh shongol ya muk pay*	41 m, 15 s[5]	1 h, 37 m
collecting *bobae* sapling at Tagiy for *pongow* jamming sticks	*bobae pongowom waeray muk pay*	8 m	
collecting cane grass stems for external walls[6]	*dol sezel gaimb lor kaeray*	3 h, 0 m, 10 s	2 h, 9 m
collecting cane grass stems for internal walls[7]	*waenuw sezel gaimb lor kaeray*	1 h, 39 m, 30 s	35 m
collecting *paym* pandan leaves[8]	*paym shor kaeray miy*	3 h, 12 m, 30 s	
fetching previously collected *paym* pandan leaves from house	*paym shor aenda bort muwk pay*	5 m, 15 s	
collecting moss for sealing eave gaps[9]	*kwimb muk maembow pesom bay*	53 m, 30 s	
Total times		**13 h, 40 m, 40 s**	**12 h, 51 m, 00 s**

1 Collected 11 saplings with steel tools (7 from Kondorow and 4 from Honael) and 13 with stone (all from Kondorow); average length 457 cm and circumference 7.5 cm.
2 Peeled 14 saplings using steel tools and 10 using stone.
3 Collected 10 lengths of vine totalling 8,601 cm using steel tools and 10 lengths of vine totalling 7,171 cm using stone tools.
4 Collected 5.78 kg of bark from 3 *waen*, 3 *inom* and 3 *wenet* trees using steel tools and 2.72 kg from 3 *waen* and 2 *inom* trees using stone tools.
5 Also collected 1 *gwai* sapling during this time for *gamonk* battens.
6 Made 7 trips for stone side and 9 trips for steel side.
7 Made 3 trips for stone side and 1 trip for steel side.
8 Collected 34 leaves at Maip & Kwimb, weighing 10.56 kgs.
9 Collected 0.68 kg of moss at Taeziyba.

Table 6.5 Preparation of other materials for walls.

next stage, *bor mor homaebay*, involves making a series of axe cuts into the trunk, breaking through the brittle outer bark to give it a degree of pliability for the next stage. The succeeding step is *buwk lay* 'bark sheet hit', which involves beating the pandan trunk with club-like pieces of wood to pulverize and soften the fibrous inner pith and further increase the malleability of the bark. The trunk is then *puw henday* 'cut open lengthways' with a single, straight axe cut, and prised open with a pair of sharpened stakes to reveal the inner fibrous pith. This is pulled out – *buwk iy muw tuwae* 'bark-sheet excreta take throw-away' – using an axe to cut it away from the bark. The bark is then pressed flat to produce a sheet, which is finally rolled up carpet-like for transportation to the house site. The preparation of pandan-bark sheets involves relatively little axe work and we might anticipate a relatively small time difference between those prepared using stone and steel tools. In fact, the work undertaken using stone axes to prepare bark sheets for the Honael house was completed more quickly; the time discrepancy being largely due to the different times spent searching for suitable trees. The variations in the times taken to complete the same tasks on different occasions are salutary reminders of the range of time to be expected in undertaking any such work, of which the data here are an indication.

The time it took to collect and prepare the other wall materials are also approximately comparable for both stone and steel (Table 6.5), except for the cutting of cane-grass stems, which was completed more quickly with a steel bush knife. The saplings used for battens are common and soon collected and easily split lengthways in two. Similarly, the strips of bark used as lashing materials are readily gathered. It takes longer to find sufficient rattan palms in the forest, but the preparation of vine for lashing is quick and easy, and is usually undertaken as required during construction work. It entails splitting the flexible stems lengthways, the split often initiated with the teeth, and sometimes includes the discarding of the softer *hobor* 'inner pith' to leave the corneous *shongol* 'outer skin'. The dead *paym* pandan leaves are likewise collected from forest locations. The total time spent amassing and preparing the raw materials for the walls of the house was 74.08 hours using stone tools and 61 hours using steel tools, which gives a ratio of 1:1.22 – that is, for every hour worked using steel tools an additional 13 minutes was necessary to accomplish the same output overall using stone tools.

Task	Wola term	Time spent	
		Stone-tool side	Steel-tool side
Ridge-poles			
searching for tree at Honael	*aend bez iysh ez biy*	22 m	8 m
felling tree	*iysh bway*	4 m, 30 s[1]	50 s[2]
clearing round tree	*iysh tay bort payay*		55 s
cutting off branches	*pəga payay*	1 m, 30 s	1 m
carrying trunk into clearing	*iysh tor tok pay*	1 m, 30 s	
cutting trunk to required length	*bokor liy*	2 m	2 m
sharpening end of pole	*bez waendiyay*	12 m, 30 s	3 m, 30 s
peeling bark off trunk	*humbiy togay*	6 m	1 m
carrying pole to Honael	*bez aend tay tok pay*	6 m[3]	14 m, 30 s[4]
pinning pole to ground to straighten	*bez tonga mbabtuw duwn bay*		22 m
Rafter poles			
felling & trimming rafters[5] at Momiyaek	*polpen bway*	2 h 19 m	1 h 49 m, 45 s
peeling off bark and moss (and trimming some ends)	*shongol dokor tuway diy kwimb muw tuway diy*	1 h 30 m	1 h 4 m
tying up bundles of rafters	*polpen tongay*	9 m, 45 s	23 m
carrying bundles of rafters to Honael	*polpen tok pay*	1 h 10 m, 30 s	2 h 18 m, 45 s
Other roofing materials			
collecting bamboo for *ka* roof battens[6]	*taembok kaembay*	47 m, 15 s	47 m, 15 s
collecting *shonon* saplings from Tetaenda for *iybhedla* eave battens[7]	*shonon huwniy muk pay*	28 m	28 m
peeling bark off *shonon* saplings (by rubbing on fence stake)[8]	*shonon huwniy shongol iyziylay*	20 m, 30 s	23 m, 30 s
collecting casuarinas fronds	*naep shor aend salom miy*	3 m, 30 s	12 m, 30 s
collecting *gwai* sapling for front eave *sal* arch	*gwai huwniy muk pay*		30 m, 45 s
total times		**7 h 44 m, 30 s**	**8 h 51 m, 15 s**

1 *Bort* tree, circumference 37 cm.
2 *Waen* tree, circumference 40.5 cm.
3 Length of ridge-pole 666 cm
4 Length of ridge-pole 610 cm.
5 27 rafters with stone tools and 30 rafters with steel: average circumference of rafters 15 cm, average length of 335 cm.
6 Collected 3 lengths of culm with steel tools and 2 lengths with stone: average circumference 16.5 cm, average length 640 cm.
7 Collected 6 saplings with steel tools and 12 saplings with stone: average circumference 4.5 cm, average length 305 cm.
8 Peeled 9 saplings each on both steel and stone sides.

Table 6.6 Preparation of roofing materials.

The roofing materials, in comparison, take considerably less time to collect (excluding the thatch) and the stone:steel tool times are near parity (at 1:1.14), with the work taking marginally longer with steel tools (Table 6.6).[15] Trees of suitable dimensions for the pair of *aend bez* 'ridge-poles' are fairly common and men soon fell and strip them of bark. Sometimes, if the tree trunk is not quite straight, house-builders will pin it on a level area of ground to straighten it while the green wood has some flexibility in it. Another technique is to suspend the pole between two stakes about 30 centimetres above the ground, arranged in such a way that gravity may straighten any bend in the timber. Young trees suitable for *polpen* 'rafters' are likewise relatively common, particularly in some previously cultivated areas that have been abandoned for a few years. They are soon felled, trimmed and carried back to a house site. The other roofing materials are also easily collected and prepared, with *taembok* 'bamboo' cultivated around some house yards and saplings suitable for lathes common in regrowth adjacent to settlements and easily debarked. If the split aerial roots of the *paym* pandan are used for lathes they may take longer to collect, as these large screw-pines are not common, although their whereabouts in any locale is well known to residents.

15 The difference is attributable to the longer time taken to transport the rafter poles to the house site.

Plate 6.6 Cutting and levering open pandan trunk.

Plate 6.7 Pulling fibrous interior from flattened pandan trunk to give buk 'bark sheet.'

The thatching material, unlike the other raw materials, is collected when the framework of the house is completed and it is ready for its roof covering (Table 6.7). The collection of thatching material involves cutting down cane or pulling up grass or collecting pandan leaves and transporting them to the house site. If using cane grass, preparations may include the making of semicircular gutters from lengths of pandan trunk, which are used to contain the cane thatch at the eaves during thatching. Some smaller materials are similarly collected as needed during house construction, such as bark sheets for the eaves, timber for fireplaces and drainage ways, and moss or leaves to plug gaps around the eaves (Table 6.8). Sometimes men use leftover materials around the site for some things, such as offcut pieces from stakes for *pongow* 'jamming sticks' and pandan pith to plug gaps along eaves.

Material agency and animism

According to the materiality argument this account of the preparation of materials to build a house is cock-eyed in focusing on how the house-builders make ready the materials for use in construction, albeit it encompasses affordance issues. Instead it should consider what the materials do to the builders, not only in respect of how their properties-cum-affordances set certain limits to what they can do but also in respect of how the 'actions' of materials affect the men and their relatives, which returns us to the

Task	Wola term	Time spent Stone-tool side	Steel-tool side
Thatching Preparations			
collecting previously used pandan gutters from elsewhere	*aenk tor tok obay*	45 m	
felling pandan for new gutter and carrying longitudinal ½ trunk back to house site[1]	*aenk bway tokor gənk mond tok obay*		13 m, 30 s
removing pith from inside pandan trunk	*iy tuway*		8 m
collecting saplings for stakes to hold gutter around eaves	*aenk torom iysh huwniy payay*	21 m, 30 s	3 m, 15 s
Thatching			
finding suitable cane-grass clumps	*gaimb ez bay*	49 m, 15 s	11 m, 15 s
cutting cane-grass thatch[3]	*aend gaimb payay*	12 h, 21 m, 35 s	6 h, 55 m, 15 s
carrying cane-grass bundles to house site[4]	*aend gaimb tok pay*	9 h, 3 m, 10 s	14 h, 21 m, 30 s
total times		**22 h, 58 m, 0 s**	**22 h, 15 m, 15 s**

1 Pandan felled at Tetaenda.
2 84 bundles cut using steel tools and 58 bundles cut with stone.
3 Men cutting cane-grass thatch: Sorliy and Balen (steel), Kot, Pes, Neleb, Balen, Piyp and Sorliy (stone). For relationships between named persons, see Figure 7.6.
4 Women carrying cane-grass bundles (both steel and stone cut): Lenday, Hundbiy, Piriyn, Horshiyow, Tindiy, Gənk, Dorwa and Kwalten. 83 bundles steel-cut (average journey time per bundle approx 10 m, 15 s) and 61 bundles stone-cut (average journey time per bundle approx 9 m, 17 s).

Table 6.7 House-thatching preparations.

Task	Wola term	Time spent Stone-tool side	Steel-tool side
collecting log for step across front of house[1]	*wil aend dombep payay*	38 m, 55 s	
collecting log for under-eaves drainage channel[2]	*iyb gən dombep payay*		12 m, 15 s
searching for and felling pandan for fire screens	*aenk ez bubtuw bway*		54 m, 10 s
cutting incisions in pandan trunk, beating and removing pith	*bor mor homay diy buwk liy diy iy muw tuway diy*	49 m, 35 s	58 m
collecting saplings for fire-screen frames	*aend kor tor iysh huwniy payay*		6 m
collecting log for footing of pig stalls[3]	*showmay kuwl dombep payay*		28 m, 30 s
felling trees for logs to roof pig stalls[4]	*kat tor iysh bway*	5 m, 15 s	5 m
collecting timber for making doors[5]	*tiriyau haekomb iysh payay*		21 m, 15 s
collecting log for sitting-room fireplace surround[6]	*wil aend taenktay dombep iysh payay*		10 m, 45 s
collecting log for men's sleeping-room fireplace surround[7]	*uwpuluwk aenda taenktay dombep iysh payay*		25 m
collecting logs for women's room fireplace surround[8]	*tenda taenktay dombep iysh payay*	9 m, 45 s	
ferrying various prepared materials to house site (stakes, vine, rafters etc.)	*aend iysh haek aenda munk bort tok pay*	3 h, 10 m, 15 s	
total times		**5 h, 16 m, 00 s**	**3 h, 18 m, 40 s**

1 Collected *henk* log 412 cm long by 34.5 cm circumference.
2 Collected *huwshiy* log 615 cm long by 30.5 cm circumference
3 Collected *keret* log 361 cm long by 21.5 cm circumference.
4 Collected 2 timbol tree trunks 141 cm (steel) and 106 cm (stone) circumference.
5 Collected *sabok* log 163 cm long by 79 cm circumference.
6 Collected *henk* log 432 cm long by 28 cm circumference.
7 Collected *keret* log 366 cm long by 21.5 cm circumference.
8 Collected 2 *keret* logs 269 cm long by 23 cm circumference, and 229 cm long by 24 cm circumference.

Table 6.8 Preparation of other materials (e.g. for internal house fittings).

Plate 6.8 Carrying polpen 'rafter poles' and other materials to house site.

curious argument about inert substances having agency (Chapter 5).

The attraction of the materiality approach to archaeologists is understandable, with no actors to ask about the properties of things let alone enquire into symbolic associations (Chapter 4), for in the final analysis objects can hardly substitute for living persons, not even in Melanesia where the so-called 'new ethnography' talks of people's behaviour this way. But this is what some seem to suggest, focusing on the 'socialness' of things, pushing the materiality turn beyond the 'humanistic' interpretative stance, by arguing that material things can have a social effect beyond serving as symbols. In going further they attribute agency to things. According to Were (2013:584), for instance, among the Nalik 'materials are intimately linked to action through their capacity to instigate social change'. Regarding houses, Fahlander (2008:128) tells us that the 'built-up environment is as much an active generator of social behaviour as it is constituted by it'. In attributing agency to materials, this version of materiality parallels cognitive archaeology's use of the extended-mind thesis (see Chapter 5). The material world, most broadly the environment, unarguably impinges on the social world. It is the implication that things have agency just by being there that causes problems, which ranges from 'determinist techno-fetishism to less radical notions of the importance of the non-human in social studies' (Fahlander 2008:128).

The fallacy of material determinism was asserted as the social sciences emerged; for example, Durkheim noted that material things may affect 'social evolution, whose speed and even direction vary according to the nature of these elements; but they contain nothing of what is required to put it in motion... As an active factor, then, the human milieu itself remains.' (1964:113). Regarding Were's contention above, for instance, it is not materials[16] *per se* that contribute to social change, but the things that humans can produce using them (such as tools). The attribution of agency to humans collectively is no reason, of course, to overlook the material aspects of life, as the social sciences largely did subsequently (Sillitoe 1988:5), to the vexation of archaeology (Schiffer and Miller 1999), as illustrated in the following comment: 'Just as archaeologists find it difficult seeing through the material to the social, so it seems the ethnographer or sociologist struggles to see through the web of social relations to materials and their properties.'

16 It is unclear what Were had in mind, perhaps the invention of steel or plastic.

(Knappett 2007:21–2). Although analysis of social actions that ignores material context is undeniably partial, it is even more questionable to attribute agency to things, which only have a potential to open opportunities and set limits on behaviour. The latter is apparently what some commentators intend by 'agency', advocating that we define it as 'providing affordances and constraints for thought and action', although it is unclear why we cannot simply talk of 'effects' if we are referring to 'agency (effects) of things on people' (Tilley 2007:19). The pseudo-agency of materiality is further evident when we consider how human actors can vary in their intentions using the same stuff. It seems that recognizing that things feature in social action (for in the hands of actors they may contribute to the initiation and maintenance of social processes – albeit tacitly, in ways not always evident to them) is not so much a step forwards as an exercise in running on the spot, amounting to little more than an obscure statement of the obvious, or nonsense if taken literally to indicate that inanimate things can take action.

The bestowing of agency on things and imagining that they can interact with persons and induce them to behave in certain ways has ethnographic parallels in animist beliefs, which since Victorian times have remained in the closet, together with 'primitive artefacts', as embarrassing cultural legacies. Until recently that is, with currently popular Amazonian cosmology (Santos-Granero 2013) supplanting the Melanesian gift as the ethnographic *motif de jour*. And material-engagement theorists make the association openly, as mentioned in the previous chapter: 'the closest parallel to the idea of a material agent that we have is not that of a person but rather that of a fetish' (Malafouris 2013:135). Recent efforts to square animist discourse about things having life with our scientific understanding of the world suggest that 'the spirit that enlivens them is not in but of matter ... the power of agency lies with their materiality itself', and according to this interpretation 'things are in life rather than ... life is in things' (Ingold 2007:12). While it is an intriguing exercise in counter-ethnocentricity to argue that animistic beliefs offer a novel way to further our understanding of the world – after the Victorian evolutionists (e.g. Tylor 1871) had consigned them to the bottom of the intellectual heap as spurious primitive belief systems compared to science's rigorous understanding of the world – it seems unlikely realistically to make much headway. It is improbable that an animistic perspective could ever overturn the Enlightenment informed scientific view of planet Earth (particularly given the technology that it has spawned and the ability of humans to intervene in nature).

The animist view of the world certainly does not apply to the New Guinea highlands. My friends in the Was valley are unsure what animating force to attribute to other animals, which they recognize as *enjay* 'alive' like humans, though they do not know if this is due to them having a *wezow* 'soul force' like them. As one friend responded to a question on the matter with respect to birds: '*aib hendaybay, nau sor inj kau*', 'who knows, we're not birds I tell you?' And questions such as 'are trees *enjay*' sound ridiculous. They are not; while they may grow, they cannot, after all, move around and do things (Sillitoe 2003:116). A growing plant is *kahagay* 'raw-green' and one that is no longer doing so is *kabiy* 'dead-dry'. To ask if a stone is *enjay* is equally, if not more, ridiculous; it just *wiy* 'is'. When I have observed that people associated certain *hungnaip* 'ritual-stones' with the *towmow* 'ghost-spirits' of ancestors, and some deep pools of water and deceased persons' skulls with them too, individuals have patiently explained that this does not mean that they consider them *enjay* 'alive' in any sense. Rather, they are places where they believed *towmow* sometimes took up residence for a while, and where they made offerings to them in times of sickness, when divination indicated that they were there, to persuade them to desist from attacking their kin and making them ill (Sillitoe and Sillitoe 2009). To insist on materiality's agency argument regarding local perceptions would be an ethnocentric imposition.

It is strange that those who draw inspiration from phenomenology, philosophy's response to the death of god (as discussed earlier, Chapter 2), should themselves turn to animism and smuggle supernatural intercession back in. While it may appeal to those of a metaphysical bent, who are happy to deal in ideas with no empirical confirmation, it is unlikely to impress those with a scientific background who demand factual evidence, nor in truth many persons inculcated into a non-animistic belief system whose life experience tells them that things are inanimate. In this respect, the discipline of archaeology shows dissociative-identity-disorder-like symptoms, with archaeometrists working in the natural sciences unable to see how a fetishist perspective can apply to their research that contributes so significantly to our knowledge of past environments (e.g. those engaged in luminescence dating of minerals in pottery shards or identification of ancient plant communities from pollen). When I asked such an archaeological colleague on my corridor what he made of the agency view, he quipped 'do they expect us to talk to knapped tools and pot shards!' This is a graphic example of the tensions mentioned in Chapter 2 and caused by the deep intellectual divide in Western thought traceable back to Platonic idealism and Aristotelian realism.

It is arguable that this is the wrong way to think of animism. It is possible that interpreting what people say literally, like the Victorian evolutionists, misrepresents their ideas and does their belief systems an injustice (Harvey 2005:179 *passim*; Santos-Granero 2013). It was Lévy-Bruhl (1926) who pointed out that so-called 'primitive' thought that focuses on the engagement of people with things highlights joint involvement in the world, wherein both human and non-human parties are caught up in the same processes. This subsequently led to the suggestion that that there are, at root, two human views of reality: one involves causality and the other participation (Tambiah 1990). According to the former perspective, humans are separate (epitomized in the nature and culture distinction) and act on the world; whereas in the latter view, humans are joined with it (epitomized in the relational ontology of animism) and share in the world. If we put animistic ideas within the context of wider socio-cultural practices, we see how local views of relations between persons and material things inform perceptions of place and extend to ecological perceptions of the environment (Bird-David 1999). In this event, those who subscribe to animistic ideas do not necessarily believe that things are literally alive but have a conception of the world more akin to Gaia, that all things are related in the greater global system.

An acknowledgement of the 'active power of materials' in this sense leads the 'ecological' approach to materiality to propose an even more radical interpretation of the agency of materials, seemingly combining a phenomenological worldview with the second law of thermodynamics (Ingold 2007, 2012 – citing Deleuze and Guattari 2004; Gibson 1979; Simondon 2005). This approach, focusing on the properties or affordances of materials, argues that 'Things are alive and active ... because the substances which they comprise continue to be swept up in circulations of the surrounding media that alternately portend their dissolution ... Spirit is the regenerative power of these circulatory flows.' (Ingold 2007:12). Process philosophy likewise argues against the idea of inert matter, pointing out that all entities are caught up in spatial and temporal processes. Its distinction between 'pan-experientialism', the idea that all things go through an existence, and 'panpsychism', the idea that all matter has conscious agency (Griffin 2001:97), avoids the accusation of subscribing to animistic-like beliefs.

A focus on the process of becoming as opposed to structure -- that is, that nothing is fixed -- became prominent in later twentieth-century Continental philosophy (e.g Deleuze and Guattari 2004; May 2003). From this perspective 'Bringing things to life, then, is a matter ... of restoring them to the generative fluxes of the world of materials in which they came into being and continue to subsist.' (Ingold 2007:12). It advocates an 'ontology that assigns primacy to the processes of formation as against their final products, and to the flows and transformations of materials as against states of matter' (Ingold 2010:92). This formulation seeks to override, with its Gaia-like living-planet focus, the distinctions between animate and inanimate, human and non-human or in previous formulations, culture and nature. It approves of Heidegger's (1971:165–82) distinction between object and thing (Knappett 2011), advocating the latter as necessary in the investigation of the material world. An object is stationary, something made and thought complete, defined by its 'overagainstness' (Heidegger 1971:167) or separateness from its natural surroundings, which we can handle but cannot engage with processually. A thing, on the other hand, comprises matter on the move such that we can engage with its flow or 'thinging' and the ongoing process of its 'becoming', with our movements in tune with its component elements through time (Harman 2010).

Arguing for the agency of materials beyond the agency of human actors further aligns the materiality approach with the 'incompatibilist' position of 'causal determinism' in the philosophical debate over free will, which maintains that the idea that humans are free to choose what they do is an illusion. What happened in the past, combined with the laws of nature, determines the course of the future; there is only one way things can go since the Big Bang when atoms and their particles (some of which comprise us) were set off on an unalterable course through space-time, which determines everything that has and will happen to both living and lifeless matter (Honderich 1990; Pereboom 2001; Strawson 1986). Further, adopting an animistic-like perspective to the agency of materials recalls the venerable theological philosophy of 'divine determinism', which likewise argues, for instance after Augustine, that to think that humans are free to choose their actions is mistaken, as a godly power determines all that happens and we can do nothing to alter our fate. The correspondence between the material-agency approach and 'incompatibilist determinism' is puzzling given that the materiality approach draws inspiration from phenomenology, which assumes human free will – notably evident in existentialist versions, where it is a central tenet. It suggests some profound confusion.

These are stronger determinist views than those of environmental determinism, which is the deterministic approach more familiar to anthropologists and archaeologists, as noted above, which may allow for acts of human free will hedged around by the limitations set by the environment. The incompatibilist and divine determinism positions are also quite at odds with Wola views, as evident in Was valley stateless political arrangements that extend liberty equally to all (Chapters 1 and 3). They differ too from the ethno-variety of determinism, which allows for the agency of humans and the idea of free will, and acknowledges that we are morally responsible for our actions. Ethno-determinism's holistic focus on cultural milieu, the seat of shared moral values, ensures that the morality of behaviour is a central issue. In the Was valley a code of reciprocity features prominently, where the equal exercise of free will is a key feature of acephalous political-economic relations, and power is uniformly distributed through the community.

From material flows to ecosystem cycles

In respect of houses, the materiality perspective comes down to distinguishing between 'objects' buildings and 'things' materials. The latter may comprise the former for an interval, until they decay and either fall down or are demolished. So what is new? It is obvious that buildings are in 'process', that they decay and eventually collapse, that they are

> part of the world, and the world will not stop still but ceaselessly unfolds along its innumerable paths of growth, decay and regeneration, regardless of the most concerted of human attempts to nail it down ... it is the process that claims their [builders and residents] energy and attention, more than the product. ... [in] their daily struggle to limit the damage inflicted by invasions of insects, and rodents, the rot of fungal infestation and the corrosive effects to the elements
>
> (Ingold 2013:48)

Many centuries ago, Alberti (1988:27) made a similar observation in his classic architectural work:[17] 'rain is always prepared to wreak mischief, and never fails to exploit even the least opening to do some harm: by its subtlety it infiltrates, by softening it corrupts, and by its persistence it undermines the whole strength of the building, until it eventually brings ruin and destruction on the entire work'. Furthermore, the suggestion that material culture studies do not encompass the cycle from raw material to manufactured object to discarded rubbish is also open to challenge. In some regards, these stages reflect the distinction between anthropology, which focuses on the making and use (or consumption) of things (Knappett 2005; Lemonnier 1986), and archaeology, which focuses on things when abandoned and in various stages of decomposition (Pollard 2004; Rathje and Murphy 1992).[18]

From the materiality viewpoint, we need to consider the life cycle of a building, which is 'always work in progress', focusing on its materials from the 'time it took for the trees to grow, the time to let their wood dry, the time to build with them and ... [when] the building has reached the end of its useful life ... the time spent putting away, taking out and pulling down' (Ingold 2011b:212, 217). The dense philosophical pitching of the argument hides the apparently unacknowledged geo-ecological-cycling focus. For instance, we are told that our concern should be 'matter in movement, in flux, in variation' and that 'this matter-flow can only be followed' (Deleuze and Guattari 2004:451), which amounts to 'a simple rule of thumb: to *follow the materials*' (Ingold 2011b:213). This affords an interpretation of Heidegger's (1971:181) obtuse reference to 'thinging from out of the worlding world' as alluding to the 'interweaving of materials in movement', which is part of 'the world's becoming ... energised by the flows' that feature 'processes of respiration and metabolism' (a nod finally towards ecology). In material culture studies it boils down to the investigation, in ecological context, of persons making 'objects' from 'things' or (raw) materials: 'neither brick nor mortar ... are object. They are materials' that people 'follow ... weaving their own lines of becoming into the texture of material flows comprising the lifeworld. Out of this emerge the kinds of things we call buildings.' (Ingold 2011b:214–15). What else are books such as *Made in Niugini* and this one about?

In pointing out that materials are subject continually to change, the 'ecological' approach to materiality comes down to tracking the shifting interactions between the constituents of things, in the final analysis down to the bioscientific level of molecular interactions.[19] This applies to animist ontology too, which is arguably analogous to the ecological model of the cycling of matter around an environment. Indeed it is plausible that a belief in 'spirit' forces animating the entire world equates with the scientific idea of 'energy' forces driving the universe. Physics defines energy as the capacity to do work (cause actions), and conceives as constant (while one form of energy may transform into another, none can be taken away or added, which suggests that it is eternal), and while we can identify different energetic interactions with matter (mechanical, chemical, nuclear etc.) and observe the outcomes, we can no more sense energy than we can a spirit. In technological contexts we firstly have persons fabricating, that is changing, materials into objects; and secondly those objects inevitably deteriorating with use; until thirdly abandoned to disintegrate when their constituents may transform into something other than when selected as raw materials: all transactions that involve energy.

In this approach to materials 'nothing is ever finished: "everything may be something, but being something is always on the way to becoming something else" ... Materials ... are substances-in-becoming (Barad 2003:822):

17 As Ingold 2013:49 notes himself.

18 And occasionally we link the three stages, particularly in using ethnographic evidence to further understanding of use life and discard practices, which are central to the accumulation of the archaeological record (Shott and Sillitoe 2001, 2004, 2005). In respect of houses, such research requires building a house and then waiting for it to fall down and rot, and then excavating the remains to obtain information such as presented in Chapter 4's seminar's 'excavated' house plan; the findings will differ depending on the environment, notably climatic, topographic and edaphic conditions.

19 This view of materials will hardly come as a surprise to those involved in conservation – for instance of ethnographic collections in museums – who continually battle to hold 'flow' in check.

They carry on, overtaking the formal destinations that, at one time or another, have been assigned to them.' (Ingold 2012:435). After reading this I thought of our pine kitchen table – at which we have entertained Tim Ingold and his family – made by joiners at Durham Pine Limited in the nearby village of Sherburn, which one day will be too battered to be serviceable and I may recycle by chopping up for firewood. The trees in the woodland adjacent to our home may take up some of the CO_2 released to produce new timber that may in turn serve as roof joists when felled and processed. Also, the porridge that I ate for breakfast this morning sitting at the table may contribute to the joists via the CO_2 that I am exhaling as I metabolize the oat carbohydrates to get the energy to live and write these thoughts.

The polished stone axe used by men in the New Guinea Highlands (Sillitoe 1988a:43–53) affords a particularly apt example of an object that exists in the 'process of becoming', its manufacture occurring in use, starting out as a roughed-out blade from a quarry where suitable metamorphic stone occurred and ground into a fully polished axe blade over time, often as it passed from hand to hand (both in trade and as an object transactable in socio-political exchanges). A stone axe was subject to further modification, even when a fully polished final blade, especially when the cutting edge was accidently chipped, demanding further grinding to restore it to usability (the same applies to the set of chisels that I inherited from my grandfather, which with carborundum honing are shorter than when he purchased them a century ago). This process, combined with the gradual erosion of stone in use, resulted in blades changing in shape and size until after many years of use they were too small to be serviceable any longer, when they were discarded for nature to finish off the erosion process, ultimately reducing the stone to mineral molecules for environmental recycling.

We may think of physical entities as both phenomenological 'objects' and 'things'. It depends on context, and to rule one out would impoverish our ability to comprehend and communicate about stuff. If we take something in the present, such as a Was valley house, we can speak of it as an 'object', complete and standing out there, which is what I understand by my friends' comments and behaviour when they refer to an *aenda* 'house'. But they are aware that over time, in the manner of a 'thing', it will change; a new *aenda* becomes increasingly tatty, pest infested and leaky, until too *mondowgay* 'dilapidated' to be habitable any longer. Life's experiences soon teaches people that artefacts can 'flow' or change over time from useful 'objects' today to useless 'things' tomorrow – a bow can break, the split-cane 'string' doing so regularly, and an arrow when fired may be lost in the bush and rot there. The fact that many of the artefacts documented in *Made in Niugini* are no longer in use underscores the obsolescence aspect of objects. Something exists right now as an 'object' or separate concrete phenomenon – the fact that it will change and become a 'thing' may only be an issue occasionally during its use life, probably at the end. If someone is shot with an arrow during a skirmish, for instance, it presents itself only too painfully to him as an 'object', the archer giving the palm-wood-arrow point the necessary energy/'agency' to intrude into the flow of his life, literally embedding itself in his living flesh. The wounded person will perceive of it as a brute foreign 'thing' to be removed immediately without any consideration for its 'thinginess' quality, which may involve it 'becoming' C-rich ash and CO_2 gas if thrown into a fire on removal,[20] something irrelevant to him at the time of undergoing agonizing battlefield surgery.

It is debatable what this entropic natural-cycling view of materials has to contribute to the ethnographic study of material culture. It is certainly alien to my Was valley friends who talk about raw materials in respect of the *maen* 'properties' that suit them to the objects they intend to make, whose thoughts may include a concern for *buriysha* 'durability' (as with timber for *polpen* 'roof rafters' not susceptible to *mbwiy* 'woodworm' attack), for they are well aware that objects become *hunjiy* 'old' and *korsha* 'wear-out' and have to be replaced. This is why they rebuild their houses every five years or so, as they become *modowgay* 'rotten' and too dilapidated to be habitable. But no one has spoken about the recycling of decayed materials, about them 'becoming something else'; if asked, they are more likely to talk of them becoming *buwmhaez* 'lost'.[21] By 'lost' they do not imply 'disappear', for people are aware, for instance, that rotting plant material accumulates to comprise *waip* 'litter' that contributes to the build up of *suw pombray* 'black soil'. The consequent high organic content of this dark top soil makes it rich in *suw iyba* 'soil grease' that contributes to soil fertility and good plant growth (Sillitoe 1996:165). We might arguably construe this as intimating a notion of recycling, if we think of potential future building materials such as bamboo growing in the soil.

If we follow what the materiality argument implies and adopt an ecosystem-cycling approach, as with Was valley farming practices for instance (Sillitoe 1996), we impose an alien intellectual construct on people's behaviour and interaction with the environment, in much the same way as we do in employing the idea of society (see Chapter 3), again appearing to know more about their place than they know

20 Another 'thing' fate for an arrow tip removed from a wound, on its 'flow' through the universe, is for it to be pushed into a piece of pig fat put in a short bamboo tube, from which it emerges as the fat rots puss-like, so encouraging the build up of puss around any splinters left in the wound to expel them (palm wood often splitting and leaving fragments).

21 Unless they salvage them still usable from an old house and reuse in building a new one.

themselves; although concerning phenomena 'out there', such understanding is open to objective confirmation. Nonetheless, it is legitimate to use the ecosystem-cycling approach to explain, in our scientific terms, how people interact with their environment, so long as we do not conflate it with local ideas. But we have yet to see studies applying the materiality approach, as opposed to discoursing on it, tracking the movement of the same molecules and elements through a series of cycles, which, as my mention of CO_2 indicates, would pose considerable methodological challenges.[22] There are other attempts to follow cycles in human ecology, notably focussing on livelihood regimes. Such an ecosystem-cycling approach features, for instance, in my enquiry into how people manage in the Was valley to keep some gardens under crops for decades in a nominally shifting-cultivation regime. The farming-system account presents the ecological approach that focuses on soil-plant-human relations (including the nutrient cycling that occurs with material growth and decay) in parallel with an ethnographic one that concerns Wola practices and ideas about their cultivation regime. It even includes an investigation of the remains of abandoned houses, the sites of which, when structures collapse or are burnt, frequently support small gardens. And the analysis of these places extends to the elemental level, as the 'ecological' approach to materiality intimates, documenting and discussing their nutrient status (Sillitoe 1996:392). Furthermore, these sites sometimes support clumps of bamboo, the split culms of which may find their way into houses of the future. And so on, with the elements continually cycling through the world.

Sustainability of resource use

In a macrocosmic sense, the Wola have an awareness of the cycling of matter, with respect to building materials such as lengths of timber and bark sheets. While some houses in the Was valley are abandoned and left to fall down and rot away, others are dismantled with a view to rebuilding on the same site. Some of the materials are durable and may last for many years (as elsewhere, Saini 1966). In this event it is not necessary to collect all the raw materials needed to build the new house. When they take down the old house – *aend kundiyay* 'house dismantle' – residents do so with a view to recycling whatever materials are salvageable. It took Mayka Pes and his wife Nonk (see Chapter 7 and Figure 7.4 for more on these persons) about three hours to dismantle their old *enza tenda aenda* 'men's and women's house'. They estimated that the old house had stood at Tombem for 5 years and they took it down because it was no longer sufficiently weatherproof and was infested with fleas.[23] The family had moved out of the old house about 3 months previously, after Pes built a single roomed *momaenda* 'family house' nearby – this house subsequently became a pigsty and day room for his wife and children. They carefully laid out the reusable materials around the site, while burning all that was rubbish, such as rotten cane-grass thatch (Meggitt 1957:168 and Steensberg 1980:146 also give examples of Enga and Duna reuse of building materials). They salvaged and reused the following raw materials, which comprised much of what was required for the new house:

- *aend dol* 'wall stakes';
- *aend buwk* 'bark wall sheets';
- *aend gamonk* 'wall battens';
- *haeluwp ya* 'rattan vine';
- 2 *aend bez* 'ridge-poles';
- 24 *aend polpen* 'rafter poles';
- *paym koma* 'roof lathes' of pandan roots;
- *paym shor* 'pandan leaves' for wall lining;
- *wenet humbiy* 'engel bark' for the eaves;
- *gaimb* 'cane grass', some salvageable roof thatch.

Pes had to cut some new wall stakes to replace rotten ones, collect some new rattan and bark-lashing materials, make some new roof lathes and arrange for some relatives to help him cut and collect some fresh cane-grass thatch.

The recycling of materials relates to the sustainability of resource use. Until recently, the residents of the Was valley were self-sufficient. All constructions were built entirely from locally available renewable materials, and their use was compatible with the region's montane ecological rhythms. We arguably have some important, indeed urgent lessons to learn from such people's relations with their environment (Bay and Ong 2006; Heath 2009). When abandoned or demolished, structures rot away entirely, as noted with respect to the materiality-cum-affordances debate. They feature no foreign non-biodegradable materials and decompose back into the natural surroundings with no pollution effects or irreversible damage. It is common, as commented in the opening chapter, for vernacular building traditions to be similarly in tune with local environmental cycles. The clay-brick buildings of Mali, for instance, which are often taken as archetypical vernacular architecture, which 'when left to decompose … almost poetically, return to the land from which they came' (Marchand 2009:42). While this is good for the environment, it is not so good for archaeology, to harp back to the previous chapter's

22 After the biochemistry of photosynthesis, for instance, which is among the most elegant of scientific findings (the Calvin cycle up there with the Krebs cycle and DNA double helix), contributing to a powerful explanation of the cycling of material elements – such as the CO_2 molecules from my kitchen table turned to firewood.

23 People show considerable tolerance of *erel* 'fleas', but are less tolerant of *kiyaem* 'bedbug' infestations, which are likely to prompt the abandonment or dismantling and rebuilding of houses.

discussion, leaving little in way of occupational evidence other than 'soil shadows' where post-holes were (Joyce 2012) and the occasional material remains, often refuse (Rathje and Murphy 1992).

The question of human-environment relations in the New Guinea highlands is a fraught one. Since their arrival, some outsider commentators have argued that the local population is degrading the environment, particularly with population growth and intensification of farming practices (Allen 2001; Allen and Crittenden 1987; Brookfield 2001; Brookfield and Brown 1963; Hanson, Allen and Bourke 2001; Wohlt 2004). While it is beyond question that human activities have impacted on the region's environment for millennia, as they have across the planet (for instance, there is evidence of manipulation of the highlands' environment through proto-farming management of natural resources 10,000 years or so ago), it is debatable whether such environmental modification to meet human needs, continued currently by the heirs of this ancient farming tradition, is resulting today in degradation, as opposed to a managed alteration of the land and resources to supply people's wants.

It is usual to equate degradation with unsustainable activities damaging environments irreversibly, and not with human actions that modify them, some change being inevitable, as with any other animals' activities. We are not somehow apart from nature, whatever our causality view of reality may lead us to think. The issue is the extent of the change and whether it imperils natural processes such that not only are they unable to function as previously (with the possible extinction of some species etc.), but also that their dysfunction degrades the system's productivity, particularly with respect, from a human viewpoint, to our needs. An ecosystem is sustainable for the foreseeable future so long as the relationships between components, such as their cycling around it, remain in equilibrium; that is, it will continue as it is without structural change due to depletion or degradation of resources. To paraphrase the widely cited so-called Brundtland Report (World Commission on Environment and Development 1987), an unsustainable arrangement is one where current activities compromise the environment such that future generations cannot meet their needs.

It is questionable whether current human activities in the Was valley are so degrading the environment that they threaten to compromise the natural resources that future generations will inherit, depriving them of the wherewithal to meet their needs (Sillitoe 1996; 2010a). But amassing data on human-environment relations to support this in-ecological-balance viewpoint is difficult. It requires assessment of the limitations that the environment sets regarding present and future needs in the context of current technology and social arrangements. One approach is to calculate carrying capacity, a concept that comes from ecology, which indicates the maximum density of population that available natural resources can support using current technology without degradation and loss of productivity long term (Dewar 1984; Ohtsuka 1995). The core of the carrying-capacity idea is that an ecosystem is sustainable so long as population remains within certain limits, first hinted at by Malthus and now central to the popular ecological footprint metric (Wackernagel and Rees 1996), which is a version of carrying capacity widely used in discussions of sustainability (to give attention-grabbing headlines that some believe necessary to reverse despoliation of the planet, such as we need five planet Earths if all of humanity is to live at the level of US citizens currently).

The accurate calculation of carrying capacity has however proved difficult in the Was valley, various aspects of cultivation practices compromising measurement of the environmental and demographic variables (Sillitoe 2010a:196–215). In view of the problems that attend its calculation, such as assessing fallow intervals, I have suggested that the availability of certain key raw materials used in the manufacture of things might serve as a proxy for carrying capacity (Sillitoe 2010a:244). Tracking the fate of marker species that are in heavy demand, such as rattan, may be a sounder approach to assessing sustainable use of resources and environmental limits. This is to draw on the ecological concept of 'keystone species'. These are species that play a significant role in maintaining the balance between all the other species that comprise an ecological community, disproportionately influencing community structure compared to their abundance. The concept is popular in conservation, where particular efforts are made to protect such species in threatened communities in view of their effect on species diversity. While its validity is debated in ecological contexts (the complexity of the structure of many natural eco-systems such as rainforests making it difficult to identify such keystone species with confidence), we can apply it in a modified form to the availability of resources used by humans because people can tell us what resources are important to them and they can assess their scarcity, something that we can subsequently measure through surveys of plant/animal communities.

It is possible to assess the sustainability of building practices where all the materials traditionally used in house construction are locally available and renewable. We can compare the supply of widely used or key raw materials in the Was valley with their rates of collection and use, which will relate to population and demand for houses, and so assess the environmental impact of building activities. The data in Table 6.9 indicate the availability of the raw materials used in house construction. They give the occurrence of materials in the various vegetation successions outlined previously, which mirror Wola plant-community categories (Sillitoe 1996:182–200). These data show the commonness

Wola name	Common name	Botanical name	Rainforest Mean plants 100 m^{-2} ($quadrat^{-1}$)	Secondary woodland Mean plants 100 m^{-2} ($quadrat^{-1}$)	Cane grassland Mean plants 100 m^{-2} ($quadrat^{-1}$)	Gardens[1] Mean plants m^{-2} ($cast^{-1}$)	Abandoned gardens[1] Mean plants m^{-2} ($cast^{-1}$)	Wola assessment
PANDANS & PALMS								
aenk	karuga screw-pine	*Pandanus julianetti*	0.64	0.05	0.85			abundant
goiz	palm	*Gulubia* sp.						scarce (rare in gardens)
keret	screw-pine	*Pandanus adinobotrys*	2.6					occasional
mbet	betel palm	*Areca* aff. *macrocalyx*						occasional
nort	screw-pine	*Pandanus concavus*						
paym	wild karuga screw-pine	*Pandanus antaresensis*						occasional (rare in grassland and houseyards)
pundin	screw-pine	*Pandanus archboldianus*	1.52					common
GRASSES								
gaimb	cane grass	*Miscanthus floridulus*	0.04	1.1	20	0.19	0.1	abundant (scarce in forest and swamp)
gaimb henj	cane grass	*Miscanthus floridulus*						
senz	kunai grass	*Imperata conferta*			6.2[2]			scarce (abundant in some abandoned gardens)
taembok	bamboo	*Nastus elatus*						scarce
OTHER								
haeluwp	rattan	*Calamus fuscus*	0.08					common
hathaen	stones	Limestone						
henk	tree fern	Cyatheaceae (various species)	19.44	6.4	6.9	0.55	0.08	common
homb	sphagnum moss	*Sphagnum nov-guineense*						occasional
kwimb	moss	Frullaniaceae (various species)						abundant
tedbel	gingerwort	*Alpinia* sp. (sect. Dieramalpinia)	3.72					

1 Where trees occur in gardens they are seedlings and where they occur in abandoned gardens they are saplings.

2 This occurrence related to rockland communities in the survey of the composition of plant communities.

Table 6.9 Occurrence of raw materials

Wola name	Common name	Botanical name	Rainforest Mean plants 100 m^{-2} (quadrat^{-1})	Secondary woodland Mean plants 100 m^{-2} (quadrat^{-1})	Cane grassland Mean plants 100 m^{-2} (quadrat^{-1})	Gardens[1] Mean plants m^{-2} (cast^{-1})	Abandoned gardens[1] Mean plants m^{-2} (cast^{-1})	Wola assessment
TREES								
bat	silkwood	*Cryptocarya* sp.	0.12					
bobae	spurge	*Macaranga* sp.	0.08	0.5	0.35	0.02		limited (common in secondary woodland)
bort	silkwood	*Cryptocarya laevigata*	0.04	1.7	0.25	0.06	0.2	limited (common in secondary woodland)
en	spurge	*Macaranga* sp.	0.68	0.25	0.1			scarce
gwai	switchsorrel	*Dodonaea viscosa*		0.85	0.1	0.11	0.04	common (scarce in grassland)
haezuwmb	white magnolia	*Galbulimima belgraveana*	0.92					occasional
hombolem	gazelle cherry	*Prunus gazelle-peninsulae*	0.6					scarce
huwshiy	cunonia	*Caldcluvia papuana*	0.36	0.2	0.1			occasional
inom	stinging tree	*Debregeasia* sp. / *Maoutia* sp.	0.08	0.75	1.2	0.17		limited (common in secondary forest)
kaeriyl	oak	*Lithocarpus rufovillosus*		0.1	0.55			common – especially where rocky
kol	spurge	*Phyllanthus* sp.	13.6	0.15	0.15			limited (rare in grassland)
morowa	cypress	*Papuacedrus papuanus*						scarce (rare in houseyards)
naep	casuarina	*Casuarina oligodon*						abundant (largely cultivated but many are self-sown)
pay	chestnut oak	*Castanopsis acuminatissima*						common (limited in rainforest)
pel	southern beech	*Nothofagus starkenborghii*	5.04	0.1	0.1			abundant
romiya	melastome	*Astronia* sp.	0.44					
sabok	sabia	*Meliosma pinnata*			0.15			scarce
saemow	mimosa	*Albizia fulva*						scarce (common in secondary forest)
serep	balubug	*Platea excelsa*	0.72	0.05				
shongaen	spurge	*Macaranga pleioneura*	0.04	0.45	0.05	0.05		limited
shonon	chenille	*Acalypha* sp.		0.2				occasional
shuguwl	mangeao sp.	*Litsea irianensis*						
shwimb	Indian bead ash	*Elaeocarpus dolidrostylus*						common (scarce in secondary woodland, and rare in houseyards)
timbol	spurge	*Homalanthus novoguineensis*	0.44	2.6	0.35	0.07		limited (common in secondary woodland and scarce in grassland)
waen	woolly cedar	*Trema orientalis*		5.5	0.1	0.42	0.1	common (scarce in grassland)
wenet	engel	*Engelhardia rigida*						common (scarce in grassland)
wok	gamboge	*Garcinia* sp.	4.48					occasional

of construction materials, both in quantified terms and as the Wola perceive them. The former information comes from a quadrat survey of the various vegetation communities found in the Was valley, conducted to gather quantifiable data on their floristic composition (see Sillitoe n.d. for further information).[24] It shows the frequency of occurrence of the various plants, which is an indication of resource scarcities. The local people's assessments of the availability of the materials complement these plant survey data.[25] Respondents maintain that none are in short supply, which complies with the observations. While some occur rarely, they exist in sufficient amounts to meet demand, supplying sufficient materials for the construction of houses adequately to satisfy all needs for shelter. Indeed if overuse and scarcity of some raw materials was an issue, we might expect some tightening up of current relatively casual attitudes to rights of access.

We can estimate the abundance of marker species knowing their occurrence from the quadrat survey (Table 6.9) and the area of the territory exploited for raw materials by a given population (Sillitoe 1996:181).[26] If we take rattan as one of our marker species, we have a quadrat occurrence of 0.08 plants per 100 square metres, which for the territory of Aenda and Ebay *semonda*, where *iyshabuw* forest covers 32.8 km², results in a total of 26,240 plants.[27] We know the amounts of materials used in building a house (Table 6.2). The quantity of rattan used in the construction of the Honael house was 7 stems totalling 75.34 m in length.[28] We can then estimate the annual level of rattan demand from the number of houses on the *sem* territory and the rate at which new ones are built, with the average life of a house being 5–10 years. The number of houses occupied by people on Aenda and Ebay *semonda* territory averaged 112 over four censuses (Wolaland.org; Sillitoe 2010a:180–6; Chapter 3)[29] and if we take the average life of a house to be 7.5 years, this gives us a mean building rate of just over 15 houses a year at recent population levels and the associated rate of house construction. This gives a rattan exploitation rate of 105 stems annually, which is well within the forest's ability to supply rattan sustainably.

In order to have an idea of the total demand for rattan – or any other raw materials, such as target timbers – it will be necessary to document the amounts used in other activities, such as the erection of garden fences, bridge construction and the manufacture of artefacts (which previously involved the use of locally available materials largely – except for some imports from elsewhere such as seashells and stone axes). These data are available (Sillitoe 1988a, 2010a and Chapter 9); they include the lengths of fence erected around gardens (from which it is possible to calculate rates of rattan use lashing stakes together),[30] the amounts of vine required to build bridges and to make the various artefacts in which rattan features (the amounts of rattan used, the number of artefacts owned and their average use life). According to these data (Table 6.10), the total amount of rattan previously harvested and used in a year by the Aenda and Ebay *semonda* community was just over 3,750 m; these data overestimate current-day rattan demand by about ten per cent, as many artefacts are no longer made.

These combined data indicate the impact of construction, manufacturing and cultivation activities on the natural environment and resource availability, as assessed according to exploitation of the marker species rattan by a certain population occupying a known territory over a given period of time. But in order to assess the

24 The dimensions of the quadrat areas surveyed varied, as the table indicates, according to the size of the vegetation making up the communities. They were ten metres square where the vegetation included large plants (*iyshabuw* forest, *obael* woodland and *gaimb* cane grassland), and one metre square where the plants were smaller (*em* gardens and *mokombai* abandoned gardens); all of the plants occurring in the demarcated areas were counted.

25 These assessments come from some men I asked to judge the relative abundance of the raw materials according to the following six categories (Wola equivalents in brackets): 1 – abundant (*onduwp ora*); 2 – common (*onduwp*); 3 – occasional (*onduwp sha*); 4 – limited (*gənk sha*); 5 – scarce (*gənk*); and 6 – rare (*gənkden ora*).

26 This calculation assumes that the raw materials collected from territories elsewhere by persons living on the territory surveyed balance the raw materials collected by those with rights to collect materials on the territory surveyed who themselves reside elsewhere – that is, we have quits.

27 This figure of 8 plants per hectare is possibly on the low side with rattan probably occurring more frequently in remote, infrequently visited and exploited forested regions; data on rattan abundance elsewhere suggest that this may be so – for instance Siebert (1993) reports 86 and 284 plants per hectare for two *Calamus* species in Indonesia (albeit such comparative data are difficult to assess, concerning different species growing in National Parks at lower altitudes).

28 The assumption here is that this figure is representative of houses generally – it would, of course, be sounder to obtain an average by measuring the amounts of rattan used in building a representative sample of different houses, including some reuse of rattan that would reduce the harvesting rate.

29 The numbers of houses were as follows: 1973 = 107, 1976 = 111, 1978 = 132 and 1983 = 96 houses; the fluctuation over a decade indicates the variation that we can expect in resource use and the caution necessary in interpreting such calculations. In 1983 fewer families maintained second houses in remoter downstream forest regions.

30 Clearing areas for gardens is unlikely to result in any more loss of rattan – or other useful materials such as good timber – because those that occur on land cleared for cultivation are likely to be used in fence construction.

	Length of rattan used per item (cm)[1]	Constructed/made per annum	Rattan used per annum (m)
Construction activities			
house construction	7534	15	1130.10
bridge construction	25000	~1	250.00
garden-fence erection	50 per 0.50 m fence	3141 metres	1570.50
Artefact manufacture			
stone-axe binding	1197	26	311.22
ornamental stone axe	1783.5	0.14	2.50
dog's hunting collar	43	0.8	0.34
arrow ferrules	77.5[2]	6.9	5.35
shields	170.5[2]	2	3.41
head decoration	310.5[2]	0.3–5	5.05
neck ornaments	369[2]	1.3–13	31.90
limb decoration	1242.7	1.3–53	242.71
girdle-straps	42	14.4	6.05
musical instruments	506[2]	1.2–6.3	18.85
total			**3577.98**

1 Average where category includes several items.

2 Length of split-rattan strand used, not entire stem (remainder may be used for another task or discarded).

House construction, data this study. Note that house construction overestimates annual amount of rattan used because it treats all house types as if use same amount of rattan as *enzatenda*, when many are smaller and use less rattan in construction.

Bridge construction, data this study. In chapter 8 I guesstimate the vine used per annum in bridge building to be 750 metres. But some of this will comprise vines other than rattan, perhaps one-third. Also, those living on the other side of the river will contribute materials from their territory and labour too in bridge construction and maintenance; it seems fair to assume half each over time.

Fence construction, see *From Land to Mouth* (Sillitoe 2010:139–45, 271–4). The total length of *em paip* garden fence, with vine lashed stringer sapling along the top edge enclosing the cultivations of Aenda and Ebay *semonda* families surveyed in 1976 (Sillitoe 2010:178–86), was 9842 metres. The length of fence enclosing gardens cultivated for the first time was 4150 metres (which taking the average time between cultivations to be eighteen months means that men construct two-thirds of this fence in a year), the length enclosing gardens cultivated twice was 2266 metres, three times 770 metres, four times 644 metres, ≥5 times 1012 metres and ≥10 times 1000 metres (the figure in the table is calculated assuming that all the vine on a fence is replaced after every five years and so includes these last two lengths). The vine is twisted around hewn stakes and sapling stringer such that twice the length of vine is used to secure any length of fence, but this difference is offset with men splitting rattan into two or more strands (depending on stem size), and the calculation in the table assumes that length of fence equates with length of vine used to secure it. A further complicating factor is that men use other vines too, in addition to rattan, and the guesstimate used in the table calculation is that fifty per cent is rattan.

Data on artefact manufacture, see *Made in Niugini* (Sillitoe 1988).

– Arrow ferrule data calculated for plain-pointed arrow (black-palm wood), broad-blade arrow, rounded-blade arrow, fine-blade arrow, claw-tipped arrow and bone-tipped arrow (short-pointed and long-pointed versions), using both stone and steel tools in manufacture.

– Stone-axe binding estimates that entire binding is renewed every two years, and ornamental stone-axe binding (data for both stone- and steel-manufactured) on a guesstimated use life of 50 years (use life is considerable as they are stored in men's houses and only brought out for events).

– Rattan used for slings of wooden shoulder and under-arm shields, and slings and binding of bark shoulder and hand shields. Rate of manufacture and rattan use based on a community involved in armed hostilities every fifteen years (Sillitoe 2010:238–9) and half the men making and using new shields.

– Head decoration includes: burr wig, feather-circlet headdress, sprung-feather headdress and enamelled-bird-of-paradise headdress.

– Neck ornaments include: cowrie-shell necklaces (front-mounted and side-mounted) and shell disc necklace.

– Limb decoration includes: twisted armband, entwined armband, woman's armband, Huli style armband, woven armband and woven wristband.

– Musical instruments are drum and panpipes.

– Items omitted from artefacts list are fire-lighter and bow-string, on which rattan only used as a temporary measure, and cane bag, which is normally woven from split bamboo stems.

Note that many of these data are of historical interest only because people no longer produce many of the artefacts and their impact on the environment as measured by rattan use has consequently decreased noticeably. Subtracting these items from the above table reduces the total annual use of rattan by nearly 10 per cent to 3201.11 m. It is probably smaller still, with men increasingly using large nails in house construction, particularly of *mbolinaenda* style houses, in preference to rattan.

Table 6.10 Annual exploitation of rattan.

sustainability of this level of exploitation – as an alternative to carrying capacity – we need some idea of the rate at which plants replace themselves whatever the level of harvesting pressure on them. According to Bøgh (1996), three different *Calamus* species in Thailand had average growth rates of 1.24, 1.42 and 0.20 m per year; he notes that light conditions were responsible for the considerable variation. And Milner-Gulland and Mace (1998:181) report that *Calamus* canes grow 15–20 m in 10 years and can be harvested biannually from that age. Let us assume that rattan has an annual growth rate of 1.0 m, which is probably a conservative estimate with the Wola leaving the base of the plant intact to grow further stems when they cut palms down (regrowth is consequently rapid, as it is not dependent on seed propagation alone; although conversely, the palm may grow more slowly at higher altitudes). This gives us a growth rate of 26,240 m of rattan every year, and if we assume that the average length of palm stem collected is 10 m, we have 2,624 stems available annually for harvest sustainably. The impact of people on their environment as measured by rattan use is therefore relatively modest, with one-eighth of the sustainable take-off being harvested (particularly when we take various caveats into consideration, with guesstimates of usage erring on the generous side).

These findings compare with those for land use and demographic trends – assessed in the context of a discussion of carrying-capacity limits (Sillitoe 2010a:175–252). They all indicate that the exploitation of the environment is sustainable in the Was valley so long as population remains within certain limits. While the rates of renewal of many resources are unknown, for many building materials (e.g. saplings, cane grass, rafter poles, bamboo) it is likely to be in the lifetime of a house; whereas for others it will be considerably longer (e.g. the dominant southern beech, which occupies an important place in the forest's ecology, lives for 500 or more years). There are other problems with assessing sustainability this way, reminiscent of those faced calculating carrying capacity. One is that people often reuse some materials from dismantled houses when building new ones, as described above, such that their use life may extend to decades (Saini 1966). While it is feasible that we could collect data on rates of reuse, these are highly variable and would demand the survey of a sizeable sample of houses under construction to achieve a reliable estimate. Another issue that compromises such calculations is the increasing use of imported manufactured materials in house building – such as metal nails, corrugated iron, sawn timber and plywood (Chapter 8). This applies also to the making of artefacts generally, with manufactured imports increasingly substituting for things – such as items of clothing and adornment, people now preferring T-shirts, blouses, shorts and baseball caps. The fact that many things are no longer made and used significantly reduces demand for some local materials, such as rattan and bark fibre.

These technological innovations, resulting from contact and commerce with the world outside the Was valley, relate to the thesis widely attributed to Boserup (1965), that human ingenuity will overcome environmental constraints, albeit this has occurred via diffusion in the Was valley before degradation and shortages became an issue locally. The implication is not that exploitation of resources has become more sustainable; rather, viewed globally, the Wola are exporting some of their environmental footprint elsewhere. This relates to appraisals of the 'technological fix will save us' argument that point to disturbingly highly levels of environmental destruction caused by associated substitutable industrial technology, with possibly irreversible climate change and so on. These critiques indicate that we have something to learn about sustainability from arrangements such as feature in house building in the Was valley (where the use of non-renewable materials remains low today and there are no wasteful mounds of construction debris, causing pollution problems, with toxic matter such as asbestos and lead).

It is evident that sustainability is a complex idea, which ultimately requires us to see the world as an integrated ecosystem connected through space and time, in the way that the materiality perspective suggests. There are intriguing echoes here with the phenomenological notion of dwelling in, as opposed to on, the world (Heidegger 1993; Ingold 2000). The local problems encountered in assessing the sustainability of human-environment relations in the Was valley mirror those at the global level. Even the idea of sustainability is disputed. What timescale are we talking about, for instance, as palaeo-geographical evidence indicates that environments will change regardless of human activities, and what may be a sustainable practice today will not be so in the (distant) future. While the calculations are somewhat gross, making assumptions based on averages drawn from small samples of data, they give some idea of the Was valley ecological footprint. After all, the global footprint calculations, while good for attention-grabbing headlines and provoking discussion among students, are based on considerably more dubious data (Franz and Papyrakis 2011).

7

House construction

A how-to-do-it account and critique

When house-builders have amassed all the materials they judge they will need, the next task is to prepare the house site prior to starting construction.[1] Site preparation may involve some considerable work if they are building at a new location where no house has stood previously. The preparation of such a site may demand heavy earth-moving work, as the sloping terrain of the Was valley often makes it necessary to excavate and move considerable volumes of soil to produce a level platform (albeit men will only seek to level land on relatively gentle slopes, steeper ones being liable to landslips in heavy storms; see Chapter 3). Flatter land is usually too low-lying and often waterlogged. They are likely to have to cut down and clear the vegetation covering such a virgin site to start with, doing so in much the same way as they chop down vegetation when establishing a garden (Sillitoe 2010a:261–6). Today, they excavate and level a house platform – *aend suw tombay* 'house soil dig' – using spades (previously they used large *olon gəmb* 'man's digging sticks' – Sillitoe 1988a:74–6). They may use their hands to pick up clods and clear them away. The usual procedure is to deposit excavated soil behind you, so to speak, that is downslope and thus build up part of the platform in addition to digging it out of the slope. The sticky clay subsoil lends itself to the building up of such areas, with men consolidating the clod-built part of the platform by stomping on the clods at intervals to compact them down. The intention is not only to produce a level and clear surface to build on, but also to excavate an area down to the firm clay subsoil so as to ensure a sound foundation for the house.

The house that we built at Honael (which supplies the quantitative data on the logistics of house construction discussed in this chapter) required the levelling of a new site. We agreed the area that we needed to clear and even up, and I marked it out, dividing it in half with a length of vine and stipulating that on one side the builders could work with spades but that on the other they could only use digging sticks and hands (Figure 7.1). After burning off the sun-dried cane and fern vegetation previously cut down on the site, and levering out the roots,[2] they proceeded to level the area. They excavated 13.22 m^3 of soil out of the slope and dumped the spoil downslope to build up a house platform of 8.24 × 7.02 m (Figure 7.2).[3] Finally, they carefully levelled the site by skimming off thin layers of clay – referred to as working the *waenuw* 'forehead'. The time efficiency of steel spades compared to wooden digging sticks was 1:1.38 (Table 7.1); that is, for every hour worked using spades it was necessary to work an hour and twenty-two minutes with digging sticks to prepare the same area (1.8 m^2) of house site.[4]

If men rebuild a house on a previously occupied site, which they often do (it is probable that some house sites have been in use for generations), then preparation of the site prior to starting construction may entail relatively little work beyond clearing it up. They will ensure a clean and even surface by lightly skimming off the top layer of clay with a spade or, previously, a large digging stick. They commonly deposit the soil, mixed with the layer of organic refuse that built up inside the previous house, to one side, adjacent to the site, often combining it with the ash resulting from

1 For comparative ethnographic data on house building in the mountains of New Guinea, see Clarke 1971:100–6; Heider 1970:252–71; Meggitt 1957; Pospisil 1963:257–63; Steensberg 1980:124–84.

2 Maenget Saendaep, whose family was to occupy the house when built, had previously cut down the *gaimb* 'cane grass' and *saezuwp* ferns covering the area when he cleared a nearby garden (the likely time taken to clear the area is calculated using data collected at Ganonkiyba during a time-and-motion study of garden establishment – Sillitoe 2010a:261–6).

3 They excavated the wedge of soil upslope from the dotted line A–B and deposited the soil downslope, to give the platform area. The wedge drawn in dotted lines represents soil excavated outside the markers (that is, surplus to requirements) before I spotted what was happening (in other words they dug out 0.62 m^3 of extra soil on the spade side).

4 Discounting for the time spent digging out the extra 0.62 m^3 of soil using spades makes a trivial difference to the comparative efficiency, increasing the ratio to 1:1.41, which is one hour twenty-five minutes of digging-stick work to achieve one hour of spade work.

Plate 7.1 The Honael house site before any levelling (surrounding cane grass cut down in clearing a garden).

Plate 7.2 View of Honael house site during levelling work.

Plate 7.3 Levelling the Honael house site.

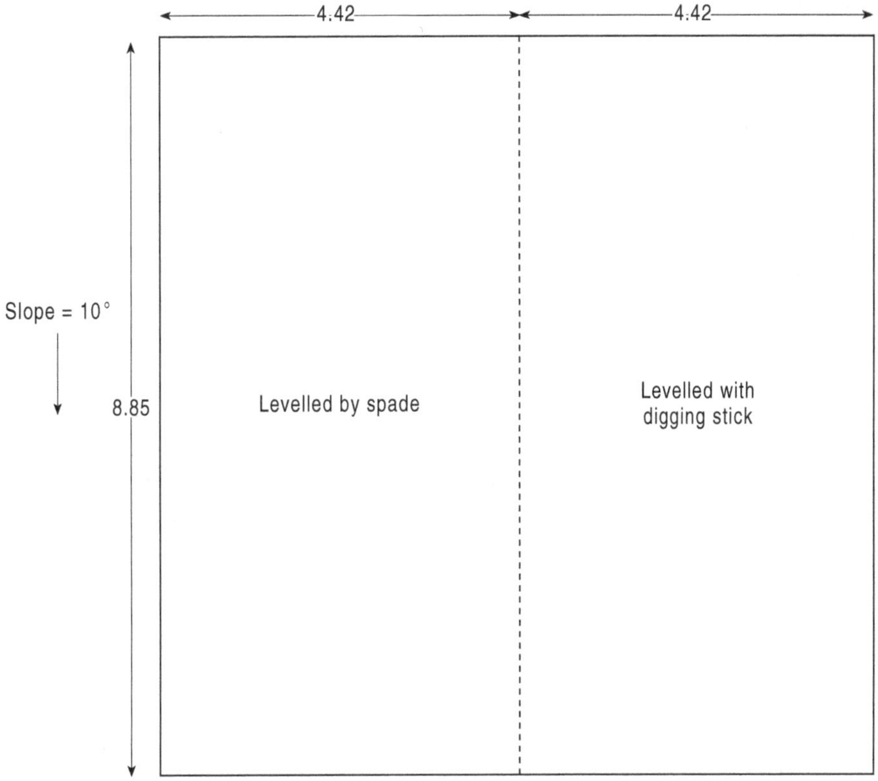

Figure 7.1 The area levelled at Honael for house site (m^2).

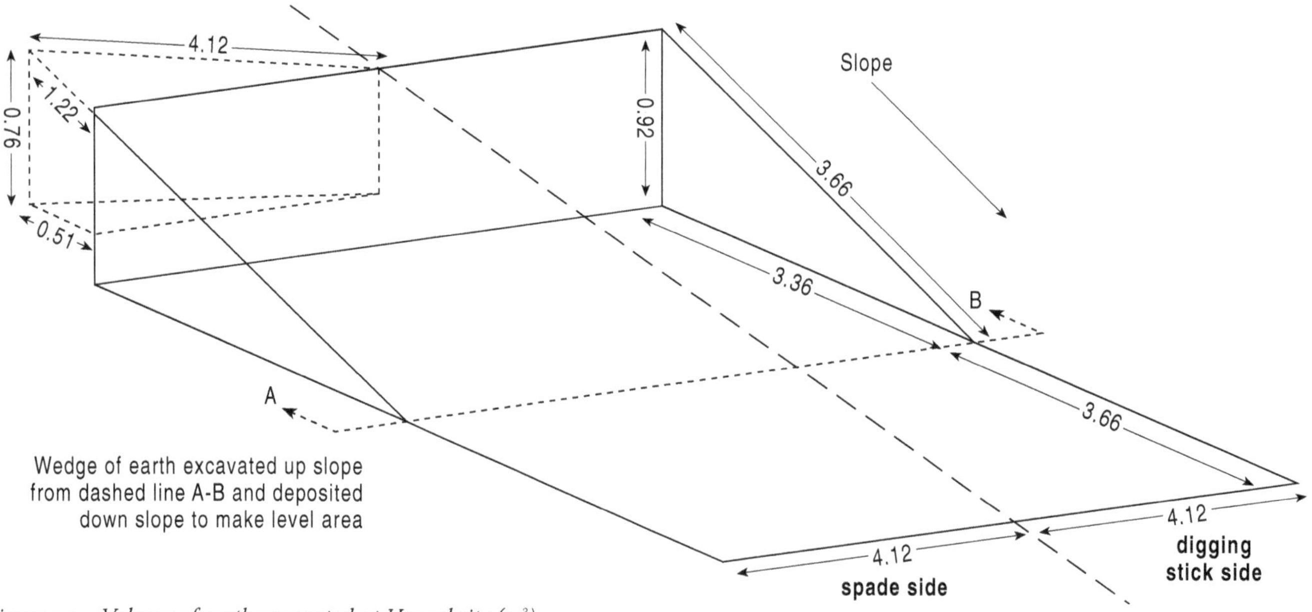

Figure 7.2 Volume of earth excavated at Honael site (m³).

		Time taken[1]	
Tasks	**Wola term**	**Stone tools**	**Steel tools**
cutting down cane grass (*gaimb*) and ferns (*saezuwp*)[2]	*aendtay karayay*	4 h, 8 m, 5 s[3]	1 h, 10 m, 55 s[3]
burning off dead ferns and cane grass	*gaimb haeray*	1 h, 17 m, 30 s	1 h, 21 m, 30 s
levering out roots and	*gaimb tay aezay*		
feeding fires with rootstocks and	*imbil haeray*	3 h, 6 m, 30 s	4 h, 59 m, 45 s
throwing other roots downslope	*tay tuway*		
digging out site[4]	*suw tombay*	8 h, 29 m, 45 s	4 h, 0 m, 30 s
final levelling of house site prior to construction	*waenuw lor payay* [or *waenuw waeray* or *waenuw tombay*]	4 h, 43 m, 15 s	4 h, 13 m, 30 s
total time		**21 h, 45 m, 5 s**	**15 h, 46 m, 10 s**

1 The work detailed on this table took place over 11 days.
2164.84 m^2 cleared (of which 78.14 m^2 was the ground area of the house); half area calculated as cleared using stone tools and half with steel.
3 Estimated times from Ganonkiyba garden clearing data (Sillitoe, Stewart and Strathern 2002) where 1 m^2 was cleared in 3.01 minutes using stone tools and 0.86 minutes using steel tools.
4 Volume of soil excavated on stone side 6.3 m^3. Volume of soil excavated on steel side 6.92 m^3.

Table 7.1 Clearing house site.

the burning of the rotten and un-reusable materials from the old house to create a fertile bed for an *em gəmb* 'small mixed vegetable garden' (Sillitoe 1996:392). These plots are productive and support a range of crops, such domestic refuse producing a good friable tilth. If, on the other hand, they abandon a house to rot and collapse, or burn it down without rebuilding on the site, they may likewise establish a garden on the organic rich remains (Sillitoe 2010a:16–17).[5]

Cooperation in house building

In preparing the site for a house that he or his family will occupy, a man often works alone – for instance, if building a *tenda* 'women's house' for his family to occupy, which is the concern of no one else, as opposed to an *enza* 'men's

5 Not all ash promotes soil fertility. People refrain from putting the *mem* 'ash that builds up in domestic fireplaces' onto gardens, instead tossing it away into surrounding undergrowth, because they say that the *buriy* 'strong' hard wood burnt on fires results in 'strong' ash, which mixed with soil makes it dry and hard, whereas *tomiy* 'soft' plant material such as grasses and cane give 'soft' ash that improves soil conditions and benefit crops.

house' that several male relatives may reside in. It may take several weeks to level a house site working alone, with men engaging in the heavy work periodically in bursts. Similarly, an individual may collect together the materials needed to construct a house working alone, again possibly doing so over an extended time period. Such self-sufficiency is common and a prominent feature of Wola life. It occurs also in the cultivation of swidden gardens and the manufacture of many artefacts. In discussing the trend in *Made in Niugini* (Sillitoe 1988a:563–5), I refer to it as the 'user-makes' principle, to convey the image of families able independently to produce all they need to subsist as comfortably as anyone else, structured according to the sexual division of labour arrangements (see Chapter 10). It is a well-known arrangement in subsistence communities, for instance Adam Smith (1993:188) noted long ago that 'in a rude society there is a good deal of variety in the occupations of every individual... Every man does, or is capable of doing, almost every thing which any other man does, or is capable of doing', although he was largely unaware of the implications.

The independent 'user-makes' ethic relates to equality that is a central feature of the constitution of the stateless political-economy (see Chapters 3 and 10). Every family has the opportunity and capacity to supply itself with the things that it needs, having equal access to all the materials, necessary skills and technical wherewithal required to meet its livelihood wants, such as houses to accommodate its members. These autarkic arrangements, where households are able to supply themselves adequately, are integral to the egalitarian order in which none can exploit more resources than others, control labour or access more means and so control the 'economy' and accumulate extra assets and power (Sillitoe 1988a:40–2). Furthermore, the things that circulate in the socio-political exchange arena (where persons compete for social status and influence) are disassociated from livelihood production activities, further subverting opportunities for exploitative relations. The egalitarian arrangements of the Wola political economy prevent any seeking to control resources and production (Sillitoe 2010a) by effectively isolating the subsistence economy (featuring a household mode of production – where means are sufficient to meet demand and all have the same access to resources to supply their needs) from the socio-political domain (featuring delayed reciprocal transactions – where scarcity applies by definition, with demand for valued exchangeable things exceeding supply).[6] While this disjunction of the domain of material well-being from that of political manoeuvring is central to the stateless political-economic order – bringing about the relative political autonomy of the individual integral to acephalous arrangements – it takes us away from the issues of immediate interest.

While men are able to construct houses single-handedly, and on occasion do so, more often than not they receive help from a few relatives, particularly when it comes to thatching the building. Some may give considerable help, some just lend a hand for a morning or afternoon, and others may drop by and offer advice more than assistance. When Mayka Pes built an *enzatenda* 'mens'-womens' house' at Tombem, for instance, he levelled off the area working alone, which took him about 'one moon'. He also collected all the raw materials himself, without help, which he estimated took him 'about two moons'. When he built the house, three relatives helped him (Figure 7.3 – he could not recall those in the thatching party); they did so for no reward. When he rebuilt the house on the same site some years later, he received help from three other relatives and a considerable number assisted with the thatching (Figure 7.4); again for no reward except for a share of the *nay saway* 'meal of earth-oven cooked vegetables' prepared for the thatching party. When Maenget Hiyp built a *momaenda* 'family house' at Haelaelinja for his wife, children and pigs, he also received help from three men (two of whom also helped him collect raw materials) and an even larger number of relatives assisted with the thatching (Figure 7.5).

The group of men who co-operated in building the *enzatenda* 'mens'-womens' house' at Honael (Figure 7.6), documented in detail here, is somewhat atypical in composition, with nine individuals helping to collect and prepare the building materials and eight assisting in the construction work. My demands to observe house construction and the rewards I offered influenced its composition. It was necessary for political reasons to agree that it was the turn of certain men to help me and receive the payment on offer.[7] It is unusual for those who assist in building a house to demand any payment, unless the help they give is substantial. When Mom, for example, single-handedly helped Tensgay (his elderly ZHF)[8] build his *tenda* 'women's house', he received a K20 cash reward. It is similar to assistance with collecting building materials, as noted in the previous chapter, where if persons receive anything it is to recompense them for their efforts and not due to any scarcity in the supply of materials or skills.

While the Honael team may be somewhat irregular, it illustrates (together with the other genealogies showing those co-operating in house construction) the network

6 While there are connections between production and transaction, these domains are effectively disassociated from one another; things from the productive sphere that find their way into the exchange one, or finance the securing of other scarce things from elsewhere for use in it, are first divested of their production associations.

7 Note that nobody came from elsewhere to help, not only because of my attenuated network but also due to the desire of those living locally to corner the rewards.

8 See Figure 7.5, Mom is Duw's elder brother.

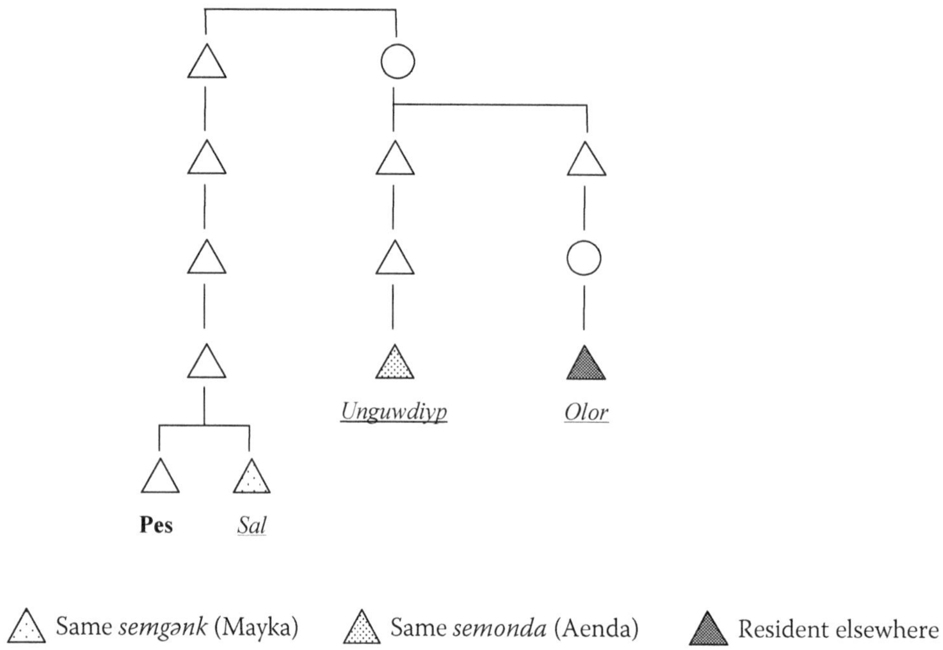

Name in bold (e.g. **Pes**) is house owner. Names in italics and underlined (e.g. *Sal*) helped build and thatch.

Figure 7.3 Genealogy of first Tombem house-builders.

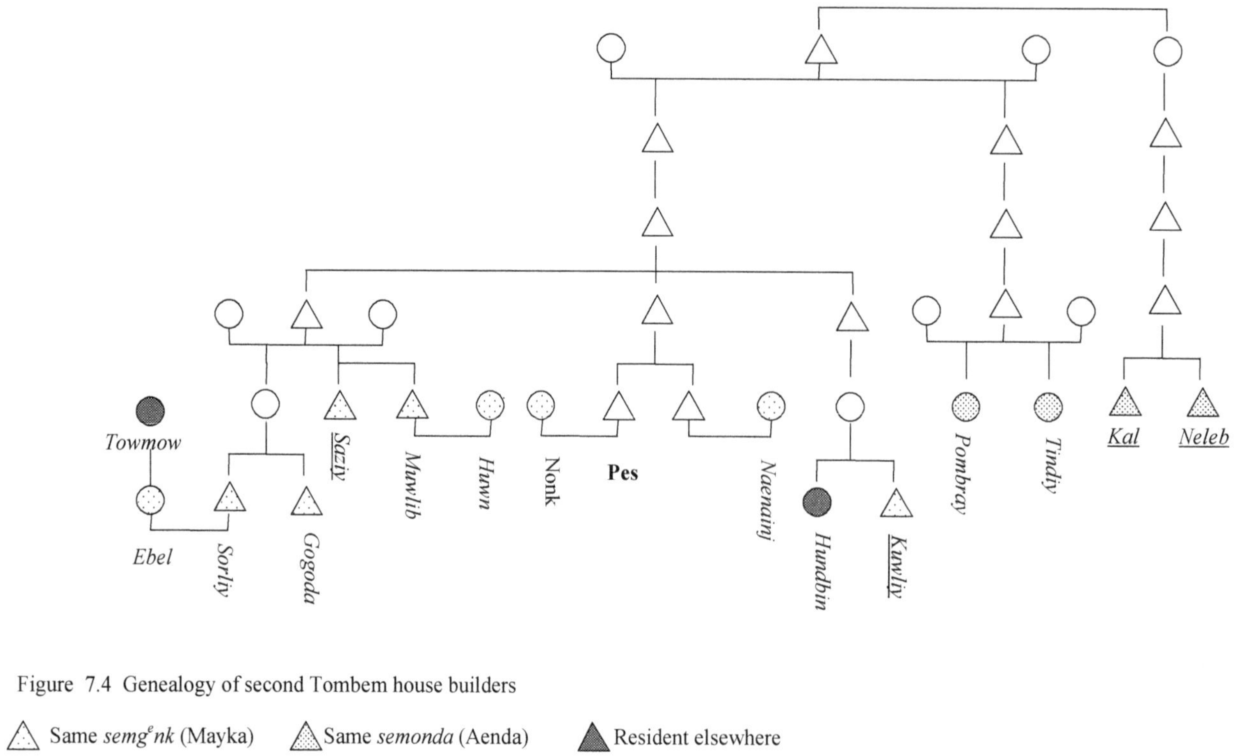

Name in bold (e.g. **Pes**) is house owner. Names in italics (e.g. *Ebel*) helped with thatching only (together with four young children). Names in italics and underlined (e.g. *Kal*) helped build and thatch.

Figure 7.4 Genealogy of second Tombem house-builders.

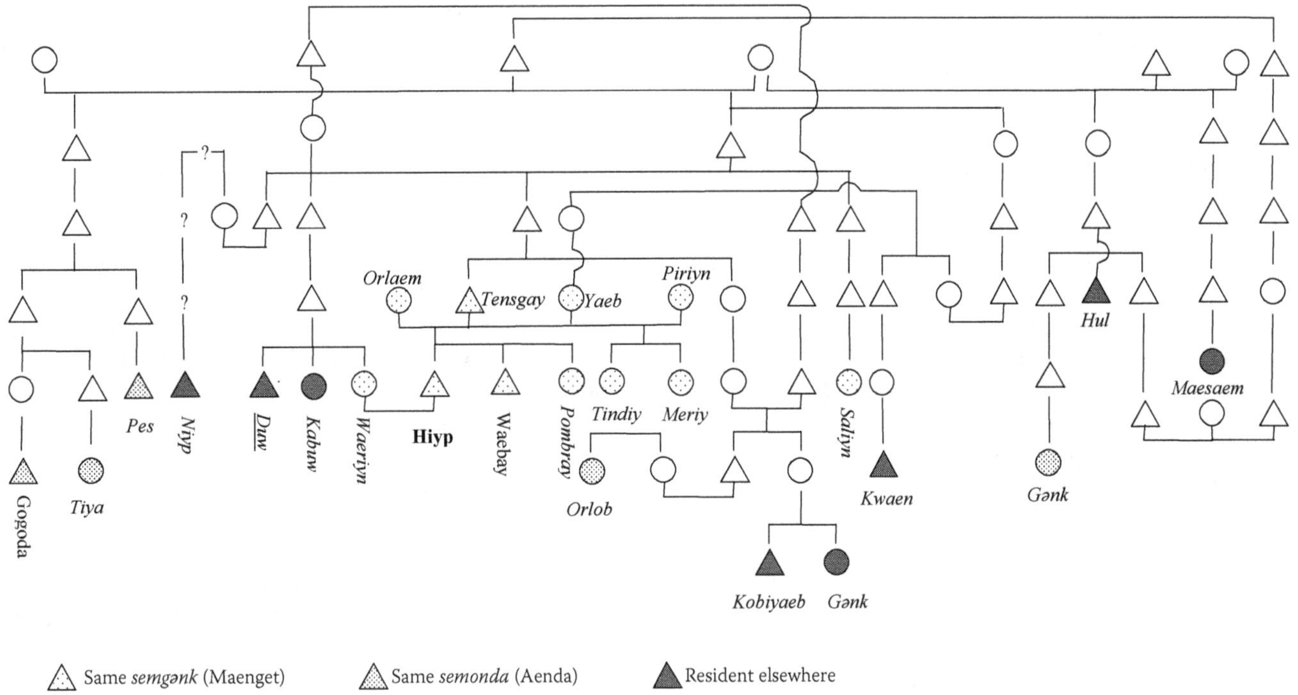

Name in bold (e.g. **Hiyp**) is house owner. Names in italics (e.g. *Tiya*) helped with thatching only (and some young children). Names in italics and underlined (e.g. *Duw*) helped build and thatch. Others named (e.g. Waebay) helped collect materials, build and thatch.

Figure 7.5 Genealogy of Haelaelinja house-builders.

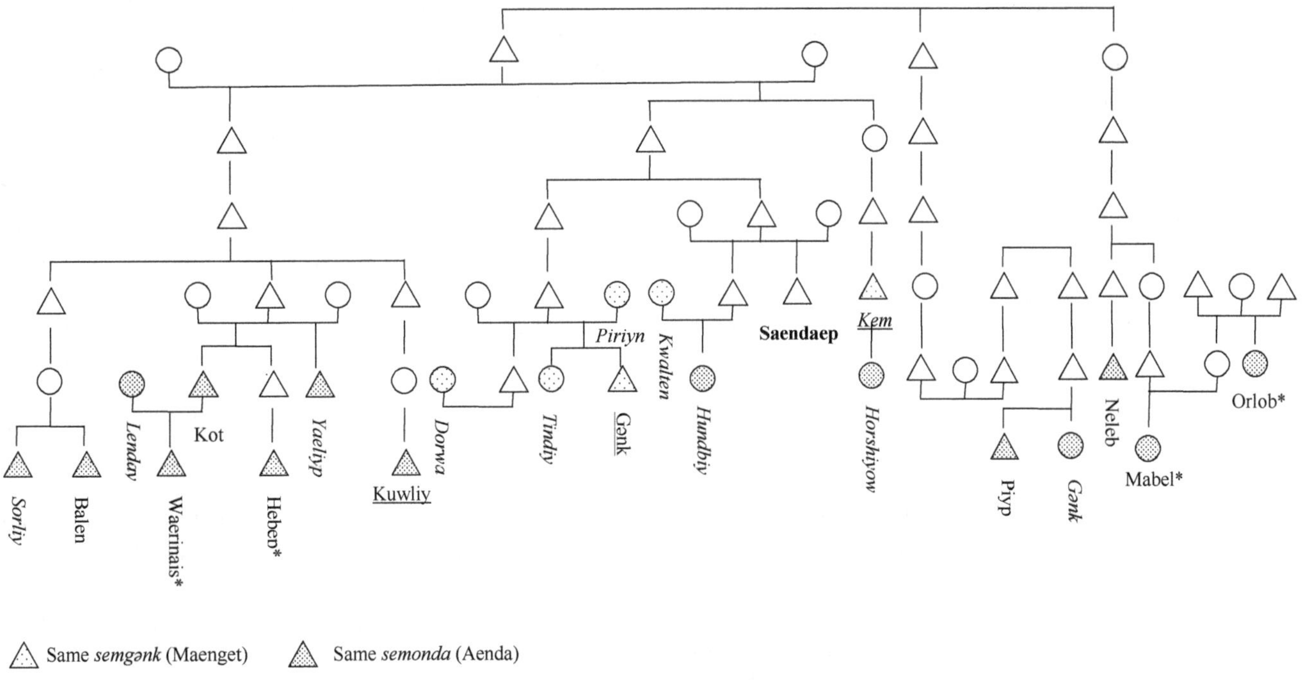

Name in bold (e.g. **Saendaep**) is house owner. Names in italics (e.g. *Lenday*) helped with thatching only. Names in italics and underlined (e.g. *Kem*) helped build and thatch. Names underlined (e.g. Kuwliy) helped collect raw materials and build. Names with asterisk (e.g. Hebep*) helped collect raw materials only. Others named (e.g. Kuwliy) helped collect materials, build and thatch.

Figure 7.6 Genealogy of Honael house-builders.

character of action-set recruitment and group activity in the Was valley, and the prominence of individual agency. In order to co-operate in any activity (such as building a house) individuals have, of course, to come together. They unite, when personal and joint interests correspond, via the ramifying networks of social relations that connect them together. All persons are embedded in overlapping networks, connected through kinship largely, which include and radiate out across Wolaland from the neighbourhoods where they reside (Sillitoe 1979a:86–104). Neighbourhoods, focusing on *sem* 'families', constitute loosely knit communities of a few hundred or so residents (see Chapters 1 and 3). While individuals have kin connections with several such neighbourhoods, it is through residence at particular locales that they determine the 'family' of relatives they interact with regularly and with whom others identify them, and who likely assist them.

The 'network' image differs significantly from that of 'structure', with no jural norms guiding group allegiance decisions and connections being continually negotiated, adjusting to ongoing social process. The scope afforded individual action thwarts recruitment through actors recognizing an obligation to support fellow members of corporate kin groups (constituted by descent, alliance or whatever) to which they acknowledge responsibilities. While action-sets sometimes mirror neighbourhoods closely, this does not imply that they are political corporations that demand the support of those associated with them. Rather, it is proximity of residence that is key: someone is more likely to co-operate with those s/he interacts with daily, in building a house for instance, than with those living elsewhere. Individuals are free to decide according to their personal interests – which are, of course, culturally informed – whether or not to co-operate with relatives in an activity. The interests that prompt them to join in or not as they think best may include such things as a wish to repay previous help, show good feeling or out of sympathy for a relative struggling with a task, oblige someone to assist them later, or because of a concern for their reputations as reliable persons, and so on.

The collectivities that come together to undertake activities are not entirely random, as kin interaction informs their make-up. They coalesce through the network extending out from the person instigating an activity, such as building a house. He can look to some relatives in his neighbourhood to support him, with whom he interacts regularly, and some persons from elsewhere with whom he maintains meaningful social relations may join too. An individual may possibly join in activities at several places where he keeps up networked relations, that is recognizes social responsibilities to some residents. These action-sets are temporary; one with the same membership may never form again, as the two parties that worked on Pes's two Tombem houses illustrate.

While undertaking a group activity presumes some kind of agreed organization, it is so informal in the Was valley as to be scarcely recognizable. A pursuit such as house building or bridge construction does not feature any sort of leadership to co-ordinate actions. No one oversees what others do, they all know what needs to be done and have the necessary skills to get on with it while co-operating informally (as the film associated with this book shows).[9] That there is no call for overt direction is another aspect of tacit knowing. If it is necessary to reach a decision over some matter, the participants will discuss it until they reach a consensus. All those involved have a right to participate in the 'talk' and to try to persuade the others to act in accord with their views. While some men, notably respected *ol howma* 'men of the clearing' (Chapter 1; Sillitoe 1979a:106–11), may have more influence than others and hence a better chance of shaping opinions, they cannot compel others to follow their wishes. In a manner reminiscent of everyday interaction, network members continually influence one another's behaviour through their values, expectations and interests as they cooperate, reaching compromises as necessary.

Construction of exterior walls

When the house site is level, clean and firm, the first task in *aend bay* 'house building' is marking out the plan (Table 7.2). The builders use lengths of cane, sapling and vine to *maenjet may* 'mark out' the position of the walls and fireplaces (Figure 7.7), pegging these markers in place with small pieces of wood (the Enga and Dani similarly mark out a house plan after clearing the site – Heider 1970:261; Meggitt 1957:167–8).[10] The length of a house varies depending on the style. A three-roomed *enzatenda* 'men's-women's house' such as we built at Honael is some six metres, whereas a one-roomed *tenda* 'women's house' may be half this length. The width of a house is approximately twice the height of an adult. The builders may test the dimensions by lying on the ground across the width of the house with their heads close to the outside wall and feet over the fireplace;[11] this is the usual arrangement when sleeping, with legs slightly flexed and feet pointing towards the fire. A house-builder may use a length of wood or cane to measure the distance from wall to fireplace and then use this to check the distance at other points so as to ensure a symmetrical ground plan. This is important because an asymmetrical house will be

9 See the book's website at www.community.dur.ac.uk/paul.sillitoe/wola/houses.

10 See Clarke 1971:105; Meggitt 1957:163–4; Pospisil 1963:259; and Steensberg 1980:130, 136, 142, 173 for ground plans of houses built elsewhere in the highlands, which although they vary in detail, follow a broadly similar design.

11 Similarly with house-builders in Mali, 'The human body was the yardstick for many building dimensions.' (Marchand 2009:135).

Plate 7.4 Marking out the ground plan of house.

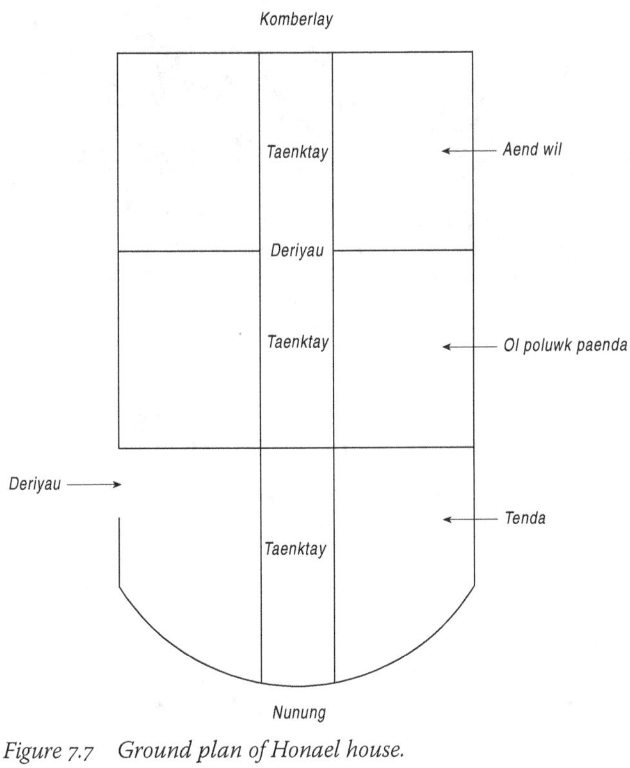

Figure 7.7 Ground plan of Honael house.

structurally unsound. Other than this, house-builders work without any formal plans, although they have a good idea of the finished structure, following tried-and-tested templates, albeit they may decide details of construction during work, these emerging as they respond to properties of particular raw materials and unforeseen site issues.[12]

Once we had the Honael house plan marked out, I divided it longitudinally in half down the centre, stipulating that those working on the house could only use steel tools on one side and stone tools on the other – the left-hand side, looking into what would be the *aend wil* 'men's sitting room', was for steel, and the right-hand side for stone.[13]

When they have the house plan marked out to their satisfaction, the builders proceed to drive in the exterior *aend dol* 'wall stakes' that will comprise the sides and rear of the house. They start at one end of the side wall and work backwards towards the rounded enclosed part of the house plan along the lines pegged out. While the stakes may vary in length by some centimetres, they drive them in level so

12 As did medieval masons (Ingold 2011b:211–12).

13 Note that the times taken to complete various tasks recorded in the following tables do not include any time spent resting, smoking and eating.

Plate 7.5 Starting to drive in wall stakes down sides of house.

Plate 7.6 Driving in wall stakes, approaching the rounded end of house (note the men sitting and sighting along the wall line).

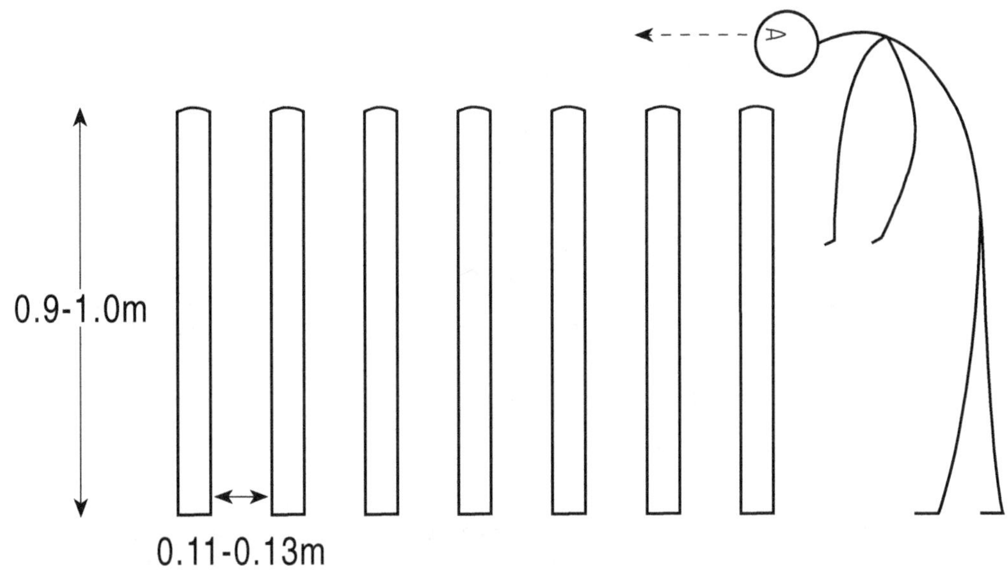

Figure 7.8 Sighting along wall stakes.

that their tops are at the same height. They leave a space of about twelve centimetres between each stake, which is a couple of centimetres more than the average stake width. It is important that they *aend dol aesay* 'spear wall stakes' in a straight line because these comprise the footing of the entire building. If they are not true, the whole house will be askew and again structurally unsound. In order to ensure that they are straight, someone may stand and *duwn henday* 'sight along' the tops of the stakes to see they form a straight line; if working alone, a man will drive in two or three stakes at a time and then step back to sight along them (Figure 7.8).

If necessary, the builders may revise the plan, moving the wall-line markers so that they are both straight and parallel to one another. They may check the heights of stakes on opposite walls by marking a stick against a stake on one side and then comparing it with a stake on the other. As they approach the semicircular end, they gently angle the walls inwards by about ten degrees for the last fifty centimetres. In constructing the rounded end wall, they follow the arc marked out on the ground by the length of vine.

The builders continue the construction of the exterior wall by lashing *gamonk* 'battens', comprising saplings split

Plate 7.7 Care is taken to ensure straight line: correcting the alignment of a stake.

Figure 7.9 Position of exterior wall *gamonk* 'battens.'

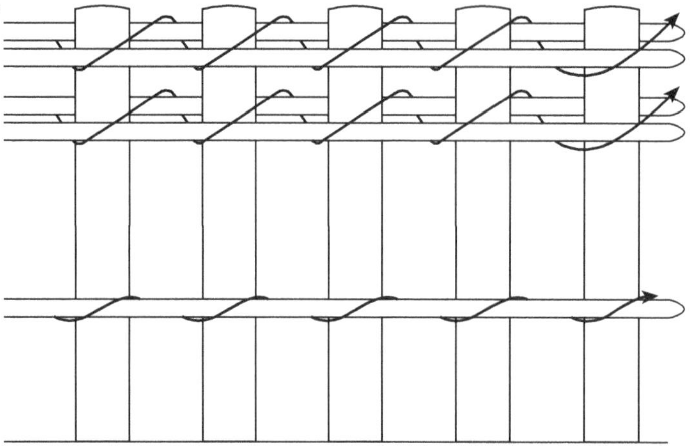

lengthwise in two, along the stakes, so securing them together. They lash a pair of battens, one on either side of the stakes, along the wall's top edge. Next they lash a single *tay gamonk* 'base batten' along the outside of the stakes some thirty centimetres or so from the ground. They may also lash a second double row of battens in place twenty to twenty-five centimetres below the top one, depending on the style of wall under construction (Figure 7.9). The initial lashing of these batten rows is temporary, simply done by firmly winding a length of vine or bark strip continuously around stakes and battens along the wall.

When the battens are in place, the builders check that the lengths of wood or cane marking the location of the cross walls and central fireplaces are correct, as these mark the positions of the *aend pinj* uprights that will support the pair of ridge-poles. These uprights are two-metre-long stakes with a slight scoop cut into their top edge to accommodate the ridge-poles. Again, men may use a length of wood or cane to measure and mark the distance to the off-centre point for each upright. When satisfied, they drive in the uprights, one pair on the front-wall line of the house and another pair towards the rear, on either side of the fireplace line. They judge their heights to be the same by standing next to them, using their own stature as guides, each upright driven into the ground to about a man's height. Next they lift the ridge-poles into position, their sharp pointed ends towards the front of the house, and then they try a few rafter poles in place, placing these between ridge-pole and wall-top battens. When arranged, they stand back to judge the line of the roof and check the symmetry of the house. If walls or roof look askew, they reposition the offending stakes, which is a skilled task

Plate 7.8 Securing the top row of *gamonk* 'battens.'

Plate 7.9 Lashing the bottom row of gamonk 'battens' in place.

usually involving the shaving of a small sliver of clay off the side of the existing posthole to increase the height of a stake, or the driving of the stake further into the soil to decrease its height.

When they are satisfied that the structure's proportions are correct, the builders remove the rafters and ridge-poles and put them to one side. They then permanently lash the single base batten in place around the wall using strips of bark, running these on continuously and winding them tightly around each stake two or three times (Figure 7.10). Next they position the sheets of *aend buwk* 'pandan bark' around the inside of the walls with the outside surface of the bark facing into the house. They secure these bark sheets in position by driving in *pongow* 'fastening stakes' at intervals in the interior of the house. A single sheet of bark reaches about three-quarters of the way up the wall. (Meggitt 1957:168–9 and Steensberg 1980:166–8 document the construction of similar 'packed' walls in the Enga and Western Highlands regions.)

If the wall features only the top double and base single split-sapling batten rows, the house-builders will fill in the top gap by splitting a suitably sized piece of pandan bark longitudinally off an entire sheet and arranging it to fill the gap, jammed behind the *pongow* stakes that reach to the top of the wall (Figure 7.11). When they have the pandan-bark sheeting in position, they proceed to lash the top, double-batten row permanently in place, using strands of split *haeluwp* 'rattan'. The number of strands split from a rattan stem depends on the size of the vine – it may be divided in half or quarters or even eighths if sizeable. Men join the lengths of rattan together with a sheet-bend knot. Some

Plate 7.10 Driving in pinj 'ridge-pole supports' to check symmetry of house.

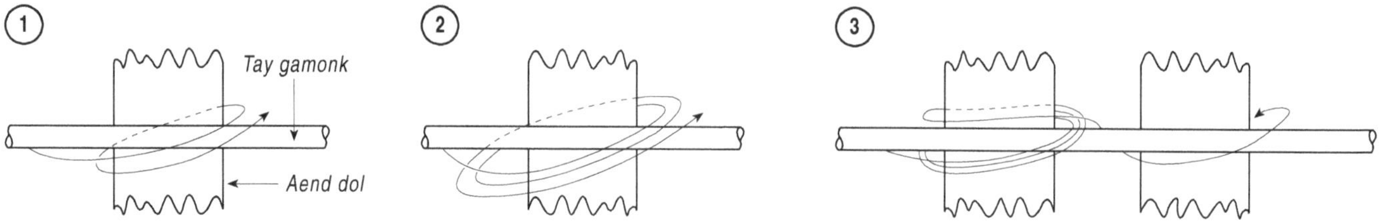

Figure 7.10 Lashing of gamonk 'battens.'

Plate 7.11 The lashing used to secure bottom gamonk 'battens' in place (start).

Plate 7.12 The lashing used to secure bottom gamonk 'battens' in place (finish).

Plate 7.13 Positioning buk 'pandan-bark sheet' against interior side of wall stakes.

Plate 7.14 Sharpening *pongow* 'stakes' to secure *buk* 'pandan-bark sheet' in place (note roll of bark by wall stakes).

Plate 7.15 Pinning *buk* 'pandan-bark sheet' in place with *pongow* 'stakes.'

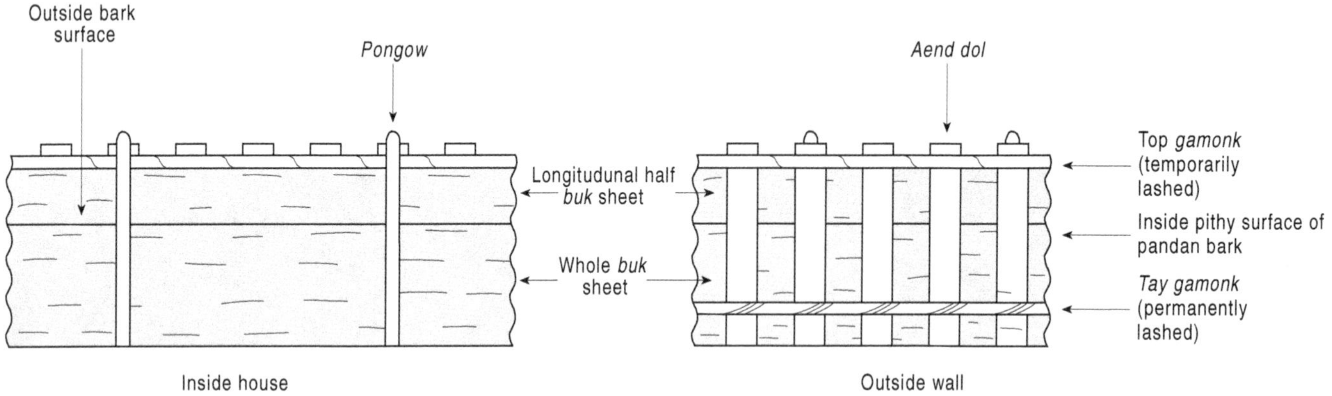

Figure 7.11 Securing of buk *'pandan-bark sheets.'*

builders simply wind the rattan once tightly around the two battens and each *dol* 'stake' for the length of the wall; while others prefer a more elaborate lashing (Figures 7.12 and 7.13). Whatever they choose, they include the *pongow* stakes in the lashing when they reach them with a turn or two around their top ends, so sandwiching the pandan-bark sheets tightly between the outer *dol* and inner *pongow* 'stakes' (Figure 7.14).

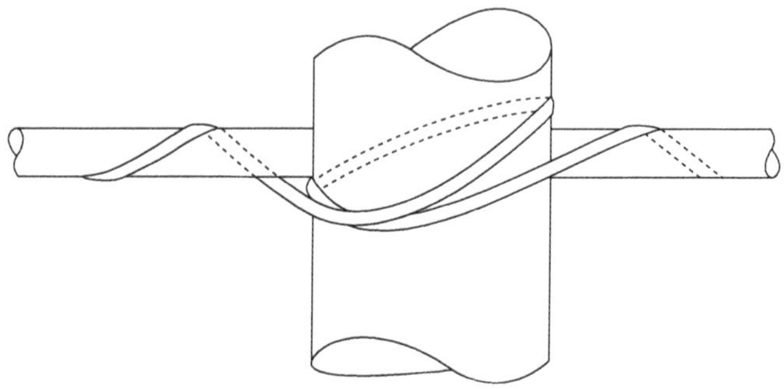

Figure 7.12 Simple lashing of gamonk *'batten'*

If the wall features the middle double row of split-sapling battens, positioned twenty to twenty-five centimetres below the top one, the single entire pandan-bark sheet will reach to here, as will the tops of the shorter *pongow* stakes used to fix it in place. In this style of wall, men permanently lash these middle battens in place with rattan, like the top battens, incorporating the tops of the *pongow* stakes as they go (Figure 7.15). This leaves a gap along the top part of the wall, which they fill with tough *aendashor* pandan leaves, after they have permanently lashed the top row of battens in place. They secure the pandan leaves in position with slender, shorter *pongow* 'fastening sticks' jammed between the top and middle rows of battens (Figure 7.16). They may construct this kind of wall in *enza* 'men's houses' or the men's part of *enzatenda* 'men's and women's houses.' Some prefer it, saying that pandan leaves pegged around the tops of walls, called the *hwimb*, give an aesthetically pleasing appearance. They do not line the top of the walls in *tenda* 'women's houses' or *momaenda* 'mixed houses' in this way, because pandan leaves are not strong enough to withstand the activities of pigs that would soon trash them. In an *enzatenda* they may continue the optional middle-batten row round the outside *tenda* women's part of the house (i.e. as a single outside batten similar to the *tay* base one), such that all of the exterior wall of the house has the same appearance, which some do for aesthetic reasons (the batten serves no technical purpose).

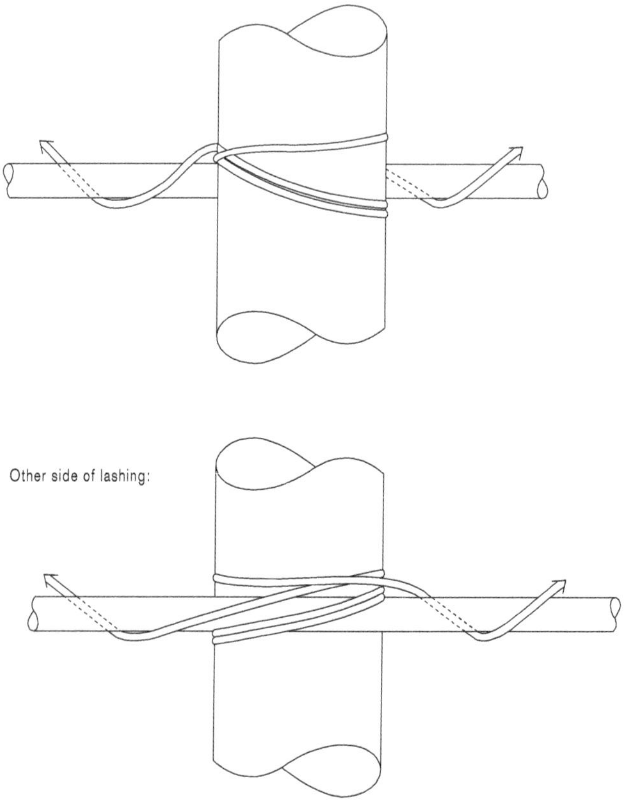

Figure 7.13 Other lashing of gamonk *'batten'*

In an *enzatenda* 'men's and women's house' such as we constructed at Honael, the builders remove one *dol* 'stake' from the wall towards the rear of the house and leave a hole in the pandan-bark sheet, to serve as the crawl-through

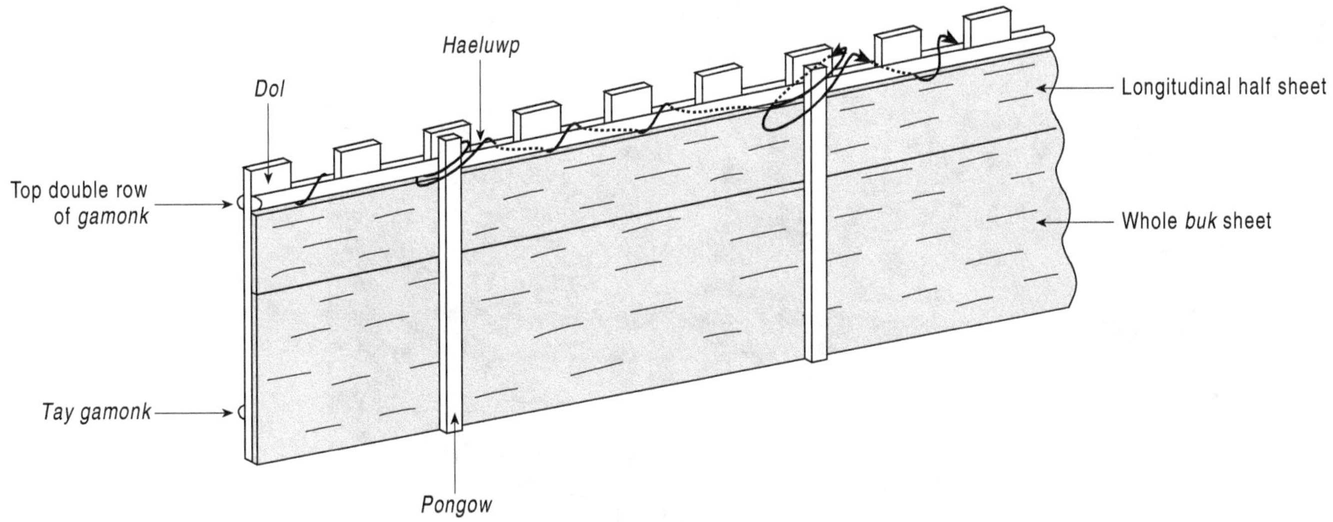

Figure 7.14 *Lashing top row of wall gamonk 'battens' and pongow 'stakes.'*

Plate 7.16 *Securing buk 'pandan-bark sheet' along top part of wall, pushing on pongow 'stake' as lashed in place.*

Plate 7.17 The lashing used to secure top gamonk 'battens' in place

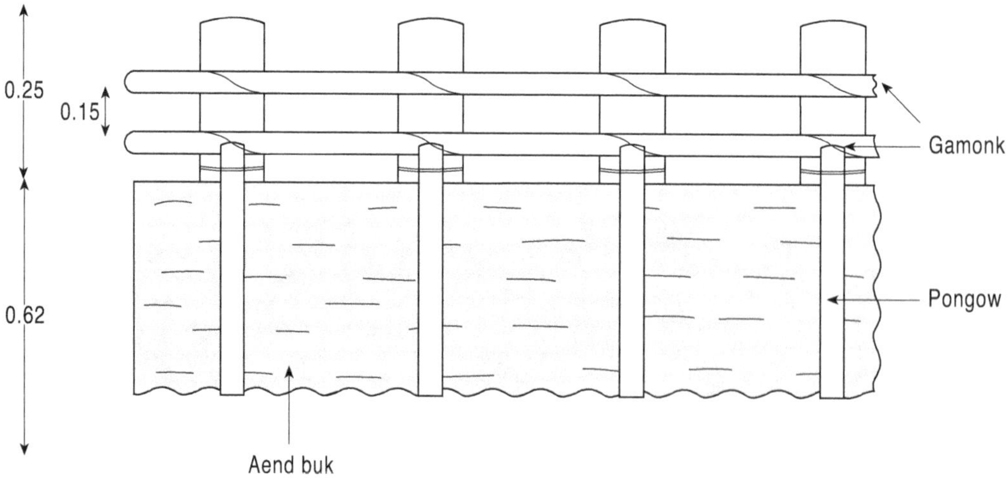

Figure 7.15 Securing partial buk 'pandan-bark sheet' wall lining (m).

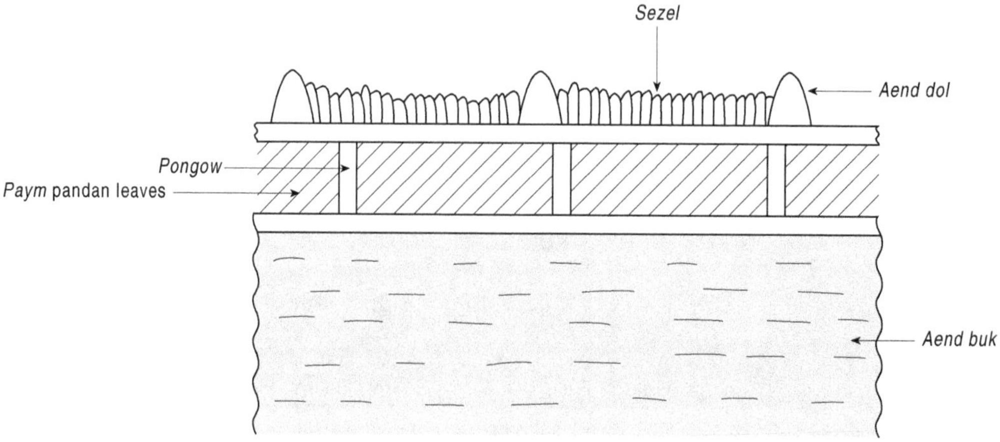

Figure 7.16 Pandan leaf lining of top of wall.

Plate 7.18 View of Honael site showing completed walls.

deriyau 'doorway' into the women's *tenda* room (Figure 7.17). If they have continued a single middle batten running round the outside of the *tenda* section of the house, this will serve as the top of the doorway with another short piece of battening inserted behind it inside. Otherwise, men may lash two short battens between the stakes on either side of the doorway to serve as the lintel. They may secure these door-top battens with a simple spiral lashing or execute a figure-of-eight weave called *tabok* round them with rattan (Figure 7.18). They may cut two or three short lengths of wood and lash above the doorway battens to reinforce the bark sheet there, and drive in some short stakes along the bottom to give a shallow step.

The builders complete the outside wall of a house by filling in the exterior spaces between the *dol* 'stakes' with lengths of *gaimb* 'cane grass' stem, concealing the raggedy pith visible on the inside of the pandan-bark sheets. They block off any tears in the bark with pieces of pandan leaves (no light should enter the house through the walls). They insert the *sezel* 'wall canes' between the *gamonk* 'battens' and push firmly into the ground, trimming off level with the top of the wall (Figure 7.19).

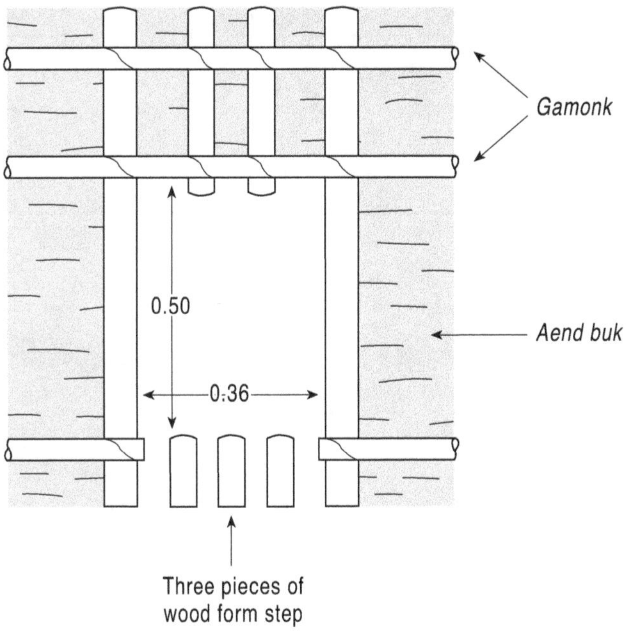

Figure 7.17 Entrance into rear tenda 'women's room' (m).

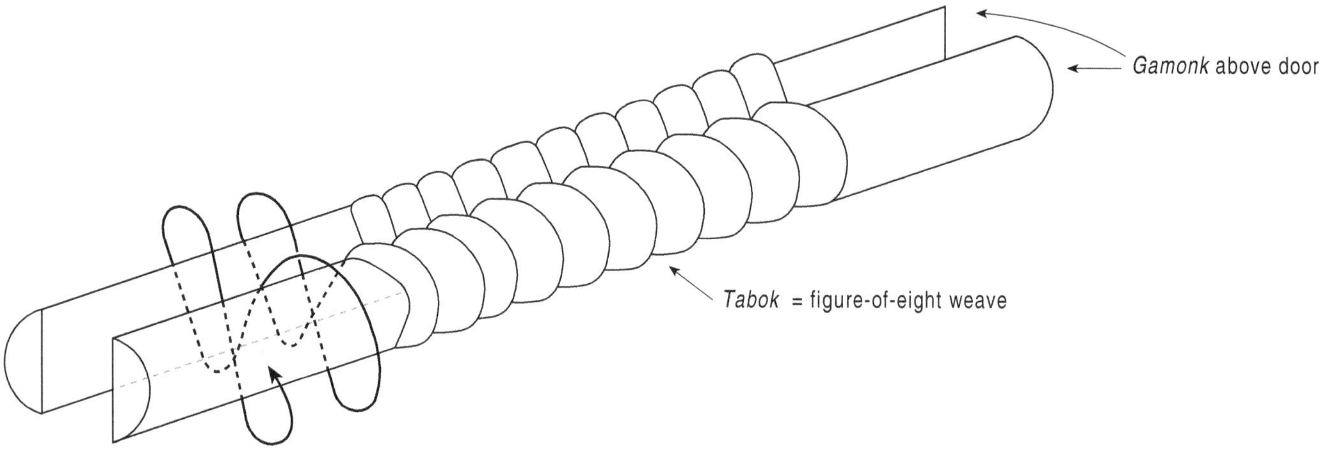

Figure 7.18 The tabok weave.

Plate 7.19 The doorway into a tenda women's house.

Plate 7.20 Carrying cane stems to house site.

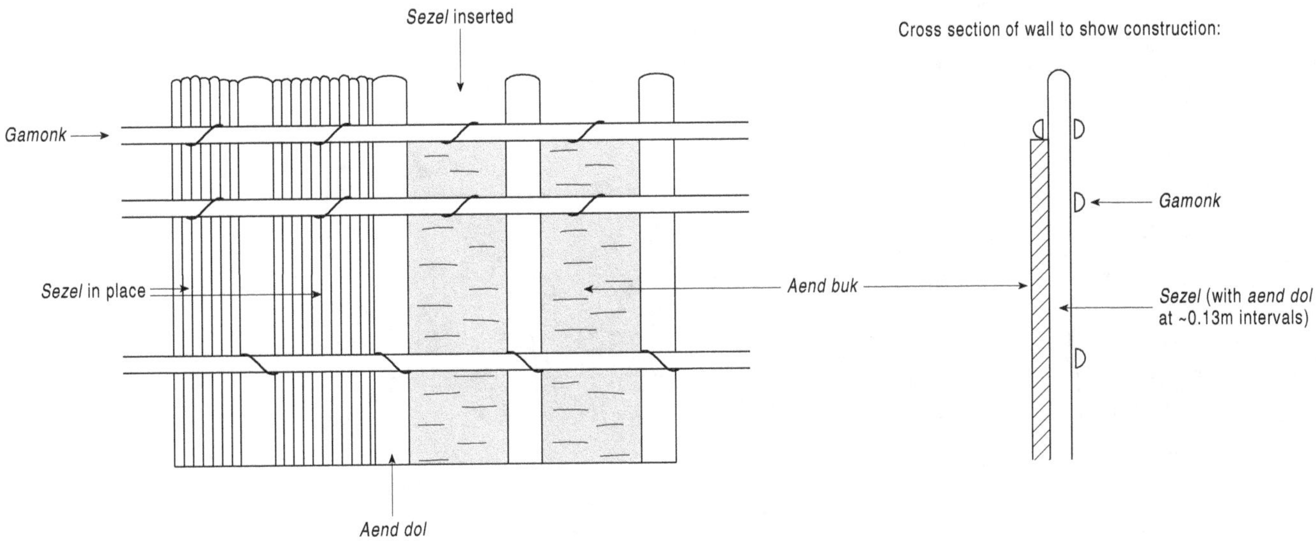

Figure 7.19 *Insertion of sezel 'wall canes.'*

Plate 7.21 *Inserting sezel cane stems along outside of wall.*

Plate 7.22 *Trimming sezel cane to top of wall.*

Plate 7.23 Completed outer wall showing sezel cane filling between dol 'stakes.'

It took nearly twenty man-hours to erect the exterior wall of the Honael *enzatenda* house (Table 7.2), which followed the full pandan-bark design throughout (there was no pandan-leaf upper part in the men's quarters). The table does not include the time spent replacing the battens and stakes pulled down during the land-rights dispute mentioned in Chapter 3's discussion of house site selection; the damage was not large. Regarding the comparative time efficiencies of using steel and stone tools, the steel side of the house took longer to construct, the ratio being 1:1.18 in favour of stone; that is, for every hour worked on the stone side they worked an hour and eleven minutes on the steel one. While the difference is perhaps unexpected, such time divergence in favour of stone tools occurs in other tasks and may be attributed partly to the better finish attainable with steel tools and the extra time taken to achieve it. Also, a considerable amount of the work undertaken in constructing a wall does not require the use of any cutting implements, stone or steel. These data show how the time taken to complete the same tasks can vary; men work at different rates, some make mistakes that have to be put right and so on.

Variability and materiality

These time differences, which recur in this account of house construction, bring up the issue of variability again. The differences in the times taken to complete the same tasks on the stone and steel sides of the Honael house, sometimes counter-intuitively favouring stone tools, indicates the extent of variation seen in house building. The 'ecological' approach to materiality (focusing on raw materials, see Chapter 6) also brings up the issue of variability. In arguing that matter is not fixed but always in flux, it highlights the continuous variability of materials: one length of beech timber is not the same as another with differences in the grain and hardness of the wood. In the context of material culture studies, Leroi-Gourhan (1993:306) noted half a century ago that 'the making of anything is a dialogue between the maker and the material employed', reaffirming what artisans have known for millennia with respect to variations in raw materials. Regarding the splitting of timber, for instance, 'The practised woodsman brings down the axe so that its blade enters the grain and follows a line already incorporated into the timber through its previous history of growth.' (Ingold 2011b:211); in other words, he 'goes with the grain', which as any woodworker knows, determines in some measure the outcome of his actions. Artisans intuitively recognize and respect such variations in the materials they work with, and strive to use them, as evident in such sayings as being 'true to the material'. Skilled persons seek to afford full expression to such material qualities, which is a key feature of tacit knowing. It is in this sense that materials exert 'agency' on any productive process.

The variation in materials relates to the variability in things made. There is no standard bow or digging stick, and no two houses are exactly the same. They all vary to some extent in shape, size, finish and so on, as we might expect for handmade things where makers go with the flow

Task	Wola term	Time spent: Stone-tool side	Time spent: Steel-tool side
marking out house plan	*aend maenjet menay or aend tomb bay*	1 h, 2 m	
driving in stakes for external wall	*aend dol aezay*	1 h, 6 m, 15 s	1 h, 34 m, 45 s
sighting along walls	*dol duwn henday*	47 m, 45 s	49 m, 45 s
splitting rattan vine for lashing *gamonk* battens	*haeluwp dekel bay*	7 m, 30 s	26 m, 45 s
splitting saplings longitudinally in half for *gamonk* battens	*gamonk dekel bay*	4 m, 30 s	5 m
temporary lashing top rows of *gamonk* battens	*gamonk maen kay*	12 m, 15 s	22 m, 30 s
lashing bottom *gamonk* battens in place	*tay gamonk tongay*	1 h, 24 m, 15 s	51 m
driving in *pinj* stakes to try *bez* ridge-pole and *polpen* rafters	*pinj aezay*	15 m, 15 s	
paring and sharpening *pinj* stakes	*pinj waendilay*	3 m, 30 s	16 m
trying *bez* ridge-pole in place on *pinj* stakes	*bez maenjet bay*	18 m, 30 s	
trying a few *polpen* rafters in place	*polpen maenjet bay*	10 m, 30 s	
taking down *polpen, bez* and *pinj*	*polpen diy bez diy pinj diy kundiyay*	7 m	
paring *pongow* stakes	*pongow waeray*	4 m, 30 s	24 m
putting pandan-bark sheets in place in *nunung* bell-end	*aend buwk nunung bort tor bay*	31 m, 30 s	
putting pandan-bark sheets in place on long side wall	*aend buwk tor bay*	2 m, 30 s	10 m
trimming end off pandan-bark sheet to fit wall	*aend buwk bokor luw tuway*	8 m, 30 s	1 m, 45 s
cutting pandan bark-sheet longitudinally in half and trimming	*aend buwk gənk pelay*	9 m, 50 s	13 m, 15 s
placing half pandan-bark sheet on side wall and securing	*buk gənk mond poror lubtuw tongay*	1 m, 30 s	36 m
placing half pandan-bark sheet in *nunung* bell-end and securing	*buk gənk mond poror lubtuw tongay nunung bort*	8 m	
driving in *pongow* stakes and lashing to secure pandan-bark sheets[1]	*pongow ma haebtuw tongay*	22 m, 15 s	10 m, 15 s
re-aligning some *pongow* stakes	*pongow waek duwn bay*		13 m, 30 s
lashing top *gamonk* battens permanently in place	*gamonk tongay*	50 m, 30 s	58 m, 15 s
undoing *gamonk* batten lashed wrongly	*gamonk korob tongaebtuw hezbiyay*		7 m
re-aligning some *dol* stakes	*aend dol duwn aezay*		6 m, 25 s
permanently re-lashing top row of *gamonk* where *dol* adjusted	*gamonk waek tongay*		5 m, 50 s
pushing *sezel* cane in around outside walls	*sezel bombay*	1 h, 12 m, 55 s	1 h, 28 m, 45 s
trimming *sezel* cane level[2]	*sezel payay*	46 m, 20 s	
sealing gap between wall and eaves in men's sleeping room with moss[3]	*uwpuluwk tubael maembow pes bay*	18 m, 45 s	
total times		**8 h, 50 m, 35 s**	**10 h, 26 m, 30 s**

[1]On stone side this includes some time paring *pongow* stakes.

[2]They trimmed the *sezel* cane level on the steel side as they went pushing it in.

[3]According to the men building the house, it would be up to the women occupying the *tenda* to attend to such finishing work as they saw fit.

Table 7.2 Construction of external walls.

of the materials.[14] There is also variation in the doing of tasks: no two strokes of an axe are identical. Artisans have to navigate 'a continuous variation of variables' (Deleuze and Guattari 2004:410). The variation intriguingly reflects the variability that characterizes social life generally in the Was valley, which challenges the inadequacies of normative assumptions about behaviour, as mentioned previously. It reflects an emerging weakness of twentieth-century social science as we increasingly seek to encompass the variability of human behaviour, with a shift of focus to social process, and the associated notion of the 'agent', whose behaviour comprises the social system (see Chapter 3). Individual variation is everywhere; we have to contend with the 'variable individual' and not only the 'moral person', something that has gradually occurred to me as I have become aware of the extent of the variation that typifies Wola behaviour and knowledge, and their capacity to destabilize categories (see Sillitoe 2010a:50–5).

While residents of the Was valley recognize broadly agreed values that inform proper ways to behave and share a commonly held body of knowledge, significant variation occurs, as pointed out previously, depending on different personal situations, expectations, experiences, individual histories and so on. They thwart generalizations about social norms, being reluctant to discuss matters such as rules of conduct, particularly in the abstract, responding to hypothetical questions about behaviour by asking to whom, where and when the event happened. And they drop no symbolic hints amenable to interpretation as suggesting collective meaning to aspects of social life. There is actual behaviour and sometimes actors comment on it, although different individuals may have different understandings of events. They are likely to respond *mbinyom konem* – equivalent to the Pidgin expression *em laik bilongen* 'it's their wish' – to questions about the way people should behave. In short, individuals have considerable latitude and free choice, which is in keeping with their stateless order where, with the stress on personal autonomy, there is no authority to legislate on proper behaviour.

The accommodation of individual variation in the study of social interaction and in actors' understandings of behaviour present considerable methodological challenges. They imply the collection of a large body of ethnographic data, ethnographic-determinism par excellence, to assess the extent of difference and to do it justice, demanding attention to detail, including statistically adequate data samples – a challenge to current reflexive methods – to detail the behaviour and views of many individuals and catch the variety in any community. It will demand, for instance, ethnographic records on the construction of several houses for different families to secure a sound range of data on variations in materials used, construction activities and time invested in house building (as I pointed out some time ago in *Made in Niugini* with respect to artefact manufacture, where accommodating variation would require the making of a dozen or more examples of an object by different persons – Sillitoe 1988a:18).

In my experience, such detailed work would demand a team, being beyond one researcher, bringing with it a host of other complications, including the possible compromising of participant-observation, which is identified as a way to tap somewhat into tacit knowing (see Chapter 11). The managing of such large volumes of data to investigate issues of socio-cultural variation is increasingly feasible with expansion in computing power and accessibility, as noted in Chapter 4, and approaches such as agent-based modelling of behaviour. The demand to sample the building of several houses or making of many artefacts underlines the contrast between the omniscient view sought by social scientists and the personal one of social actors who, not aspiring to such a perspective, have only their own idiosyncratic view. The divergence from the individually variable local level of understanding that seeking to achieve an overview entails, further questions attempts to access actors' tacit knowledge, because if what every person knows varies to some ill-defined extent, we can only meaningfully seek to know one individual's tacit knowledge, so far as this is feasible anyway, which returns us to reflexivity, albeit *à deux*.

Construction of interior walls

The next stage in house building involves the construction of *waenuw* 'interior walls', if any, to divide the house into *tubael* 'rooms'. After clearing out any refuse left over from erecting the exterior wall, the builders drive in further hardwood stakes along the marked-out lines of the house plan, to form, as with the exterior wall, the framework of the walls. They incorporate the *pinj* 'ridge-pole-support stakes' in the partition walls; if they removed these when building the exterior wall, they again reposition these stakes off-centre, using a sapling or cane 'yardstick' to measure their positions. They lift the *bez* 'ridge-poles' into position on the stakes and place some *polpen* 'rafter poles' where the interior walls are located, to show the roofline. They cut the stakes for the interior walls to different lengths so that they decrease in size from the centre of the house down to the outside walls, following the slope of the roofline (Figure 7.20).

14 This aspect of manufacture changes, of course, with mechanization (Ingold 2011b:59–62; 2013:122–4), which aims to standardize both process and product with no recognition of individual variation in materials, treating these the same way (with implications for skills with the associated shift from sexual to occupational division of labour); consider, for instance, the automated pressing out of the steel sheets into the bodywork of a vehicle, where an automobile factory would cease to function if such components were not standard, as it could assemble no cars.

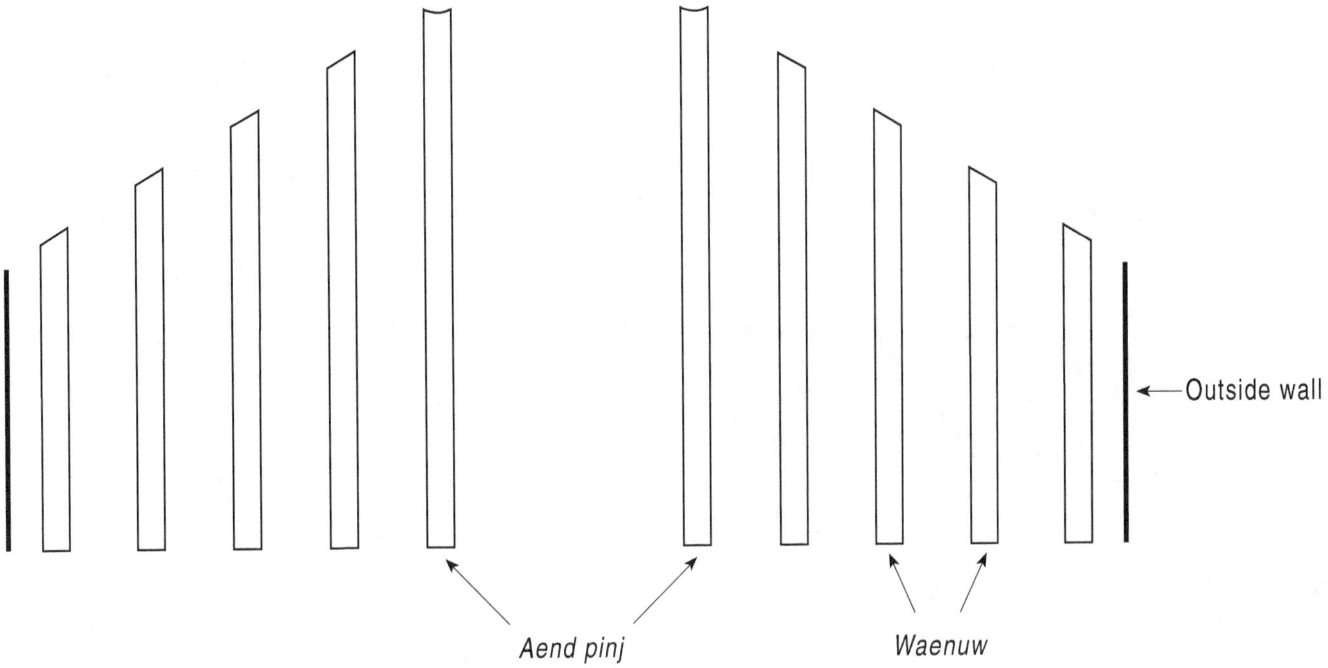

Figure 7.20 Alignment of waenuw 'interior wall' stakes.

Plate 7.24 Driving in the pinj 'ridge-pole support stakes.'

Plate 7.25 Lifting the bez 'ridge-poles' into place.

Plate 7.26 Driving in the dol 'stakes' of interior wall, following line of roof slope.

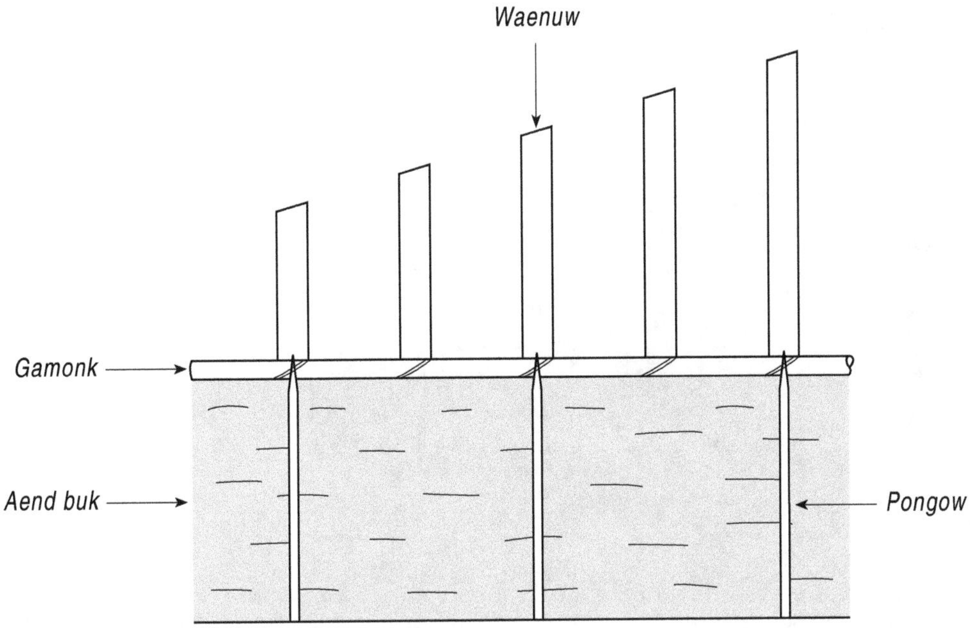

Figure 7.21 *Securing of buk 'pandan-bark sheet' on lower waenuw 'interior wall.'*

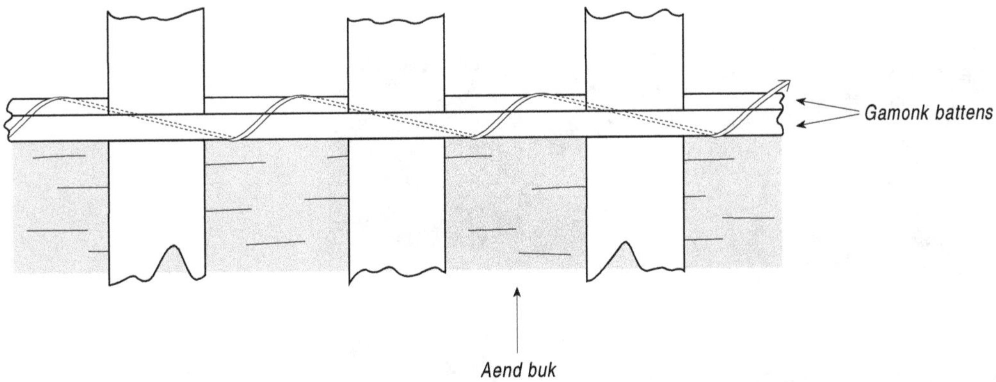

Figure 7.22 *Lashing gamonk 'battens' on waenuw 'interior wall.'*

Next men lash two split-sapling *gamonk* 'battens' horizontally across the interior-wall stakes, about sixty centimetres above the ground, for the width of the house. On the *tenda waenuw* 'dividing wall between the women's and men's sleeping quarters', they position a whole *buk* 'pandan-bark sheet' along the lower section, possibly together with a further strip to bring up to the height of the exterior-wall bark lining (sometimes they use dried pandan leaves here instead),[15] and secure in place with *pongow* stakes. On the *wil waenuw* 'dividing wall between the men's day room and sleeping quarters', they position two sections of *buk* 'pandan-bark sheet' along the lower section, leaving a gap open between the *pinj* 'ridge supports' that will serve as the crawl-through doorway into the sleeping quarters. They then lash the split-sapling battens permanently in place on both interior walls with strands of rattan, incorporating the *pongow* stakes with a turn or two around their ends, so sandwiching the pandan-bark sheets tightly against the *waenuw* stakes (Figure 7.21). Again, some men simply spiral-wind the rattan once tightly around the two battens and stakes (Figure 7.22), while others execute a more elaborate lashing. They may put the bark sheet on the women's or the men's side of the stakes on the *tenda waenuw* wall, with the pithy inner surface towards the stakes and the bark side facing into whichever room. The bark sheet should go on the sitting-room side of the stakes of the *wil waenuw* wall, unless the builders intend to face the lower part of the wall on the day-room side with *shwimb* tree bark, when they put it on the sleeping-room side (again, inner-bark side faces the stakes).

On the *wil waenuw* wall, they may again weave a *tabok* figure-of-eight pattern in rattan (Figure 7.18) around the *gamonk* 'battens' where they run between the two *pinj* 'ridge stakes'. This will comprise the top of the *deriyau* 'doorway' into the men's sleeping room, giving a pleasing

15 Pandan leaves are alright here as they do not house pigs along this wall.

Plate 7.27 Lashing pair of gamonk 'battens' across waenuw 'interior wall'

Plate 7.28 Figure-of-eight lashing.

finish, as with the women's room doorway. They may also secure a couple of dummy stakes to the battens above the doorway to fill the gap in the *waenuw* wall. And if the gap between the *pinj* 'ridge stakes' is too wide for a doorway, as in the Honael house, they insert two other stakes next to them to narrow it appropriately.

A further double row of battens called *tuwmainj*,[16] comprising several flexible *shonon* sapling withies, are lashed in an arc around the top edge of the *waenuw* stakes on all partition walls, following the contour of the roofline, and secured using a rattan strand in a simple spiral lashing around each stake. The wall builder then inserts lengths of *sezel* cane through both the *gamonk* and *tuwmainj* battens, filling in the gaps between the stakes, pushing into the soil and trimming off to the roofline.

Finally, men cover the upper part of the room partitions with dry *aendashor* pandan leaves arranged horizontally over the *waenuw* stakes. This wall surface is called the *hwimb* too. They press the leaves flat and even, and cut off the semicircular brackets that attached them to the tree. They secure the leaves in place with *pongow* 'fastening sticks' jammed vertically between the *gamonk*

16 The word *tuwmainj* is also the name of the willie wagtail bird (*Rhipidura leucophrys*) but according to Wola people there is no connection.

Plate 7.29 Completing figure-of-eight binding along top of door frame.

Plate 7.30 Lashing the tuwmainj saplings in place on waenuw 'interior wall.'

Plate 7.31 Inserting sezel cane stems along waenuw 'interior wall.'

Plate 7.32 Arranging aendashor pandan leaves over upper part of waenuw 'interior wall' stakes.

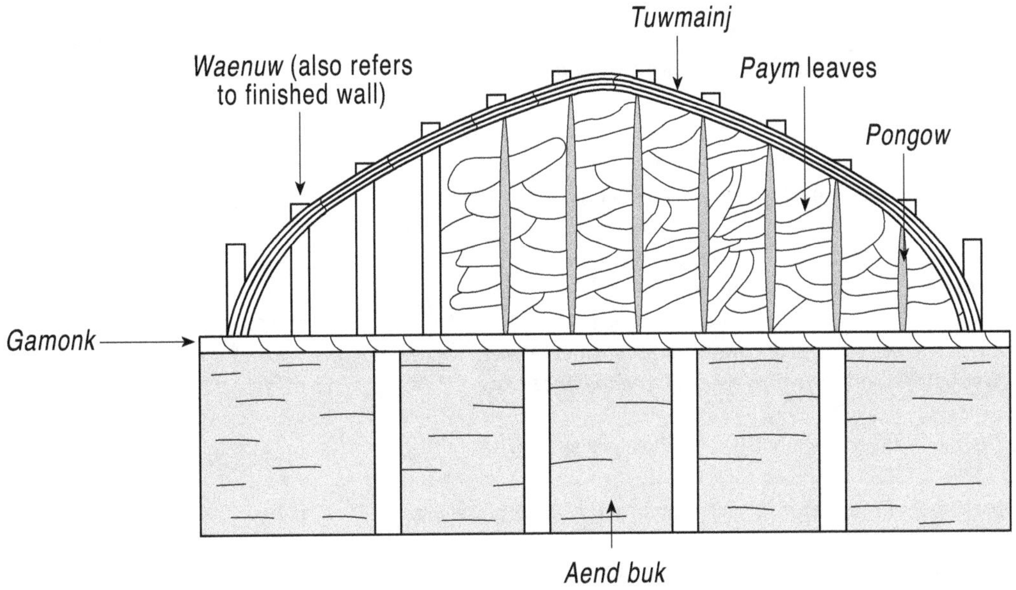

Figure 7.23 Pandan-leaf lining of upper waenuw 'interior wall.'

and *tuwmainj* battens (Figure 7.23). At either end of the wall, where the *tuwmainj* batten arch is low, they lash the tops of the *pongow* to the opposite *waenuw* stake. It is usual to place the stout pandan leaves on the *aend wil* 'sitting-room' side of the *wil waenuw* wall, regardless of the side to which the *aend buwk* 'pandan bark' is attached. Sometimes men line both sides of the upper wall with pandan leaves. It is on the *hwimb* part of *wil waenuw* wall that men display animal skulls as hunting trophies and the mandibles of pigs slaughtered, especially those from large animals with curling canine tusks (mentioned in Chapter 4's 'dig report'; Sillitoe 2003:136–7). There is no rule as to which side of the *tenda waenuw* wall they attach the pandan leaves, it may be the same side as the pandan-bark lining or not; again, they may put leaves on both sides.

It took nearly ten man-hours to erect the interior walls of the Honael *enzatenda* house (Table 7.3). Regarding the comparative time efficiencies of steel- and stone-tool use, there is no difference, which is perhaps to be expected, as the construction work features little axe work. (Note that the restriction of stone- and steel-tool use to one side of the house divided longitudinally in two obviously did not apply here; instead, steel tools only were used in constructing the rear *tenda waenuw* interior wall and stone tools the front *wil waenuw* interior wall.) Again, these data show how the time taken to complete the same tasks can vary with men working at different rates.

Construction of other walls

The front external wall is the last one erected, enclosing the, until now open-fronted, structure. It varies in construction depending on type of house. If it is a *tenda* 'women's house' or a *momaenda* 'mixed house' with a single apsidal semi-domed end, the front external wall is identical in construction to the *wil waenuw* wall described above (it is likely to be the only wall across the width of the house, as internal partitioning walls are usually absent). If such a house has two apsidal semi-domed ends, it will require no front external wall as enclosed all around with *aend dol* 'wall stakes' and accessed via doorway(s) situated on the side (Figure 7.24).

If the house is an *enza* 'men's house' or an *enzatenda* 'men's-women's house', it will feature a low *hiytpaip* 'front external wall' (literally *hiyt*-fence) of characteristic open

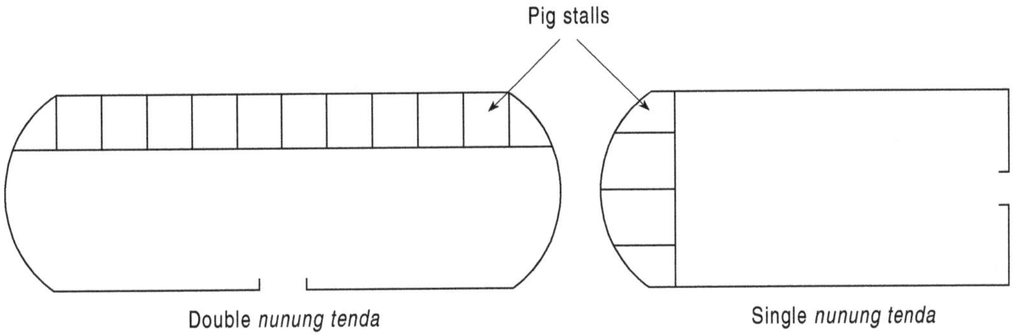

Figure 7.24 Ground plans of tenda 'women's houses.'

<table>
<tr><th rowspan="2">Task</th><th rowspan="2">Wola term</th><th colspan="2">Time spent</th></tr>
<tr><th>Stone- tool side</th><th>Steel-tool side</th></tr>
<tr><td>clearing debris out of house site in preparation for erecting pinj supports</td><td>dowhuwniy muw tuway</td><td colspan="2">2 m, 30 s</td></tr>
<tr><td>measuring position of pinj supports with cane-grass stem</td><td>gaimb hulom pinj maenjet bay</td><td colspan="2">5 m</td></tr>
<tr><td>driving in pinj supports</td><td>aend pinj aezay</td><td colspan="2">22 m, 55 s</td></tr>
<tr><td>putting bez ridge-poles in place, adjusting pinj supports and trying some polpen rafters</td><td>aend bez ma taebtuw diy polpen maenjet bay</td><td colspan="2">32 m, 50 s</td></tr>
<tr><td>paring waenuw dividing wall stakes</td><td>waenuw waendilay</td><td></td><td>22 m, 30 s</td></tr>
<tr><td>driving in waenuw dividing wall stakes[1]</td><td>waenuw aezay</td><td>23 m</td><td>22 m, 15 s</td></tr>
<tr><td>splitting rattan vine</td><td>haeluwp dekel bay</td><td>1 m, 15 s</td><td>12 m</td></tr>
<tr><td>splitting saplings for battens</td><td>waenuw gamonk dekel bay</td><td></td><td>2 m, 30 s</td></tr>
<tr><td>lashing double row of temporary battens on waenuw dividing walls</td><td>waenuw gamonk maen kay</td><td>12 m, 30 s</td><td>7 m</td></tr>
<tr><td>cutting pongow stakes to secure bark sheets against waenuw walls</td><td>waenuw pongow waeray</td><td>30 m, 30 s</td><td>16 m, 30 s</td></tr>
<tr><td>measuring off pieces of pandan-bark sheet for either side of wil waenuw doorway</td><td>wil waenuwon aend buwk maenjet bay</td><td>11 m</td><td></td></tr>
<tr><td>cutting off 2 pieces of pandan-bark sheet</td><td>aend buwk pelay</td><td>10 m, 30 s</td><td></td></tr>
<tr><td>positioning pandan-bark sheets on waenuw walls</td><td>aend buwk waenuw bort tor bay</td><td>3 m, 30 s</td><td>18 m</td></tr>
<tr><td>securing pandan-bark sheets in place with pongow stakes</td><td>pongow ma haebtuw tongay</td><td>18 m</td><td></td></tr>
<tr><td>cutting bark sheet longitudinally in half for top part of waenuw wall</td><td>aend buwk gənk pelay</td><td></td><td>5 m</td></tr>
<tr><td>positioning half sheet and securing with pongow stakes</td><td>aend buwk tor bubtuw pongow haeruw ma haebtuw tongay</td><td></td><td>16 m</td></tr>
<tr><td>flattening 2 paym pandan leaves to fill in above pandan-bark sheet (instead of half bark sheet)</td><td>paym shor ma bat say</td><td>4 m, 45 s</td><td></td></tr>
<tr><td>cutting 2 paym pandan leaves to size and jamming in place</td><td>paym shor kaerubtuw torbay</td><td>3 m</td><td></td></tr>
<tr><td>adjusting some waenuw stakes</td><td>waenuw duwn bay</td><td></td><td>2 m</td></tr>
<tr><td>lashing gamonk battens and pongow stakes in place permanently</td><td>pongow gamonk kab tongay</td><td>15 m</td><td>16 m</td></tr>
<tr><td>bending tuwmainj saplings across top of waenuw walls and lashing in place[2]</td><td>tuwmainj tongay</td><td>35 m</td><td>18 m, 30 s</td></tr>
<tr><td>pushing cane stems in along waenuw dividing walls[3]</td><td>waenuw sezel bombay</td><td>9 m, 30 s</td><td>18 m, 30 s</td></tr>
<tr><td>trimming cane level with top of waenuw dividing walls</td><td>sezel payay</td><td>3 m, 30 s</td><td>1 m, 30 s</td></tr>
<tr><td>cutting two stakes for either side of uwpuluwk tubael doorway, driving in and securing under gamonk beading</td><td>tiriyau tokol waendiyubtuw diy aezubtuw diy tongaebtuw diy</td><td></td><td>8 m</td></tr>
<tr><td>lashing figure-of-eight binding round double beading over door openings[4]</td><td>tiriyau tabok lay</td><td>36 m, 40 s</td><td>19m</td></tr>
<tr><td>cutting semicircular bracket off paym pandan leaves</td><td>paym hogor kaeruw tuway</td><td>3 m, 15 s</td><td>5 m, 30 s</td></tr>
<tr><td>flattening paym pandan leaves</td><td>paym shor ma bat say</td><td colspan="2">5 m, 45 s</td></tr>
<tr><td>cutting pongow jamming sticks[5]</td><td>hwimb pongow waeray</td><td rowspan="2">38 m, 45 s</td><td>29 m, 15 s</td></tr>
<tr><td>securing paym pandan leaves on waenuw walls</td><td>paym shor waenuw hwimb torbay</td><td>20 m, 15 s</td></tr>
<tr><td>total times</td><td></td><td>4 h, 54 m, 10 s</td><td>4 h, 54 m, 45 s</td></tr>
</table>

1 When driving in the stone stakes, some paring of stakes undertaken too. One half of the stakes on each *waenuw* dividing wall (*wil/uwpuluwk tubael* and *tenda/uwpuluwk tubael*) were prepared and erected using steel or stone tools, but subsequently (in view of the impracticability of dividing further work on these internal walls in half) all of the work in erecting the *wil/uwpuluwk tubael* dividing wall was undertaken using stone tools and all of that erecting the *tenda/uwpuluwk tubael* dividing wall was undertaken using steel tools.

2 Using stone tools, also cut two small pieces of wood to fill in gap over top of *uwpuluwk aend* men's room doorway.

3 On *enza* sleeping-room side of *tenda waenuw* wall using steel tools and on *enza* sleeping-room side of *wil waenuw* dividing wall using stone tools.

4 On steel side worked on *uwpuluwk aend* men's-room doorway and on stone one *tenda* women's-room doorway.

5 Used pieces of wood cut off when trimming *polpen* rafters.

Table 7.3 Construction of internal walls.

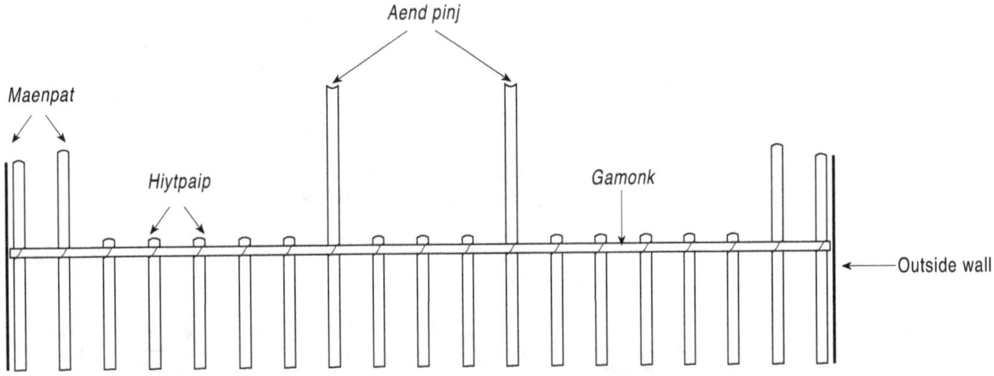

Figure 7.25 Alignment of exterior hiytpaip 'men's sitting-room-wall' stakes

design. It comprises a row of short stakes enclosing the *aend wil* room behind a low fence-like barrier over which men have to step in entering the house, leaving open the upper part, framed like a triangular window, by the roofline. In erecting the wall, house-builders first drive in the stakes in line with the front pair of *pinj* 'ridge-pole supports'. The majority are short *hiytpaip* 'low-fence' stakes, sixty or so centimetres high, while the two or three stakes at either end called *maenpat* are longer, cut to follow the slope of the roofline (Figure 7.25). They locate the line of stakes in from the end of the exterior walls, by a couple of *dol* 'stakes', some forty centimetres (Figure 7.26). When roofed, this recess will serve, porch-like over the open-fronted *aend wil* entrance, to protect the occupants from wind and slanting rain.

The builders initially secure a temporary double row of *gamonk* 'battens' along the stakes of the front wall, positioned either side of the stakes and aligned with the top of the short *hiytpaip* stakes, using a simple spiral rattan-strand lashing. Next they position a sheet of *buk* 'pandan bark' along the outside of the stakes, the grey bark surface facing outwards and extending to the height of the short *hiytpaip* stakes. They secure it again by driving in *pongow* stakes that they lash to the tops of adjacent wall stakes, so sandwiching the bark sheet between the stakes. They simultaneously fasten the *gamonk* 'battens' permanently in place, again with a simple spiral lashing. They may also add a *tuwmainj* 'batten arch' of flexible saplings bent round the line of the roof, as on other partition walls.

It is usual to cover the *maenpat* side stakes with dry *aendashor* 'pandan' leaves, secured in place with *pongow* stakes jammed under the *gamonk* 'battens' and *tuwmainj* 'batten arch' (if the arch is added to the front wall; if not,

Plate 7.33 Fashioning a short dol 'stake' for the low hiytpaip 'front wall'.

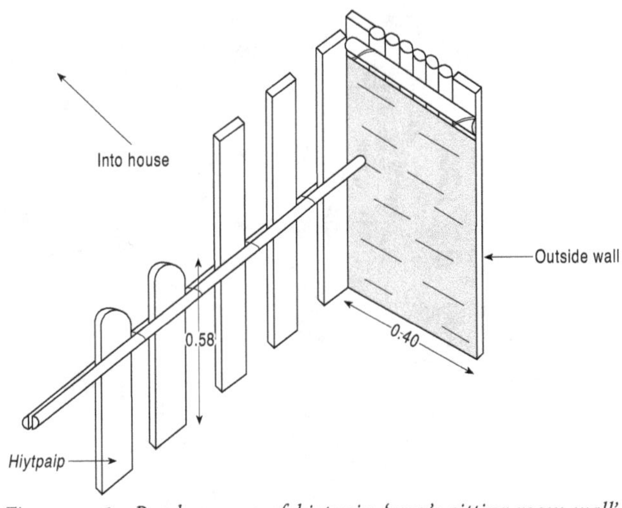

Figure 7.26 Porch corner of hiytpaip 'men's sitting-room-wall' entrance (m).

Plate 7.34 The low hiytpaip 'front wall' showing longer maenpat side stakes following slope of roof line.

Plate 7.35 Fastening gamonk 'battens' and pongow 'securing stakes' in place along hiytpaip 'front wall'.

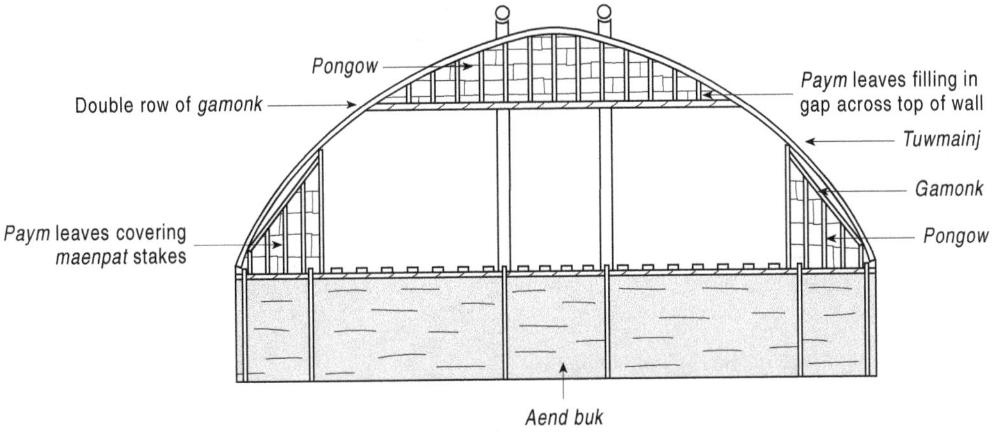

Figure 7.27 Front view of aend wil 'men's sitting-room' wall with open entrance.

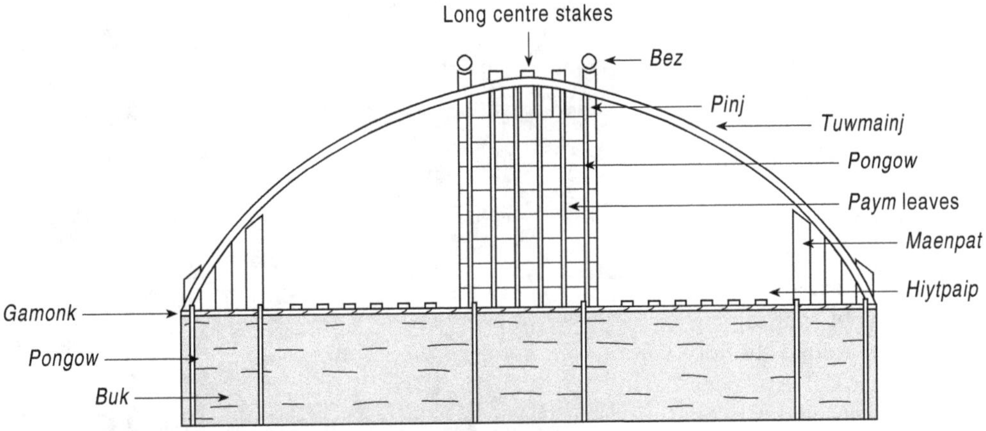

Figure 7.28 Front view of aend wil 'men's sitting-room' wall with partially closed entrance.

Plate 7.36 The lashing used to secure pair of gamonk 'battens' along top of hiytpaip 'front wall.'

men add a short *gamonk* 'batten' lashed along the top edge of the *maenpat* side stakes following the angle of the roofline). They may go further and lash a double row of split-sapling *gamonk* 'battens' across the top of the *aend wil* entrance and fill in the triangular crown space demarcated below the house ridge with dry *aendashor* 'pandan' leaves, jamming with *pongow* stakes placed on both sides under these *gamonk* 'battens' and the *tuwmainj* 'batten arch' (Figure 7.27). On some houses they fill in the front wall further by driving in long stakes between the two *pinj* 'ridge supports', reaching up to the roofline and again facing with dry *aendashor* 'pandan' leaves, secured in place with *pongow* stakes jammed vertically under the *gamonk* 'battens' and *tuwmainj* 'batten arch' (Figure 7.28). They fill in the centre of the front wall in this way, they say, if the wind blows smoke around too much inside the sitting room.

On some houses men entwine a fancy rattan trussing around the *gamonk* 'battens' along the top edge of the low *hiytpaip* wall where men climb over into the *wil* 'day room', to give a durable finish with a pleasing appearance. They produce a design, called *ndul*,[17] by twining a strand

17 Some refer to the pattern as *baerob*; both *ndul* and *baerob* refer to twilled patterns of armband weave (Sillitoe 1988a:422),

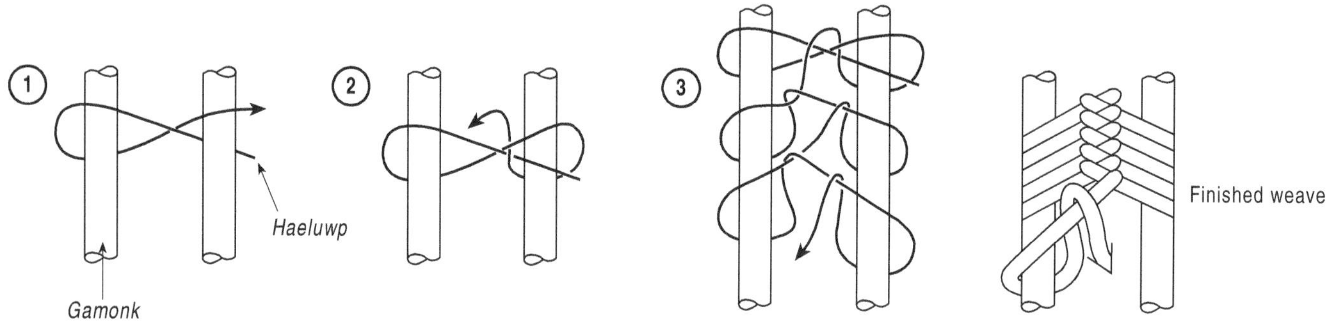

Figure 7.29 *The ndul weave on gamonk battens of hiytpaip 'men's sitting-room wall.'*

of rattan back and forth around the two battens (Figure 7.29). The practice is a relatively recent innovation that arrived sometime in the 1970s in the Was valley from the neighbouring Nembi valley. Other houses retain the simpler lashing used previously, which is the same as the twisting lashing that secures the top double row of battens on the *aend dol* exterior wall.

The final walls built are those of the *showmay kuwl* 'pig stalls', if there are any in the house. These are less walls than staked pens, being similar in construction to battened garden fences (Sillitoe 2010a:271–2). They comprise stakes driven into the ground to the height of the exterior walls, demarcating individual stalls, with lengths of split-sapling *gamonk* 'batten' lashed spirally both sides around the top edge, as on the outside walls (Figure 7.30). These pens are constructed in the women's quarters only, and may be found in *tenda* 'women's houses', *momaenda* 'mixed houses' and *enzatenda* 'men's and women's houses'. It is usual to site two or three pig stalls in the apsidal semi-domed end of such houses. In some double-apsidal-ended houses several may be erected down the long wall opposite the crawl-through doorway, particularly if the occupant customarily herds large numbers of animals (Figure 7.24).

People place heavy split logs called the *kat tor* ('*kat* structure') longitudinally across the top to roof the stalls; the logs dry out nicely and may act as a firewood store, to be replenished with new timber. Finally, they stake a *dombep* 'sill log' along the ground in front of the stalls, which acts to keep the pigs' refuse inside the stalls, preventing it spreading around the women's living area. The stalls are nothing like farm pig sties of popular imagination. They are relatively clean and do not smell strongly; the animals infrequently defecate indoors. The waste within largely comprises uneaten remains of food and foliage bedding; and women regularly clear these out, often throwing such waste on nearby vegetable gardens (this contrasts with Enga pens, according to Meggitt 1957:173, that are roughly floored to keep pigs out of their muck).

Plate 7.37 *Fastening gamonk 'battens' along top edge of showmay kuwl 'pig-pen' stakes*

which the herringbone pattern along the top of the wall resembles.

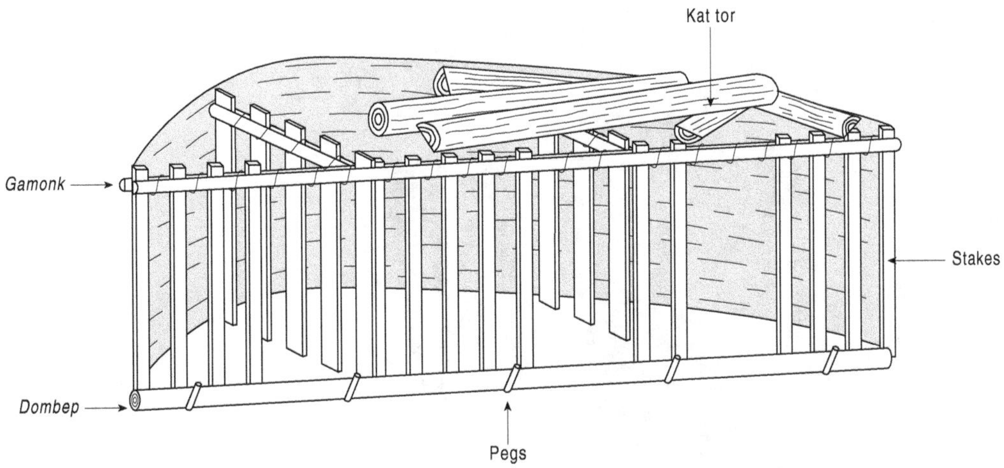

Figure 7.30 Construction of pig stalls.

Today some families house pigs in separate structures, heeding the admonitions of colonial and missionary authorities about it being dirty and unhygienic to sleep with their animals (Chapter 1, Sillitoe 2003:251, 346). It is unclear to what extent these cautions reflect the reaction of Europeans to people living with 'mucky' pigs and to what extent they represent advice based on sound medical evidence. It seems likely that Highlanders who have lived in close proximity to pigs for centuries, if not longer, would have noticed dangers to their health long ago. Whatever, women say that stalling the animals in with them assists in their management, as it ensures that the pigs in their charge 'know us'. Also, the animals' body heat contributes significantly to a room's temperature; they act like bristly radiators housed along the wall. Together with a smouldering fire on a bed of hot embers, women and children are assured warm cosy rooms during the night.

Ensuring a warm shelter is certainly an issue in the chilly mountains of New Guinea where night-time temperatures can fall to 5°C or less (Sillitoe 1996:70), unlike sultry lower and coastal altitudes, where the main purpose of shelter is to provide privacy and protection from rain, and possibly any evil spirits lurking in the vicinity. The squat Highlands' house is well designed for wet and chilly montane conditions, being waterproof, windproof and well insulated. Data on the micro-climate inside the houses of Enga speakers in the Lagaip region, who occupy a mountainous environment and squat houses similar to the Wola, show their thermal effectiveness (Sinnett and Whyte 1973:257). The mean difference in the minimum

Plate 7.38 The pig stalls showing the kat tor logs stacked on the top.

Plate 7.39 Pig standing inside stall.

temperature inside and outside houses was 11.8 °C (range inside being 14.5–23.8 °C and outside 7.0–10.3 °C) and the mean difference in the maximum temperature inside and outside was 6.6 °C (range inside being 19.5–31.2 °C and outside 16.3–25.8 °C). The need for a snug home accounts for the squat design of Highlands' houses. A cosy fug can soon be built up inside such low structures with a fire.

The downside is that they can sometimes be rather smoky, although usually only when a fire is first kindled; once established the smoke levels fall considerably.[18] Arguably, the smoky atmosphere is a negative well-being issue, possibly harming occupants' pulmonary health. For instance, Sinnett and Whyte (1973:270–1) refer to 'smoke-filled houses' in reporting on the 'progressive development of obstructive and restrictive airway disease' with age in the Lagaip region. They echo Cleary's and Blackburn's (1968) association of chronic respiratory disease with such domestic air pollution, who found that air samples from Western and Eastern Highlands houses had high levels of aldehydes, formaldehydes and particulate matter in addition to carbon monoxide. According to Master (1974), his Lufa region data unquestionably support the association. But tobacco smoking arguably contributes more to respiratory ill health than housing conditions. Data from the neighbouring Huli region support this view, where men over forty years, many of whom have smoked heavily during their adult lives (unlike women, who until recently were discouraged from smoking), suffer high rates of respiratory infection (at 17.6 days a person a year, compared to 4.5 days for women of the same age – Frankel 1986:68–9). It is a correlation that challenges the house-pollution argument that asserts that regarding 'the prevalence of lung disease, air pollutants from smoky fires in poorly ventilated huts remain the major and most preventable factor' (Master 1974:1655).

It took a little over six man-hours to erect the front wall and insert the pig stalls of the Honael *enzatenda* house (Table 7.4). The construction of the front wall accounted for 44 per cent of this time and the pig stalls the rest. Regarding the comparative time efficiencies of steel- and stone-tool use, stone proved more efficient, with a ratio of 1 for stone to 1.3 using steel; that is, for every hour worked using stone tools the builders spent an hour and eighteen minutes working with steel ones. Again, while perhaps unexpected, the time difference is not large, and is partly attributable to variations in the time men took to complete the same

18 An abiding memory of an old college friend's visit to the Highlands involves smoke levels in houses. We were sheltering from the rain in an empty *momaenda* house. While Wola friends kindled a fire, he sat upright on one side of the fireplace talking as I lay down on the other with my head supported on my forearm. The thick smoke from the newly lit fire gradually filled the roof space, descending to billow through the doorway. Suddenly my friend, who had been sitting stoically with his head increasingly enveloped in a cloud of smoke, started to cough uncontrollably and made a desperate rush for the doorway, diving through it to gasp in fresh air. He turned towards me with his eyes watering and we both burst into laughter. Soon the smoke thinned when a bed of hot embers was established and he returned to his place and we cooked a meal.

<table>
<tr><th rowspan="2">Task</th><th rowspan="2">Wola term</th><th colspan="2">Time spent</th></tr>
<tr><th>Stone-tool side</th><th>Steel-tool side</th></tr>
<tr><td colspan="4">Low front wall</td></tr>
<tr><td>paring short maenpat stakes</td><td>maenpat waendiyay</td><td></td><td>46 m, 15 s</td></tr>
<tr><td>driving stakes in and trimming to size</td><td>maenpat aezay</td><td>40 m</td><td>16 m, 45 s</td></tr>
<tr><td>cutting pongow stakes[1]</td><td>maenpat pongow waendiyay</td><td>3 m, 45 s</td><td>10 m, 30 s</td></tr>
<tr><td>splitting rattan vine</td><td>haeluwp dekel bay</td><td colspan="2">9 m, 15 s</td></tr>
<tr><td>splitting sapling in half for gamonk batten and lashing temporarily in place</td><td>maenpat gamonk dekel bubtuw maen kay</td><td colspan="2">7 m, 45 s</td></tr>
<tr><td>measuring pandan-bark sheet</td><td>aend buwk maenjet bay</td><td colspan="2">1 m, 30 s</td></tr>
<tr><td>positioning bark sheet</td><td>aend buwk maenpat bort tor bay</td><td colspan="2">2 m, 15 s</td></tr>
<tr><td>pinning bark sheet in place with pongow stakes</td><td>aend buwk pongowom ma hae</td><td colspan="2">8 m, 45 s</td></tr>
<tr><td>lashing gamonk batten and pongow stakes in place permanently</td><td>pongow gamonk kab tongay</td><td colspan="2">23 m, 30 s</td></tr>
<tr><td>total times</td><td></td><td>1 h, 8 m, 45 s</td><td>1 h, 38 m, 30 s</td></tr>
<tr><td colspan="4">Pig stalls</td></tr>
<tr><td>collecting timber[2] for stall stakes</td><td>iysh pel maenmaen bay</td><td></td><td>4 m, 45 s</td></tr>
<tr><td>paring stall stakes[3]</td><td>showmay kuwl paip waendiyay</td><td>46 m, 25 s</td><td>42 m, 15 s</td></tr>
<tr><td>marking out plan for 3 pig stalls</td><td>showmay kuwl maenjet bay</td><td colspan="2">8 m</td></tr>
<tr><td>passing stakes into house</td><td>paip iysh aendon kalay</td><td></td><td>1 m</td></tr>
<tr><td>driving in stall stakes</td><td>showmay kuwl aezay</td><td colspan="2">28 m, 30 s</td></tr>
<tr><td>splitting saplings in half for gamonk battens and lashing along top of pig-stall stakes[4]</td><td>maenpat gamonk dekel bubtuw diy tongaebtuw diy</td><td colspan="2">53 m, 30 s</td></tr>
<tr><td>splitting tree trunks for stall roof[5]</td><td>iysh dekel bay</td><td>1 m</td><td>2 m, 30s</td></tr>
<tr><td>placing logs on pig-stall roof[6]</td><td>iysh kat tor wiy</td><td>1 m</td><td>5 m, 30 s</td></tr>
<tr><td>measuring footing log, cutting and pegging in place</td><td>dombep lay</td><td></td><td>11 m, 30 s</td></tr>
<tr><td>total times</td><td></td><td>1 h, 34 m, 55 s</td><td>1 h, 54 m, 00 s</td></tr>
</table>

1 3 *pongow* stakes cut with stone tools and 3 with steel.
2 Southern beech split previously for wall stakes.
3 12 stakes pared using stone axe and 12 stakes with steel axe.
4 Also trimmed a few stakes in the process.
5 Stone-felled tree trunk split into 2 pieces and steel-felled one into 7 pieces.
6 The steel time includes finding 4 more logs laying around house site and putting on top of stalls.

Table 7.4 Construction of other walls.

tasks, with a considerable amount of the work undertaken not requiring any axe work, stone or steel. When they did use these tools, they may sometimes have chosen to work on easier materials with stone axes (such as smaller timber pieces requiring less work to split and dress), leaving the harder materials for steel tools, which they could also have worked on longer to achieve the better finish possible.

Breaking the chain?

The next step in house building is roof construction, although it is usual to assemble some of the roof framework before completing the internal walls, as some of the illustrations show. House-builders routinely proceed in this interdigitated way, going from one activity to another and back again during house construction, which somewhat compromises their presentation according to different stages. It is an aspect of the flowing approach to work noted previously, artisans intuitively following the 'flow' of variations in materials, responding to their qualities as 'they work with the grain', which can result in them toing and froing between tasks, not completing them in a sequence as in this book's recounting of house construction. This prompts criticism of the associated *chaîne opératoire* approach (see Chapter 2) for not apparently realizing that 'we have here an unbroken, contrapuntal coupling of a gestural dance with a modulation of the material' (Ingold 2012:434); albeit Leroi-Gourhan, the coiner of the term *chaîne opératoire*, was apparently aware of the issue, as the earlier quote shows. The responsiveness of skilled persons to materials is an aspect of matter having a 'life proper to matter' (Deleuze and Guattari 2004:454), which the blueprint approach purportedly hides in seeing raw materials as inert matter shaped according to a preconceived image. But an account arranged according to a series of defined technical operations does not preclude the likelihood of toing and froing, as in the construction

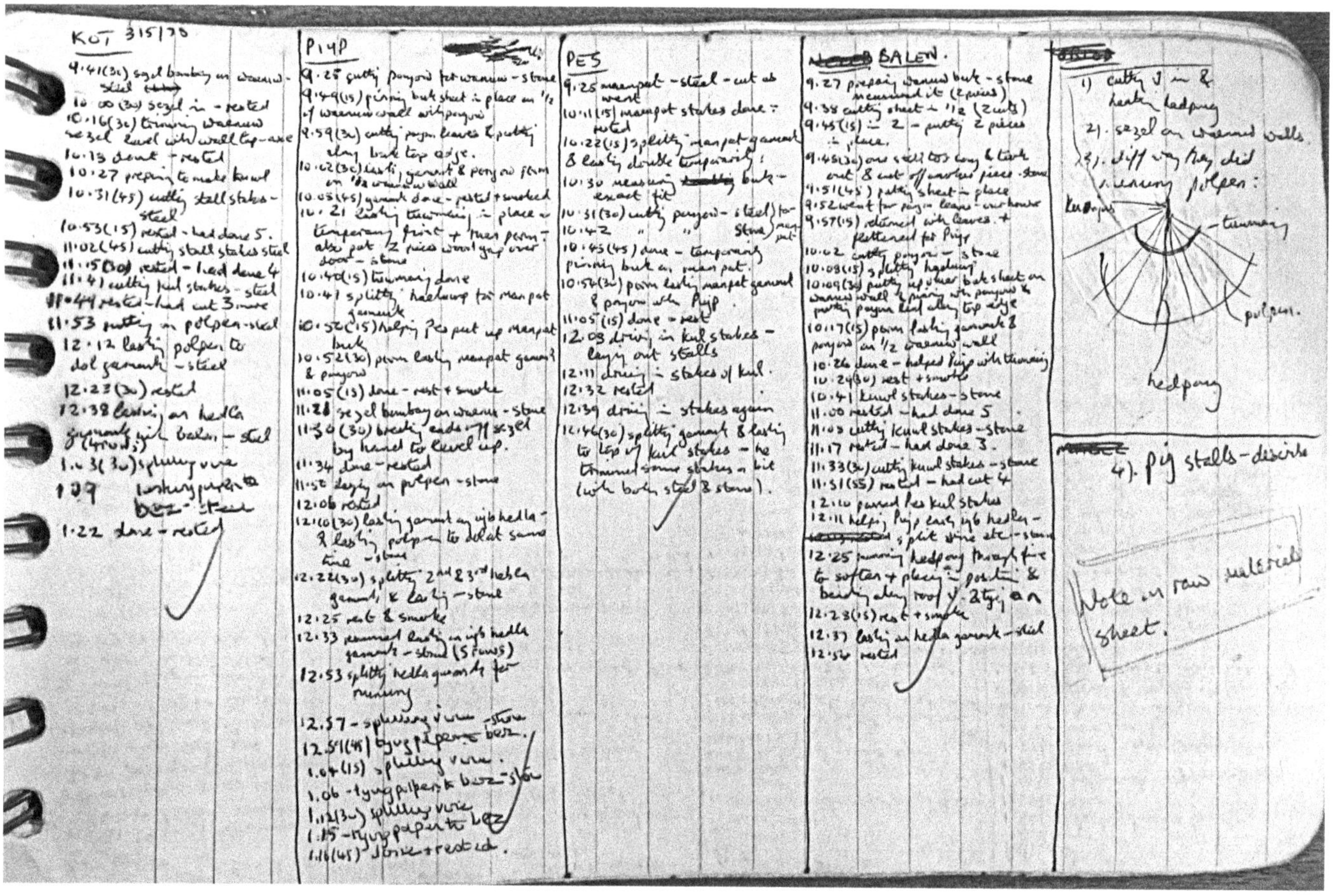

Plate 7.40 Page of field notebook documenting activities during Honael house building.

of the Honael house. The problem is how to reproduce such hopping from one task to another and back again; the only ways I can think of would be either to publish my chaotic field-notebook records, which would give a difficult to penetrate record, or to film the entire process of house building, which would result in several days of movie to view.[19] This seems less a criticism than an affirmation of the impossibility of catching actual experience on paper.

Furthermore, regarding variability, we can anticipate that no two lots of toing and froing will be the same in constructing different houses. Individuals vary in their skills – some more handy than others – and differ in the way they tackle tasks. The same persons may even vary in the way they do things from one occasion to another. If we seek to engage with the actual variation that characterizes life and document, for instance, the construction of several houses, we shall have a very complex, if not confusing, account of many different sequences of actions. It is necessary to standardize such a record in some way to make sense of what is going on. It is more economical, and certainly easier for the reader, to present the work as a series of stages, after the convention in generations of technical and craft books (see Chapter 2), and followed by the *chaîne opératoire* approach, without implying that they are followed rigidly; although presumably there has to be some sequence, for you cannot thatch a house before erecting the walls. This approach complies with local Was valley practice and understanding, my Wola friends identifying and having terms for various stages, as the tables in this book testify, which in turn presumably inform their tacit knowing to some extent.

The toing-and-froing point relates to the notion of agency, over which there appears to be some confusion (see Chapter 5). The attempt to overturn the Cartesian inner and outer worldview threatens to throw out the bath along with proverbial baby and water. Both human actors and the materials they work on may have agency from the materiality perspective. It is necessary to engage with materials to know them. They respond to every action, and skilled persons react spontaneously in turn when working on them, exploiting their potentials to achieve their ends. This is tactile knowledge. According to this view of materiality, craftspersons do not impose their ideas on matter, but rather they emerge from working closely with

19 Both of these alternatives would make our job easier by averting the need for any analysis and editing etc., which was demanding and time consuming for the house-construction data.

materials; their ideas are not separate from the inherent properties of the materials but come out of practical engagement with them. But the outcome does not reside in the affordances of materials alone. Raw materials are not only directly perceived, as the 'ecological' approach implies, but are also indirectly known, situated in a socio-cultural world of ideas, as the 'humanistic' approach to materiality points out (Knappett 2007:22). People relate to the properties of materials, as practically experienced during any process, both objectively as tangible physical reality and subjectively as intangible cultural conception. Both inform their actions, having material in their hands and an image in their heads of what they plan to make, of the 'form' they intend to 'impose' on the materials.

A person making something surely has a blueprint in mind, knows what s/he expects to make. [20] The way that men in the Was valley mark out the plan of a house with sticks and vine demonstrates it. No one starts out netting a string bag and ends up hewing a digging stick. As Marx (1974:174) noted long ago,

> A spider conducts operations that resemble those of the weaver, and a bee puts to shame many an architect in the construction of her cells. But what distinguishes the worst architect from the best of bees is this, that the architect raises his structure in imagination before he erects it in reality. At the end of every labour-process, we get a result that already existed in the imagination of the labourer at its commencement… Besides the exertion of the bodily organs, the process demands that, during the whole operation, the workman's will be steadily in consonance with this purpose.

While an artisan inevitably engages with the materials he uses, moderating his actions in response to their emergent qualities in his hands; these do not determine the thing made so much as possibly modify somewhat its final form (Ingold 2013:115); for instance, an unforeseen knot in wood may demand a change in shape. He selects the materials after all for their suitability to make the thing he conceives of; a straight, strong trunk for a house ridge-pole or a pandan trunk for heavy sheets of bark.

What differentiates the agency of persons from that of the materials they work on is that they have a '"sense" or an "experience" of agency' such that it is 'not causality but consciousness that differentiates the human sense of agency' (Malafouris 2013:215). In intuitively following the 'flow' of material variations, craftspersons recognize that they are implicated in the processing and 'becoming' of the material, at the time they are working on it anyway. On the one hand, an artisan making something exercises agency, imposing a design on what is taken at the time as inert material, even if its qualities oblige a certain way of working with it. On the other hand, the agency of the material extends beyond its current qualities, being caught up in inevitable ecological processes that lead to its change-cum-degradation as an object over time. We see that there are two senses to the proposal that objects have agency, one consequent on the 'vitality' of the materials that comprise them and the other on their 'agency' as things that interface with users. They are not mutually exclusive, as some arguments suggest. As materials, they may be subject to biophysical processes independent of human beings, as evident in the deterioration of goods over time (see Chapter 6). And as objects, they may influence, when made and used, the behaviour and capabilities of humans, even affecting the plastic neural networking of the brain (see Chapter 5), which in turn feeds back to how persons engage with, or act on the material world, both intentionally, with chain-like steps in mind, and unintentionally, with the properties of materials in hand. This latter distinction recalls the 'subsidiary' and 'focal awareness' one (Chapter 2), reflecting connections to tacit issues, which are apparent in how craftspersons skilfully adjust their tool-using movements to the material worked on, responding to its qualities, with a rhythmic union of perception and action (Ingold 2011b:58–61). It also connects with the phenomenological aspect of the 'ecological' approach to materiality, seeing matter as an 'active participant in the world's becoming' (Barad 2003:803).

Roof construction

Returning to roof construction in the Was valley, the builders lift the *bez* 'ridge-poles' back into position, having usually removed them during the erection of the walls. They place some *polpen* 'rafters' in position, resting on the ridge-poles and the *dol* exterior wall, and stand back to check again that the roof is symmetrical in shape. If it appears out of line, they pull up the offending *pinj* 'ridge support' stakes and realign as necessary (it is for this reason that they usually construct the roof framework before completing the internal walls). They rest the ridge-poles in the curve cut into the ends of the *pinj* 'ridge supports' (Figure 7.31). They are balanced there, not lashed in place.

In the event of an earthquake, the ridge-poles – not being secured to the *pinj* uprights on which they rest – can move freely to and fro, together with the entire structure. The Enga house design similarly allows for movement, featuring a single ridge-pole either resting on a horizontal beam lashed between two upright main posts or resting on top of a single dished post, similar to a Wola ridge-pole (Meggitt 1957:169–70). There is no danger in this region

20 Even critics of the *chaîne opératoire* approach, it turns out, admit to having some stepwise blueprint in mind: 'before setting out I need to have arrived at some overall conception of the task to be performed – of what is to be done, how to do it' (Ingold 2011b:54).

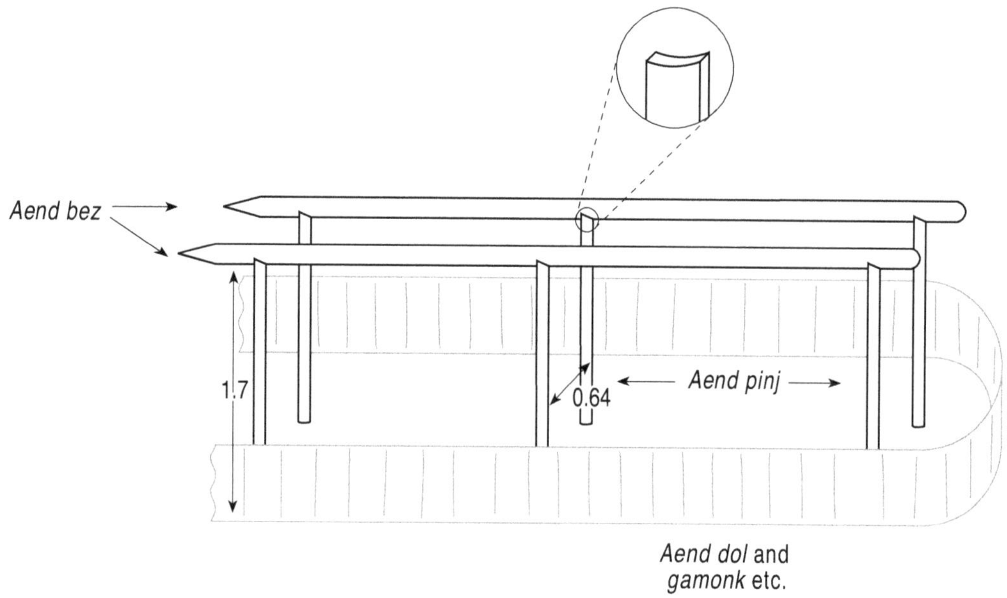

Figure 7.31 Positioning of bez 'ridge-poles' (m).

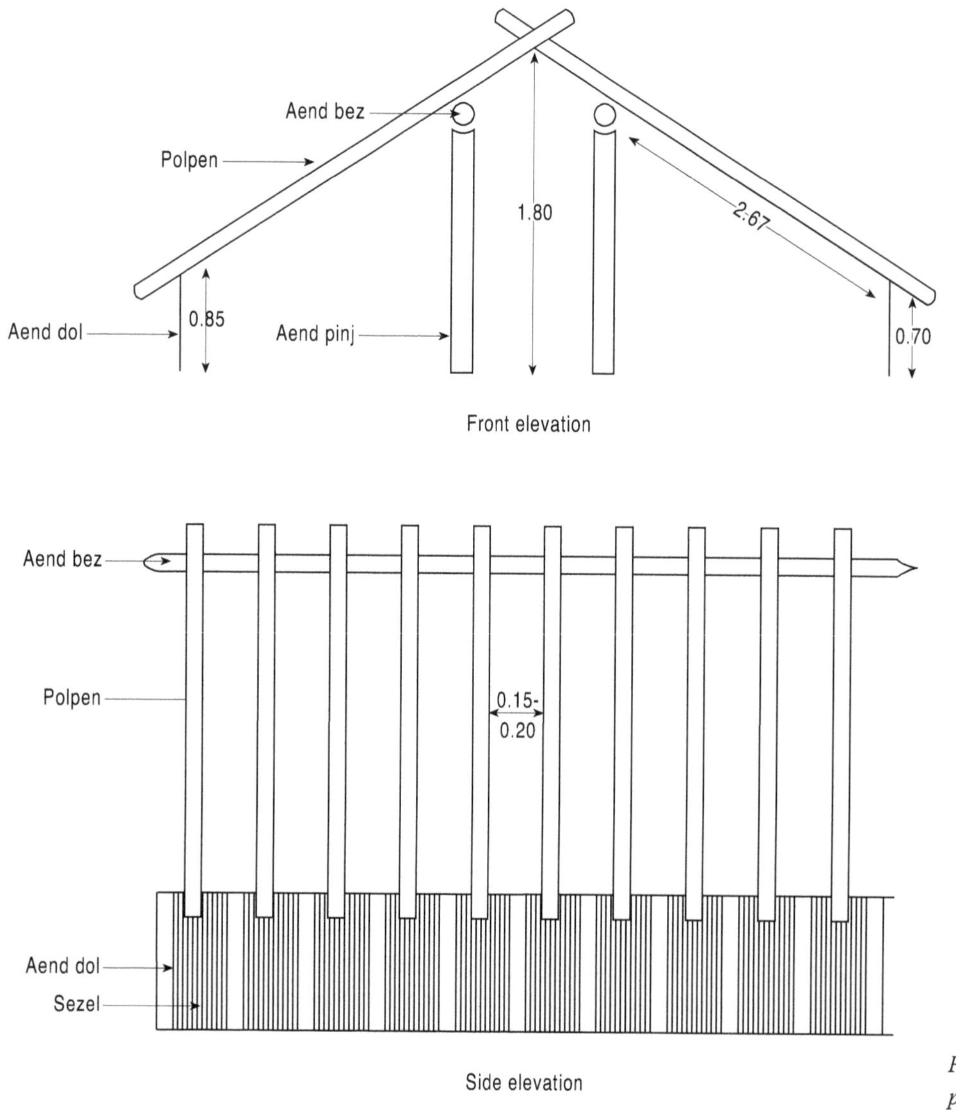

Figure 7.32 Positioning of polpen 'rafters' (m).

Plate 7.41 Placing polpen 'rafters' in position.

Plate 7.42 Lashing polpen 'rafters' to gamonk 'batten' running along the wall.

of being trapped in a collapsed house, as can tragically happen in some earthquake-prone regions of the world where occupants are buried alive under fallen masonry and reinforced concrete. Even if a quake was large enough to shake the ridge-poles off their posts, the roof would not collapse immediately (although it would obviously demand repair). Nobody has ever accounted for house design in this way to me, by referring to structural movement during earthquakes. But other tectonic activity does come up in discussions of roof construction. People talk about the need for a strong roof in case a *dorpind* 'sky-fall' event occurs. It is probable that accounts of *dorpind* refer to pyroclastic and ash fallout from volcanic eruptions, and although no such eruption has occurred for many generations – unlike *duwlbaeray* 'earth tremors' – the possibility of such an event and its dangers are passed on in myth-like accounts (see Sillitoe 1996:131–3).

When they are satisfied with the roof alignment, men put all the *polpen* 'rafters' in place at regular intervals down either side of the house, resting on the outside wall and a ridge-pole (Figure 7.32). They place a rafter between most, if not all, of the *aend dol* 'wall stakes', which gives a gap of fifteen to twenty centimetres between them. They secure the *polpen* 'rafters' in place by tying to the outer *gamonk* 'batten' running along the top of the wall (Figure 7.33), using a simple spiral lashing taking a turn or two around each rafter (Figure 7.34). Then they secure the rafters to the

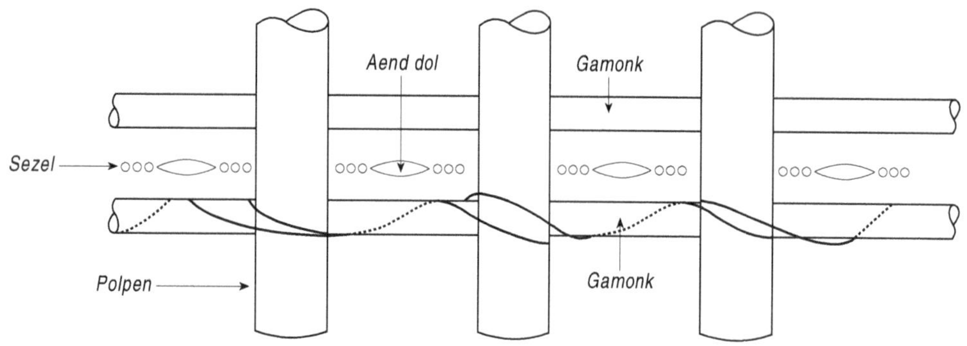

Figure 7.33 *Securing of polpen 'rafters' along wall line.*

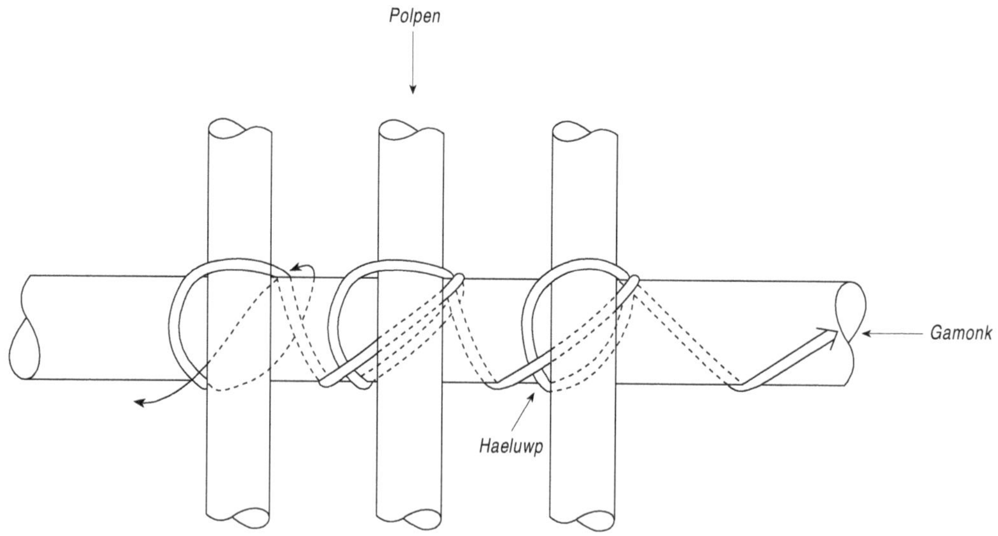

Figure 7.34 *Lashing of polpen 'rafters' to gamonk 'batten'.*

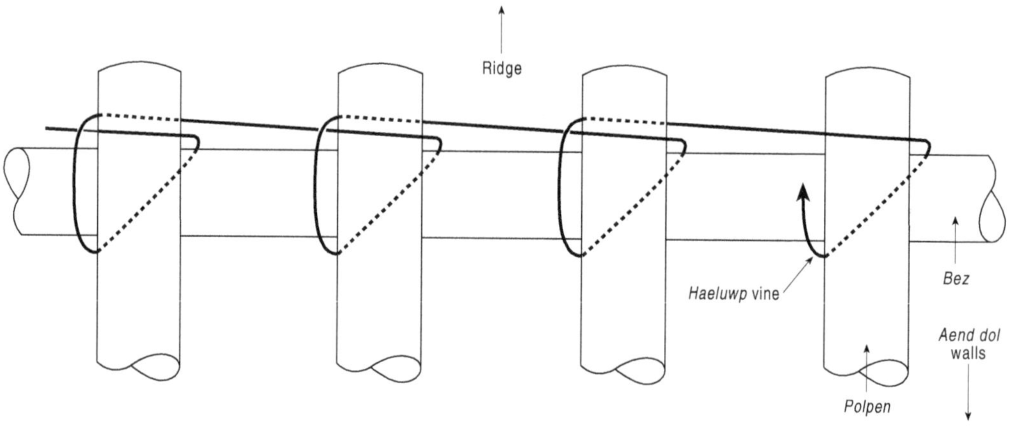

Figure 7.35 *Lashing of polpen 'rafters' to bez 'ridge-pole'.*

ridge-pole on which they rest, using a simple lashing pulled tight around each rafter (Figure 7.35). (See Steensberg 1980:143–5 for details of similar lashing in the Duna region.)

When the rafters are secure, the builders fasten a long stout sapling called the *kuwngis* in the 'V' crotch formed by the meeting of opposite *polpen* 'rafters' at the ridge of the house, using a simple spiralling lashing around each pair of rafters and the sapling (Figure 7.36). Before lashing in place, they measure where the *kuwngis* sapling reaches the final pair of rafters at the apsidal semi-domed end of the roof and cut a small nick in it ('A' in Figure 7.37). They run the nick through a fire to soften the wood and then they bend the sapling here to form an angle (if they just bent the sapling, they say it would probably break in two). The angled part of the sapling, called the *hedpong*, runs from the crotch of the rafters down to the centre of the rear semicircular wall, where it is tied to the outermost *gamonk* 'batten' running around the top of the wall.

The stout *hedpong* sapling comprises part of the semicircular roof structure at the *nunung* 'apsidal end' of

Plate 7.43 The lashing used to secure polpen 'rafters' to bez 'ridge-pole' (start).

Plate 7.44 The lashing used to secure polpen 'rafters' to bez 'ridge-pole' (finish).

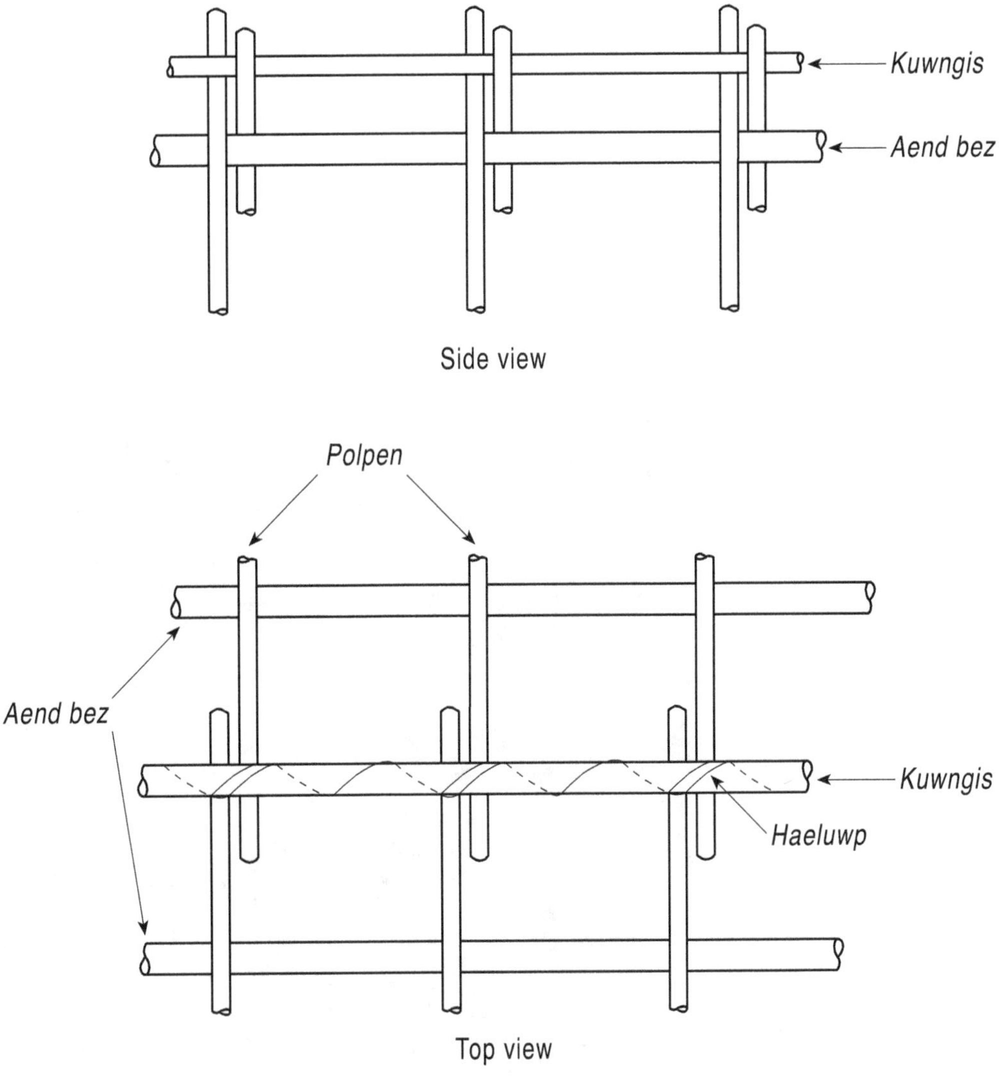

Figure 7.36 Positioning of kuwngis 'ridge tie sapling' in 'V' crotch of polpen 'rafters'.

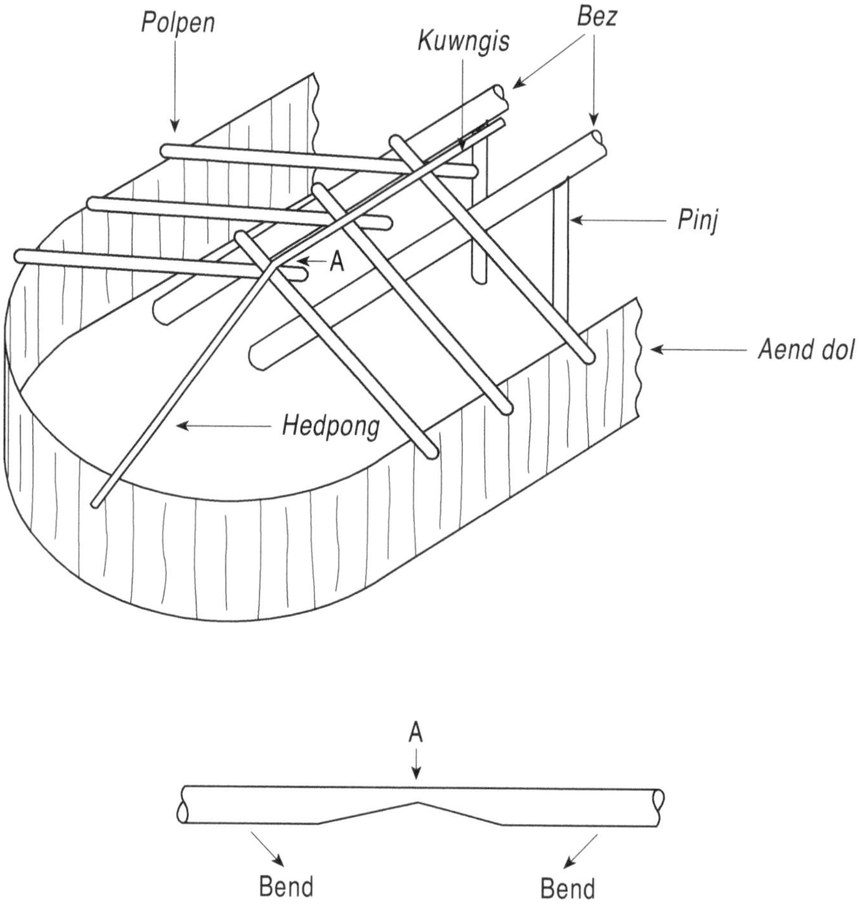

Figure 7.37 Positioning of hedpong 'apse end tie sapling'.

Plate 7.45 Tying the kuwngis 'ridge sapling' in place (note the hedpong running down to rear wall).

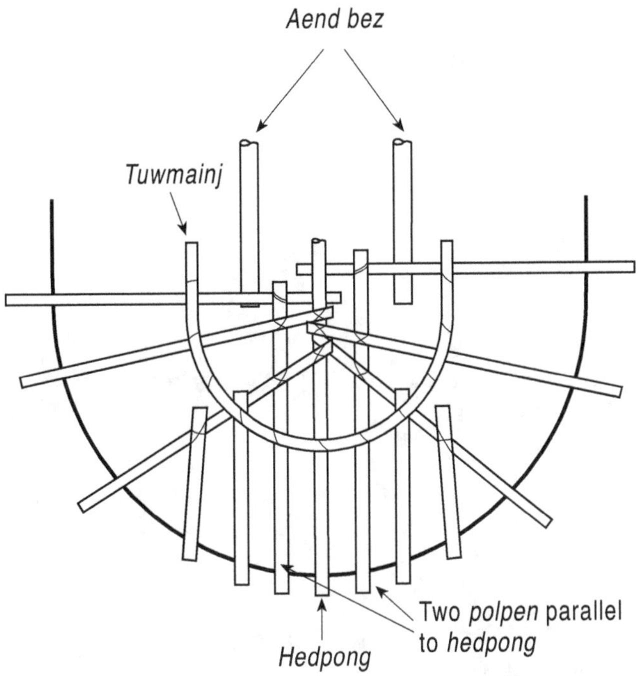

Figure 7.38 Construction of parallel-rafter apse roof.

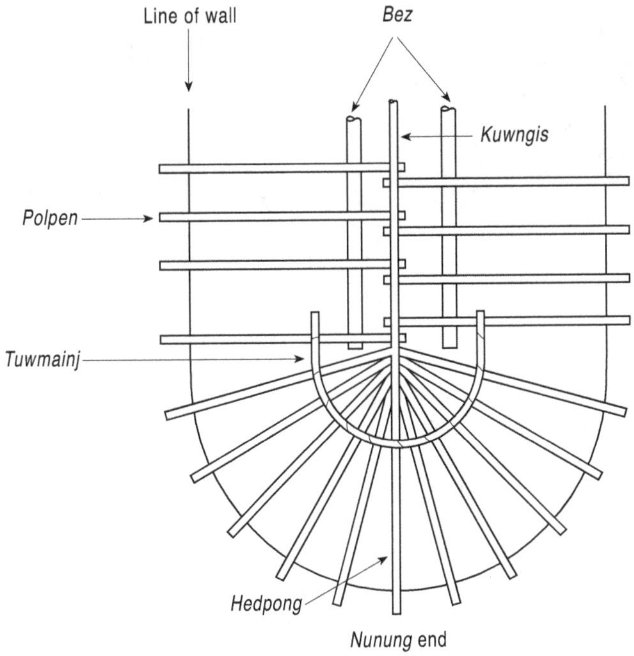

Figure 7.39 Construction of radial-rafter apse roof.

the house. The arrangement of rafters here varies. In one pattern, men lash two *polpen* 'rafters' on either side of, and parallel with, the *hedpong* sapling (Figure 7.38). They then arrange four to six further *polpen* 'rafters' at intervals around the semicircular end, bevelling their ends to fit with the roofline where they rest on the *hedpong* sapling and two parallel *polpen* 'rafters', to which they are lashed. They

Plate 7.46 Placing polpen 'rafters' in position around the semicircular roof end.

Plate 7.47 Securing the tuwmainj 'withy arc' to rafters forming semicircular end of roof.

fill in the spaces with shorter rafter lengths lashed in place. They again lash the rafters to the outermost *gamonk* 'batten' running around the top of the exterior *dol* wall. They complete the apse framework by lashing a pair of flexible *shonon* saplings together, one on the inside and the other the outside of the rafters, using a simple spiral binding. This *tuwmainj* 'withy arc' follows the rounded contour of the roof and further secures the rafters. Another apsidal roof-end design arranges all of the *polpen* 'rafters' in a radial pattern, their bevelled ends meeting at the roof ridge where the *kuwngis* bends to become the *hedpong*, resting there without lashing (Figure 7.39). It is the *tuwmainj* 'withy arc' – again comprising a pair of flexible saplings sandwiching

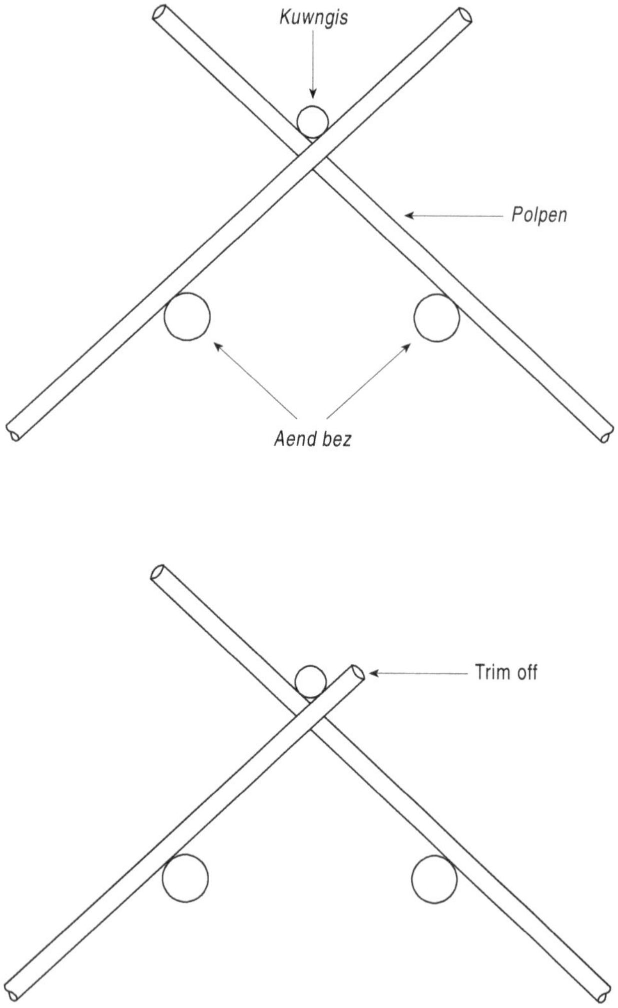

Figure 7.40 Trimming of rafters along roof ridge.

Plate 7.48 Cutting the protruding ends off polpen 'rafters' along ridge

the rafters, lashed following the roof's contour just below the ridge – which secures the rafters in place, together with the lower ends of the rafters again lashed to the outermost *gamonk* 'batten' running around the top of the exterior *dol* wall.

The house-builders do not measure and trim the rafters to size as they arrange and secure them in place along the sides of the roof. They are invariably longer than required and are positioned so that they extend up beyond the *kuwngis* ridge line of the house (Figure 7.40). A man climbs onto the ridge when all the rafters are in place and trims them to the ridge line, leaving the *kuwngis* sapling in a neat crutch. Someone will also trim the *bez* 'ridge-poles' to final size, having arranged these so that surplus pole pokes out at the rear *nunung* 'apsidal end' of the house.

When the *polpen* 'rafters' are all fastened in position, the next task is to secure horizontal battens in place across them, running parallel to the exterior walls and at ninety degrees to the rafters. The battens comprise two kinds. Along the eaves, where the rafters extend out over the exterior walls, men attach *iybhedla gamonk* 'eave battens' (Figure 7.41). These comprise *shonon, gwai* or *bobae* saplings, stripped of their bark like all saplings used in house construction. It is necessary to have strong saplings along the eaves to stop the thatch from subsequently sagging between the ends of the rafters. They lash four or five *iybhedla* rows in place over the eaves and exterior wall, using a simple spiral binding either running a rattan strand along each batten and around the rafters or the reverse, by running the binding down every fourth or so rafter and around the battens.

The remaining battens, called *ka*, which are secured over the main part of the roof, comprise lengths of split *taembok* 'bamboo' culm, or split *aendashor* 'pandan' aerial roots, or stems of *gaimb henj* 'cane grass' (Figure 7.42). The *ka* battens made from the first two materials are usually long enough to run the length of the roof in one stretch. The purpose of the *ka* is to support the thatch. In preparation

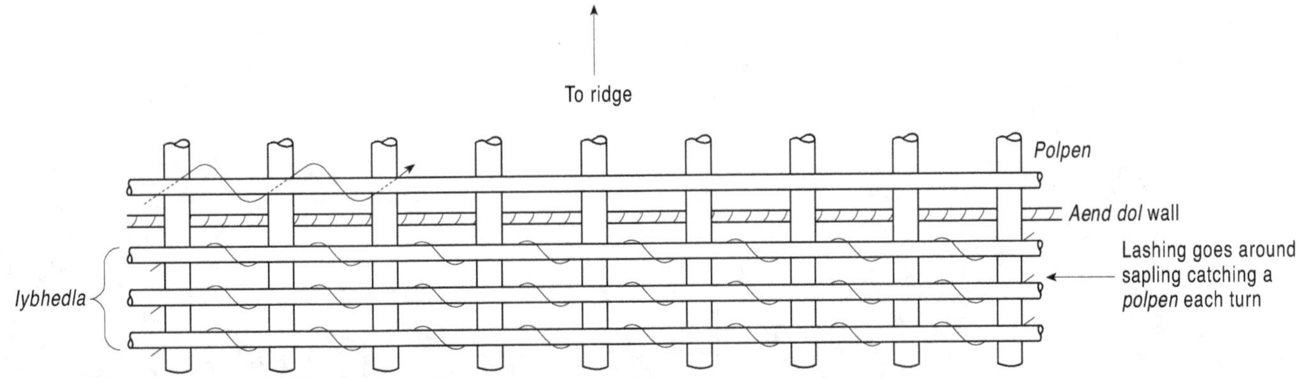

Figure 7.41 *Lashing of iybhedla 'eave battens.'*

Plate 7.49 *Lashing iybhedla gamonk 'eave battens' along the eaves.*

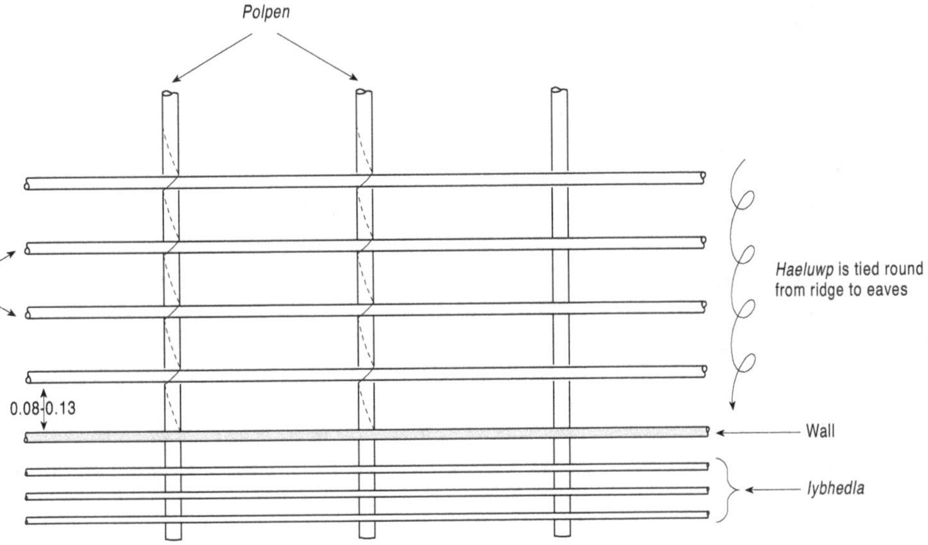

Figure 7.42 *Securing of ka 'roof battens' (m).*

Plate 7.50 Arranging ka 'roof battens' along roof.

Plate 7.51 The spiral lashing securing ka 'roof battens' to polpen 'rafters.'

for securing them in place, the roofers tie lengths of rattan strand to the *kuwngis* along the ridge adjacent to the rafters, which hang down streamer-like inside the house while they arrange the *ka* 'roof battens', ready to wind around the rafters. They position the *ka* battens ten centimetres or so apart down either side of the roof and again secure with a simple spiral twining down alternate *polpen* 'rafters'.

At the semicircular *nunung* 'apsidal end' of the roof, they bend the *ka* 'roof battens' around such that they run at about thirty degrees and not at right angles to the *polpen* 'rafters'; reminiscent of a 'duck's rear', as I described it in my field notes (Figure 7.43). The *iybhedla* 'eave battens' continue to follow the contour of the exterior wall. The house-builders may place sheets of bark from trees such as casuarinas, cypress and engel over these eave battens before they start thatching, especially if they intend to thatch the house with kunai grass or 'tile' it with pandan or gingerwort leaves, and only occasionally if they use cane grass (Figure 7.44). The purpose of this *iybhedla humbiy* 'eaves bark' – laid with the outer rind uppermost around the eaves and extending over the exterior walls – is to form a waterproof seal, ensuring that rain runs off and does not percolate through the thatch when it reaches the bottom of the roof. They may fill in the gaps between the odd-shaped bark slabs with pieces of tough *aendashor* 'pandan' leaves.

Before proceeding to thatch a house, the builders may construct a *sal* 'gable fringe', particularly where a house has

Plate 7.52 The ka 'roof battens' at the semicircular nunung 'apsidal end' of the roof.

an open-fronted *wil* 'day room'. The fringe comprises a trim of fronds arranged around the gable end of the porch, which further serves to protect those sitting inside from wind and driving rain. The fronds are either the feathery conifer-like foliage of the *naep* 'casuarina' or leafy lengths of *tedbel* 'gingerwort' stem, which men insert between partially split saplings or lengths of split *aendashor koma* 'pandan aerial root'. They arrange these saplings or aerial roots, which are the *sal gamonk* 'fringe battens', in an arch around the gable end, pushing their sharpened ends (which are left whole) into the ground just beyond the exterior wall eaves (Figure 7.45). They sandwich the ends of the protruding *ka* 'roof battens' between the *sal gamonk* 'fringe battens' and secure with a simple loose spiral binding around these latter battens. They bind these loosely to allow them to push the casuarina or gingerwort fronds between the split in the battens to form the *sal* 'fringe' hanging down around the *wil* 'day room' entrance. They may place some lengths

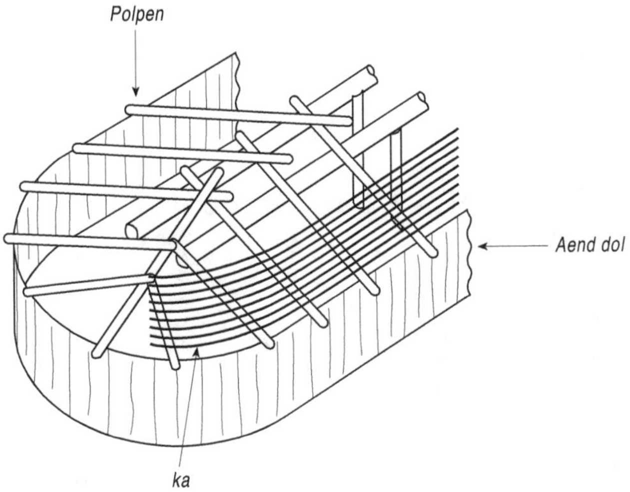

Figure 7.43 Positioning of ka 'roof battens' around apse end.

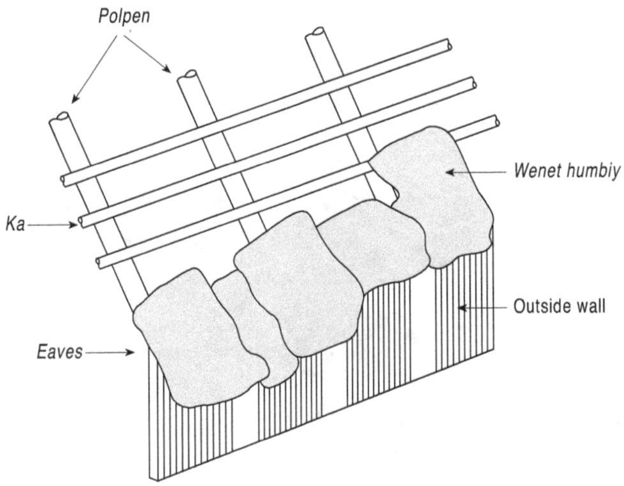

Figure 7.44 Positioning of bark sheets on eaves.

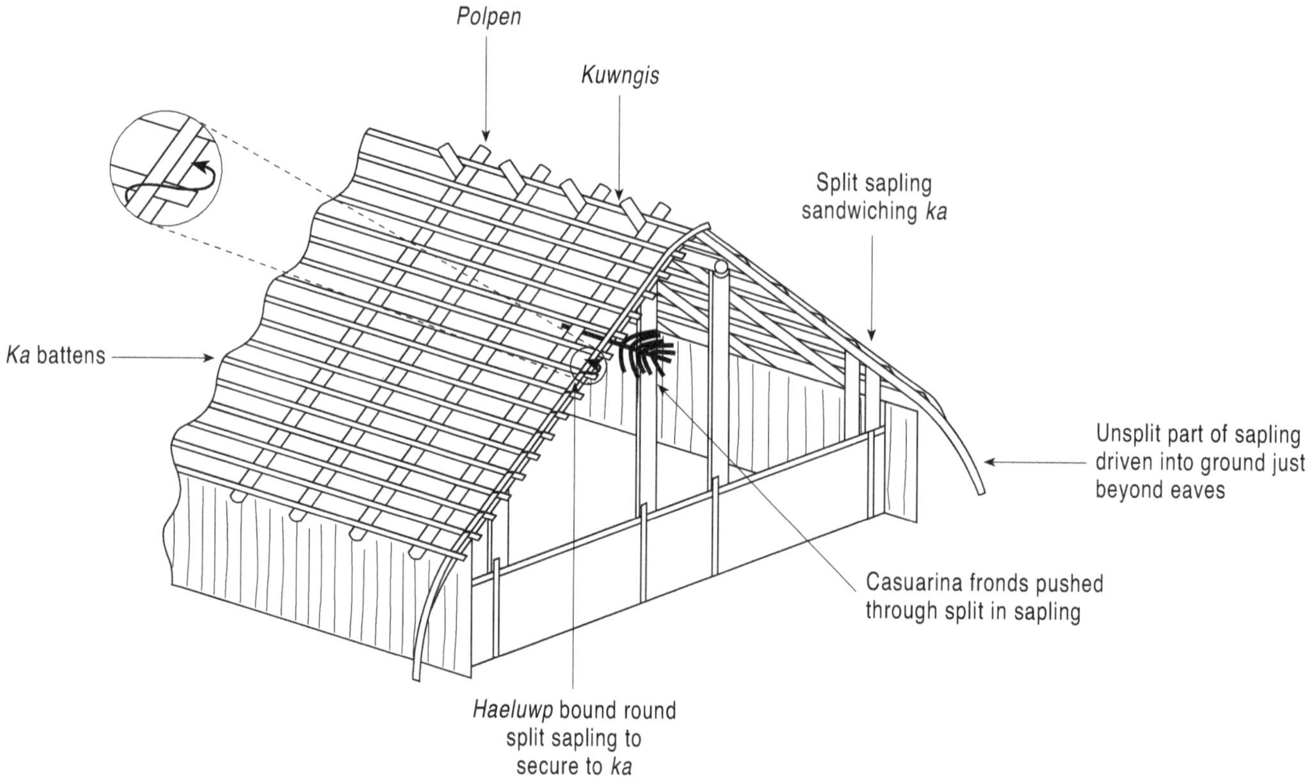

Figure 7.45 Construction of casuarina-frond porch fringe.

Plate 7.53 Securing the arched sal gamonk 'fringe battens' around gable of house.

Plate 7.54 Pushing casuarina fronds between battens to form sal 'fringe' around house entrance.

Plate 7.55 Securing stake to eaves to support the aenk tor 'pandan gutter'.

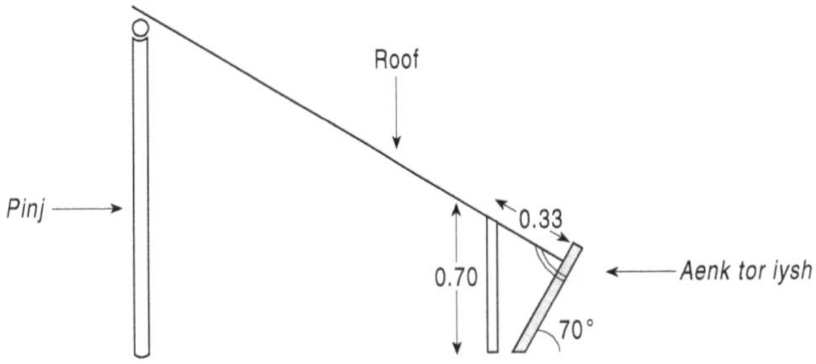

Figure 7.46 Positioning of thatching gutter stakes at eaves (m).

of timber on the fronds, where they stick through into the roof framework, to hold them in place temporarily until the thatch is on. Sometimes they drive in further saplings or aerial roots and bend these underneath the fringe as the *sal aengaybayay* 'fringe hold-up', to support the fronds jutting out over the entrance. They may insert other withies, and bend over the top of the fringe, securing it to those holding it up from below to give a stiff fringe. They finish off by trimming the foliage even around the gable front.

The construction of the Honael house roof took a total of nearly eleven and one-half man-hours (Table 7.5). Regarding the comparative time efficiencies of steel- and stone-tool use, there is scant difference (the ratio is stone 1 to steel 1.04, which amounts to a mere two minute's difference in an hour, which is within the error margins of the time recordings). This again reflects a situation where construction work features little use of these tools. These data continue to show how the time taken to complete the same tasks can vary with men working at different rates, making mistakes that they have to put right and so on.

Thatching

When they have the *ka* 'battens' and *sal* 'fringe' in place, the roof is ready for thatching, which house-builders are keen to proceed with promptly. They may use *gaimb* 'cane grass', *senz* 'kunai grass' or *aendashor* and *nort* 'pandan' leaves or *tedbel* 'gingerwort' leaves to thatch a house, as listed previously. They have similar affordances, supplying a waterproof layer when arranged appropriately on a roof, although their properties vary. The materials they use largely depends, as noted, on their availability locally. The most commonly used in the Was valley is *gaimb* 'cane grass', followed by *senz* 'kunai grass'.

When thatching with *gaimb* 'cane grass', as on the Honael house, men firstly erect a gutter around the eaves comprising semicircular (i.e. unflattened) lengths of hollowed out *aenk* 'pandan' trunk, which serves to hold the cane grass in place during thatching and to give an even edge to the thatch. They secure the *aenk tor* 'pandan gutter' in place by *aezay* 'driving' in stakes at an angle at intervals around the walls adjacent to the eaves (Figure 7.46). The

Task	Wola term	Time spent	
		Stone-tool side	Steel-tool side
Roof construction			
cutting ends off ridge-poles	*bez kaembay*	5 m, 55 s	1 m, 55 s
putting rafters in place	*polpen haeray*	16 m	19m
splitting rattan vine	*haeluwp dekel bay*	27 m, 30 s	53 m, 30 s
lashing rafters to *gamonk* beading along tops of side walls	*polpen dol bort tongay*	26 m	11 m, 30 s
splitting saplings for *iybhedla* battens and placing over eaves	*iybhedla dekel bay*	2 m, 30 s	
lashing *iybhedla* battens to rafters on eaves of side of house[1]	*iybhedla horay*	20 m	41 m, 30 s
lashing rafters to ridge-poles	*polpen bez bort tongay*	12 m, 4 5s	13 m
trimming rafters to size along ridge	*polpen nibdiy kaembway*	16 m, 30 s	4 m
preparing and positioning *hedpong kuwngis* sapling	*hedpong haeray*	3 m, 15 s	
lashing *kuwngis* sapling in crutch of rafters	*kuwngis tongay*	20 m, 45 s	
putting rafters in place round apsidal end (and trimming to size[2])	*polpen nunung bort haeray*	11 m, 15 s	
cutting ends off ridge-poles to fit apsidal-end roof line	*bez kaembay nunung bort*	4 m, 5 s	1 m, 20 s
lashing rafters to wall round apsidal semi-domed end[3]	*polpen nunung bort tongay*	7 m, 15 s	18 m, 15 s
splitting saplings for apsidal end *iybhedla* battens	*nunung iybhedla dekel bay*	4 m	
lashing *iybhedla* battens on to rafters over apsidal end eaves	*nunung iybhedla horay*	21 m, 15 s	9 m, 45 s
lashing *tuwmainj* saplings around top of rafters in apsidal end	*nunung tuwmainj tongay*	12 m	
splitting bamboo for lathes[4]	*taembok dekel bay*	49 m	27 m, 15 s
positioning lathes and tying in place[5]	*ka bay*	36 m, 45 s	34 m
spiral-lashing lathes to rafters	*ka tongay*	41 m	58 m, 45 s
knocking in some *waenuw* wall stakes below roof line	*waenuw lay*	3 m, 30 s	
splitting saplings for *sal* and lashing across front eaves[6]	*sal gamonk dekel bubtuw tongay*	8 m, 15 s	18 m, 45 s
pushing in casuarina fronds for fringe across front eaves	*sal say*	11 m	12 m, 55 s
Thatching			
sharpening stakes to hold gutters	*tor iysh waendilay*	12 m	26 m
driving stakes in along eaves[7]	*aenk tor bubtuw iysh aezay*	7 m, 30 s	
lashing stakes along eaves	*aenk tor tongay*	11 m, 30 s	13 m
positioning pandan-bark gutters	*aenk tor bay*	27 m, 30 s	11 m, 30 s
adjusting gutters	*aenk tor duwn bay*	9 m, 30 s	
passing up thatch	*gaimb kalay*	44 m, 15 s	
thatching roof	*aend tinjiyay*	1 h, 28 m, 15 s	1 h, 21 m
passing thatch up and placing horizontally along ridge[8]	*aend homb lay*	53 m, 15 s	
treading thatch down along ridge	*homb gaimb haelambay*	12 m	
removing pandan-bark gutters	*aenk tor hezbiyay*	2 m, 45 s	1 m
times taken		**9 h, 10 m, 40 s**	**8 h, 56 m, 15 s**

1 5 rows of *iyb hedla* battens were lashed to rafters on stone side of house and 4 rows on steel side.
2 Trimming done with steel tools
3 The steel column includes time splitting saplings for *gamonk* beading, and stone column some time trimming ends of *polpen* rafters.
4 2½ culms with stone and 2½ using steel tools.
5 On stone side includes some splitting of rattan.
6 Lashing on steel side was incorrectly done initially and had to be redone.
7 Steel-made stakes driven in as sharpened.
8 The stone side of the house was thatched with cane cut using stone axes and the steel side with cane cut with bush knives. The third *homb* 'ridge' layer was thatched with cane cut using bush knives only.

Table 7.5 Construction and thatching of roof.

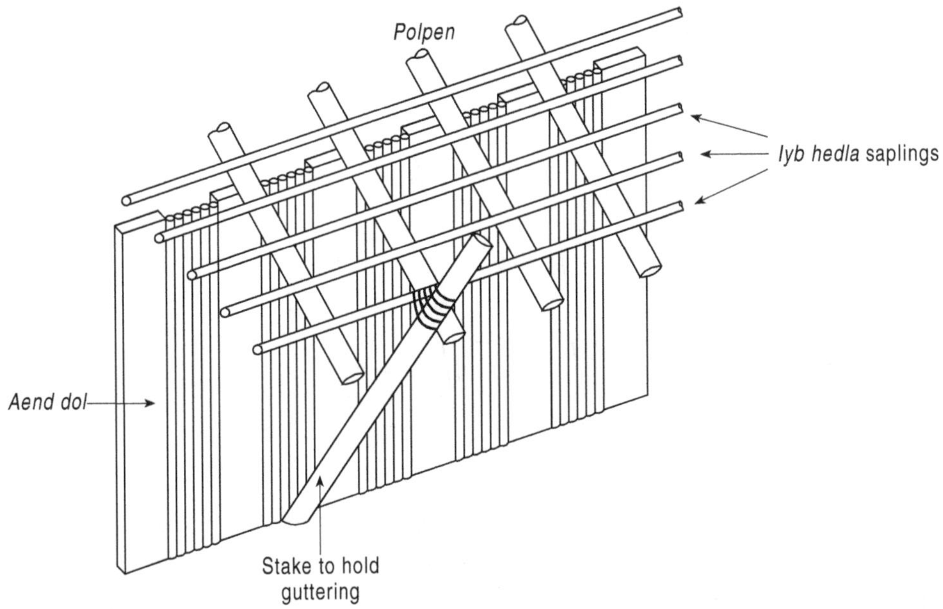

Figure 7.47 Lashing of thatching gutter stakes to eaves

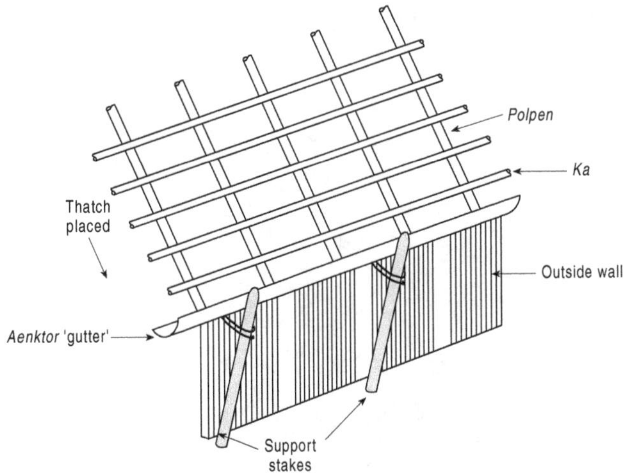

Figure 7.48 Positioning of thatching gutter along eaves.

number of stakes depends on the size of the house; four or so down either side wall and seven or so stakes around the semicircular end. They secure the stakes to the eaves, either by lashing tightly to the *iybhedla* battens (Figure 7.47) or loosely to them, arranged to give a vine cradle (Figure 7.48). The latter arrangement ensures that the cane thatch covers the eaves entirely, whereas with the former the ends of the rafters will protrude. They arrange the lengths of pandan-trunk-bark guttering, which vary from two to five or more metres long, around the eaves supported by the stakes. The gutters are semi-communal property, it being common practice for men to pass them on to others to use – everyone knows who recently built a house in their locality and is likely to have some gutter lengths lying around their house yard that they may ask to use – until too damaged to reuse that is, when they make new ones as necessary. Sometimes men drive in some long poles vertically across the gable end

Plate 7.56 Arranging the aenk tor 'pandan gutter' in position along eaves.

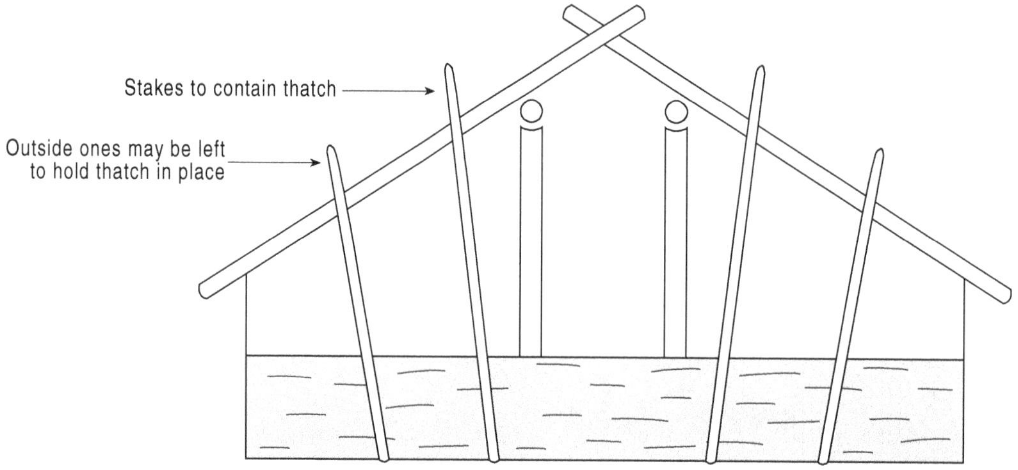

Figure 7.49 Gable-end poles to contain thatch.

of a house, extending above the roofline, to contain the cane grass there too during thatching operations, which they usually remove subsequently when the thatch has settled (Figure 7.49).

When thatching a house with cane grass, house-builders aim to complete the task in a day and they usually rely on the help of some relatives and friends to achieve this (Figures 7.4 to 7.6). While some of the men in the work party cut *gaimb* 'cane grass' stems, women ferry bundles of the leafy cut cane to the house site where they pile it up (Table 6.7). Meanwhile other men see to the *aend tinjiyay* 'house thatching'. If they are using some thatching materials reclaimed from a dismantled roof, they put these on first, followed by freshly cut cane stems. The thatchers arrange the cane in three layers on the roof. The first layer comprises bundles of cane placed vertically in the same direction as the *polpen* 'rafters', following the pitch of the roof (Figure 7.50). They bang the cut ends firmly in the *aenk tor* 'pandan gutters' at the eaves, to ensure that the thatch has an even edge. The second layer comprises bundles of cane placed running in the same direction but arranged such that it reaches and covers the ridge of the house and overlaps the lower first layer. Some men may climb up on to the roof to arrange the bundles of cane, which others pass up to them. The third layer comprises bundles of cane grass placed horizontally along the *homb* 'ridge' of the house and trodden firmly in place. The thatchers may further secure this ridge layer in place by bending up a few cane stems from the second thatch layer and pushing them into it (Figure 7.51). When they have finished the thatching, they remove the *aenk tor* 'pandan gutters'.

Thatching the roof of the Honael *enzatenda* house (excluding cutting and ferrying cane to the site – see Table 6.7) took a total of nearly six and three-quarter person-hours (Table 7.5). Regarding the comparative time efficiencies of using steel and stone tools, there is again little difference; although here it was in favour of steel tools, as expected throughout the house-building work

Plate 7.57 Carrying a bundle of cane-grass thatch to house site

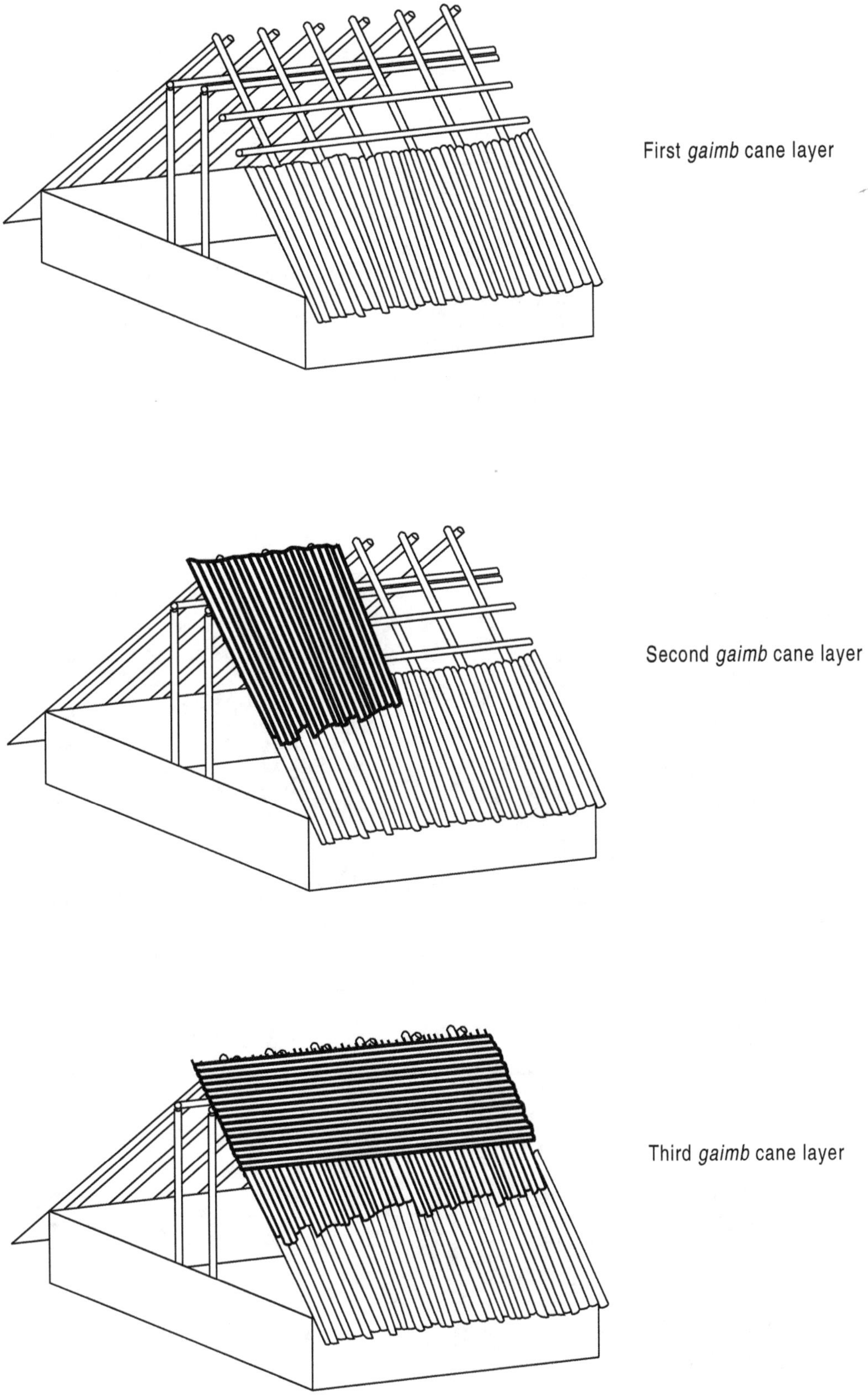

Figure 7.50 *Thatching with cane grass.*

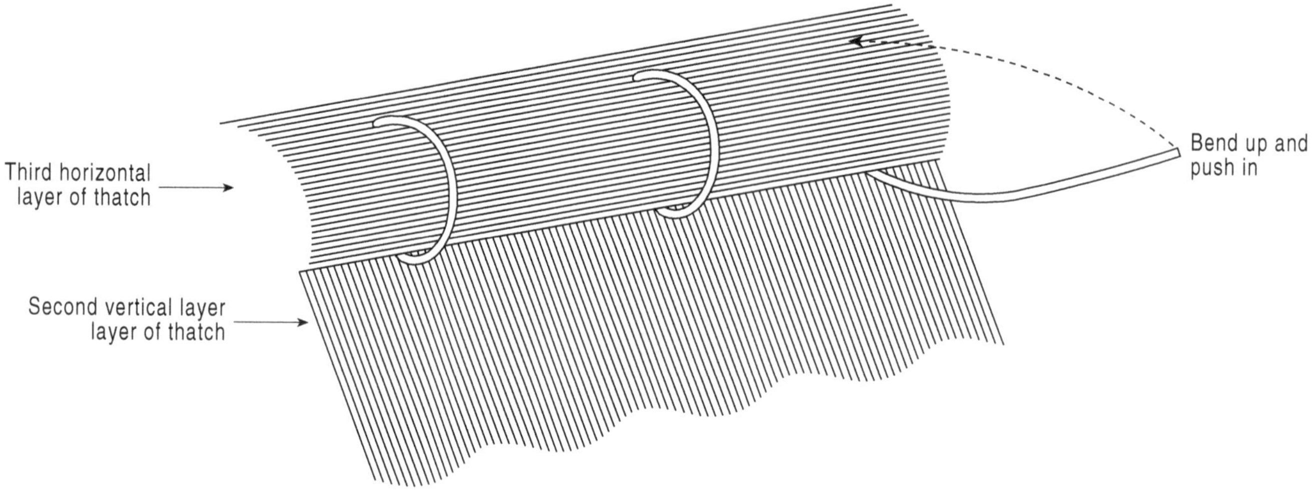

Figure 7.51 Securing horizontal thatch layer along ridge.

Plate 7.58 Delivering a bundle of cane-grass thatch.

Plate 7.59 Thatching house with first layer of cane grass.

Plate 7.60 Thatching house with second layer of cane grass.

Plate 7.61 Completing thatching along the homb 'ridge' of the house (note smoke of fires lit inside house to start drying out green materials)

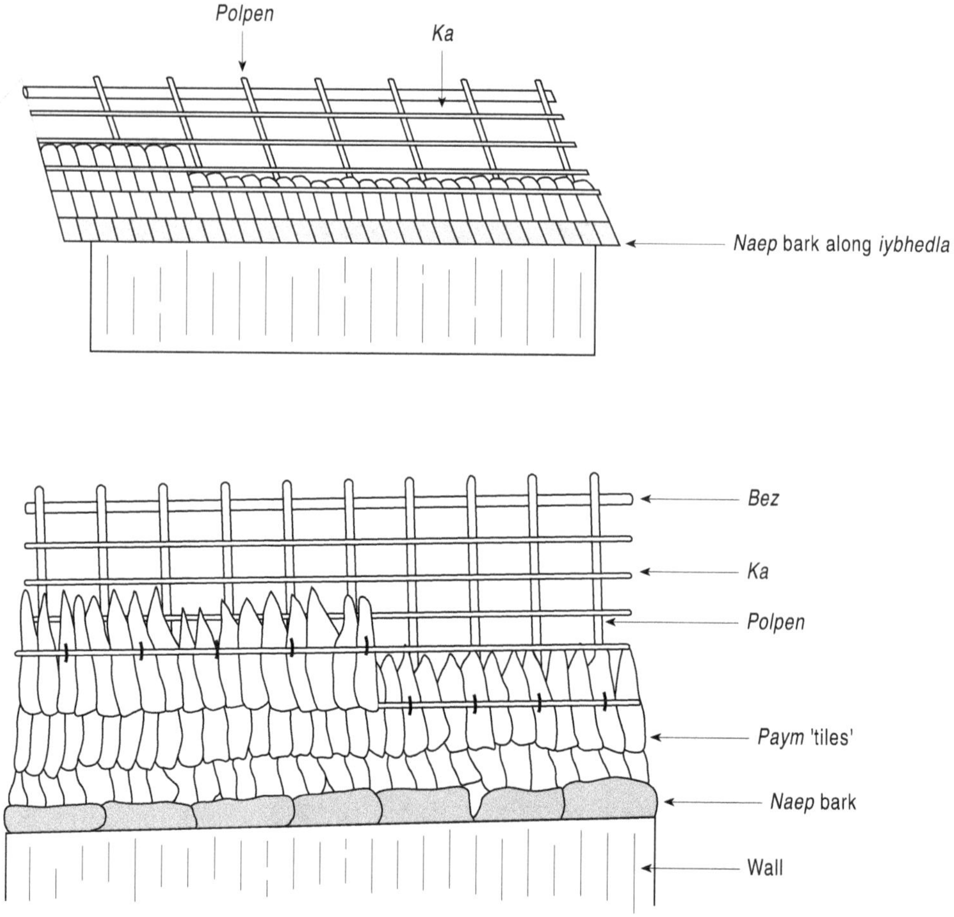

Figure 7.52 Roofing with pandan leaf 'tiles'.

(the ratio is steel 1 to stone 1.14, which amounts to nine minute's difference every hour). The differences between these tools are largely due to work undertaken erecting the *aenk tor* 'pandan gutters'. The actual thatching times are nearly the same, which is as expected for a job that does not feature the use of these tools and where men are working together co-operatively on the same task such as to blur any differences in work rates.

The thatching of a house with *senz* 'kunai grass' takes longer, albeit there is no call for *aenk tor* 'pandan gutters'. According to data collected during the building of other houses, the rate of thatching with *senz* 'kunai grass' is 1.95 m^2 of roof per hour, compared to 5.27 m^2 of roof covered per hour with *gaimb* 'cane grass' (including the time spent erecting the thatching gutters), which is 2.7 times faster. It is usual for people to collect kunai grass over several days and pile it up, hayrick-like, adjacent to the house. They tie in small bunches as they *senz kobay* 'pull up grass', winding a blade of grass around the base of each bundle and tucking the end in. When thatching, men take these small bundles of grass and lash them tightly one next to the other along the *ka* 'battens'. They start thatching around the eaves, after placing sheets of tree bark over the *iybhedla* 'eaves battens' as described, and work their way up to the *homb* 'ridge', a row of thatch at a time, overlapping each row of grass bundles with the previous one below (Enga thatching follows a similar pattern – Meggitt 1957:174–5).

The 'tiling' of a roof with large *aendashor* or *nort* screw-pine or *tedbel* gingerwort leaves can also be time consuming. Again, the roofers collect the leaves they estimate they will need before they start. They begin tiling once more along the eaves, after placing sheets of tree bark over the *iybhedla* 'eaves' battens, and work their way up to the ridge. They overlap each leaf with its neighbour and over the row of leaves below, after tiles. They secure the leaves by either passing them under a batten such as a *shonon* chenille sapling (running across the roof like the *ka* battens) with their narrow end towards the ridge – called *gamonk shor hauway* – or by folding the leaves over the sapling – called *gamonk shor maeray*. When they have several leaves in place, they lash the sapling to the adjacent *ka* batten by twisting a rattan strand around both, so sandwiching the leaf 'tiles' in position on the roof (Figure 7.52). When they have the leaves in place, they secure *daengbiyp* 'anchor logs' along the bottom edges of each row to stop them blowing up in the wind. They weatherproof the ridge line of the roof by folding large overlapping leaves over it, again held in place with a pair of logs on either side,

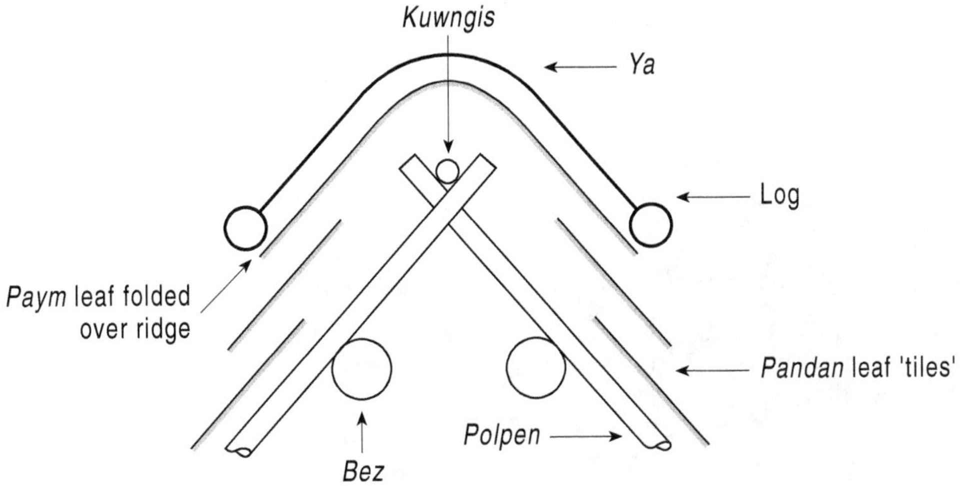

Figure 7.53 Cross-section of leaf-'tiled' roof ridge.

secured with lengths of vine passing over the roof crest (Figure 7.53). They sometimes secure the ridge line of roofs thatched with *senz* 'kunai grass' in this way too.

The thatching of a house brings together the largest group that cooperates in a productive activity in the Was valley (Figures 7.4 to 7.6), which is noteworthy in view of the autarky that households largely pursue in respect of supplying their material requirements (above; Sillitoe 2010a). While a couple may thatch a small house alone – complying with the usual self-sufficient family productive arrangements – it is common for some other relatives to give a hand. It is usual for future occupants of a house to co-operate in building and thatching it, particularly an *enza* 'men's house'. They work together without anyone giving directions. People explain that these work bees are popular when thatching a house because it is necessary to complete the task quickly to ensure that the structure is warm and dry. They say that if they dragged thatching out over several days then the house would become cold and damp. Both men and woman collect thatching materials. Regarding cane grass, the sexual division of labour requires men, who handle axes or bush knives, to cut stems down, while women ferry the thatching materials to the house site. Both may collect and transport kunai grass or pandan leaves.

People help one another with no expectation necessarily of a reward. Although by helping others thatch their houses, they ensure that kin and friends will be ready to help them thatch theirs in turn, albeit people deny that they keep a close eye on who helps whom. The reciprocity is an aspect of the informal give and take of everyday life. It is usual however for the occupiers of the new house to prepare a substantial meal of earth-oven-cooked vegetables, called the *nay wem saway*, to feed the helpers. It is not an obligation or a payment but a sociable act; it is an opportune moment to share a meal with several relatives collected together. They may sit around chatting for some time while the food cooks; on one occasion we sat around for three hours while the occupants of a newly thatched house prepared an earth oven and we waited for the vegetables to cook. While it is poor etiquette to turn up at such a meal if one has not helped with thatching, such an uninvited person will receive some food if s/he arrives when it is distributed. The invitation extends to individuals who may have helped with the house construction in other ways than thatching, as I discovered when I made an acerbic comment about an old man suddenly turning up as we opened an oven and was told to be quiet as he had supplied some of the *aenk tor* 'pandan guttering' used to contain the cane grass during thatching.

Other fittings

Finally, there are the fittings and furnishings of a house, although Was valley houses feature few of these. The principal fittings are fireplaces and doors.

A fireplace comprises two components. There is the *mem tay* or *taenk tay* 'ash source' hearth. This is situated in the centre of a room and comprises four *dombep* 'sills', which are straight lengths of timber fastened to the ground with pairs of pegs knocked in either side at intervals. Two of the pieces of wood run the length of the room, approximately demarking the space between the *pinj* 'ridge-pole supports'. The other two short pieces, cut to fit across the demarcated space, are pegged a metre or so apart between the two long pieces. The central enclosed rectangular area is the fireplace. Residents excavate a shallow scoop here in the hard, clay subsoil where a bed of ash will build up (if the soil is soft they may not bother, as it will mix with the bed of ash that builds up and not impede them pushing sweet-potato tubers into the hot embers to bake). They line the inside edge of the wooden framework surrounding the fireplace scoop with flat stones called *momberael* to protect it from the hot embers (the limestone mentioned in Chapter 4's 'dig report'). The enclosed rectangular areas on either side of the fireplace may serve as storage places for implements,

Plate 7.62 Positioning *aend kor* 'fireguard' in roof space.

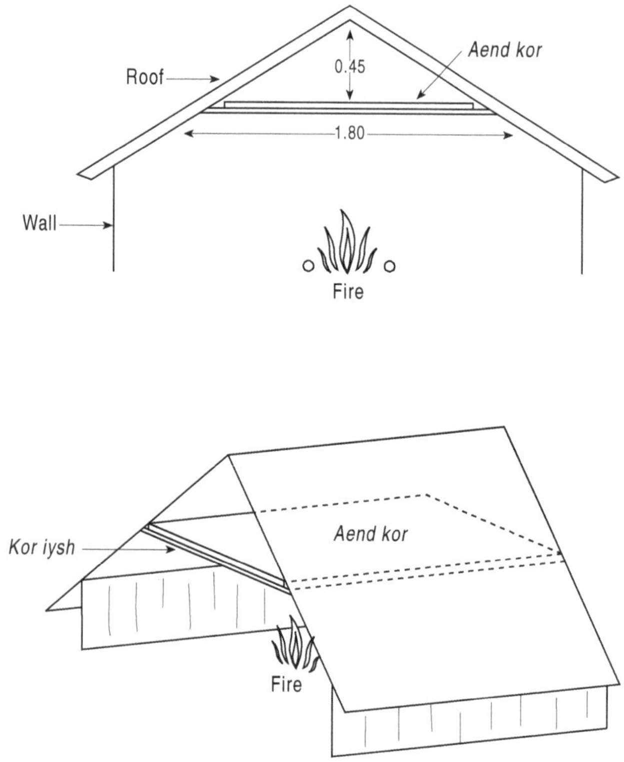

Figure 7.54 Position of *aend kor* 'fireguard' (m).

such as split-sapling cooking tongs, and food, both raw and cooked. Adjacent to doorways, they demarcate the crawl-through way.

The other fireplace component is the *aend kor* 'house fireguard' installed in the roof space above a fire, which serves to protect the thatch from flames and sparks (Figure 7.54). It comprises a sheet or two of *buk* 'pandan bark' supported by two lengths of timber, the *kor iysh* 'fireguard supports', lashed to the rafters (Figure 7.55). It is usual to install the baffle-boards before thatching the roof. They also serve as a dry storage space, for instance of seeds, albeit things placed there are soon smoke-stained and sooty. Each room with a fire has a baffle-board; so the three roomed Honael house had three of them.

The doors comprise slats of wood that slide one above the other between two upright jambs on either side of the doorway, described earlier. The *togol* 'jambs' are short stakes driven into the ground to the height of the *deriyau* 'doorway' and lashed at the top to the *gamonk* 'batten' running lintel-like along the top (Figure 7.56). The door-fitter arranges the jambs with a space between them and the wall stake behind sufficient to slide the door slats through. He positions the *togol* 'jambs' of the interior doorway from the *wil* 'day room' on the *uwpuluwk* 'sleeping room' side of the *waenuw* 'interior wall', and the jambs for the exterior doorway into the *tenda* 'women's house' on the outside of the *dol* 'exterior wall. He splits the *deriyauhaekomb* 'door slats' from logs, measuring against the doorway for size and carefully paring into flat boards. Occupants close a door by

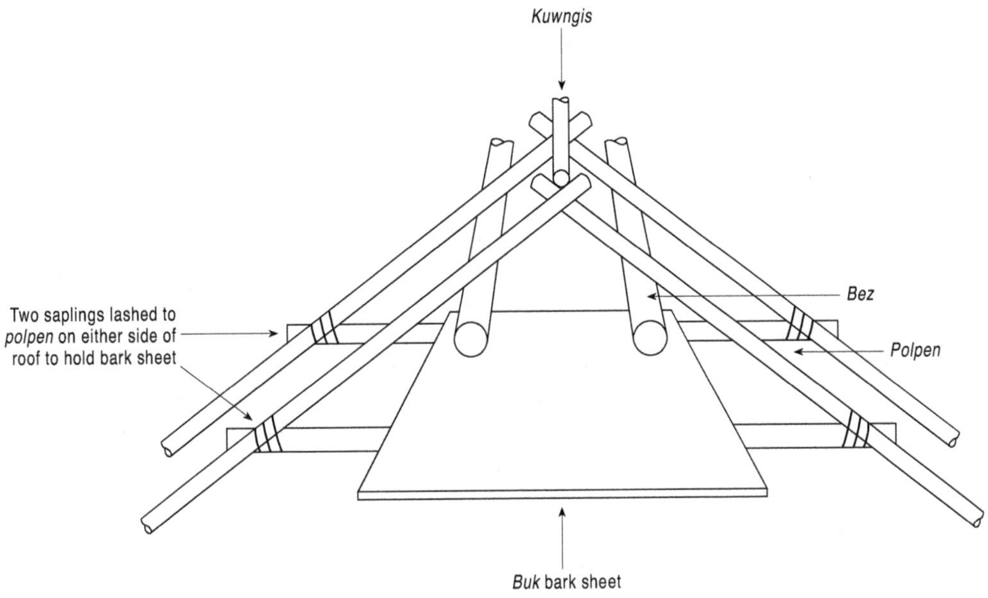

Figure 7.55 Structure of aend kor 'fireguard'.

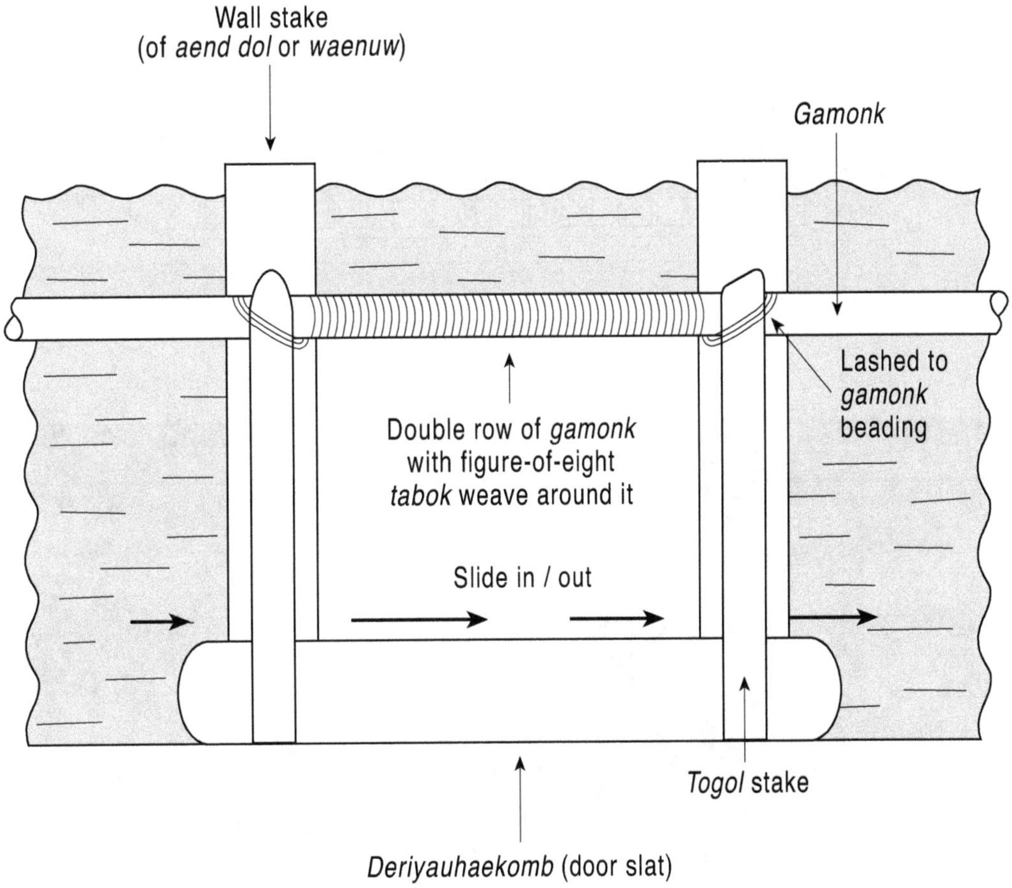

Figure 7.56 Structure of doorway.

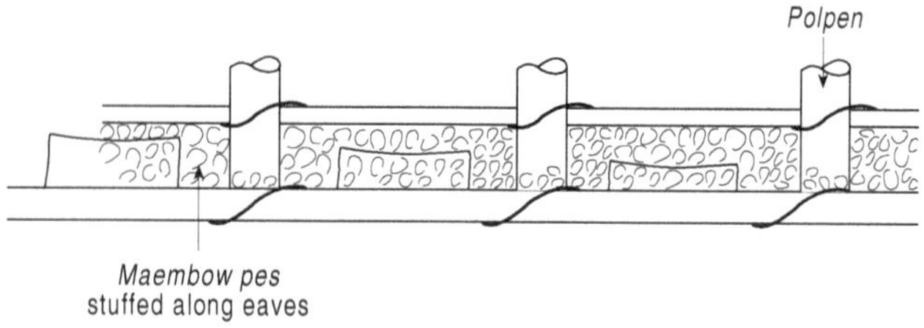

Figure 7.57 Wadding pushed in under eaves.

arranging several slats horizontally across the doorway, one above the other, securing in place by jamming one or two longer slats diagonally across them.[21]

Other interior work includes blocking the gap between the tops of the *aend dol* 'exterior walls' and the *iybhedla* 'eaves' to stop cold draughts (Figure 7.57). Men may use *homb* 'ground moss', *kwimb* 'tree moss', *gwai* 'switchsorrel' leaves or pith from inside *aenk* 'pandan' trunks (left over from beating bark wall sheets) as *maembow pes* 'eave wadding'. They stuff the wadding material into the gap in the eaves and may jam in place with small sticks.

21 Steensberg 1980:131–2 reports on similar fire baffle-boards, fireplaces and doorways in the Duna region.

Looking up into the roof of the porch of the *wil* 'day room', men often fill the gaps between the *polpen* 'rafters' that are visible there with *aendashor* leaves. They may fill the gap adjacent to the *sal* 'fringe' batten like this too. They secure the leaves in place with short *pongow* 'jamming sticks' pushed under the rafters (Figure 7.58). This lining of the porch roof hides the thatch underneath from view.

If the builders have attached the *buk* 'pandan-bark sheet' on the sleeping room side of the *wil waenuw* interior wall (in an *enza* 'men's house'), they face the *wil* 'sitting-room' side of the wall with *shwimb humbiy* 'Indian bead ash bark', securing with *pongow* 'stakes'. They may also peg a *dombep* 'sill' log, such as a length of palm-tree trunk, along the bottom of the wall, and along both of the *aend dol* 'exterior walls' inside the sitting room. They position these sill logs some thirty centimetres or so out from the bottom of the walls. They used them to prop up exchangeable things such as pearl-shells in displays during socio-political transactions.

Further furnishings are non-existent in many houses. In some, men construct themselves *paenda tor* 'beds', which are low platforms comprising a rough rectangular wooden frame that has slats of cane-grass stem lashed across it.

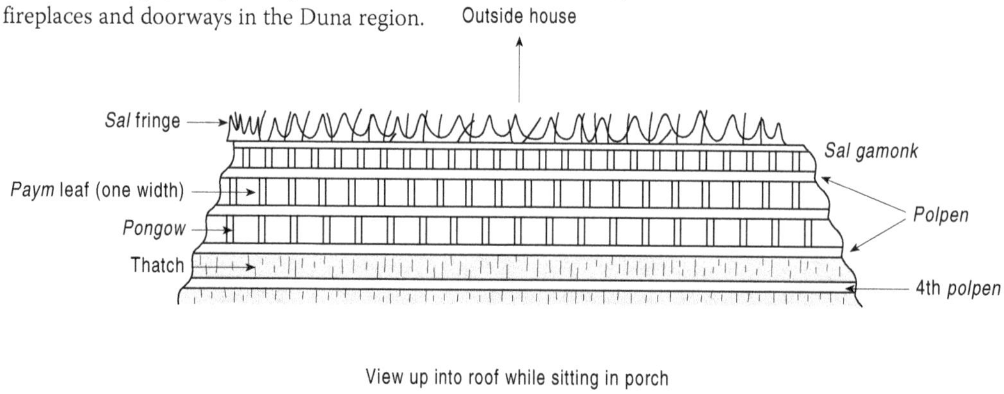

Figure 7.58 Pandan-leaf lining of porch roof.

Plate 7.63 Collecting up refuse from building work and tossing on fires to hobol kay 'dry out' house.

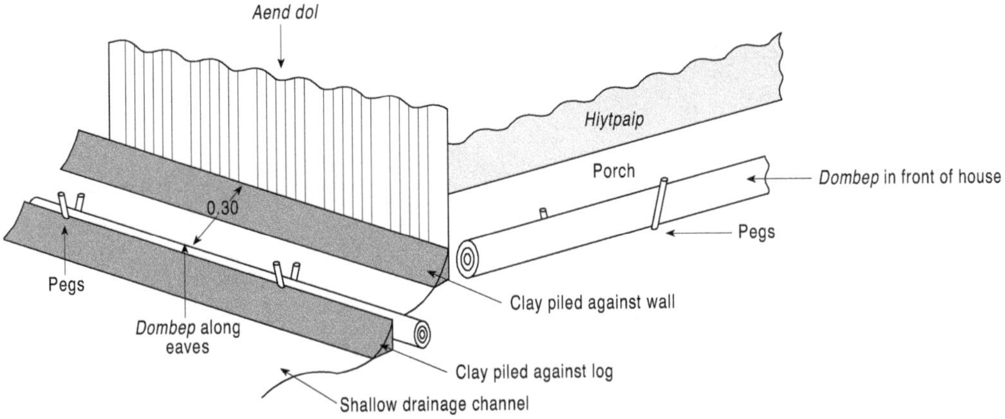

Figure 7.59 Drainage way under eaves.

Previously, some men slept curled up on large wooden shields (Sillitoe 1988a:163–6). Many do not bother to make beds at all, unless the house becomes so infested with fleas that they are an intolerable nuisance and stop them sleeping soundly. They may decide to wait until a newly built house has dried out somewhat before making beds. The clay floor is cold and damp and to sleep above it on a bed would be chilly and uncomfortable, even debilitating according to some, and likely to make you ill. Until the floor has dried out sufficiently, they will sleep on beds of dry leaves that insulate sleepers from the floor.

Newly built houses are *kahagay* 'raw' and occupants are anxious to dry them out and warm them up. They light fires in all the rooms soon after they have constructed the fireplaces to start the *aend hobol kay* 'house drying out' process. They may feed these initially with some of the refuse lying around from the construction work, including pieces of wood, vine strands and cane-grass stems.

Other than clearing up the waste left over from the construction work, either burning it or tossing it away, the only remaining work involves the arrangement of exterior-drainage ways. Men peg *dombep* 'sill' logs, eight centimetres or so in diameter, under the eaves of the house, running parallel with the walls and some thirty centimetres away from them. They then dig a shallow scoop, two to three centimetres deep, along the outer edge of the log, directly below the eaves, putting the excavated clay along the side of the log and against the bottom of the house wall and treading firmly down (Figure 7.59). The resulting shallow *iyb gən* 'water channel' acts as a gutter, directing run-off around the house and preventing it from entering. The occupants may only construct these drainage ways on the uphill side of a house to deflect the considerable run-off arriving downslope there, particularly during storms, whereas on the downhill side there is only the water that drips off the roof, which will naturally run away.

Men may also peg a *dombep* 'sill' log in place across the front of a house, parallel with the front wall and running between the ends of the exterior side walls that stand out to create a recess. The thirty-centimetre-or-so space between the log and the front wall, which has the thatch overhanging it to give a gable-end like porch, often serves as a firewood

Plate 7.64 View of completed Honael house.

Task	Wola term	Time spent: Stone-tool side	Time spent: Steel-tool side
Fireplaces			
lashing supports in roof space and positioning pandan-bark sheet fireguards[1]	*aend kor bay*	11 m, 30 s	26 m, 15 s
cutting wood for fireplace frames and fitting[2]	*dombep kaeray*	11 m	28 m, 45 s
cutting pegs to secure frame logs[3]	*iysh hul waeray*	10 m	3 m, 45 s
pegging frame logs in place[4]	*dombep lay*		11 m
Doors			
cutting log in two	*iysh bokor lay*		8 m, 30 s
splitting two logs into rough slats[5]	*iysh dekel bay*	15 m	11 m, 30 s
measuring slats against doorway	*deriyauhaekomb maenjet bay*		2 m, 30 s
shaping slats[6]	*deriyauhaekomb waeray*	13 m, 15 s	51 m, 15 s
cutting door frame stakes, driving in and lashing to wall[7]	*tokol waerubtuw ma hae*	5 m, 30 s	8 m
sliding slats into place to close door	*tirabo*	1 m	2 m, 15 s
shaping diagonal slats to jam door closed	*deriyauhaekomb waeray*		8 m, 45 s
fitting long diagonal slats[8]	*deriyauhaekomb bo*		2 m, 15 s
Drainage			
pegging tree trunk along line of drainage channel[9]	*dombep lay*	5 m, 15 s	5 m, 45 s
digging drainage channel[10]	*iyb gaen may*	9 m, 45 s	23 m, 15 s
total times		**1 h, 22 m, 15 s**	**3 h, 13 m, 45 s**

1 Used stone tools to construct *tenda* fireguard (supports cut from two leftover rafters) and steel tools to construct *wil* and *uwpuluwk tubael* fireguards (supports cut from four saplings; used piece of pandan-bark sheet leftover from walls on *wil* fireguard).
2 Fitted fireplaces in *wil aend* and *uwpuluwk tubael* using steel tools and *tenda* using stone tools.
3 Used pieces of Southern beech lying around site left over from making stakes.
4 Steel column includes some time cutting length of fireplace frame (from piece of Southern beech lying around site) and pegs.
5 Made *tenda* door using stone axe and *wil* door using steel axe.
6 Made 6 slats for *tenda* door and 7 slats for *wil* door, plus a further 4 diagonal jamming slats.
7 Used discarded timber collected for supporting thatching gutters.
8 No diagonal slats made for *tenda* door, as security is not an issue (men store valuables in *uwpuluwk tubael*).
9 Cut pegs from discarded pieces of wood around house site, using stone axe on front log and steel axe on side of house log.
10 Dug trench along front of house with digging stick (3.7 m long) and trench down side of house with a spade (4.9 m long), piling excavated clay against walls.

Table 7.6 Internal fittings and drainage.

store with logs piled up there. The occupants may dig out a gutter scoop along the outside of the log to take water coming from the side of the house and channel it away from the entrance.

The completion of these various fittings and furnishing tasks on the Honael *enzatenda* house took a little over four and one-half person-hours in total (Table 7.6). It is tricky to compare the comparative time efficiencies of steel and stone tools because of the differences in the work undertaken, such as constructing two fireplaces using steel tools and one using stone. If we compensate for the differences in work done (e.g. by dividing steel fireplace making time in half, discounting work only done with steel axe in making doors and correcting drainage-way work for different distances constructed), it took longer to complete similar tasks working with steel tools. The ratio is 1 to 1.61 in favour of stone, such that for every hour worked using stone axes, an hour and thirty-seven minutes was worked using steel ones. These data illustrate how the time differences in favour of stone tools are partly attributable to the better finish men can achieve with steel tools and the extra time they take to do so, for instance in shaping the *deriyauhaekomb* 'door slats'.

Total inputs: stone versus steel

We are now in a position to review the total inputs required to build a house. Although the three-roomed Honael house was larger than many dwellings – at 6.29 m long by 3.76 m wide and 2.16 m high at the ridge (see Figures 7.60 and

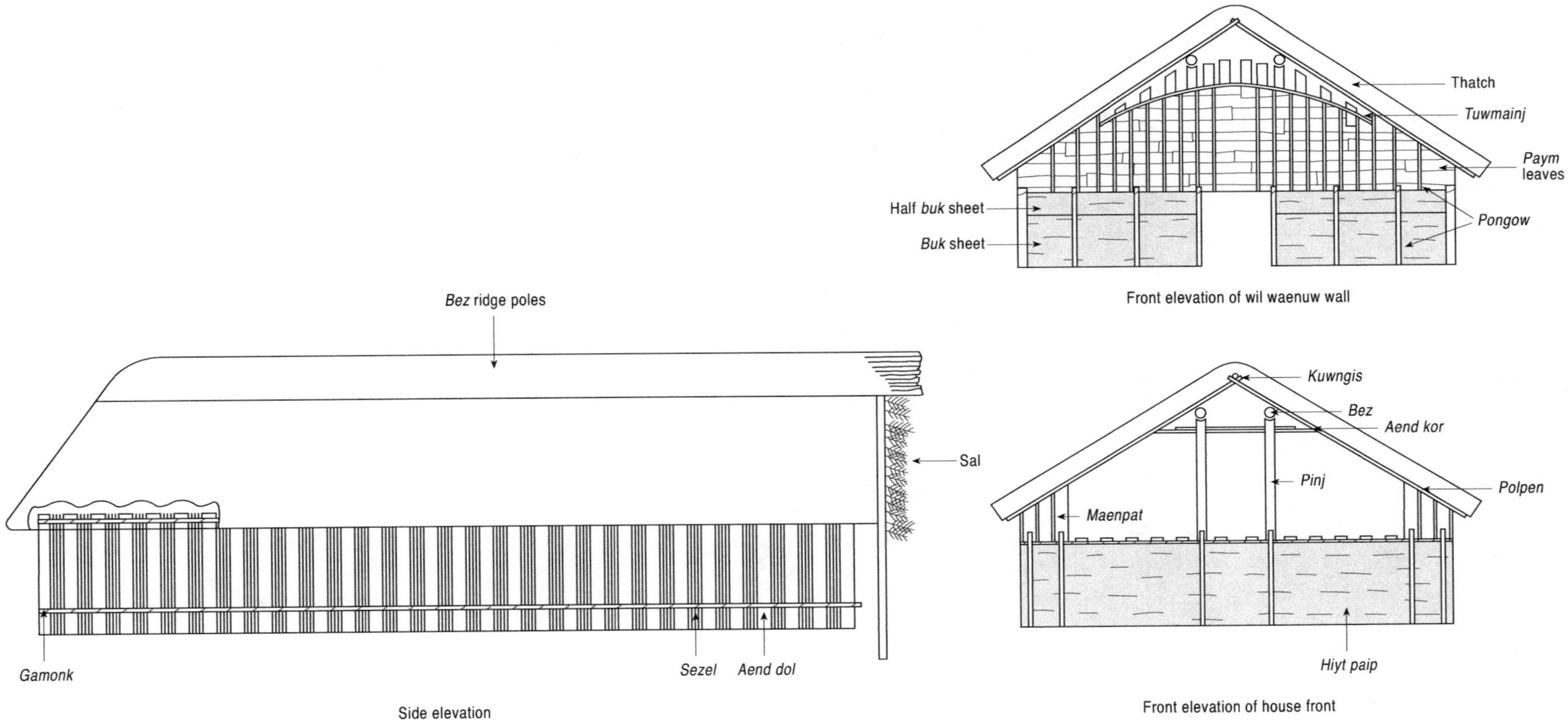

Figure 7.60 *Elevations of Honael house.*

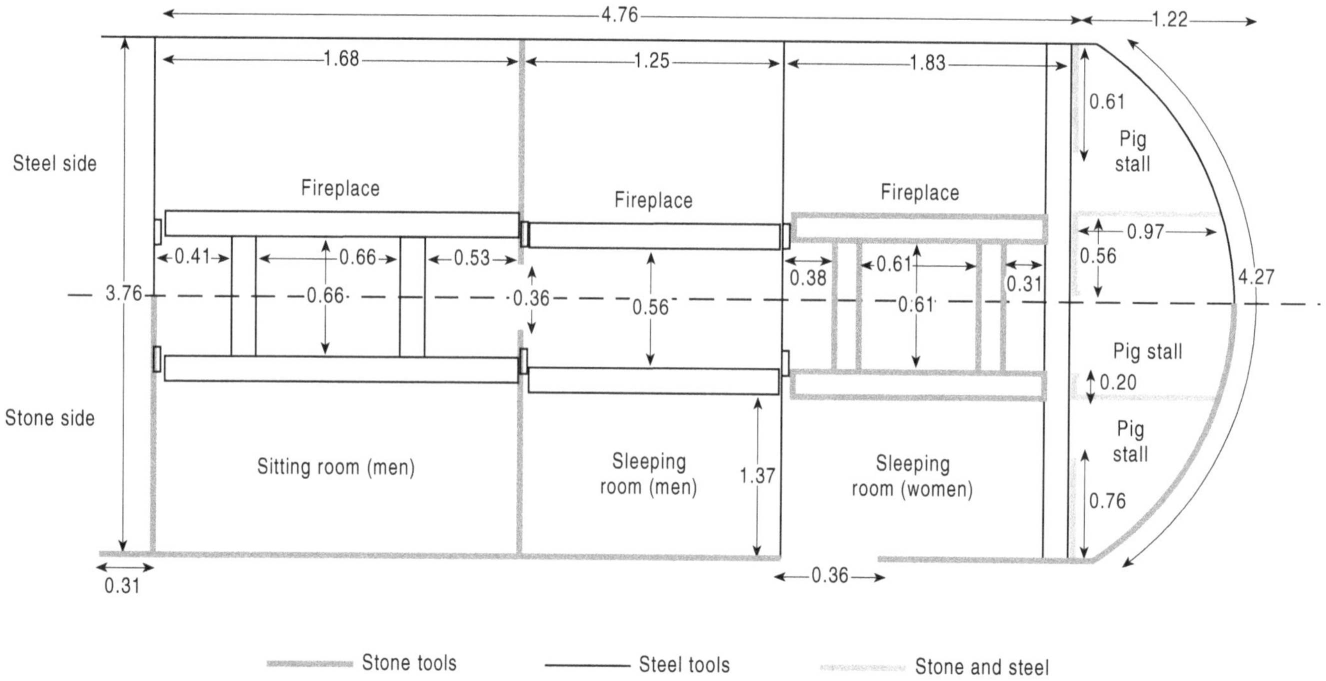

Figure 7.61 Ground plan and dimensions of Honael house (m); showing stone- and steel-tool use..

Plate 7.65 The Honael enzatenda 'men's-women's house.'

Materials	Number
aend dol stakes in straight walls	50
aend dol stakes in *nunung* 'apsidal end'	21
sezel canes between *dol* stakes (average; range 5 to 12 canes)	8 per space
buwk pandan-bark sheets	10
pongow stakes holding *buwk* to outside wall (stone side)	8
pongow stakes holding *buwk* to outside wall (steel side)	8
pongow stakes holding *buwk* to *nunung* 'apsidal end' wall	5
waenuw stakes in *wil* wall (2 above door opening)	17
waenuw stakes in *tenda* wall	15
sezel canes between stakes in *waenuw* walls	5 to 8
pongow stakes holding *buwk* on *wil waenuw* wall	8
pongow stakes holding *buwk* on *tenda waenuw* wall	5
pongow stakes holding *aendashor* leaves on *wil waenuw* wall	20
pongow stakes holding *aendashor* leaves on *tenda waenuw* wall	21
maenpat stakes	6
hiytpaip stakes	14
pongow stakes holding *buwk* on *hiytpaip* wall	6
gamonk split-sapling beading	22
tuwmainj saplings on *wil waenuw*	10
tuwmainj saplings on *tenda waenuw*	8
pinj ridge supports	6
bez ridge-poles (5.79 m long)	2
polpen rafters (av. 2.67 m long) on sides of roof	44
polpen rafters in *nunung* 'apsidal end'	13
ka batten rows (steel side)	17
ka batten rows (stone side)	14
iybhedla batten rows (steel side)	4
iybhedla batten rows (stone side)	5
showmay kuwl pig-stall stakes	36[1]
aend kor fireguards and *taenk tay* fireplaces	3

1 Some of these were discarded pieces of wood lying round site picked up and used to make stalls.

Table 7.7 Material specifications of Honael enzatenda house (refer to Figures 7.60 and 7.61)

7.61 for the dwelling's detailed specifications)[22] – the data presented in this chapter give a reasonable indication of the materials used, the tasks and work undertaken, and time spent overall constructing a house. The houses built in the Was valley do not vary greatly in size and structure (see next chapter), and this account, focusing on the construction of an *enzatenda* 'men's-women's house', includes the range of structural features found in all houses, both men's and women's.

Other house designs basically differ in their ground plans and hence construction. An *enza* 'men's house' comprises the front two *enzatenda* rooms with an *aend wil* 'foyer' and *uwpuluwk tubael* 'sleeping room'; the latter has a rounded *nunung* 'apsidal end', being the rear of the house. A *tenda* 'women's house' truncates the front rather than the rear of the *enzatenda*, having no open-fronted foyer room and usually comprising one room with no internal dividing wall; the *waenuw* wall with the central crawl through doorway comprises the front outside wall of such a house (Figure 7.24). A *momaenda* 'family house' is built according to the same design and only differs in having men, women and children living together in it. A house with two rounded *nunung* 'apsidal ends', usually a *tenda* 'women's house', will have crawl-through entrances in the side wall like the *tenda* room of the Honael *enzatenda* house (Figure 7.24); such houses are often longer and may have a central solid internal dividing wall and two entrances to give accommodation for two women and their children.

The builders used most of the materials they collected (Table 6.2) in constructing the various parts of the Honael house (Table 7.7). The stuff left over was trivial, comprising some strips of bark, lengths of rattan vine, some cane-grass stems and small pieces of wood. It is testimony to people's powers of estimation and again reflects tacit know-how

22 Compare Clarke 1971:105; Meggitt 1957:163–7; Pospisil 1963:259; Steensberg 1980:130, 136, 142, 173, for similar ground plans and elevations of houses built elsewhere in the Highlands.

Tasks	Time spent Stone-tool side	 Steel-tool side
travelling and searching for materials	10 h, 18 m	
making wall stakes	35 h, 37 m, 23 s	30 h, 18 m, 17 s
preparation of pandan-bark wall sheets	14 h, 20 m, 55 s	17 h, 51 m, 20 s
preparation of other wall materials	13 h, 40 m, 40 s	12 h, 51 m, 00 s
preparation of roofing materials	7 h, 44 m, 30 s	8 h, 51 m, 15 s
house-thatching preparations	22 h, 58 m, 00 s	22 h, 15 m, 15 s
preparation of other materials	5 h, 16 m, 00 s	3 h, 18 m, 40 s
clearing house site	21 h, 45 m, 05 s	15 h, 46 m, 10 s
construction of external walls	8 h, 50 m, 35 s	10 h, 26 m, 30 s
construction of internal walls	4 h, 54 m, 10 s	4 h, 54 m, 45 s
construction of other walls	1 h, 34 m, 55 s	1 h, 54 m, 00 s
construction and thatching of roof	9 h, 10 m, 40 s	8 h, 56 m, 15 s
internal fittings and drainage	1 h, 22 m, 15 s	3 h, 13 m, 45 s
grand totals	**152 h, 24 m, 08 s**	**145 h, 46 m, 12 s**

Table 7.8 Total house-construction times

in operation. When asked how they knew when they had enough materials readied, the house-builders simply signalled that this was just another of my inappropriate, even daft questions. The amassing of the raw materials also showed again the informal leaderless character of group co-operation, because there was no discussion between the men about the amounts of materials needed, they just set about amassing them; although they did sometimes discuss the suitability of some materials, such as certain pandan-tree trunks for making bark wall sheets.

It took a total of just over 298 person hours to build the house, including the time taken to collect all the necessary raw materials (Table 7.8). This is a combination of the time spent working with both stone and steel tools. According to these data, working with steel tools only, it would have taken about 291½ person hours to build the entire house, whereas using stone tools only, it would have taken about 304¾ person hours to do so. Overall the time difference is small, with the stone:steel ratio coming out at 1.05:1, which means that for every steel hour worked they had to work 1 hour and 3 minutes using stone tools to accomplish the same work. If the comparison is restricted to the work in which these tools were used continuously, the house-builders completed such tasks 1.43 times faster using steel tools.[23] These overall statistics confirm the point made several times previously, that the relative time efficiency of steel compared to stone tools is not of the order that many have estimated elsewhere (for instance, see Godelier and Garanger 1973; Hampton 1999; Pétrequin and Pétrequin 1993, 2011; Salisbury 1962; Steensberg 1980:147; Townsend 1969).

It is difficult in some regards to evaluate the overall impact of steel tools, which arrived with Europeans in regions neighbouring the Was valley, but these data on house building (which agree with data on other axe-using activities, Sillitoe 1979c, 1988a, 2010a) suggest that it was probably not of the magnitude widely thought – in this part of the Southern Highlands anyway. Direct comparisons between use of stone and steel tools are not necessarily possible. They require different techniques, as described. And comparative tool efficiencies vary depending on the materials worked on. Also, men tackle tasks with steel tools not previously attempted with stone ones, such as felling large, hard wooded trees to fashion wall stakes. Undertaking such heavy work reduces the comparative time efficiency of steel axes. Time alone is consequently an inadequate measure of comparative efficacy. It also misses the more proficient use of resources achievable with steel tools; such as the better utilization of timber in producing walling materials – albeit the consequences may be fairly small, particularly as resource scarcity is not an issue and hardly repays the effort of such improved exploitation. Time comparisons overlook the better quality of work attainable too. Furthermore, not all men previously had access to stone axes of the same quality, which were not of a standard size or material; some even shared an axe, which presumably impacted on their work in some way.

The schedules do not include the time spent maintaining axes, which would increase the differences somewhat, the elaborate hafting of stone axes demanding more attention (see Sillitoe 1988a:47–55). The inexperienced younger men arguably used the stone axes less efficiently, as they were unfamiliar with the requisite technique, differing in swing and delivery from steel (see Chapter 2), which may have increased chances of damage, demanding more attention to bindings than necessary in practised hands, in addition to

23 With a stone: steel ratio of 1.43:1, they worked 1 hour and 26 minutes using stone tools to complete the same work as achieved with steel tools.

them taking longer over tasks. A stone axe blade was also more prone to chipping, which if serious would demand much honing to restore the cutting edge (likewise a steel axe blade, when occasionally damaged). The cutting edge of a stone axe is less keen than a steel one anyway. The obverse is that men consequently spend longer today sharpening steel axes to maintain their sharp edge.[24] Indeed, the cutting edges of the stone axes we used had not been attended to for twenty years or so (left unused for this time their owners retrieved them from various places to sell to me for deposition in museums). Also, the house-builders may have used the axes in a more cavalier way than their owners previously, as they did not belong to them, so decreasing their efficacy further.

In discussing stone-tool use, this book is not trying to recreate some bogus 'stone-age ethnographic present', engaging in dubious salvage anthropology or cultural reconstruction. It seeks to place discussion in historical cultural context, not freeze the Wola in some questionable 'traditional' past, as the ethnographic-present-is-historical-fiction critique argues (Crittenden 1982:234; Smith 1980:85–90). The ethnographic moment here is a complex blend that situates today in respect of the past, with a view to the future, as discussed later (see Chapter 8) in reviewing the changes to house design following extensive contact with the world beyond the Was valley. This chapter gives comparative data on stone- and steel-tool use, including the time differentials in their use – part of a long-term research project – to further understanding of the extent of associated technological change and the wider implications (Pétrequin and Pétrequin 2011). The New Guinea Highlands was one of the last places, as noted in Chapter 4, where such enquiries were possible: I had the fortune to know individuals who had lived in stone-using times for a goodly part of their lives and, recalling them well, could discuss and demonstrate how they used the tools. The objective is to further understanding of the implications of stone-tool use, notably the time repercussions of their substitution by steel tools, on the assumption that we can only fully appreciate change and current practices if we understand what happened in the past.

A narrow focus on the present, ethnographic or otherwise, misses deep principles and assumptions that are rooted in the past, which to some extent continue to influence current attitudes and behaviour. The forces of globalizing change do not disconnect people from previous times, as pointed out by those who advocate that anthropology needs to consider, or even is, history. Furthermore, attempts to understand something about technological orders independent and unaware of industrial society, before they were significantly impacted by such forces, may prove of value not only in advancing appreciation of humanity's rich variety but also potential – for instance, in respect of balanced relations with the natural environment. They give a needed historical perspective to, and counters today's focus on, technologically affected globalization. It is partly in response to this burden that ethno-determinism advocates attention to the collection of ethnographic data, absence of which diminishes the possibility of future cross-cultural enquiry furthering our understanding of the human condition, which will likely involve advances in cognitive and neuroscience, as noted in respect of comprehending tacit aspects of life.

The extent to which the Honael data represent house construction generally relates to the wider question of variability. While the data are indicative of the time invested in house building, there is inevitably some variation between the construction of one house and another. We cannot be sure where the Honael house would fall on a bell curve compiled from data on the construction of a representative sample of houses, as suggested previously (if feasible). But observations on the building of other houses suggest that it is representative and not an outlier. It took a little over 94 person hours, for instance, for Maenget Pes helped by three relatives (Figure 7.4) to rebuild his family's *enzatenda* house at Tombem,[25] spread out over four days like the Honael house[26] (Table 7.9). It took less time overall to build than the Honael house, as it was situated on a previously levelled site that only required clearing off. Many materials were also re-used from the house dismantled there, including some of the cane thatch (see Chapter 6). The difference between the two houses was 7 hours 16 minutes, after subtracting the time spent collecting and preparing building materials for the Honael house and that spent on tasks not recorded at Tombem, such as making interior fittings.[27] These data compare with the time spent building other houses; for instance, Pes recalled that it took him and three relatives (Figure 7.3) three days to build the original house and one day to thatch it, while it took Maenget Hiyp and three relatives (Figure 7.5) three days to complete a *momaenda* house at Haelaelinja, after he and two helpers had spent two weeks previously amassing the necessary materials and levelling the site (these times compare with those for the Enga – Meggitt 1957:162). Sometimes men spread the work

24 No time was spent repairing serious damage to either stone or steel tools during house construction.

25 I observed the building of this house as a preliminary study, to familiarize myself with building practices prior to undertaking the detailed documentation of house construction at Honael.

26 The fifth day working on the Honael house was spent completing internal fittings and drainage, processes not documented for Tombem house.

27 Time spent clearing and levelling both sites is also omitted as not comparable, reducing Tombem house time to 85 hrs, 10 mins. It took 92 hrs, 26 mins to complete equivalent construction work on the Honael house over four days.

Task	**Time taken**
levelling house site prior to construction	9 h, 0 m
marking out house plan	30 m
dressing and driving in stakes for external wall	3 h, 40 m
splitting saplings longitudinally in half for *gamonk* battens	2 h, 25 m
lashing *gamonk* battens in place	1 h, 50 m
paring and sharpening *pinj* stakes	35 m
trying *bez* ridge-pole in place on *pinj* stakes	15 m
paring *pongow* stakes	20 m
securing pandan-bark sheets on walls	5 h, 30 m
pushing *sezel* cane in around outside walls	2 h, 30 m
securing *paym* pandan leaves on *waenuw* walls	30 m
construction of internal walls	11 h, 15 m
constructing low front *maenpat* wall	40 m
putting rafters in place	1 h, 0 m
lashing rafters	4 h, 55 m
lashing *iybhedla* battens to rafters along eaves	2 h, 30 m
lashing *kuwngis* sapling in crutch of rafters	50 m
positioning lathes and tying in place	6 h, 0 m
lashing supports in roof space and positioning bark fireguards	2 h, 30 m
splitting saplings for *sal* and lashing across front eaves	2 h, 15 m
putting bark sheets around eaves	55 m
attaching stakes along eaves and positioning pandan-bark gutters	1 h, 10 m
cutting cane-grass thatch	10 h, 40 m
carrying cane-grass bundles to house site	16 h, 55 m
thatching roof	5 h, 30 m
total time	**94 h, 10 m**

Table 7.9 Tombem house construction.

of house building over several days, especially if working alone; for instance, it took Maenget Korobol eleven days, working sporadically, to build his wife's new *tenda* house at Haelaelinja and another five days to complete the internal fittings.

It is possible to go further and estimate the labour a community invests overall in its housing stock, using census data (Sillitoe 2010a:217–20) that include information on number and types of houses occupied by a population and the time since the houses were built (Table 7.10).[28] According to these data, a community of about 350 persons (adults and children) builds 17.3 residential houses (that is *enza, tenda, enzatenda* and *momaenda*) per annum (Table 7.11), which works out at 1 new house per 20.2 persons a year.[29] If all these houses were new build, this would represent a communal labour investment – using the Honael house data – of some 5,155 hours per annum, which works out for this Was valley community at 79.8 hrs per adult male[30] and 6.7 hrs per adult female spent on house construction each year.[31]

28 The housing-stock data distinguish between houses in the Haelaelinja region – in the heavily occupied part of the Was valley nearer Nipa administrative outpost in the Nembi valley – where the majority of the population reside most of the time, and houses in the downstream forested region where many families maintain a second homestead occupied for varying periods.

29 This annual rate of house construction is about half that reported by Meggitt (1957:162) for the Enga, at 14 per cent per annum compared to 26 per cent of total housing stock.

30 Adults building houses are taken to be persons between 20 and 50 years of age; women contribute 7.9 per cent of labour time.

31 These times are considerably lower than those recorded for construction work in a time survey (by a factor of four, Sillitoe 2010:341–6). If we allow that some houses are rebuilt, the difference increases even further; for instance, if we assume that half the houses are new build and half rebuild, this is a communal labour investment (using the Tombem house data) of some 3,392 hours per annum (10.3 per cent of which is contributed by women), which works out for this Was valley community at 39½ hrs per adult male a year and 3¾ hrs per adult female. But the time survey included activities other than house building under the construction category. Also, the construction times recorded in this chapter only cover periods of actual work and omit (the fairly considerable) periods of time spent resting (smoking, chatting and idling) that are included in the time-survey estimates.

House type:	Enza	Tenda	Enzatenda	Momaenda	Mbolinaend	Showmayaend
Housing stock, survey December 1976 (for population of 347 persons)						
Haelaelinja region >3.5 yrs	13	11	5	2	2	1
2.5–3.5 yrs	1	5	1	3		
<2.5 yrs	3	16	4	13	5	5
total houses (Haelaelinja)	17	32	10	18	7[1]	6[2]
downstream forest region	2	7	12	12		
total houses (both regions)	**19**	**39**	**22**	**30**	**7**	**6**
Housing stock, survey June 1978 (for population of 351 persons)						
Haelaelinja region: >5 yrs	6		2			1
1.5–5 yrs	3	15	2	6	4	5
<1.5 yrs	7	27	6	7	3	9
total houses (Haelaelinja)	16	42	10	13	7[3]	15[2]
downstream forest region >1.5 yrs	2	13	9	13		
<1.5 yrs	4		1	9		
total houses (downstream)	6	13	10	22		
total houses (both regions)	**22**	**55**	**20**	**35**	**7**	**15**

1 Includes 6 'trade store' houses.
2 Include 2 abandoned houses used as pigsties.
3 All 7 'trade store' houses.

Table 7.10 Housing stock, according to age of houses.

House type:	Enza	Tenda	Enzatenda	Momaenda	Mbolinaend	Showmayaend
	2.5	7.9	1.9	5.0	1.0	2.0

Table 7.11 Average annual rate of house construction (calculated from Table 7.10 data).

These data on the number of houses built annually by a population occupying a generally known territory[32] also afford an opportunity to review further the cumulative impact of house building on natural resources and environment, as discussed in Chapter 6, which explored the idea of using key raw materials such as rattan to gauge pressure on resources and sustainability of use, in view of problems surrounding carrying capacity calculations.[33] According to these detailed housing-stock data, a community builds slightly more houses a year than estimated in Chapter 6, somewhat increasing pressure on resources. For instance, exploitation of rattan goes up by some 16 stems (173.28 m) per annum, which is well within sustainable harvesting limits for this key-species measure of environmental impact. While assessment of the effect of house-construction activities on many resources is difficult to quantify, this key-species-measure agrees with observations that suggest that the community is not noticeably depleting resources currently, certainly not to the point where people are voicing concern about shortages of building materials.

32 I say generally known in view of the problems encountered drawing boundaries around territories (Sillitoe 2010a:87, 207).

33 Although rights to resources are not confined to the territory of community of residence only, we can assume a swings-and-roundabout effect, as noted previously, with what comes from elsewhere balanced by what others take off to other places over the same extended period of time.

Energy in houses

We can use the Honael house-construction data to probe the materiality argument further. The 'ecological' approach points out that matter is active in the sense that it is subject to change over time – consequently the constituents comprising materials endure, not the things made with them. Ultimately it is the atoms (tiny packets of mass and energy) that comprise the molecules making up the materials that go on. This is what the 'becoming' of matter comes down to, according to the entropy principle that the energy in the material world disperses irreversibly, planet Earth heading constantly towards the lowest energy configuration, but for plant life using solar energy to put molecules back together for us to use and disperse again in our actions. Besides burning molecules metabolically, our energy-using activities include manipulating materials to produce objects, when we increase their energy levels for an interval, by rearranging the materials in ways that puts energy into them, thus increasing their potential to 'flow'. It is arguable that people 'grow' artefacts in this

Potential energy (PE) calculation (on Honael house data), using formula:
PE (joules) = *mgh*, where *m* = mass of object (kg), *g* = gravitational acceleration (9.8 m/s²), and *h* = height[1] of object (m).

mass of 2 *bez* ridge-poles (58 kg) at height 2 m =	116.0
mass of 57 *polpen* rafters (208 kg) at height 2 m =	416.0
mass of 31 *ka* battens (14 kg) at height 1.5 m =	21.0
mass of 9 *iybhedla* battens (3.5 kg) at height 1 m =	3.5
mass of 210 *dol, waenuw, pinj, maenpat* & *hiytpaip* wall stakes (900 kg) at height 1 m =	900.0
mass of 10 *buwk* pandan-bark sheets (110 kg) at height 1 m =	110.0
mass of 472 *sezel* canes (19 kg) at height 1 m =	19.0
mass of 18 *tuwmainj* internal wall and roof saplings (4 kg) at height 1.5 m =	6.0
mass of 26 *gamonk* bead saplings (38 kg) at height 1 m =	38.0
mass of 17 *haeluwp* vine stems (4 kg) at height 1.5 m =	6.0
mass of 144 *gaimb* cane-stem bundles (1470 kg) at height 2 m =	2940.0
mass of *aendkor* fireguard saplings and pandan-bark sheets (15 kg) at height 1.5 m =	22.5
total *mh* =	4598.0
PE (mgh) = 4598 x 9.8 = 45060 joules[2]	

1 The heights are maximums, for instance of rafters and thatch, which probably results in a somewhat inflated PE calculation.
2 Note these calculations assume that the materials express their full PE, making the fall to earth cleanly, whereas in reality this is unlikely with vegetation probably growing up around the house and becoming tangled with materials, hindering their fall; also the materials that collapse first will break the fall of those that do so subsequently.

Table 7.12 Potential energy stored in a house.

way, like organisms use energy to grow. For instance, in building a house, the Wola increase the potential energy of the materials they use. The erection of vertical walls, the horizontal ridge-poles supported some two metres above the ground, and the rafters resting on these, all store some energy in these physical configurations. This energy is released kinetically when the materials of the house 'move' and the structure collapses to the ground, the cave-in being proceeded by a series of many small 'moves' from the time the house was built. (The subsequent decomposition of the materials in the soil-litter layer slowly releases further energy contained in the chemical bonds holding together the constituent molecules, or alternatively it may be released quickly if the collapsed materials are fired or, worse, the house burns down.)

We can calculate the potential energy stored in the materials when erected into a house (Table 7.12), which is some forty-five kilojoules. The calculation underlines the repeated criticism of the 'ecological' approach to materiality, that its narrow focus on the 'activity' of matter prompts it to overlook human activity. We can in part address this criticism from an ecological perspective, by calculating the energy that people expend in constructing a house. The collection of such energy-expenditure data and their accounting analysis (although conceptually straightforward) feature several methodological challenges. While it is relatively easy, with the necessary technical equipment, to calculate the energy an individual expends in a regular activity, when several persons are engaged in an activity we have to allow for individual differences in work rates and energy expenditure (according to age, physique etc.). The tendency for persons to switch about between tasks when building houses increases problems, their effort varying irregularly: at one point driving in wall stakes, for example, and next splitting rattan vine before lashing on a sapling beading, and then driving in more stakes and so on. The recording of such varied activities by different actors demands not only time data on various tasks but also that subjects perform these activities for an interval while wearing respirators or other equipment to measure energy expenditure. While we do not have such detailed data, we can work out approximately the energy expended in building a house, using the Honael house time-and-motion data and published data from elsewhere on energy expended in various tasks.[34] According to these calculations, the builders expended a total of 324,109 kilojoules constructing a house (Table 7.13). The results of these calculations, regardless of their gross nature, parallel

34 The data on energy expenditure according to tasks come from the Lufa (an Eastern Highlands population of similar build and genetic stock to the Wola) and was measured by indirect calorimetry using a Max-Planck respirometer (Norgan *et al.* 1974). All of the energy values are averages expressed in terms of the sample group's mean body mass (57.5 kg for Lufa men) – Norgan *et al.* 1974:332. Hipsley and Kirk 1965:95–107 report energy expenditure data for a small number of activities from a Chimbu population that have a range similar to the Lufa data.

The energy values used in calculations (Norgan et al. 1974) are as follows:

Construction activity	Energy expenditure (kJ) Using stone tools	Using steel tools
preparation of materials		
wall stakes	39205	37627
wall bark sheets	17911	21929
other materials for walls	16689	15498
roofing materials	9949	11319
thatching materials	21435	22392
other materials	4223	6745
construction work		
clearing house site	28185	20249
erecting external walls	6665	8080
erecting internal walls	4421	4553
erecting other walls	2743	3494
roof construction and thatching	7807	7683
making internal fittings	1573	3734
totals	160806	163303

Table 7.13 Energy expended in house construction.

(note 34 continued)

Activity	Men kJ/min.	Women kJ/min.
clearing ground	20.5	
digging ground	24.7	
shovelling - roadwork	20.9	
cutting tree	24.3	
chopping firewood	21.8	
collecting rattan	17.6	
cutting pitpit	13	
splitting wood for posts	18.4	
sharpening posts	17.6	
putting in fence posts	19.2	
tying fence posts	13.8	
weaving pitpit wall	7.9	
roofing house	13	
sharpening axe	7.5	
standing	6.15	5.4
sitting	5.73	5.18
walking uphill	25.1	21.3
walking downhill	13.8	11.3
walking uphill & downhill average	19.5	16.3
walking with a load uphill	25.7	23
walking downhill with load	19.7	17.6
walking uphill & downhill with load average	22.7	20.3
walking uphill & downhill with (50 per cent) and without (50 per cent) loads average	21.1	18.3

the timing data with respect to the near parity of energy expended working with stone or steel tools; the extra time using stone tools is not spent on much more energy-demanding tasks.

We can compare the results of these two calculations. An evaluation of the energy that persons expend in building or making things, with increases in the potential energy of the materials that they work on, is certainly a novel use of energetics data. (These more usually feature in comparisons of the calorific value of actors' diets with the energy they expend in their daily activities, arranged on a balance sheet to assess gains against outlays, notably on production activities, which relates to the security of subsistence regimes – Ellen 1982; Hipsley and Kirk 1965; Sillitoe 2002; Ulijaszek 1995). But it begs the question: why bother? The fraction of the energy that the Wola expend in house building that is stored as potential energy in the materials they use, at a mere 0.01 per cent, is miniscule. The intention is not to create an energy store, after all, and so pushing this approach to materiality to its logical conclusion – in the context of human activities, as opposed to physico-chemical material science, so addressing Knappett's (2007:23) query about how it might be taken further – raises the question as to its aims from an anthropological perspective. On the upside, these calculations, indicating the amounts of energy people expend in rearranging natural materials to serve their purposes, afford an objective way of comparing different building traditions – such as local timber-and-thatch dwellings in the New Guinea Highlands versus concrete-and-steel high-rise structures in Europe – and so offer a further way to assess relative ecological footprints and sustainability.

The energy approach may have more relevance from an archaeological perspective; as Hodder (2012:67–8) indicates in deploring that 'the chemical and physical composition of things – what we might call their archaeometry – rarely is part of discussions of materiality', arguing that there is a need 'to explore the full entanglements and entrapments into which humans are drawn as things run out and fall apart – these entanglements are irreducibly both physical and material as well as social and cultural' and that we have to 'break the prejudices that still separate the scientific study of matter from the interpretation of materiality'. Indeed, according to Knappett (2007:23), the idea of 'the coming into being of materials will be music to the ears of those who engage in experimental archaeology, as it hints at an often lacking philosophical background or framework to such study'. It is a line of enquiry that this branch of archaeology could surely further contribute to, collecting data on the energy persons expend undertaking various tasks in manufacturing things from the past and present; albeit to be true to possible energy transactions previously it may be necessary for experimenters to live on a diet from those times (this caveat applies equally to the suggestion that experimental archaeology might contribute to our understanding of tacit knowledge – see Chapters 4 and 5). Such enquiries introduce behavioural data from living humans on the handling of otherwise currently inert – whatever materiality advocates argue – materials and objects found in the archaeological record. And measuring the energy that humans transfer to materials, which contributes to their 'flow' or distance they have to travel down the entropy slope to lower energy levels, also has some relevance in archaeological contexts regarding the decay of deposited objects.

The analysis of energy expended in house construction is certainly not intended to suggest that the Wola think of their activities in this way; I employ energy accounting in the same manner as I use chemical and physical analyses of soil to further understanding of farming practices, without implying that people express any ideas comparable to nitrogen fixation or bulk density (Sillitoe 1996). While they are aware of the *buriy* 'effort' that they expend in constructing a house, individuals do not think of it in terms of energy, but rather experience it though their own sweat and muscle exertion. It is an aspect of their tacit knowledge, which the deployment of this foreign ecological strategy to further appreciation of their activities does not get us any nearer to knowing, such as their tactile going with the flow of their materials. The energetics calculations illustrate vividly the above point about the wide gap between the all-knowing perspective of academic analysis and the local-knowing one of actor experience, deploying a foreign strategy to further appreciation of peoples' activities, inevitably distorting understanding of their behaviour and knowledge, which is the subject of enquiry. '*The cybernetic nature of self and world tends to be imperceptible to consciousness* insofar as the contents of the "screen" of consciousness are determined by considerations of purpose.' (Bateson 1972:444 – original italics). If anything, the imposition of this foreign view on their activities, quantifying the effort involved in terms familiar to us, only exacerbates the gap, and certainly makes it more obvious. It is ironic that following the phenomenologically inspired materiality approach to its logical entropic conclusion should take us further away from an understanding of embodied behaviour.

8

Some other structures

Tradition and change

A noteworthy feature of the Was valley housing stock is its uniformity. All houses are built along the same lines and feature relatively few structural differences, which, as pointed out in the previous chapter, allows us to use data on one house to represent the material and labour inputs needed to build houses generally – albeit as an *enzatenda* the Honael house was larger than many. Variations in construction depend primarily on the gender of residents, differentiating between men's and women's accommodation, as noted in the opening and previous chapters. It is possible, for instance, to distinguish men's accommodation by the open-fronted day room; although some houses with this feature – like the *enzatenda* 'men's-women's house' – include women's accommodation too.

While all houses are traditionally built according to the same general design and constructed in the same way, they vary with respect to dimensions and details, such that no two houses are ever identical. This variation around the standard plan relates in part, as noted, to available raw materials and their properties, and it intriguingly pulls in the opposite direction to the adherence of residences to the same blueprint, underlining the point that equality does not equate with sameness (Sillitoe 2010a:49). This variation on a theme echoes the way that persons subscribe to acknowledged social conventions, having been socialized to observe them from birth, yet no two individuals behave identically, being free to interpret the norms as they chose within accepted limits. While people can describe, for instance, the sequence of transactions that characterize a socio-political exchange (Sillitoe 1979a), these vary with respect to actors and occasion. It is doubtful, even, that rituals repeatedly followed the same formula.

Such variability around a customary agreed core is characteristic of a stateless order, where no authority exists to legislate over what comprises proper procedure. While individual ability is acknowledged, and sought, in practical and other matters, some persons being recognized as more knowledgeable and skilled than others at particular things (such as hunting, making some things or contributing to public events), any influence they have only extends to these limited spheres. Consequently, the nominal 'big men' – *ol howma* 'men of the clearing' – of the Was valley derive no authority, as noted previously, from the transactional skills on which their social status depends, being no more than transitionally first among equals (Chapters 1 and 3; Sillitoe 1979a). Ability and knowledge are inevitably unevenly distributed, but this does not translate into some parties systematically monopolizing knowledge and resources necessary to the lives of others, dominating them politically, threatening their individual sovereignty and the egalitarian ethos. The implications for anthropology of the inevitable associated behavioural fluidity are considerable, as noted previously, with much of the ethnographic record comprising normative accounts that such variability challenges.

What houses signify

The housing stock of the Was valley reflects these egalitarian political-economic arrangements. The variation signals little about the social order beyond relations between women and men. While intentional variation relates to the gender of a house's occupants, serendipitous variation shows no social pattern. Differences in the size and style of houses relates in no way to the personal identity or social status of the occupants. It is not possible, for instance, to tell if a successful *ol howma* 'man of renown' lives in a house because it is larger, built to a different design or features better materials or fittings. (There is perhaps a tendency, in places where *gaimb* 'cane grass' predominates, for men's houses to feature *senz* kunai-grass thatch more than other houses, which relates indirectly to status in that it shows that the occupants can afford the extra payment necessary to cover the additional work required to gather and transport the kunai thatch; there may also be a certain aesthetic consideration, although as people point out, the resulting roof is no different functionally, such as being more weatherproof.) That the relatively minor differences in dwelling design and construction communicate nothing beyond occupants' gender (which applies only to some houses) accords generally with Was valley houses apparently having scant symbolic associations, an issue picked up later.

The type of house families occupy is a matter of personal choice – *mbinyon konem* 'their thoughts' – with no social precedents evident beyond complying with gender arrangements. Some families opt for males and females living in separate men's and women's houses, and others may decide to live together in the same house, sometimes in separate rooms and other times in the same one. The decision-making is difficult to follow, depending to a certain extent on family dynamics, previous events, location and current housing stock. A man whose brothers live in a longstanding *enza* 'men's house', for instance, may decide to reside with them and build a separate house for his wife and children. If he is polygynously married, his wives may demand that he builds a house for each of them and refuse a larger house divided into separate rooms – particularly if they are on bad terms, otherwise differences between them may make life difficult for everyone.

All houses are similar overall in material construction and room size. In keeping with the egalitarian political-economic order, all persons enjoy the same level of accommodation and material living standards. There are no hovels and mansions; variations in design and construction are minor. The egalitarian ethos may account for why houses in the Was valley are not more varied with anyone seeking to improve on traditional house design and construction. This relates to the intriguing question, which I raised somewhat tentatively in *Made in Niugini* (Sillitoe 1988a:3): why is it that many populations remain satisfied with technologies that came in the eyes of Europeans to define them as primitive? The expectation in egalitarian socio-political contexts that individuals will not differ greatly from one another may partly answer this complex question. It arguably inhibits necessary experimentation and innovation.[1] Anticipating that all will enjoy more or less the same material standard of living may deter persons from seeking to advance things, possibly improving their position above others. The implication is not that architecture did not undergo endogenous change (as Kembol *et al.* 1976 describe, for instance, for the Enga region, where people adopted some features of Melpa men's houses), but that it was gradual and accessible to everyone. The only innovations that the community might tolerate are those that all can adopt and benefit from, which implies relatively small incremental changes that may freely catch on; whereas it will thwart any persons innovating in ways, beyond the reach of others, which advance them in a manner that the rest likely sense threatens domination, probably associated with other aberrant behaviour seeking to extend control over wider aspects of community life.

This may seem paradoxical to those living in capitalist states, who likely assume that support of individual political sovereignty should prompt constant innovation, with persons free to follow their own ideas; whereas the reverse pertains when the egalitarian ethic damps down such behaviour, discouraging the emergence of any significant differences that might lead to some gaining power over others and undermining the stateless order. From a wider economic perspective, the stress on egalitarian socio-political relations also arguably inhibits related accumulation of the surplus capital that is necessary for capitalist-style growth (Sillitoe 2010a:435–6). But acephalous arrangements accommodate the apparently incompatible from the perspective of Western democracy, where two opposed views, one championing individual liberty and the other equality,[2] endlessly slug it out on the assumption peculiar to hierarchical state orders that increasing one diminishes the other. Whereas they are mutually reinforcing in the Was valley. These arrangements also further challenge normative sociological interpretations of behaviour in small-scale societies promoting social solidarity, which cannot consequently, it is argued, feature individualistic action because they socialize persons to put community interests above their own (Sillitoe 2010a:36–7). In the Was valley the values that inform behaviour, inculcated from infancy onwards, afford considerable scope for individually calculated behaviour (as pointed out in Chapter 3). They promote personal sovereignty and equality according to a cultural code that simultaneously allows for orderly social interaction (Sillitoe 1979a), while inhibiting any self-interested drive for material advantage, extending in probability to equality-undermining power over others.

The absence of an ethic of experimentation and innovation characterizes social orders that condition people to think that current ways of doing things, passed on from generation to generation, are adequate. This attitude relates to ideas of time, which I have characterized in the Was valley as resembling a tightly coiled spring, the generations cycling around with relatively little change evident between each (Sillitoe 2010a:340). Until, that is, the outside world entered the valleys of the Highlands, introducing momentous changes, including new house designs, discussed below. This view of time lends further support to the previous chapter's promotion of comparative enquiries into the use of stone and steel tools, situating today with respect to yesteryear. Stone tools were part of the 'timeless' past with, from a local view, its unchanging technology – as seen for example in house construction – which was an integral aspect of the ongoing egalitarian

1 This is notably different to building traditions elsewhere that encourage innovation, such as that of Mali, where with 'experience ... master masons ... creatively expand the existing repertoire of built forms and decorations in a way that is deemed both traditional and innovative by their fellow colleagues and public' (Marchand 2009:223).

2 Broadly represented in the UK by the Conservative and Labour parties and in the US by the Republicans and Democrats.

Plate 8.1 A mbolin aenda 'white-skins' house' (heating stones for an earth oven)

stateless order. People's expectation of doing things in the way of previous generations again relates to the issue of tacitness. Actors do not think consciously about this, but accept what they do as normal; they just get on with things, for instance, building houses as they have always done. This underlines the point made several times previously that this account presents what I think is the upshot of the values and behaviour that I observe in the Was valley – it is an imposition, none of my Wola friends have ever talked to me in this way.

The disinterest that people show in what appear to be ancient house sites (see Chapter 4), may relate in part to the expectation that architecture is going to remain much the same through the generations given the inhibition of revolutionary innovation. Why recall such places and the structures erected there, if they were much the same as today? The disinterest in prehistory also reflects people's conceptions of time, in which life went on much the same from generation to generation with imperceptible changes (until, that is, recent colonial and postcolonial events), with its passage akin to a spiral rather than a line (as pertains in Europe where there is considerable interest in the sites of ancient buildings; a fascination informing vernacular architecture's appeal – Chapter 1). It is arguable that this disinterest in the past is of a piece with a lack of interest in ancestry generally, as evident in shallow genealogical memories (encompassing some six or seven generations). It also accords with the flexible land-tenure arrangements (Sillitoe 2010a:85–125). Those who occupied locales long ago may not be ancestors of current land-rights claimants, such that recalling long-past land users would not only contradict today's tenure but also inhibit the operation of the flexible land-tenure system. Furthermore, recalling previous distant generations would suggest descent-structured organization of the sort that previously befuddled New Guinea ethnography; the absence of such a record supports arguments about its inappropriateness.

Awareness of how prioritizing individual sovereignty and egalitarian relations contributes to the continuance of customary technology – by setting limits on how far persons may seek to advance themselves materially above others – reveals the error of equating level of technology with social development and thereby concluding that a lesser level indicates a primitive society. If you value liberty and equality, such stateless orders are more advanced socio-politically, affording expression of these values way beyond anything imaginable in a nation-state. It affirms criticism of the assumptions of cultural evolution (see Chapter 4). Another issue to consider is the environmental impact of different technological regimes. The so-called simple technologies, such as Wola building practices, are often more sustainable and less destructive of natural resources (as mentioned in Chapter 6). In the long term, being in balance with the physical environment is likely to prove the less primitive option, as worries over industrial impacts show, such as current ones about global climate

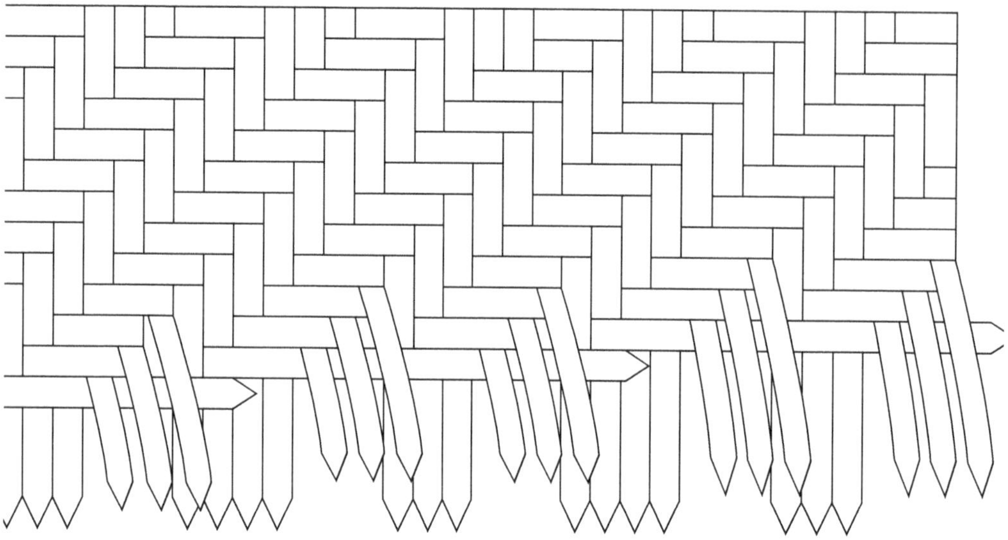

Figure 8.1 Pattern of gaimb waip 'cane-grass weave'.

change. As the evocative Native American saying has it: 'When the last tree has been cut down, the last fish caught, the last river poisoned, only then will we realize that one cannot eat money.' (Speake 2009:177). But in the Was valley today attitudes are changing, even coming into conflict, and challenging egalitarian values.

New house designs and social change

The momentous change that steel tools heralded with the outside world's appearance in the New Guinea Highlands (Connolly and Anderson 1987; Schieffelin and Crittenden 1991) included new ideas about house design that would lead to previously unheard of innovation in building, together with a certain socio-cultural upheaval. The arrival of *mbolin aenda* 'white-skins' house' architecture opened up considerable new possibilities; albeit the extent of the change is modest so far, in comparison to some other regions (see Rensel and Rodman 1997). Initially, people saw large, strangely styled houses, erected adjacent to the airstrips established by the Australian colonial authorities at administrative posts, that used foreign materials – such as metal, glass and plastic laminate. These authorities also required people to build and maintain rest houses in their settlements for the use of visiting patrols,[3] not according to their traditional squat, crawl-in design but high enough to walk into and stand upright inside. And soon afterwards missionaries arrived and arranged for communities to erect buildings of similar dimensions for use as churches. Subsequently, certain individuals have gained access to some of the imported manufactured materials and have used these in building their own houses.

Other than houses of larger dimensions, the other noticeable change has been in construction, with the innovative use of a locally abundant raw material to produce woven wall- and floor-coverings. These are made with *gaimb* 'cane-grass' stems, which people split open longitudinally and flatten out, and weave together into sheets by passing 'weft' stems over and under two or three 'warp' stems (Figure 8.1). They call the resulting woven-cane sheets, after the cane grass, *gaimb waip* 'cane-grass weave'.[4] The Wola learned to make these from the fellow New Guineans who accompanied patrols, such as policemen, and they assume that such weaving is a *nembis* 'coast' custom (Steensberg 1980:166, 170 describes similar 'pitpit matting'). Sometimes house-builders today face the walls of traditional-style houses with such cane-grass weave, notably the low front wall of *enza* 'men's house' foyers.

The building of *mbolin aenda* 'white-skins' houses' is otherwise similar to traditional house construction, especially where it features local raw materials – such as hewn wooden stakes and battens, poles for ridge and rafters, and kunai-grass thatch. After levelling a site and marking out the house plan with vine and cane stems, builders drive in substantial *dol* 'wall stakes', some two metres or so in height, leaving gaps about three-quarters of a metre wide at entrances. They secure the stakes together with battens either lashed in pairs as in traditional house construction, or today they use a *dolbez* 'wall-ridge-pole', which comprises a substantial timber nailed in place horizontally along the tops of the stakes. They may fill in the spaces between the stakes by pushing in green leafy cane-grass stems. They arrange sheets of *gaimb waip* 'cane-grass weave' vertically around the wall stakes, both inside and outside, propping in place temporarily with lengths

3 Our initial field house was such a *kiap* 'patrol officer' rest house (Sillitoe and Sillitoe 2009:176).

4 *Waip* refers to marks made in carving on objects and here it refers to the pattern of the weave; it is also the word for writing. Another term for these woven sheets is *plaen*, which is locally considered to be a Pidgin word.

Plate 8.2 *Splitting cane-grass stem to use in weaving.*

Plate 8.3 *Weaving split-cane-grass stems.*

Plate 8.4 Ground plan of mbolin aenda house with dressed stakes ready to drive in.

Plate 8.5 Driving in long wall stakes for mbolin aenda house

Plate 8.6 Nailing dolbez timber in place along tops of the stakes, held aligned by temporary lashed battens.

Plate 8.7 Sheet of woven cane grass propped in position on external wall.

Plate 8.8 Weaving flattened cane grass around window; person sitting is bashing cane grass stem nodes to split them so that he can open the stem up and flatten.

of timber pushed in at an angle against the ground. They secure the woven sheets in place with battens, preferably made from strips of pandan bark, which, if they have no nails to fasten in position (the preferred option), men lash in place with rattan-vine strands, poked through the weave and taken around the stakes. They cut out entrances and any window gaps, securing the exposed edges of the cane weave with battens.

If the house is gable-roofed, which likely if grass thatched, the builders, when initially putting in the wall stakes, also drive in a pair of substantial *pinj* uprights in the centre of the two shorter gable-end walls, branches trimmed to give a v-crotch one end and the other end sharpened to drive into the ground. They place a single stout *bez* ridge-pole horizontally between these two uprights, supported in the v-crotches, and arrange *polpen* rafters along it sloping down to rest on the top edge of the two long walls, as with a conventional house. They lash the rafters in place along the ridge-pole, and either lash or nail (if they have sufficiently long nails) along the wall-top batten. They complete the roof in the same way as on a traditional thatched dwelling, arranging and lashing *iybhedla* and *ka* battens across the rafters and securing small bundles of kunai grass to these, starting along the eaves, each row overlapping the one below as they work their way to the ridge. They fill in the gable-wall ends with a framework of saplings which they likewise thatch over or to which they attach triangular-shaped sheets of *gaimb waip* weave. If they have sheets of corrugated galvanized roofing iron, house-builders normally erect a structure with a single roof slope – one of the longer (usually front) walls higher than the other – as they commonly have problems securing a watertight ridge line. They space the *polpen* rafters so as to

Plate 8.9 Nailing pandan-bark battens to hold woven cane-grass sheet in place.

run along inside the corrugations, so far as possible, nailing the sheets in place.

When house-builders have access to sawn timber – missions, for instance, sometimes use portable sawmills – they make doors from planks arranged lengthwise with short lengths across the rear nailed top and bottom, with perhaps a diagonal running between them. Otherwise they make doors from hand-hewn planks. They prefer heavy-duty T-hinges and padlockable bolts to secure doors if they can obtain them. Windows may be open, with stout timber secured bar-like across to deter intruders; some people use transparent plastic to fill them, glass being scarce in rural areas and glazing products more so. They commonly cover the earthen floor with a sheet of *gaimb waip* weave, and if there is a fireplace, it is usually a shallow flat-stone-lined depression situated in the centre of a room. Other internal fittings are usually sparse; sometimes there is a slatted sitting-cum-sleeping platform along one wall. If the building is to be a *storaenda* 'trade store', the owner may erect a plank-topped and cane-weave-faced counter across the room, and also secure some shelves along a wall to display his wares.

Other than obliging people living in certain locales to build *haus kiap* 'patrol officers' houses' and *lotu* 'churches', the colonial authorities did not try to persuade them to change the style of their houses to comply with what these foreigners thought were better accommodation; unlike in some regions, such as the Massim, where administrators and missionaries sought to improve housing by insisting that local populations construct 'Fijian-style' houses and rewarded those who did so with various prizes (Macintyre 1989). Families in the Was valley that build in the *mbolin* 'white-skins'' style of architecture do so under their own volition. The structure that both government and missions sought to force households to adopt was the pit latrine, with prosecutions and fines for those who failed to comply. The pit should be a couple of metres or so deep with a shelter of cane grass on a simple wood frame, similar to a cane-grass lean-to (see below), erected over it for privacy. But residents of the valley were never convinced of the soundness of the colonial authority's argument about sanitation and health. To their minds, collecting excreta and urine in a pit was an unsanitary and foul thing to do! By defecating in secluded spots, individuals explained, they distribute their faeces, which are soon washed away into the soil or foraged on by pigs.

While some families experiment with lofty *mbolin aenda* 'white-skins' house' architecture, many continue to favour their ancestors' squat design. The occupation of such a spacious structure demands blankets for sleeping at night, when temperatures in the mountains can be chilly (Chapter 7; Sillitoe 1996:69–70), whereas low houses remain warm and cosy with a fire, rekindled periodically on cold nights.[5] The converse is that higher houses are less smoky and so arguably better for respiratory health. Not only is supply of bedding a problem for many families, but blankets can soon become infested with irritating fleas that can disturb sleep – flea infestation sometimes prompts, as noted, the dismantling and rebuilding of houses. Rats can be another pest, taking up residence between woven-cane-sheet walls, even building nests and raising young in the dry vegetation inside walls.[6] I can testify to the nuisance of rats, my sleep regularly disturbed by them scuttling about at night, crawling up posts within walls and along beams and running around the floor. They may pose a health hazard too, scavenging food remains (in the absence of which they

5 Even where people have houses constructed entirely from imported materials and built to Western specifications with electricity and water laid on, supplied as part of the compensation they receive when mining activity disrupts their lives and results in their relocation – as in the Porgera region – many prefer to occupy 'bush houses' rather than these 'relocation houses' because they are easier to heat and more cosy (Golub 2014:140).

6 The rats that infest houses include *Rattus ruber, R. exulans* and *R. verecundus.*

Plate 8.10 Some rafters positioned on a single-pitch roof.

will chew other things such as net bags made of bark fibre, even bars of soap in my experience). The *mbolin aenda* style of house is also not as robust as the traditional one and can sway alarmingly to and fro when *duwlbaeray* 'earth tremors' occur (Sillitoe 1996:134).

Since the first arrival of *kiap-* and *lotu-*style buildings, some people have gained access to a wider range of construction materials, such as manufactured doors and window frames, fibre board, plastic laminates and corrugated iron. If they can obtain machine-sawn timber, men may use it extensively, as in cladding exterior walls with horizontal planks. Some individuals have travelled elsewhere in Papua New Guinea or overseas, where they have seen buildings that have further widened their architectural horizons and aspirations; some have relatives living in towns, including Lae and Port Moresby (where many self-build accommodation in unplanned urban settlements, often ingeniously using, or recycling, discarded materials such as wooden packing cases and scrap vehicles – see Plocki 1975 on Port Moresby's squatter architecture). Consequently, the potential for innovation has increased considerably and some are experimenting with new styles of building construction and design. They do not necessarily use manufactured building materials. For instance, a friend constructed a modified *mbolin-aenda-*style house with a raised floor made of palm-wood slats, such as he had seen at Lake Kutubu, where the air circulation helps cool inside; however, this is not desirable at higher altitudes, and the design proved uncomfortably cold and the modification unsuccessful.

Plate 8.11 Securing corrugated-iron sheets in place on rafters.

House design may change with social relations over time, involving not only building construction but also social arrangements and dynamics between occupants. In introducing an edited collection on housing in the Pacific, Rensel points out that houses may 'connote the dynamic nature of human interaction with domestic space' (1997:9). Even if they appear to remain the same, subtle changes are likely to occur between generations, with the domestic-cycle process (Goody 1958) if nothing else. While it is arguable that availability of new building materials may not only result in different styles of houses but also promote changes in social relations within homesteads – as, for instance, Boivin (2004:66–7) maintains has occurred in Kenya and Somalia – it is possible to reverse the emphasis and argue that changing social situations may inform changes in architecture, builders using new materials accordingly. The trend in the Was valley towards more *momaenda* 'family houses' possibly indicates changing gender relations.

The cultural significance of houses may change too. The construction of new styles of house, likely featuring some manufactured materials, relates to personal identity and social status. In constructing such houses, some families are indicating something about their aspirations to be modern, to change their way of life and shake off backward reputations as – to use the vivid Pidgin term – *bus kanaka* 'jungle natives'. But it is more than a symbolic issue of signalling standpoint. It is also a practical matter of increasing one's material standard of living, of desiring a more comfortable existence. It relates to potentially revolutionary political-economic change consequent upon the arrival of the outside world in the Was valley, featuring the possible spluttering emergence of a social hierarchy. After all, constructing a larger house with imported building materials is an expensive undertaking, which only a few can aspire to. The missions give some of their local pastors a leg up by constructing sawn-timber dwellings with several rooms and various manufactured domestic fittings for them. Others benefit from relatives in paid employment elsewhere, whose remittances fund house construction. Those in government employ, such as health workers and school teachers, may also reside in imported prefabricated dwellings. And those involved in national and provincial politics are able to fund the construction of large, even ostentatious buildings (notably those who hit the jackpot of election to the National Parliament, where the rampant corruption in Papua New Guinean politics today affords them access to large sums).[7]

Attempts to establish *storaenda* 'trade stores', invariably built in the modern style, are a local response to these changes, often by those without access to external resources who are seeking to get in on the new order and opportunities. Few succeed for any time. When they fail, the buildings may be turned over to some other use, or be dismantled and the materials used to construct something else. They fail for a range of reasons. There are stocking problems; these include the distances owners have to transport goods, often on foot (for instance, carrying a case of tinned fish or sack of rice 15 miles or more). The absence of any discount retail outlet obliges would-be traders to purchase goods from local suppliers, such as mission or township stores, at the same prices as everyone else, which poses problems selling them on marked up to make a profit. The trouble many have calculating profitable margins sufficient to generate income to invest in further stock and allow their enterprise to grow compounds their problems. On top of this, they have to contend with the demands of kin for goods on credit, which they sometimes never pay for, perhaps intending to reciprocate in some other way. Such demands are hard to refuse in a face-to-face community that places a high premium on sharing. It illustrates how others, willy-nilly, bring down those who may try to advance themselves above their fellows.

Being cut down to size is an ever present threat where behaviour challenges local egalitarian values and expectations, as when would-be store owners profit at the expense of others or, more blatantly, dishonest politicians and government employees embezzle public funds. The innovation and variation seen in housing, with scope today for considerable differences in accommodation between some families, runs the risk of provoking such action as people sense, albeit maybe vaguely, the power implications and challenge to local values that such differences in living standards represent. In times of conflict, for instance, it is common for European-style houses built from imported building materials for *bisnisman* 'entrepreneurs' and *memba* 'politicians' to be targets for destruction. Such cutting down of innovating poppies that get too tall is seen as an obstacle to economic development; for instance inhibiting the accumulation of necessary capital for investment. Yet a certain discord is expectable when development presumes transformation to a market state; the capitalist ethic of which openly confronts local values in advancing some economically and politically over others. The imposition of a national government has created conditions, exacerbated, as pointed out, by widespread corruption after independence, for the tentative emergence of a class-like hierarchy, with a few rich and powerful, and many poor and powerless.

It is widely accepted that an awareness of social arrangements before such momentous events is necessary to appreciate contemporary transformations and their

7 The corruption scandals surrounding Philemon Embel, the 2007–12 Member of National Parliament for the Nipa Kutubu Electorate, include misappropriation in 2006 of K165 thousand for a new church, involvement from 2005–11 in receiving K51 million of Planning Ministry funds, and being responsible in 2011 for K1.2 million of 'missing' public funds (*Post Courier* 2011a and 2011b).

implications, as these continue in part to inform the axioms and values that underpin social life, and may reassert themselves in other guises, as evident in the increasingly overt syncretism between Christianity and local beliefs. Substantial change has occurred in the Was valley in the past generation or so, but it has not wiped out the pre-existing institutions of the subsistence economy and acephalous polity. At first, people adjusted, somewhat bemusedly, to the appearance of the politically irresistible colonial state and the evangelical missions with their all-powerful deity. Arguably, some of the confusion apparent today in the Was valley reflects people's attempts to work through challenges to fundamental values, which they sense are under threat. Confusion is inevitable: a stateless nation is a contradiction in terms, after all. But the changes have not disrupted the integrity of the local order, acephalous political ways and accompanying violence (aggravated by '*raskol*' criminals and the presence of guns) reasserting themselves following the collapse of state authority after independence.

To appreciate such responses to 'modernization' by setting them in the context of deep-rooted cultural assumptions returns us to the issue of the ethnographic present. It is a complex moment, as argued in the previous chapter, situating today in respect of the past, in aligning present day behaviour with historical socio-cultural circumstance, where current morality originated and which continues, through processes of 'social reproduction', to exercise some sway over everyday life. This ethnographic present inevitably conflates the present with the past to some extent because such an amalgamation comprises the current day. However, the historical view that we need to understand what pertained in the past to appreciate what is happening today is arguably only pertinent following the intrusion of the outside world into the Highlands region and the consequent tremendous change, which also brought and made relevant a Western linear perspective of time. An atemporal ethnographic present is a truer representation of the previous conservative cultural perception of time as cycling around from generation to generation, with relatively little change evident between each – in housing arrangements, for instance – signalling an egalitarian ethos in remaining constant.

Shelters

The *aenda* 'houses' discussed so far are occupied largely in the late afternoons and during the night. They are commonly empty through the main part of the day, unless residents judge the weather too inclement to venture out, or someone is sick and stays in, resting, or an event occurs, such as a marriage or funeral, with associated exchange activity. During the daytime people often use smaller *pəgaenda* 'shelters' that are of varying permanence and design, ranging from lean-to structures to gable-roofed huts and tepee-like stooks of cane grass (Figure 8.2). They are

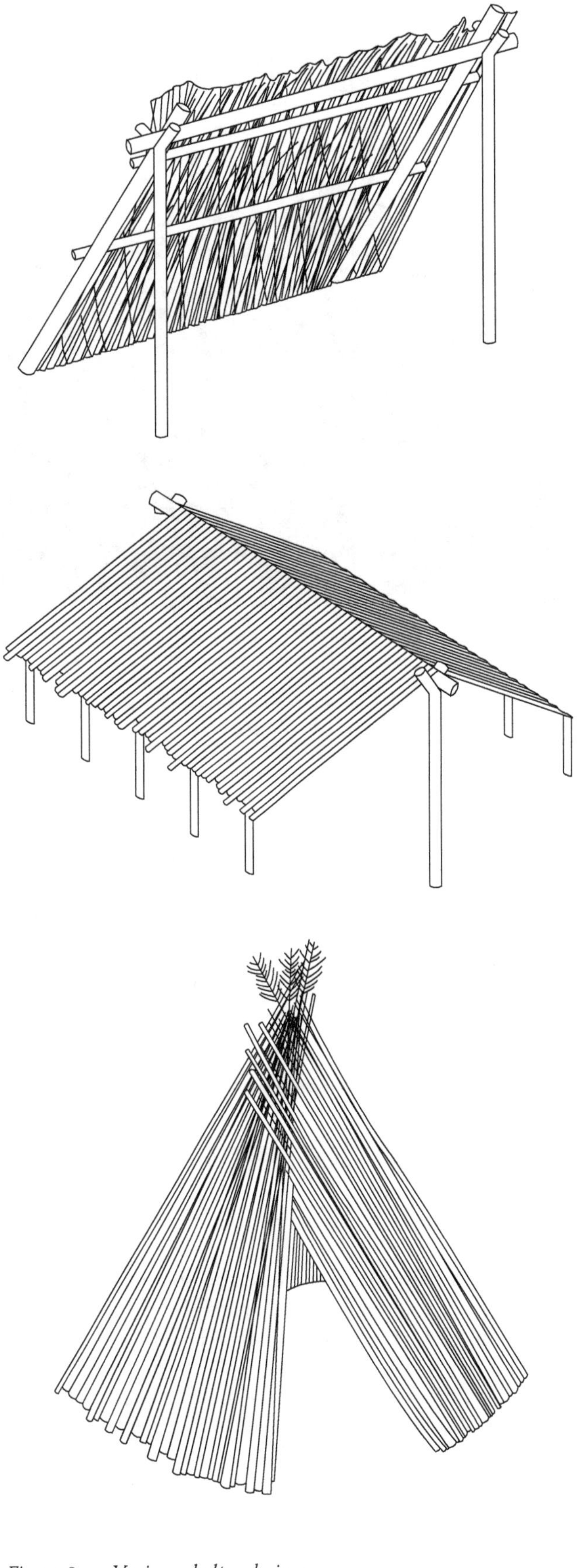

Figure 8.2 *Various shelter designs.*

Plate 8.12 A lean-to pəgaenda shelter.

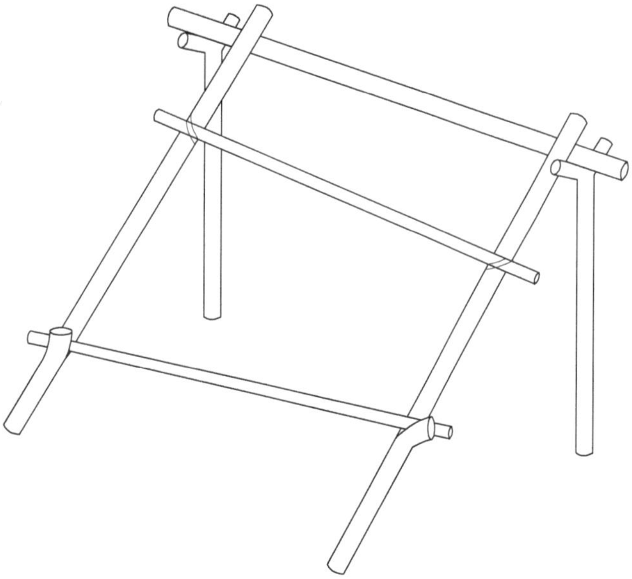

Figure 8.3 Framework of lean-to shelter.

often located adjacent to gardens, and some families have them adjacent to their houses as part of their homesteads (Meggitt 1957:175–6 describes similar shelters in the Wabag region). They invariably have a fireplace where one can sit and smoke; often people cook in these houses and some have piles of stones for use in *saway ongat* 'earth ovens'.[8]

The *pəgaenda* shelters vary in design from simple and quick to erect, to structures that are more complex and time-consuming to build. The lean-to version comprises a straightforward timber framework. This consists of two *iysh shoba* 'fork-ended uprights', fashioned from straight lengths of timber with branches trimmed to give a v-crotch one end, the other end sharpened to drive them into the ground, two to three metres apart. The builders place a stout ridge-pole horizontally across the two uprights, supported in the v-crotches, and rest two or more rafter poles on it, sloping downwards at an angle to the ground, which they may lash in place with vine. Sometimes they also erect a horizontal pole supported on two short forked poles less than one metre high at the rear of the structure to support the rafters off the ground there. Next, they place several lengths of wood or cane at right angles across these sloping rafter poles, lashed in place or resting in the crotches of any lopped-off branches (Figure 8.3). These serve as battens to support the cane-grass thatch heaped onto the structure's sloping roof. It took one-and-three-quarter hours to cut the timber and build the frame for a lean-to *pəgaenda* shelter adjacent to a garden at Ganonkiyba (Sillitoe, Stewart and Strathern 2002) and two hours to thatch it with cane grass.

The time taken to build gable-roofed *pəgaenda* depends on the elaborateness of the structure, some of which can approach a standard *aenda* 'house' in design, enclosed by walls and with a thick thatch supported on rafters and lathes. They are usually open-ended and open-sided. If the sides and ends are enclosed at all, it is commonly with rough walls of cane grass or stacked up logs. The roof usually comprises a single ridge-pole supported in the v-crotches of two *iysh shoba* 'forked uprights', with rafters lashed to it and resting on low walls that often consist of horizontal poles at eave height supported on a few stakes.

8 *Mumu* is the Pidgin word for this form of cooking with hot stones in earth-oven pits.

Plate 8.13 *Inside a pəgaenda shelter, men eat a meal.*

Plate 8.14 *A temporary tepee-like pəgaenda shelter in a garden.*

The rafters have battens lashed to them like a standard house, but they generally support a thinner covering of, usually cane-grass, thatch.

The cane-grass stook structures are erected quickly, often from cane already cut down in clearing an area for cultivation, and are intended as temporary shelters for only a few weeks; whereas other shelters, notably those built in house yards, may last for years. When they have finished with shelters, people just abandon them, as they sometimes do other houses, which may be left *aend hunta* 'house empty' until they become *aend mondowgay* 'house rotten', that is decay and collapse, as described in the last chapter; although sometimes people may again burn down dilapidated structures, enclose the area and plant a variety of crops in the ash-enriched soil.

If a family decides to house pigs in a *showmay aenda* 'pig sty' separate from a *tenda* 'woman's house', then men usually build lean-to shelters; some use old abandoned houses, if available, as sties. They increasingly keep pigs in separate structures following messages from various quarters about

Plate 8.15 A showmay aenda 'pig sty' with animals in stalls.

the health hazards of living together with animals (see Chapter 7). They erect *kuwl* 'stalls', as documented for the Honael house, in which to house animals overnight. They erect similar free-standing pens to cage cassowaries in house yards, with stakes closely arranged to prevent birds escaping and a secure roof of cane-grass thatch (as do the Enga, Meggitt 1957:176).

Large communal structures

In contrast to these small shelters, the Wola occasionally construct large buildings that require the co-operation of considerable numbers of people. Unlike everyday houses, these buildings varied in design – like the events for which they are erected, which feature novel, fashion-driven activities that arrived periodically in the Was region from elsewhere. The latest such arrival is Christianity, which came with the outside world from *ples bilong Jisas* 'Jesus' place'. The missionaries who introduced Christianity told people that they had to meet regularly to worship Jesus and ensure their well-being and place in heaven. To this end they encouraged communities to build *horaenda* 'churches' (*lotu* in Pidgin), which are similar to large *kiap*-style rest houses when built from local materials. The missions endeavour to help communities to convert such bush-material churches into larger, more permanent and imposing structures by assisting with the supply of imported building materials such as corrugated iron and machined timber, and even occasionally arranging for enthusiastic young *mboliy* 'white' volunteers to 'show' people how to construct suitable buildings. These introduced, gable-roofed, usually windowless wooden buildings resemble small village halls.[9] The churches built of bush materials are starkly empty, whereas those built of imported materials sometimes have a mission-supplied, commonly bare, table-cum-alter at one end and occasionally a churchy image on the wall, but little else. They are not the first large structures seen in the Was valley, just as Christianity is not the first cult to come from elsewhere, being the most recent in living memory of several previous such arrivals, albeit arguably the most sensational, its evangelical scorched-earth policy subsequently scuttling all previous beliefs and practices (Sillitoe and Sillitoe 2009:159–60). While its *horaenda* 'churches' are today's versions of previous cult houses, some of these forerunners featured equally large or larger structures.

The *timp* cult that immediately preceded Christianity featured a large ritual structure called the *timp aenda* (Ryan 1961:265–87; Sillitoe and Sillitoe 2009:151–8). Some men in a locality would agree to sponsor the cult, a few of them becoming spell 'owners'. They had to give valuables – such as seashells, cosmetic oil and salt bundles –to their neighbours who had hosted the cult previously, in exchange for the ritual knowledge and spells; transactions that took place over several months as the new hosts prepared to take on the cult. Incidentally, in respect of equality being central to Was valley life, such individuals having knowledge of certain rites and spells above others did not threaten to compromise the egalitarian order. They used the knowledge to help relatives and friends – for instance, to promote

9 While space considerations prevent discussion of the details of the construction of these introduced buildings, any readers interested should consult one of the many books on constructing such wooden buildings, from small sheds to large halls (e.g. DeCristoforo 2007; Haun 2008).

Plate 8.19 The large central pinj posts supporting robust bez 'ridge-pole'; and low stake wall with two crawl-through entrances at rear of timp house (photo courtesy of A.L. Crawford).

Plate 8.20 Roof construction of polpen 'rafters' and ka 'battens' with bark-slab tiles (photo courtesy of A.L. Crawford).

Plate 8.21 A slat-door entrance into a timp aenda cult house (photo courtesy of A.L. Crawford).

Plate 8.22 Pig stalls adjacent to timp cult house (photo courtesy of A.L. Crawford).

slaughtered during the cult were clubbed over the hole so that the blood spilt dripped into it. Occasionally, the man who knew the spell associated with slaughtering pigs poured some *wombok* 'cosmetic oil' into the hole too, for the *towmow* 'ancestor spirits' to 'eat'. Both pigs' blood and cosmetic oil, being 'soft' runny liquids, render the 'ghosts' mouths soft', making their sickness and death causing 'bite' ineffective. The participants also constructed lean-to pigsties adjacent to the large *timp* cult house, containing several pens in a row, erected to house and show off valuable animals before their slaughter. While the cult sought both to weaken dangerous spirit forces and appease them with the taste of blood spilt and the odour of singeing bristles from offered animals, a further significant aspect was the exchange of pork during the cult's periodic events.

The killing of pigs usually accompanied any large ritual, and exchanges of meat consequently figured centrally, together with other exchangeables such as pearl-shells transacted as men cleared their debts to entitle them to slaughter animals. Following the arrival of missionaries and suppression of traditional ritual activities, such large communal occasions have focused entirely on secular socio-political exchange aspects. The *sa* sequence was such an event (Sillitoe 1979a:271–6), which involved the construction of special buildings, the *sa aenda* and *pokaenda*. The *sa aenda* was a longhouse[11] built in a way similar to an ordinary house, only varying in its dimensions, being considerably longer. The sponsors of the *sa* event usually arranged for the construction of two longhouses, running in parallel and facing one another across a *howma* 'grassy clearing' where they staged pig-kill festivals and dances. The length depended on the number of men who agreed to take responsibility for building a section of the structure, each approximating in size to an ordinary house. The sponsors of the *sa* cycle did not build the houses themselves but contracted others to do so in return for an *aend senz liy* 'house-thatch hit' exchange payment, an occasion when the hosts donned colourful self-decoration in celebration of these opening transactions, which were the first of several large socio-political exchange sequences that characterize the *sa* cycle.

The *sa* longhouse may feature interior walls dividing up the sponsored sections into separate rooms, otherwise it comprises a single long room with several doorways and central fireplaces (see Loupis, Ruff and Ruff 1982:149–50 for photographs of longhouse interiors). The use of the rooms varied, sometimes the persons sponsoring their erection resided there with members of their family and other relatives for a period, usually fairly brief; other times kin visiting from elsewhere, coming for instance to attend one of the *sa* sequence events, occupied the house; and other times, usually most of the time, it stood empty. The sporadic use of the house prompted some modifications to its structure, particularly by those commissioned to build it, looking for shortcuts and knowing that the structure would have a limited life compared to other houses. For instance, they might support the roof on single *pinj* poles, sometimes comprising lengths of timber with branches trimmed off one end to give a v-crotch to accommodate the single *bez* ridge-pole, other times with a cup-shape cut in the end to hold the pole with a robust strap lashed around it to secure in place. The single ridge-pole arrangement made it easier to construct a straight structure with different builders adding their sections. Two purlin poles might be lashed to the underside of the rafters about halfway down and running the length of the roof to give it further stability (see

11 As the Plate 8.23 shows, though the terminology is the same, the building is quite different to the lowland New Guinea form of 'longhouse'.

Plate 8.23 A sa aenda longhouse.

Plate 8.24 A pokaenda structure built during the sa cycle.

Loupis, Ruff and Ruff 1982:142–65 for series of *sa aenda* illustrations).

The building of the *sa* longhouse illustrates the way that individualism intriguingly pervades the co-operation that occurs during collective activities in the Was valley (Sillitoe 2010a:30), which may seem somewhat contradictory. When those living on the territories of neighbouring *sem* 'family' communities agreed to co-ordinate their efforts and host a *sa* cycle, joined perhaps by a few interested kin residing elsewhere, they did so as socially networked persons, not as members supporting these *sem* as corporate groups (see Chapters 3 and 7). The individualistic ethos seen in house construction, where men often build houses alone, extends to the erection of the paired longhouses to accommodate those attending *sa* dances and exchanges, and other structures for collective events. Each participant accepted responsibility for sponsoring the collection of necessary materials and erection of a section of the *sa* longhouse; oft-times staggered, some sections erected before others, sometimes leaving disconcerting gaps in the structure (occasionally unfilled if a sponsor fails). After the event, individuals may dismantle the structure to reuse the materials elsewhere (e.g. in house building, garden fencing etc.), and anyone helping himself without permission to materials from a section sponsored by another person could be accused of theft.

The other building constructed during the *sa* cycle, the *pokaenda*, was an open-ended structure two to four metres high at the ridge and about three metres square in ground plan, which we have likened to a carport in design

Plate 8.25 A row of pens making a shiyort aend 'cassowary house.'

(Sillitoe and Sillitoe 2009:166–7).[12] It comprised four or more substantial tree trunks sunk into the ground, serving as the posts supporting the building. A gabled roof was erected resting on these posts, comprising two frameworks of *polpen* 'rafters' and lashed *ka* 'battens' secured to a suspended *bez* 'ridge-pole' and supported on substantial *dolbez* horizontal beams running between the corner posts (again two purlin poles might be lashed about halfway down along the underside of the rafters to give the roof added stability). The two sides may be enclosed with a screen of cane-grass stems pushed into the soil and secured together by three rows of paired *gamonk* battens secured top, bottom and centre. The *pokaenda* featured in some of the dances that accompanied the *sa* cycle, with men parading through it and sometimes swinging axes at the tree-trunk posts on the way.

Another characteristic structure of the *sa* cycle is the long *shiyort aend* 'cassowary house' built on a *howma* 'grassy clearing' to display the cassowaries slaughtered during one of the *sa*-sequence stages and carried on men's backs in a celebratory dance.[13] The 'house' comprises a row of small pens-cum-cages made of sharpened *dol* wooden stakes with *gamonk* beading lashed along the upper part of the stakes, and roofed with heavy logs placed across the top; similar in construction to the *kuwl* 'stalls' for pigs, described in the previous chapter. It was of shoddy construction, men using any suitably sized timber available to build it, as it only housed the birds on display for a few days before their slaughter and the dance.

Other ritual structures

Other rituals performed in the Was valley featured the building of various structures that are now defunct, knowledge of which derives largely from participants' sometimes sketchy recall. The *saybel* ritual involved the construction of an idiosyncratic structure, the *masolaenda* 'long-neck house', so called because of its tall spire-like roof (Sillitoe and Sillitoe 2009:138–9). It was unique in being a round house. Those living in the *sem* 'family' community sponsoring the festival co-operated in building it, relying on the knowledge of older men about the somewhat eccentric construction, as the festival is staged only once a generation. They arranged it to please the *towmow* 'ghosts' and persuade them not to attack their kin,[14] and also to advance the success of participants in coming exchange transactions (Sillitoe and Sillitoe 2009:129–45). The circular walls comprised *dol* wooden stakes, which according to a structure seen at Humbara in the Mendi valley that friends said resembled a *masolaenda*,[15] were lined on the outside

12 See also Loupis, Ruff and Ruff 1982:167 for a dilapidated *pokaenda* photograph.

13 This dance features as the climax in the 'Bird of the Thunderwoman' film produced by the Australian Broadcasting Corporation (see www.abc.net.au/nature/thunder.htm).

14 The *towmow* 'ancestor spirit' attacks are random; people deny that there is any moral dimension, or that the ancestors intervene, as a reviewer suggested, when a descendant behaves immorally and demand offerings of pigs in atonement (Sillitoe and Sillitoe 2009).

15 Although the Humbara spire house featured in a ritual to appease a white-skinned-woman spirit during times of particular hardship (Mawe 1982).

Plate 8.26 A masolaenda-like 'long-neck house' (at Humbara in the Mendi valley).

Plate 8.27 View inside muwnaenda 'muwn house' showing the hungnaip 'ancestor stones' along rear (reconstruction).

with slabs of bark secured in place with flexible *gamonk* beading. The building had a single long pole in the centre, which supported *polpen* rafters that radiated out to the top of the wall and had flexible *ka* battens lashed around them to take the roof covering, the lower part of which featured bark shingles with some thatch, while the steep spire was entirely thatched with kunai grass. The structure of the tall steeple is unknown (see Loupis, Ruff and Ruff 1982:98–108 for series of illustrations of the Humbara house and conjectured projection of the building's construction).[16] It 'pleased the ghosts' and also announced the coming event to people living in the region. Inside the house, at the rear, was a large earth-oven pit and pile of stones for 'cooking' the marsupials offered during the 'long-neck house' stage of the *saybel* ritual. The house remained empty until the day of the rite. When the distribution of cooked meat and vegetables was over, some men lit a fire against the central roof-supporting pole and burned the *masolaenda* 'long-neck house' down, for reasons that no one could explain beyond saying that it was customary.

The *towmow* 'ancestor spirits' were believed to be nomadic, residing in various places. One of these was with *hungnaip*, prehistoric objects such as mortars and pestles that were occasionally found in the region (see Chapter 4). The *sem* 'families' maintained *muwnaenda* '*muwn* house' (or *hungnaip aenda* or *seblaenda*) lean-to shelters, similar to *pəgaend* lean-to structures, to keep these dangerous objects in. They were arranged along the back wall in a makeshift cabinet structure, reminiscent of a row of miniature pig pens, lined with fern fronds. Nearby was the *mashoraenda* 'taro-leaf house', which was a small, open-ended, ridge-roofed structure, with an earth-oven pit in the centre where they cooked certain cuts from pig offerings, including back, belly and stomach. The first stage of the *saybel* ritual involved the rebuilding of the *muwnaenda* that housed the sponsoring *sem* 'family's' prehistoric ancestor stones. Sometimes, while pork cooked, men painted their ancestor stones black

16 People living at Humbara refused to allow me to look inside the building, unless I arranged for them to receive an exorbitant sum to pay for its conservation. According to Loupis, Ruff and Ruff 1982:104, there is a small central room under the spire.

Plate 8.28 Laying the fire to heat stones for earth oven at a pig kill, with tok pole for display of pork and table with fronds and leaves

and red, using charcoal mixed with cosmetic oil and red ochre. The skulls of deceased kin were another residing place of ancestor spirits, and were housed in small shelters comprising raised roughly thatched cabinets of hewn timber (Sillitoe and Sillitoe 2009:150, see *olmol aend* coffin below), where offerings were periodically made when divination indicated that it was the abode of an attacking ghost.

The *sor kem* was another communal ritual that featured the construction of a special building in which the person who knew the ritual procedure and associated spells performed some tasks, and where two young men who played a central role lived in seclusion for about one 'moon' (Reithofer 2006:133–45; Sillitoe and Sillitoe 2009:24–31). The *sor kem aenda* 'palm-cockatoo house' was a cane-grass-thatched structure built adjacent to the *howma* 'communal clearing' of those sponsoring the ritual. It was higher and larger than a normal house, of sufficient size to accommodate several men and an earth-oven pit, but it was only temporary and was hurriedly constructed. Its unique structural feature was a pandan leaf arranged gutter-like through one wall and pointing downslope, which the two secluded young men used during the daytime when they needed to urinate, because subject to a strict taboo for a few days at the end of the ritual not to go outside, because the 'sun must not see them', nor should other persons, particularly women. If they broke the taboo, not only would it nullify the ritual's beneficial effects in pleasing the *towmow* 'ancestor spirits', but it would provoke them to attack and make the boys sick, even kill them, while possibly visiting leprosy on relatives.

The ritual, like others, included a pig kill and exchange of pork (Sillitoe 1979a:256–69). These recurrent festivals follow a similar pattern. Arrangements include the erection of a long horizontal *tok* pole for the display of the pork that is to be distributed, supported on posts about a metre high, branches trimmed to give a v-crotch to hold the pole. Before the festival, firewood is stacked against it for heating the stones for the earth oven. Behind the pole, running parallel with it, men erect a rough slatted table of split timber supported on a pair of horizontal poles that are half a metre or so apart and resting on stakes with branches cut to give a v-crutch. The temporary table holds the tree-fern fronds and banana leaves used in cooking, and lengths of bamboo that will serve as cooking tubes (Sillitoe 1988a:195). In front of these structures men amass a mound of stones and dig a long trench for the oven.

The *iyshpondamahenday* was yet another ritual that required the building of several structures. People resorted to this ritual irregularly, when crops failed and famine threatened, so as to appease the spirit of a white-skinned woman called Howar Saliyn whom they believed responsible for the unfavourable weather causing the fall in garden yields (Sillitoe 1996:89–95). A longhouse called the *iyshpondamahenday aenda* was built at Howar in the Was valley, which is where Saliyn lived according to legend. It was similar in construction to a conventional house, only larger, being ten metres or so long, with a pitched roof of *paym* pandan-leaf thatch, the *bez* ridge around three metres high and the *dol* wooden stakes of the walls about two metres high. It had crawl-through doorways at either end in the gable walls; the participants decorated these gable ends with pearl-shells during the ritual. They divided up the floor inside the house into rectangular areas with lengths of wood pegged to the ground; these marked out sleeping areas along both walls for the men coming from the various *sem* 'families' participating in the ritual who stayed in the house, undertaking certain ritual responsibilities on behalf of their kinsfolk. A walkway was left down the centre, running between the two doors, with fireplaces along it at intervals. Two ordinary houses were also built, one adjacent to the longhouse for experienced participants to use when not engaged in ritual activities, and the other some way off for those taking part in their first *iyshpondamahenday* ritual. When they came to Horwar the participants withdrew from everyday life; others should not meet them because their engagement with the white-woman's spirit made them dangerous and contact with them could lead to sickness and even death.

After the Horwar rites, the participants pursued a nomadic existence, visiting some of the places where Saliyn is believed to have rested, where she left behind pools of water. The communities near to these pools constructed *komay* 'leaf-screen' enclosures there. These were rectangular in plan, fenced with fronds on three sides and left open on the other. Any saplings and branches could be used to build the fence, and left with their leaves attached they were driven in close together to give an impenetrable hedge through which you could not see. The participants lined the inside of the screen with *shiyp* meliad and *bat* tree[17] leaves, poked behind horizontal spars lashed across the fronds, the pale undersides of the leaves facing outwards to produce a vivid barrier. In the screen opposite the enclosure's open side there was a crawl-through door-like opening that led to a conventional house built on the other side. When the *iyshpondamahenday* participants arrived and occupied the house, men put pearl-shells on the screen; this show of wealth was to urge the white-woman's spirit to stop her life-threatening persecutions and behave morally in view of the pig offerings being made to her.

A similar, smaller *komay* 'leaf screen' was built, and hung with exchangeables such as pearl- and cowrie shells, for the *hwiybtowgow* ritual performed to cure sick persons attacked by *saem* 'forest demons' and *towmow* 'ancestor spirits'. The leafy barrier separated the participants from the lethal spirit force during the rite. A small *saem aenda* 'lean-to shelter' was also erected adjacent to the screen with an earth-oven pit excavated inside for cooking the pig offered during the rite (Sillitoe 1996:210–13). And a *hongok* tree-fern[18] stump was pushed into the ground by the leafy barrier, hollowed out to act as a receptacle for cooking some of the pig's internal organs with hot stones.

17 *Chisocheton ceramicum* and *Bubbia* spp..
18 *Cyathea* spp.

Plate 8.29 Securing the iysh doi plank-like sides of a raised coffin (reconstruction).

Coffins and graves

In the Was valley they had a tradition of disposing of the dead in a range of *olmol* 'interments'. These were not places where offerings were routinely made to ancestor spirits. It was, and continues to be, usual for two men, not close relatives of the deceased, to accept responsibility for undertaking an interment. These ranged from simple disposal of corpses at rock-face necropolises, where they were left exposed on ledges and wooden-slat stretchers jammed in place on forked-stick legs (Sillitoe and Sillitoe 2009:53), through to elaborate above-ground coffin structures and burial in fence-enclosed graves.

A popular above-ground version comprised an enclosed coffin raised some two metres off the ground. The coffin-makers started by driving two pairs of straight poles into the ground, called the *kuwgis*, each pair a shoulder-width apart and with a body length between the two pairs (Figure 8.4). They split the top of the poles in half lengthways to accommodate two hewn struts of wood, which they lashed in place between the poles on either long side of what will comprise the coffin. They cut a series of matching grooves in the top edge of these at about five centimetre intervals

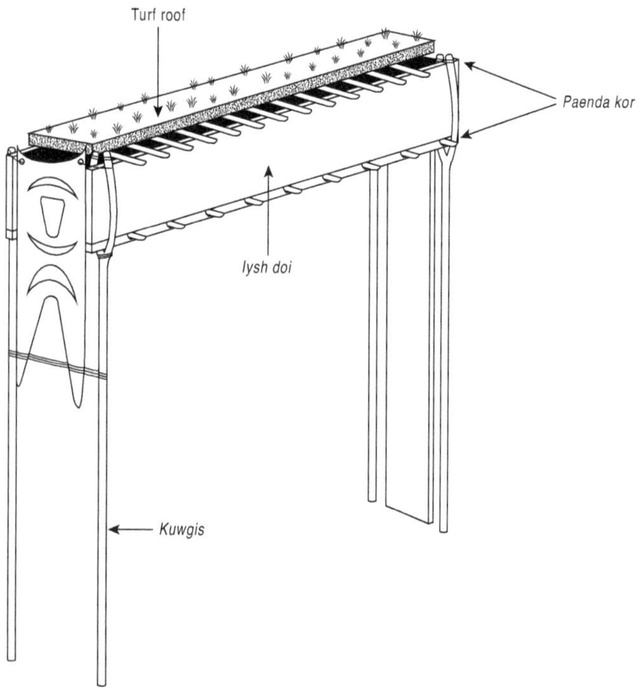

Figure 8.4 *Structure of a raised coffin.*

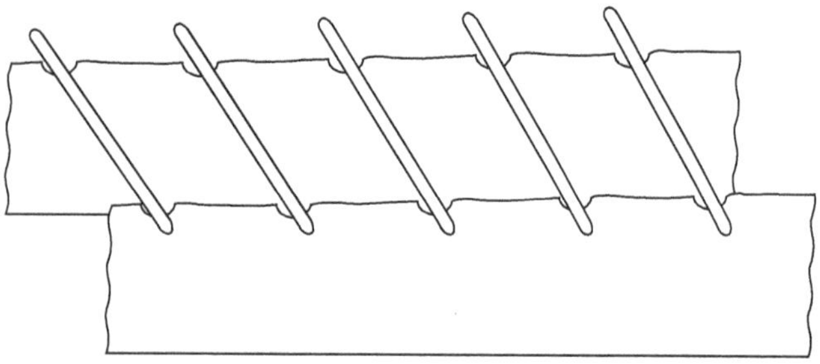

Figure 8.5 *Slat paenda kor structure of raised coffin floor and roof.*

and placed *paenda kor* 'bed rods', which are whittled spindle-like pieces of wood, between them to give the floor of the coffin (Figure 8.5). Next the coffin-makers hewed the *iysh doi* 'pared timber' plank-like sides and ends of the coffin, and lashed these in place. The side *doi* 'planks' they arranged between the split-sapling *kuwgis* legs of the structure on top of the grooved struts, sandwiching the *kor* 'spindles' in place. They cut a further series of matching grooves at close intervals along the top edges of these long plank sides and put more whittled *paenda kor* 'bed rods' in place to serve as the coffin roof.

The plank-like ends varied in dimensions, sometimes short and lashed in place to seal the coffin, box-like, others long with one end driven into the ground between the *kuwgis* supporting poles. Some had holes cut in them to see into the coffin, through which grief-stricken kin might

Plate 8.30 *A raised coffin with pearl-shell crescent designs painted on it.*

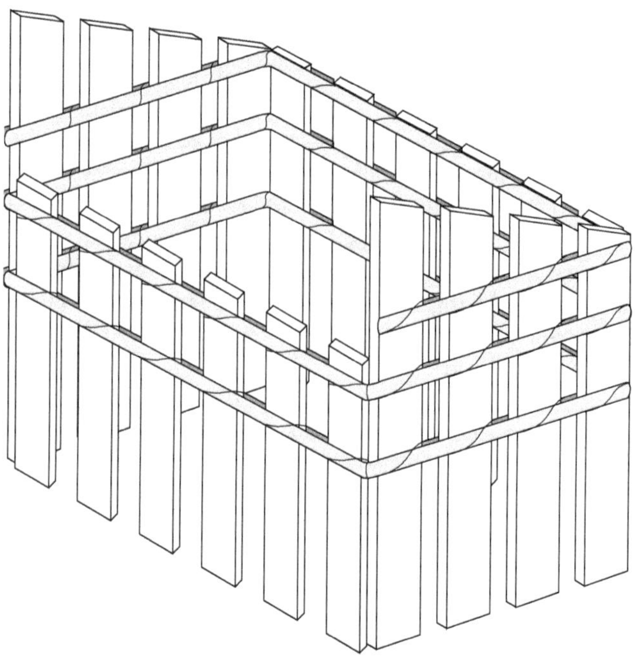

Figure 8.6 *The stakes and battens of open-fronted olmol aend 'lean-to coffin'.*

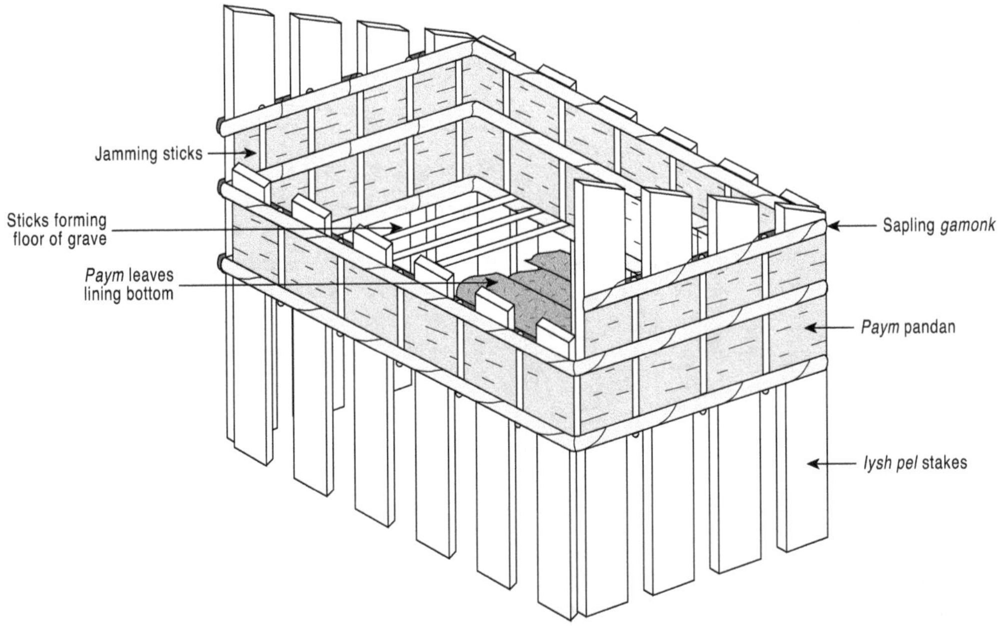

Figure 8.7 The leaf facing and floor construction of olmol aend 'lean-to coffin.'

reach inside and touch the corpse.[19] The coffin-makers lined the inside of the coffin with large tough *paym* pandan leaves, and when ready they removed the corpse from its mourning pole and laid it to rest, covering it with further pandan leaves before putting the *kor* 'spindles' in place along the top edge, and finally closing the coffin with grassy sods. Sometimes they constructed an elaborate thatched roof over the coffin. Finally, they erected a *mol kuwl* 'grave pen' of stout wooden pales around the structure to stop pigs rooting under it and, horrible thought, dislodging and eating parts of the rotting corpse.

A variation on this above-ground coffin was an open-fronted *olmol aend* 'interment house', a lean-to structure constructed along the lines of a house-cum-shelter.[20] The builders prepared beech-wood *dol* 'wall stakes' of varying lengths, which they drove into the ground to a rectangular plan, of dimensions to accommodate a human body, perhaps with legs flexed. The stakes along the *hiytpaip* 'front wall' are short (about half a metre high) and taller at the rear (about three-quarters of a metre high), and slope along the *waenuw* 'side walls', following the lean-to roof line (from about one metre at the front down to the height of the rear wall). They lashed two rows of *gamonk* 'split-sapling battens' around the entire structure, one at the height of the top edge of the front wall and the other about fifteen centimetres below it; plus a further row of such paired beading about fifteen centimetres above the front wall, along the sides and around the rear of the structure (Figure 8.6). They arranged thin sticks across the bottom row of *gamonk* beading, running from front to rear, to make the floor of the tomb, which they lined with dry, stout *paym* pandan leaves (Figure 8.7). They also arranged dry *paym* pandan leaves between the rows of beading, lining the inside and outside of the walls, and secured these in place with small *pongow* 'jamming sticks' to enclose the structure

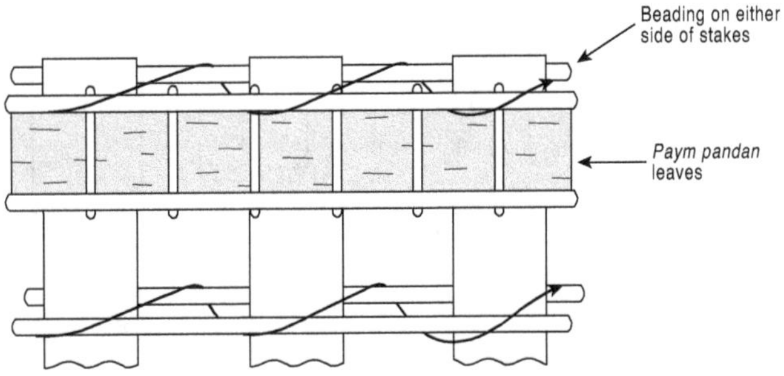

Figure 8.8 Construction of leaf facing of olmol aend 'lean-to coffin.'

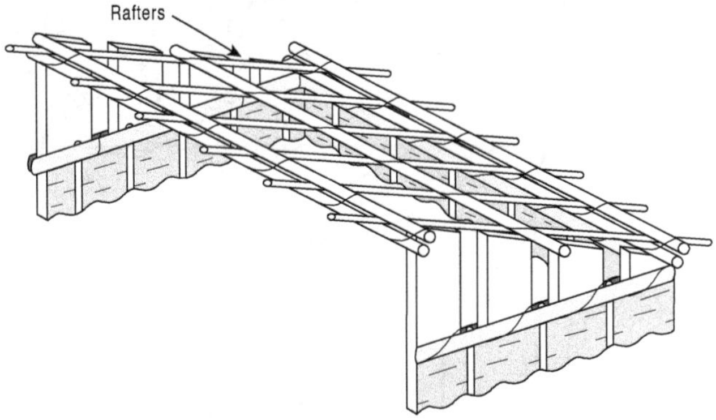

Figure 8.9 Roof construction of olmol aend 'lean-to coffin'

19 If the deceased's head was visible and he had a wig, or long hair suitable for wig making, someone might take it, the hair coming away easily after a period of decomposition.

20 See Pospisil 1963:265 for a similar grave structure in the Paniai Lakes region.

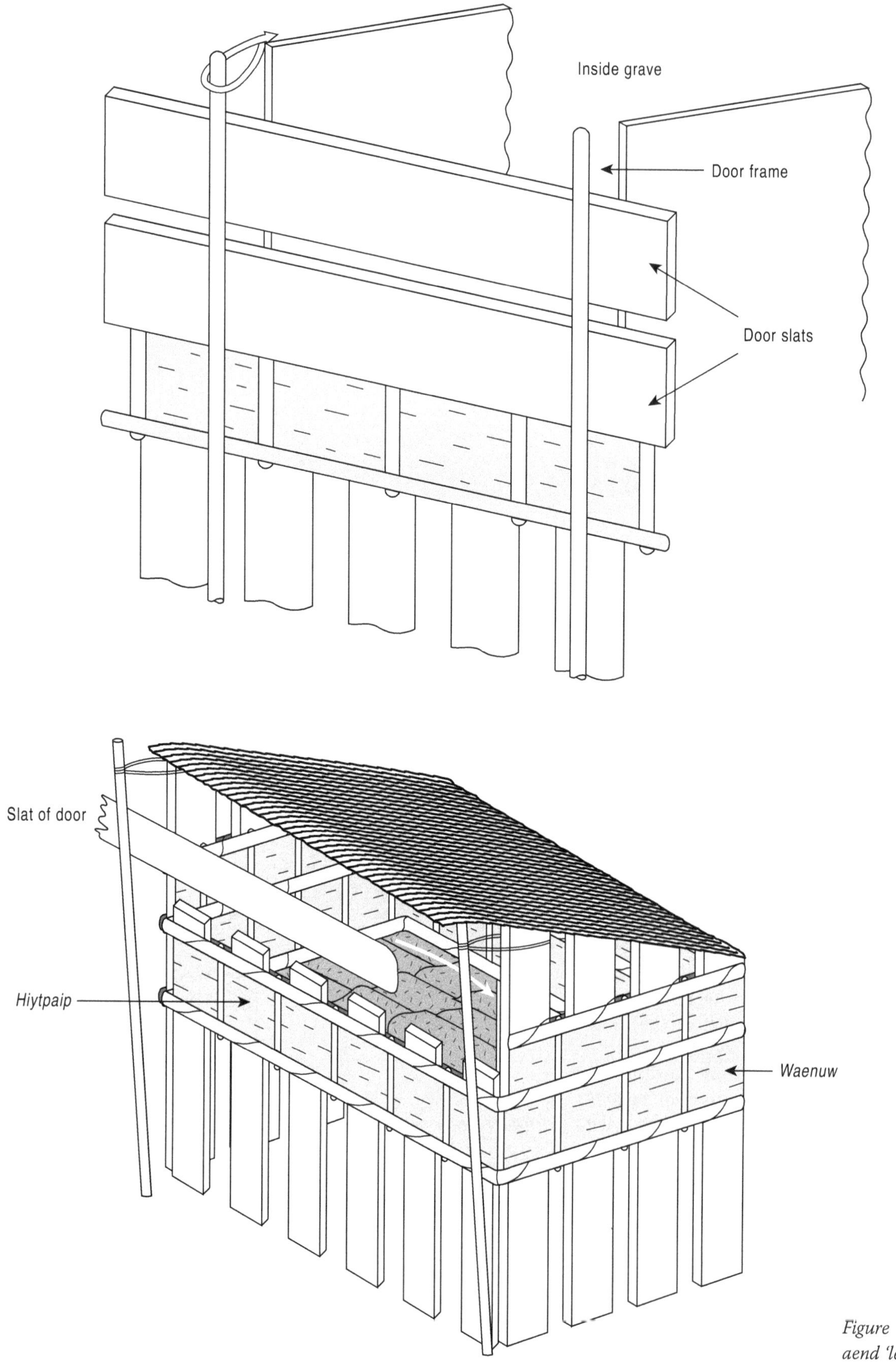

Figure 8.10 Closing of *olmol aend* 'lean-to coffin'

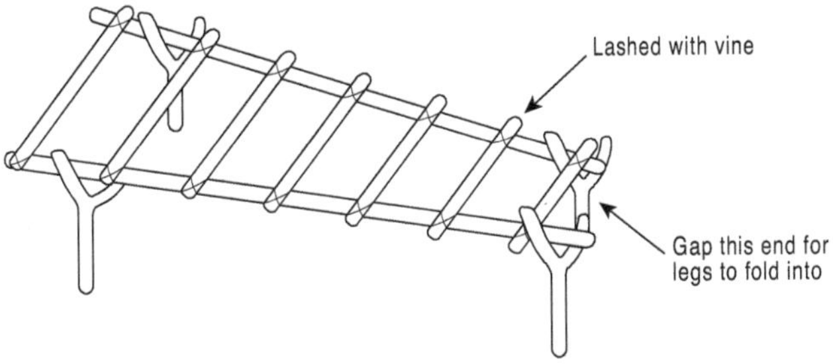

Figure 8.11 The bed framework at the bottom of a grave.

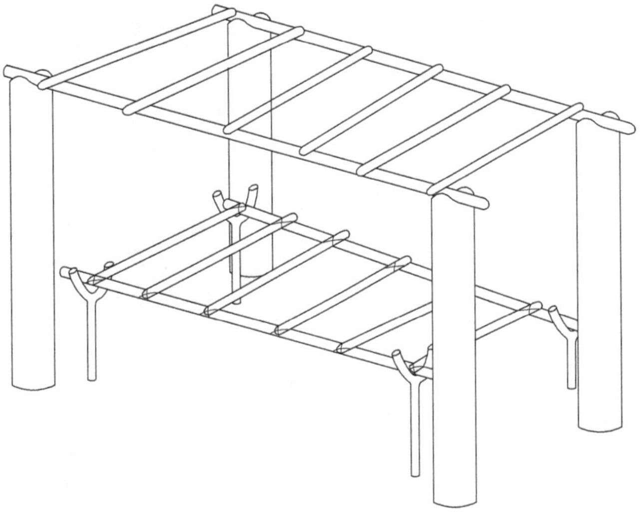

Figure 8.12 The roof framework over the bed in a grave.

Plate 8.31 Digging a grave; with iysh shoba posts and aend 'roof' planks ready.

(Figure 8.8). They roofed the tomb by sandwiching sapling rafters between two cross-members lashed across the front and rear, and reinforced with one or two further purlin-like cross-members. They then thatched the structure, usually with *senz* 'kunai grass' (Figure 8.9). When relatives placed the corpse in the tomb, they closed it with hewn planks of wood, like a house doorway, jammed between two uprights secured at both of the front corners of the coffin-house (Figure 8.10).

Today people only bury their dead in graves, after the admonitions of colonial authorities and missionaries warning them that their other practices were unhealthy, even uncivilized in their eyes. They have always buried some deceased relatives, such as those believed to have been killed in *towmow* 'ghost' attacks, as part of their strategy to ensure that the spirit responsible cannot attack again. The gravediggers are two distant kinsmen of the deceased, who dig a *suwtombay* 'hole' a metre or so deep by one-and-a-half metres long and a half to two-thirds of a metre wide; it is intentionally wider one end – the head end, to accommodate the shoulders – and tapers to the other, foot end. They construct a rough *paenda tor* 'bed frame' of wooden slats lashed in place between two longitudinal poles supported at each corner on a forked stick pushed into the ground, leaving a space at the narrower end for the deceased's lower legs and feet to rest on the ground, the corpse arranged lying face up and flexed at the knees (Figure 8.11). They line the grave with *paym* pandan leaves, securing these to the sides of the hole with pieces of split-cane-grass stem bent into a U-shape and pushed staple-like into the clay. Next they fashion four *iysh shoba* 'forked uprights' and drive these posts in at the corners of the grave to support the *aend* 'roof' (alternatively they may use stakes with a scoop cut in the top). They place two poles horizontally between these uprights, down either long side of the grave, and arrange further split-wood slats ladder-like between them (Figure 8.12). These support some plank-like pieces that the gravediggers hew to seal the roof.[21] They assemble the roof after lowering the corpse onto the 'bed' frame and then backfill the grave, stamping the clay down level and firm (Sillitoe and Sillitoe 2009:104–5). There is no graveside ceremony. Finally, the gravediggers split and shape some slender beech-wood stakes for the

21 Men prefer to use timber from the following trees for fashioning such plank pieces – *shiyp* (*Chisocheton ceramicum*), *haezuwmb* (*Galbulimima belgraveana*) or *huwgiyt* (*Toona sureni*) – because as straight-grained, they split and shape readily.

Plate 8.32 Lashing the slats of a paenda tor 'bed' in a grave.

Plate 8.33 Erecting a mol kuwl 'grave pen.'

mol kuwl 'grave pen' that they erect around the site with a *gamonk* 'split-sapling batten' lashed around the top edge of the stakes to unite them together and prevent pigs from pushing them apart.

Sometimes relatives build small *mol aenda* 'grave houses' over burials, in lieu of the fence. They construct these in the same style as regular dwellings, only in miniature, with beech-wood *dol* 'wall stakes' and *senz* 'kunai-grass' thatch roof. Occasionally, today, they innovate, using cement, for instance, to seal a grave and writing the deceased's name in it while wet, and maybe setting further decoration there too. It is intriguing to note that the accommodation of the deceased shows greater variety than that of the living. But then, dead persons cannot, after all, advance themselves materially above others or turn any differences into power inequalities. And reflecting the inverse, it would be an error to assume that only successful 'big men' warrant the more elaborate structures. The refinement of a grave seems to relate to the depth of grief felt by relatives. When a popular girl in her late teens died suddenly from pneumonia, for instance, her grief-stricken father and brothers built an elaborate small house over her grave. Whereas when an elderly man widely recognized as being of *ol howma* 'renowned status' passed away, described by one of his sons as an *ol nobtuw ol* 'man who chewed on other men' (that is of aggressive temperament), his relatives buried him in a modest and simply fenced grave. Such behaviour recommends caution on archaeologists' part, who tend to assume that more elaborate burials equate with chiefly or higher status, with significant wider interpretative implications, presupposing a socio-political hierarchy.[22] But future archaeologists working in the mountains of Wolaland may face a prior problem, namely finding graves at all; for while burials are often situated on the edge of *howma* 'communal clearings' or *aenda munk* 'house yards', they are dotted about and the surface evidence soon disappears (as have remains of previous generations' various above-ground interments).

22 And opens them up to criticism, or worse, such as Macaulay's (1979, 1980) Egyptian tomb caricature (see Chapter 4).

Plate 8.34 Constructing a small mol aenda 'grave house.'

Plate 8.35 The final mol aenda 'grave house.'

Structural symbolism

Sometimes relatives go beyond constructing interments, to embellishing them with painted designs. This was common on raised coffin graves, the planks of which may have motifs painted on them in charcoal, ochre and white clay. The relatives of an *ol howma* 'man of renown,' for instance, who was struck down by sorcery in his prime, painted some crescent-shaped symbols in orange-red ochre on his coffin. They said that they represented pearl-shell crescents. The implication was that they were marking and celebrating the deceased's social standing as an admired and much missed transactor of exchangeables. On other occasions they painted an anthropomorphic design on the large upright board extending to the ground at the foot of a coffin (Hides 1936:154), sometimes stylistic and others realistic, after the designs they painted on shields (Sillitoe 1980). If the deceased was killed violently in conflict, the painting might feature red, the same colour code used on shields to signal that the death was to be revenged, and we might interpret such decoration as communicating culturally loaded intelligence about the revenge ethic.

According to some semiotic schools of thought, investigation of such symbols associated with fixed structures might afford access to people's ideas about architecture and building construction, possibly a way into local perceptions regarding tacit issues. If so, they may allow us to get around the charge of appearing to know more about people's behaviour than they know themselves when their behaviour is the subject of our enquiries. Attention to connections between symbolic representations and observed behaviour that comprise people's lived reality, encompassing a search for connections between socio-culturally informed awareness and embodied experience, takes us in a less positivistic, even psychic direction. Indeed, exploring the relation between symbols and practice in the configuration of understanding addresses the ideological question posed previously (Chapter 2): how feasible is a synthesis of symbolical and phenomenological analysis?

The attitude of house-builders in the Was valley intimates the limitations of this approach, even challenges it. During construction work, they appear to pay no attention to symbolic or metaphysical issues, not referring to these at all. They undertake the building of a house in a thoroughly practical way, as conveyed in the previous account. And unlike some regions of New Guinea (Landtman 1920), including neighbouring Lake Kutubu (Weiner 1991),[23] or elsewhere in the world (Marchand 2009:67–81), there are no supernatural rites or spells associated with house construction. When I brought up such matters in conversation, they confused people. In constructing fireplaces, for instance, people pay no apparent attention to beliefs about *towmow* 'ancestor spirits' periodically taking the form of snakes and dwelling coiled-up under them. When I asked if the siting of houses and fireplaces had any effect on snake ghosts taking up residence, my friends said they had no idea what prompted the spirits to do so, and they knew of no steps they could take to stop them. Today, anyway, they pointed out, with conversion to Christianity, they are unsure what has happened to the snake spirits. No one pays any attention to them anymore; Christian beliefs have apparently overwhelmed them. Previously, if a divination rite indicated that a snake ghost under the fireplace of a men's-house foyer was responsible for a relative's illness, a pig might be offered there; a trident-like object comprising three wooden prongs painted with red and black stripes was driven over the animal's snout and into the ashes to blind the spirit, which, no longer able to see, could not continue to attack the sick person (Sillitoe and Sillitoe 2009:102). But even when such beliefs were alive, they did not apparently feature in building a house. For instance, residents did not line fireplaces with heavy stones to prevent snake-ghosts moving in, nor did they have protective spells or rites. The same absence of symbolic import applies to other features of house architecture, such as the variations in the construction of walls or the use of different roofing materials. The implications of their pragmatic approach to house construction are intriguing, particularly in view of the focus on idealistic issues in many of the anthropological studies that consider buildings, and further underscores the challenge of accessing unspoken concerns.

The symbolism of things associated with houses is also relatively sparse. The display of animal skulls and pig mandibles is one such class of objects, usually seen on the *waenuw* 'partition wall' in the *wil* 'men's foyer' of houses (mentioned in Chapters 4 and 7; Sillitoe 2003:136–7). The Wola, unlike some New Guinea populations,[24] do not talk of these trophy displays as depicting anything, such as resource maps of their territories, nor as works of art. Again they have no apparent meaning for actors, which they can consciously articulate anyway; it is embellishment that has no evident symbolic import for the decorators. While the skulls and mandibles may relate in part to the hunting success of the house occupants and the numbers of pigs they have slaughtered, and so advertise their achievements, the numbers of skulls and mandibles is more likely to correlate with the age of a house; the occupants of houses do not transfer skulls, usually smoke-stain blackened, from old to new houses. The way men talk about the things further suggests that they consider them more as straightforward memorabilia than trophies advertising status.

A similar absence of symbolic associations pertains to the contents of houses generally, which are sparse by materialistic Western standards. When I introduced a TV producer to an old friend at his house, he turned to me after looking inside and asked where his belongings were, and on hearing that he could see them all before him – split-sapling tongs in the fireplace, axe in one corner and water gourd standing in another, together with the contents of the man's net bag – he looked at me in surprise and commented, 'I am carrying more on me right now than he owns' (he had a large camera bag on his shoulder). People are not seeking to communicate something about themselves through the relatively few objects they have in their homes. They are not symbols of identity as the popular consumer-

23 While not associated with houses, some men in the Was valley performed rites and recited spells to attract transactables to them (Sillitoe 1979a:160–9) such as the neighbouring Foi around Lake Kutubu perform at the completion of a new longhouse (Weiner 1991:186–90).

24 Such as Ok speakers; see Bulmer 1976; Craig 1988; Hyndman 1991 on skull-trophy arrays elsewhere.

Plate 8.36 The timptinjiy disc hanging at gable end of timp cult house.

focused study of things argues for capitalist society, where individuals regularly use possessions to convey something about their status, albeit not always consciously (e.g. Miller 2001, 2010). Indeed, such behaviour is improbable in an egalitarian context, where persons do not differ markedly according to what they possess. Where no one, for instance, is materially wealthier or more powerful than others, the idea of setting out ostentatiously to advertise your assets and accompanying power makes little sense. The contents of houses tend to be similar, differentiated to some extent by occupants' gender, men and women possessing and using some different things. A survey of possessions owned by individuals revealed little pattern regarding their status, except according to gender and a lesser extent age (Sillitoe 1988a:571–9); except when a socio-political transaction was pending and men had accumulated transactables to meet exchange obligations.

Regarding the symbolism of things associated with various structures, it is possible that those found with ritual buildings may afford opportunities for interpretation, having symbolic features pertaining to beliefs; although a large part of this knowledge has disappeared with Christianity supplanting previous ideas (Reithofer 2006). The last local cult, the *timp*, featured a large circular plate-like object with a hole in the centre, called the *timptinjiy* '*timp* cover-up', made from entwined rattan. Some of the participants made it secretly inside the cult house, so that others, particularly women and children, could not see what they were doing. They painted a design on the entwined disc using charcoal and white and orange clay. It varied from place to place. One design comprised four black lines drawn spoke-like diagonally across the disc, dividing it up into eight cake-like parts, four of which were painted alternately white, two opposites red and two black. Another design comprised a black cross on a white background with target-like concentric black and orange circles around the central hole. During *timp* events the disc was suspended under the high gable end of the cult house, on the outside so that everyone could see it. In some places people also fabricated square-shaped objects, called *paranija* or *wipuwbiy* in the Mendi valley (Pretty 1969:22), made by lashing two sticks together to form a Greek cross around which they wound a ribbon of bark starting from the centre. As men entered the cult house they chewed a piece of ginger and spat onto the object as they ducked through the doorway.

Plate 8.37 The paranija squares above entrance of timp cult house (photo courtesy of A.L. Crawford).

When asked about these objects, no one could explain their meaning, saying that they 'came with the *timp*, so we did it'. There was no association with the ritual activities to suggest what these objects or their designs might signify, let alone any connection with the cult-house architecture. We can engage in speculation about possible symbolism; for example that the segments in the first design could represent the various *sem* 'families' participating in the cult, encompassed within a circle that signifies their social solidarity, or that the second design represents an ancestral eye watching that participants perform the rites properly. But without strong ethnographic grounds for doing so, such symbolic conjecture amounts to little more than an unsubstantiated flight of imagination. My Was valley friends are not alone in their denial of architectural symbolism. Regarding the decorative patterns featuring geometric motifs in brickwork on minarets, Marchand reports that Yemeni builders 'firmly maintained that there was no symbolic significance to the patterns they chose' (2001:133).[25] Similarly, he found that the motifs Malian builders carve and mould on interior house walls have no symbolic meaning for them (2009:262). But more often than not, it seems that people elsewhere see symbolism in architectural features. The disinclination of the Wola to engage in interpretation of aspects of building design or particular architectural features contrasts with the inclination of other people in New Guinea, particularly in regions where buildings and their contents apparently lend themselves to such interpretation, notably along the Sepik river (Behrmann 1950–1; Bowden 1990; Coiffier 1983; Forge 1966; Landtman 1920; Peltier 2008; Stasch 2011a; Weiner 1991:182–95), which parallel the symbolic associations identified by anthropologists who discuss local buildings elsewhere. The paucity of architectural symbolism in the Was valley is striking and calls for further reflection. A thwarted publishing episode some time ago concerning the construction of bridges – another class of fixed structures in the region, discussed in the next chapter – starkly posed the question of symbolic issues.

25 This is common across the Arab world, where much art features geometric designs, often in intricate arrangements, as Islam disapproves of figurative iconography.

9

Snakes and bridges

Social constructivism?

It was in respect of bridge construction that I became aware that giving precedence to the apparent reluctance of the Wola to recognize and engage in the interpretation of the symbolic import of fixed structures could be seen as a shortcoming. It happened when I submitted a paper on bridge building to the newly founded *Journal of Material Culture* and encountered committed 'social constructivists'. The submission was too prosaic for the then editors, heavily into the symbolism of material consumption and social construction of landscape, who invited me to speculate on symbolic issues and produce an intellectually engaging cosmological interpretation. Their letter suggested: 'you quote a fascinating myth about snakes and rivers which struck us might hint at a rather elaborate cosmological set of referents linking rivers, movement, violent waters etc. which could be explored in a very interesting way ... we would want you to make the bridges a starting point for a much wider discussion of Wola cosmology'.[1] Although I had not then coined the term ethno-determinism, the standpoint implicitly informed my approach and after briefly contemplating such an imaginative revision I rejected the idea. Appositely enough, I entitled the paper, like this chapter, 'snakes and bridges', unwittingly alluding to the delay in its appearance – missing the publication ladder in slipping down the rejection snake. It has reminded me, subsequently left languishing accusingly in my filing cabinet, of my tardiness in dealing with construction activity in the Was valley (Chapter 1).[2]

The rivers that flow along the main valleys of the rugged central cordillera of New Guinea are frequently awesome torrents of raging white water. There are several large watercourses flowing through the Wola region, such as the Was and the Lai, and bridging them is a challenge, reminiscent of tasks – for those with experience of Scouting, Outward Bound or military exercises – involving imaginary obstacles such as ravines that you have to contrive to cross using such props as some rope, poles and oil drums. The materials available to Highlanders in crossing real raging river obstacles come from their region's montane rainforest, notably vine and timber. In contriving to cross large rivers, they use rattan vines largely to construct suspension bridges, *iyb tor* 'water structures', which, swaying across churning water in the forest, have a Kipling-like *Jungle Book* quality redolent of challenging tropical ventures. The Wola bridge smaller watercourses, which are sometimes too fast flowing to wade across, with wooden walkways comprising a few logs or sometimes a single stout tree trunk, to give an *iysh tor* 'timber structure'.[3]

Bridge origins

People infrequently have reason to build new suspension bridges. There is little chance that those on popular thoroughfares will collapse through neglect, for they are subject to a process of constant renewal, individuals adding a few lengths of new vine as necessary to replace old and rotten pieces as they pass. The volume of traffic crossing most of them is high enough to ensure maintenance with relatively little effort on the part of any one individual. People say that their ancestors constructed bridges on major routes of communication long ago and that they have been constantly maintained ever since. Indeed, some of these structures hang in such precipitous and difficult locations that some persons today doubt they could replace them if they fell into disrepair. They attribute their construction to some mythical happening in which their distant ancestors learnt how to build suspension bridges, as in the following story, the aforementioned 'fascinating myth about snakes and rivers':

1 The semiotic editorial stance may account for why 'The ways in which archaeologists pay attention to the detailed recording of things rarely find their way into the ethnographic studies in the *Journal of Material Culture*'.(Hodder 2012:67).

2 Another version submitted as a chapter for an edited book, planned for publication through the British Museum, also came to nought, this time because the volume proved a ghost book (which a helpful librarian explained to me is a book that is advertised, for instance on the jacket of another, but never appears).

3 In some regions, such as the Paniai Lakes of Indonesian Papua, people construct elaborate wooden bridges (see Pospisil 1963:264).

Figure 9.1 An iyb tor vine suspension bridge across the swirling Was river

Long ago, a man called Maegai lived at Haelaelinja and one day he saw smoke curling up into the sky from fires lit by someone to burn off a new garden, at Nombra on the other side of the Was river. His curiosity aroused, he decided to go and investigate. He walked down to the river but could not cross for there was no bridge over it at that time. As he wandered up and down the bank wondering how he might contrive to reach the other side, a large black snake appeared and arched itself across the river, over the steep-sided cleft where Pelem bridge is situated today, and Maegai walked across on its back.

On the other side of the river, at Nombra, Maegai had a number of adventures with a man called Kondiyp, who lived there. He found that the latter had a pile of birds and marsupials at his house that he burnt as firewood, and Maegai put these to cook in an earth oven together with a variety of vegetables. After this, he went to find Kondiyp, who was clearing the garden from which he had seen the smoke rising earlier, and found him using his leg as a cutting block against which he was chopping cane grass. This was causing him considerable pain, so Maegai showed him how to use a log as a chopping block.[4] They then returned to the house, where he had left the food cooking. Kondiyp ate until he was bloated, but it turned out that he had no anus and was unable to defecate, throwing up his excreta instead. This prompted Maegai to erect a frame of rotten wood with a sharpened stake underneath at the place where Kondiyp customarily went to throw up, and this collapsed when he sat on it, and the stake pierced his rear so that he could defecate.

After their meal Maegai returned home, but when he came to the Was river the snake had gone and he was again unable to cross. While he walked along the bank trying to think of a way to reach the other side, the large snake came back and stretched itself over the cleft with the river boiling below, so that Maegai could cross. It occurred to him on his way over that he would wish to cross the river in future to visit Kondiyp to share meals with him, having learned about his bountiful supply of food, and he set about erecting a vine suspension bridge. The snake remained arched across the chasm throughout, allowing him to pass back and forth, 'showing him,' as reciters of the story say, the way to build a bridge. When he had completed it, the snake disappeared.[5]

4 See Sillitoe 2010a:262–5 for more on this way of clearing cane grass.

5 The Wola believe that the spirits of their ancestors sometimes reside in snakes, as mentioned in the previous chapter, and normally fear them greatly; when I asked if it was one of Maegai's ancestors manifest as a snake who 'showed' him how to build the bridge, those I questioned said they did not know, the story only

Plate 9.2 Using a tree-trunk as a bridge across the Mubi river.

Plate 9.3 Bridge across a ravine comprising several lengths of timber (note sapling handrail).

Following this, whenever Maegai saw smoke rising up into the sky from Nombra, indicating that Kondiyp was preparing an oven to cook food, he crossed the river and joined him at his meal. After a while, Kondiyp tired of this and determined to put a stop to Maegai's uninvited visits. He went to a high place, where he caught some cloud in some *wolahaeriy* bamboo tubes,[6] and returning to Nombra, released it. Maegai fell for his trick: thinking that it was the smoke from a fire he crossed the river anticipating a meal, only to find that Kondiyp had disappeared forever. But the bridge he built at Pelem is still there today.

refers to a big black snake. Although no one mentioned it, they probably have a python in mind, as the largest snake in the region, the largest of which occur at lower altitudes in the Kutubu region.

6 This cane bamboo (*Nastus obtusus*) is not customarily used to make vessels (Sillitoe 1988a:205–6).

The account, illustrating the mythical dimension to time and the past, evokes the absence of a linear notion of history concerning actual events long ago, as discussed in Chapters 4 and 8. While it is expectable, perhaps, for an oral tradition that the record of historical events is shallow and shades off into myth, such mythical conceptions, which correlate with the notion of spiralling time, clearly qualify the relevance of our ideas of history and archaeology. It is further intriguing that while the Wola attribute such constructions to mythical ancestral activity, they do not appear to show any interest in modified features of the environment that evidence the actual activities of their ancestors – such as areas of levelled land that indicate ancient homesteads (Chapter 4) – but they have an interest in some physical features that, while not evidencing any actual human interference, they attribute to legendary beings and mythical events (such as the deep pools of water in limestone sinkholes associated with Horwar Saliyn, the white-woman spirit who features in rites mentioned in the previous chapter – Sillitoe 1996:88–95). These notions relate to the idea of the social construction of place.[7]

It is difficult to see what the myth has to tell us about bridge building itself or rivers, which accords with nobody knowing of any myth about the origin of house building or, for that matter, constructions generally. Regarding the cosmological implications of the story, structuralists might seek to interpret it according to presumed universal human concerns, expressed as binary oppositions that capture the contradictions of life, such as nature versus culture or raw versus cooked, evident in Kondiyp's and Maegai's behaviour, although it is difficult to see, as with such interpretations generally, how it resolves them. Evolutionists might see in the myth an attempt to account for the development of human bodily functions or the origins of sociality, although Kondiyp's disappearance seems to undo any nascent social interaction. Functionalists might interpret the establishment of relations between the two characters in social-charter terms, although neither Kondiyp nor Maegai are named by those living on either side of the river today as distant ancestors. The problem from an ethno-determinist perspective is that actors do not offer such interpretations nor would they probably even recognize them. It is difficult to see these interpretations, as postmodernists affirm, as anything more than products of foreign researchers' imaginations. While intellectuals are, of course, free to engage in such metaphysical interpretations, it is necessary to remain aware that they are not ethnographic realities.

7 The idea that places are 'socially constructed' (as opposed to interpreted or understood) is, I think, unnecessarily obfuscatory, when the premise is that culture and history inform people's knowledge of their environment, yet purposively overlooking, as scientistic, the part natural processes play.

The idea that certain bridges, like the one featuring in this myth, cannot be replaced if they fall into disrepair, has symbolic latency for Wola speakers, as seen in the metaphorical allusion of the popular saying: '*niy* Kuwtael *pay kobaruwk*', which means literally 'I Kuwtael chop down'. People believe that the bridge at Kuwtael spans a wide stretch of river in the neighbouring Huli region and would be irreplaceable if chopped down. Married couples may use the phrase, for instance, if they separate, to indicate that they are beyond reconciliation and intend to divorce one another, that their relationship is finished like the bridge at Kuwtael would be if someone cut it down. But the Wola see that they maintain such bridges. The upkeep of bridges is casual, as noted, individuals collecting vine and adding it when they consider it necessary, unless they intend to convey a number of pigs across, in which event they may first carry out extensive repairs to ensure that a bridge is strong enough to support them and their porcine loads. The informal system of maintenance is effective, even with the considerable number of bridges that span rivers in heavily populated areas. Along 10 kilometres of the Was river, for example, there is one bridge site about every 2 kilometres (Map 3.1.2), although there is not a bridge at every site all of the time. Regardless of its age, therefore, mythical or otherwise, any bridge comprises materials no more than a few years old, with some of these possibly new and green.

Bridge construction

Occasionally men have cause to build an entirely new suspension bridge. It may be necessary to open a new route to gardens established in a previously uncultivated region, or to replace an earlier bridge, swept away perhaps by a flooding river or cut down during hostilities to prevent enemies approaching to lay ambushes. The Wola refer to the erecting of a new bridge as 'staking it' over a river: *iyb tor indiyay* (literally, water-structure stake).

The first task facing those 'staking' a new bridge is the erection of two stages, called *pok*, one on either bank facing each other across the river. These serve as the footings for the suspended structure, anchoring it firmly to either bank, and also, when completed, as the means of access to the swaying bridge. A *pok* 'stage' consists of several heavy stakes of split timber, some two or three metres long on average, driven into the bank in two parallel rows up to about a metre apart. Each row usually comprises three or four stakes, running back from the bank in a line set at right angles to it. The first stake in each row, nearest the water's edge, is the longest and this pair has two horizontal spars called *wilis* lashed across them with split-rattan strands, one near the top and the other a metre or more below it (Figure 9.1). The lower one is set at a level that will correspond with the walkway of the finished bridge, access to which will be through the rectangular gap bounded by these cross-members.

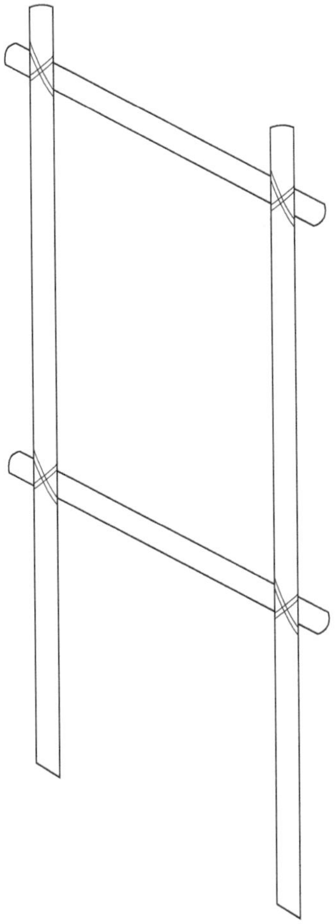

Figure 9.1 The entrance framework of the pok 'access stage' of a vine suspension bridge.

Plate 9.4 A pok structure with ladder and handrails up to vine bridge.

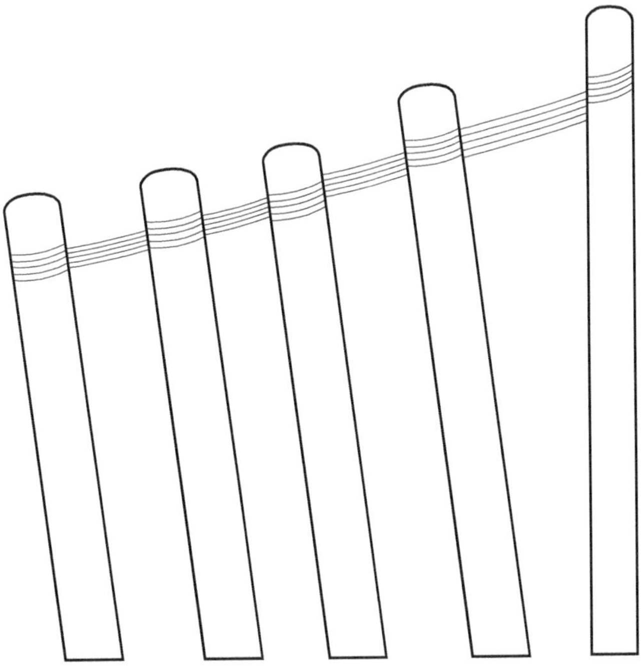

Figure 9.2 The anchor-stakes arrangement of pok 'access stage.'

The first two stakes are driven in perpendicularly, while those behind them decrease in size away from the river, and are often inserted at an angle leaning away from the water (Figure 9.2). These angled stakes serve to anchor the bridge, those in each row having their tops guyed together in series with vine, such that neighbours pull sequentially on one another to give a secure anchorage. Sometimes they have a sapling lashed between them too, which may serve as a handrail for those approaching the bridge. If there are trees on the river bank, the bridge-builders may include these in the *pok* structure for added strength. Finally, they may lash a ladder-like construction of split timber, usually of three, four or more rungs, to the lower *wilis* cross-member between the first two stakes, which running down at an angle gives access to the suspension bridge (Figure 9.3). Some *pok* structures may include small platforms jutting out pier-like over the river bank, comprising two parallel logs lashed in place more-or-less horizontally, with a slatted walkway of split wood lashed across them, similar to the ladder.

Once they have constructed the *pok* 'stage' on either bank, men are ready to erect the suspension bridge. This demands, in the first instance, getting a line across the river.

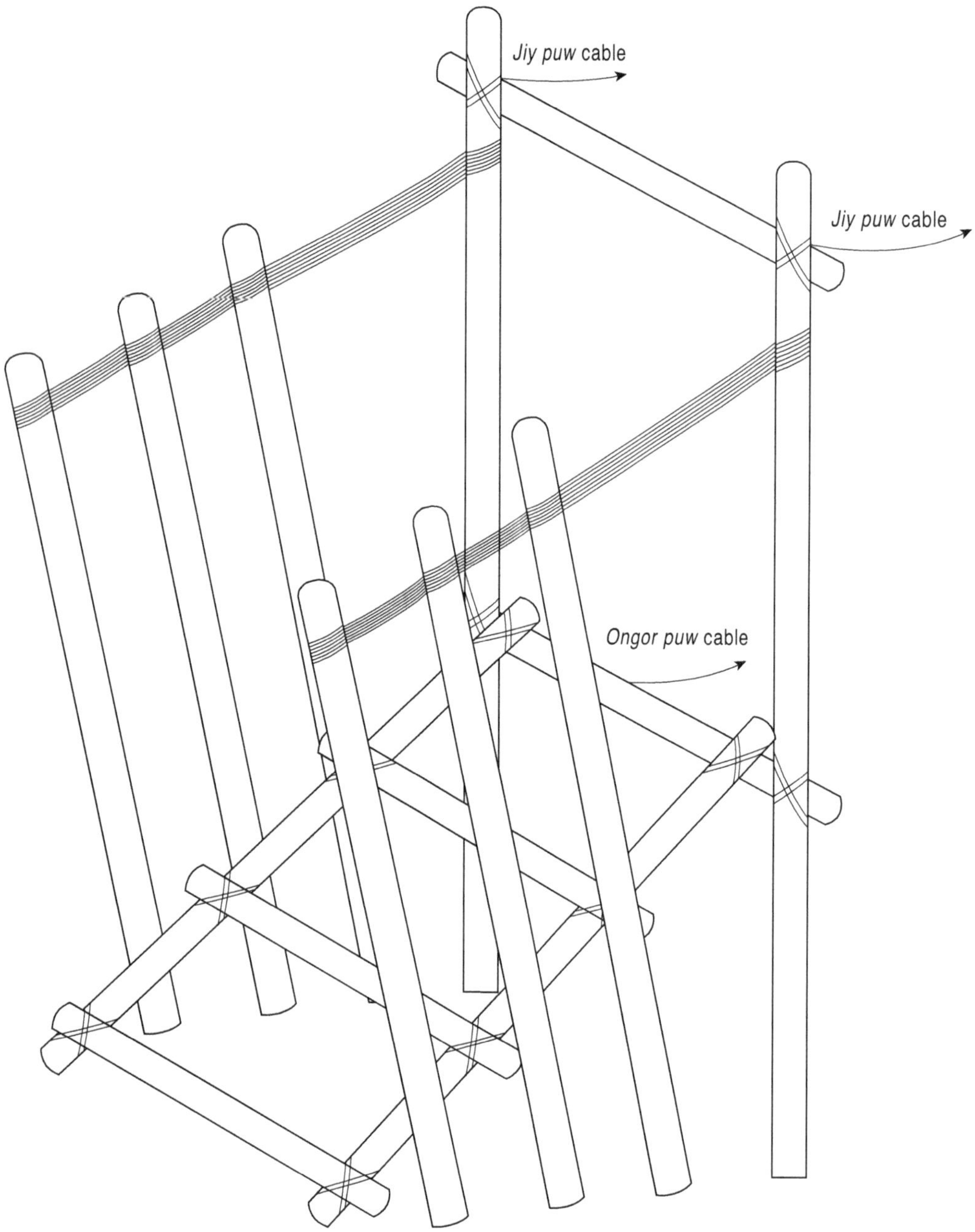

Figure 9.3 The pok 'access stage' of a vine suspension bridge showing where cables attach.

In the Was valley, they usually do this by casting a length of vine across the water, *ya luwtuw menja* 'vine cast lift'; unless the river is a small one in which the level falls sufficiently in a dry period to allow someone to wade across with the line. The casting can be a challenging task, for the major rivers of Wolaland are invariably wide and fast flowing, and bridge-builders resort to various ploys to cast the line successfully. They usually take a thin vine, or split a thicker liana in two, to give a light and easily cast line, and tie a sizeable stick to one end to act as a weight. They carefully coil up the vine line before throwing it, so that it will fly true, and then whirl the weight around and hurl it as forcefully as they can at those standing on the opposite bank. The men waiting to catch it may cut down a long branch or sapling and hold this out over the water in an attempt to snare the line cast from the other bank; on one occasion I saw a man cut down a long pole and lash a stave across one end to give a T-shaped prop for catching the line. Regardless of these various ploys, it can take many casts to land the line successfully on the other bank. Another strategy, seen in the Jimi valley, involved a vine-swing arrangement across the river, comprising a length of stout vine tied to a large branch high up a tree overhanging the river, on which a man swung to and fro until he was able to catch hold of a tree branch on the opposite bank and transfer himself and the vine across the watercourse (Vogel 1953:89).

Plate 9.5 Looking across river at refurbished pok on opposite bank with pier-like structure.

Plate 9.6 View through pok structure along vine bridge as someone crosses river.

Plate 9.7 Casting vine across river.

When they have a line across the river, men use it to haul over several lengths of stout vine, tying these onto the end with a twisted-loop knot, the *ya pap bubtu jimb* 'vine ring do join' (Figure 9.4). This can be heavy work if the river is flowing strongly, as they have to struggle against it sweeping the vine downstream. When they have dragged eight or so lengths of stout vine across the river, the builders tighten them securely in three bundles between the *pok* 'stages'. They attach two trusses of vine to either end of the upper *wilis* spars where they are lashed to the uprights of the stage, and the other one to the middle of the lower *wilis* spars. Viewed across the river at this time from the *pok* stage, the vine cables stretched across the river describe the points of an inverted isosceles triangle. The upper two bundles of vine, with others added subsequently, serve as

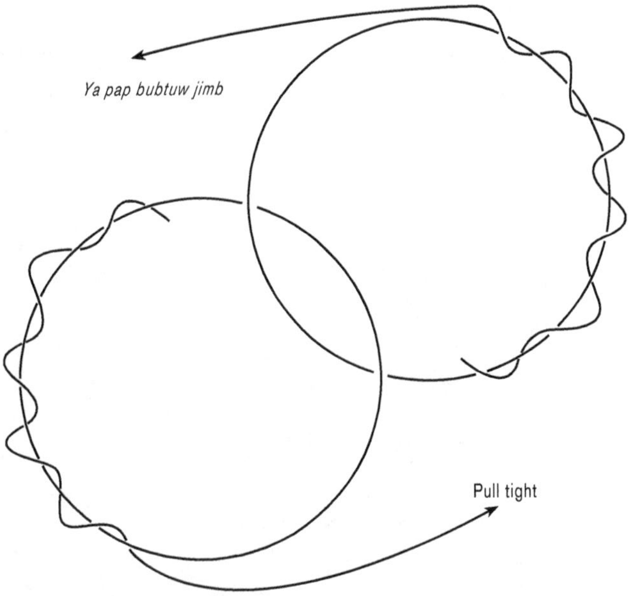

Figure 9.4 The bend used to join vine cables.

the two *jiy puw* 'hand lines' of the completed bridge and the lower one as the *ongor puw* 'foot line'.

Once they have tightened some of the *puw* cables in position, someone may cross the river with the aid of two loops of vine tied around them. He works his way across by sliding these loops along before and behind him, pulling the trusses of vine firmly together about him, preventing them from splaying apart uncontrollably and dropping him in the river. This somewhat precarious method of crossing a river is called *luwsilay*, and only a few men have the sense of balance and nerve to attempt it. Those working on a bridge may cross in this way to introduce some slack into the vine cables following their tightening to the *pok* stages, for if they remain too taut there is a danger of them snapping as they dry out and shrink. When further lengths of vine are subsequently added to the multiplex *puw* 'cables', in maintaining an established bridge, they are not stretched tight, but draped in a gentle loop along the line of the old *puw* trusses of vine. Another reason why one of the bridge-builders may shuffle his way along the positioned vine cables with *luwsilay* loops is to haul over several lengths of vine to add to the *puw* 'cables'. Alternatively, the builders may take one of the vines dragged across the river previously and use this to haul over further lengths.

When those 'staking' a bridge judge that they have sufficient lengths of vine in place for the *puw* 'cables', they start to tie short lengths of vine to the two hand lines, passing them under the foot line to establish the V-shaped cradle that characterizes the walkway of a suspension bridge, which Brongersma and Venema (1960:106) liken in shape to the neck of a fish trap. These sling-like lengths are the *porol* 'ribs', an allusion to a suspension bridge being a structure somewhat analogous to a rib cage, particularly

Plate 9.8 Knotting lengths of rattan together for pulling across river.

Figure 9.9 Pulling lengths of vine into position across river between pok 'stages.'

Plate 9.10 Crossing river using luwsilay loops.

that of an inverted snake's skeleton. They use a simple knot, the *porol bay* 'rib make', to secure these loops to the hand cables (Figure 9.5). When tying these slings in place men sometimes employ *luwsilay* loops, which allow them to secure the slings at intervals of a metre or so in the first instance, so that the bridge quickly takes shape. They discard the *luwsilay* loops when they have covered the entire bridge in this manner and proceed to fill in the gaps with further slings, until sufficient span the structure. Other men, who lack the nerve to use *luwsilay* loops, tie the rib slings close together from the start, working their way gradually across the bridge until they have completed it. They knot one end of the vine slings around one of the bundles of *jiy puw* 'hand cables', the other end dangling down to the water, which they then pick up between their toes and pass under the *ongor puw* 'foot cable', reaching down to grab it and pull tight, knotting it securely to the opposite bundle of *jiy puw* 'hand cables'. They remain upright working in this fashion, grasping the hand rails; to bend down and pull the sling lengths under the foot line would be a precarious and dangerous manoeuvre risking a

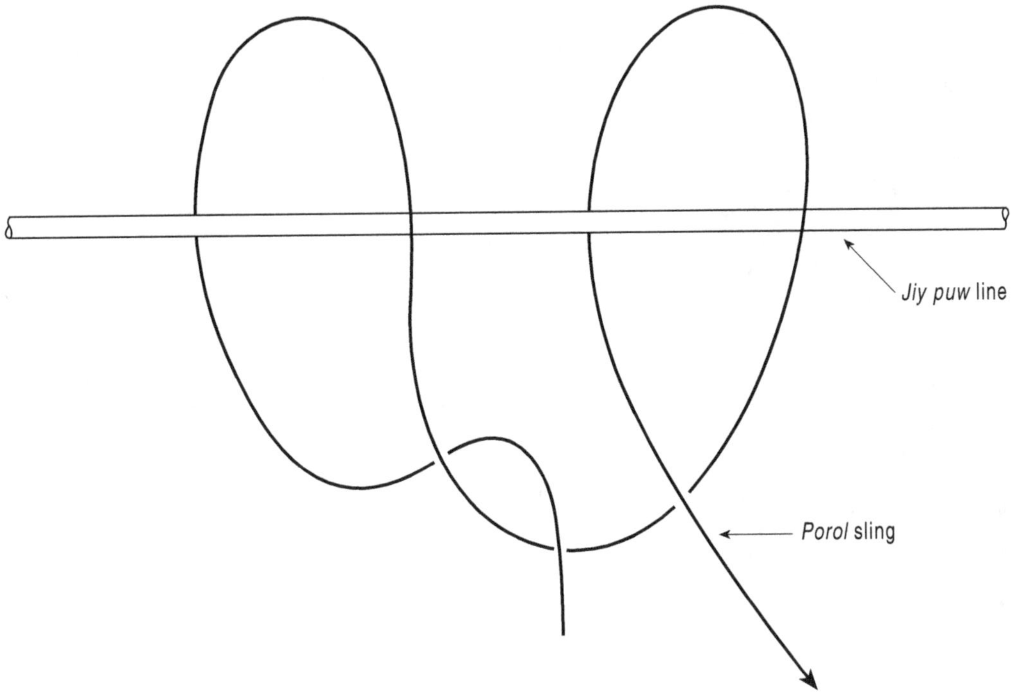

Figure 9.5 The knot used to secure porol 'rib' slings to vine cables.

Plate 9.11 Using a luwsilay loop to manoeuvre, a man secures porol slings in place.

Plate 9.12 A view across river as men work from either bank towards one another securing porol slings

headlong fall from the three unstable cable lines stretched across the water.

The *luwsilay*-loop technique is an example of tacit knowledge in action – it demanding considerable skill and nerve to keep one's balance to cross a river on the three cables and secure vine cradle strands between handrails and footway. It illustrates again the limits of participant-observation, for I have never had the courage – or been foolhardy enough – to try it for myself and learn to 'know' or embody the skill. Even if I had, I doubt I could convey it on the written page. No individual who has mastered the technique has been able to explain to me how he does it, maintaining his balance and preventing the three vine cables from splaying apart or twisting over and tipping him into the river. Those who have mastered the skill say they can 'just do it', indicating how they manage the tightrope act with the aid of the vine loops pushed along close to them. It is akin to trying to explain the skill of cycle riding, which also demands a mastery of balance and forces.

While they are working across the bridge tying the numerous *porol* 'rib slings' in position, men also rig up vine guy-lines from the *jiy puw* 'hand cables' to the branches of nearby trees overhanging the river (Figure 9.6). These *tomb* 'guys' serve to support the centre of the bridge, and prevent it swinging violently from side to side when crossed, or worse, from twisting around and tipping those on it into the river. Finally, when they have sufficient rib slings in place and the guys rigged, bridge-builders cut down a few saplings to serve as *ongor tor iysh* 'foot-structure lumber', trimming off their leafy branches and placing them along the cradle formed by the *porol* 'ribs' around the lower vine cable. These wands act to give the bridge a firmer walkway than the vine alone; they serve to weigh down the *porol* 'rib

Plate 9.13 *Sitting on a river bank, men split lengths of rattan vine for use as porol slings.*

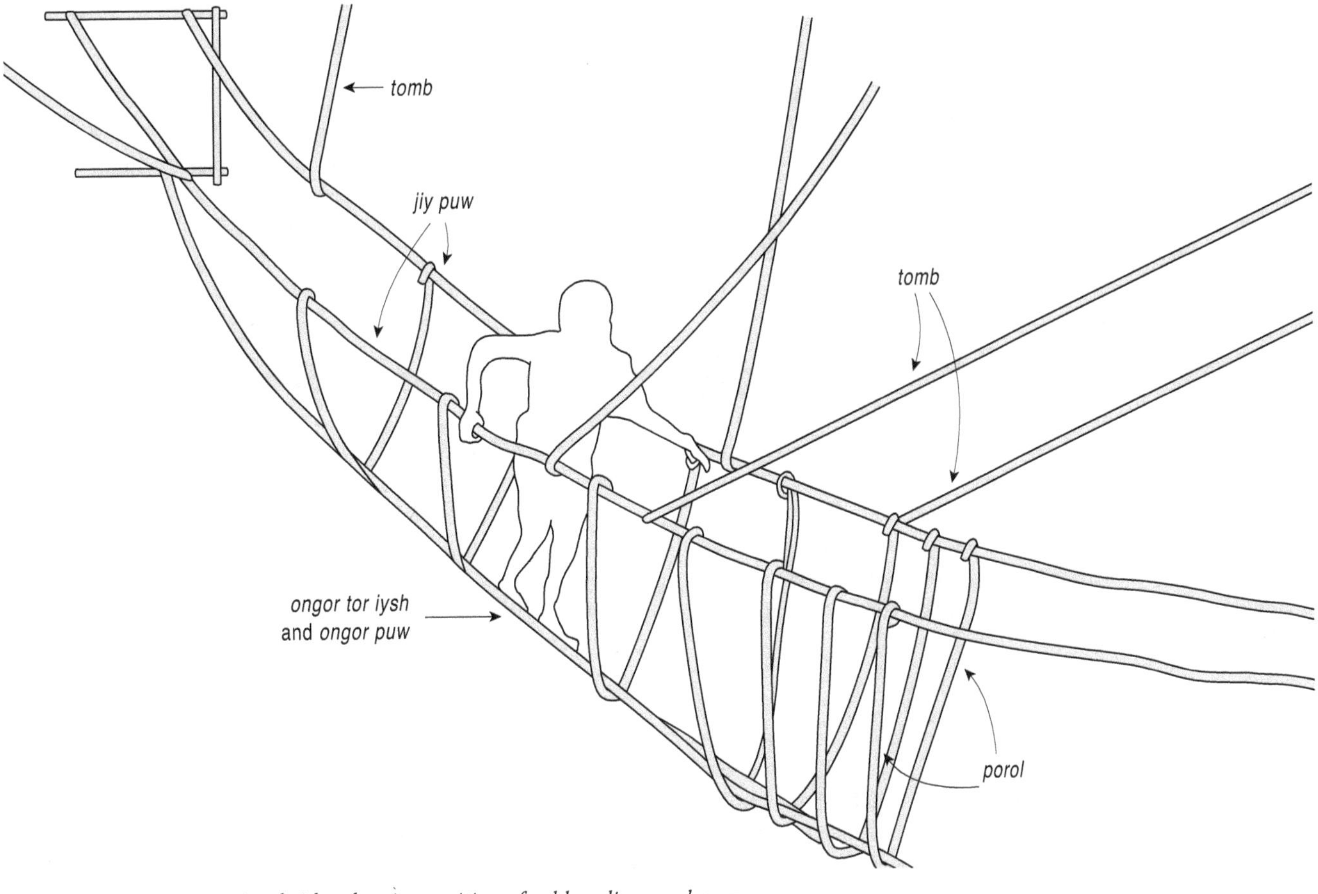

Figure 9.6 *Vine suspension bridge showing position of cables, slings and guys.*

Plate 9.14 Vine bridge showing tomb guys and ongor tor iysh walkway.

slings' so that they hang taut and provide a stable skeleton-way along which to walk.

When several men co-operate to 'stake' a bridge they can usually erect the entire structure, including the *pok* 'stages', within one day. It took six men, for example, with three working on either bank, one morning to suspend, between two extant stages, a suspension bridge that was swept away in a flood. One or two men, on the other hand, may take a week or so to erect a bridge. A man with house and gardens on opposite sides of the Was river, who single-handedly collected the materials for a bridge, took nearly a week to erect the *pok* 'stages' working alone; and a day to suspend the bridge between them, with the assistance of two relatives on the opposite bank. It is the collection of necessary materials that may take the time, not the construction of the bridge.

Bridge materials

According to Was valley residents, only certain vines are suitable for constructing suspension bridges. The strongest is *haeluwp* rattan (*Calamus fuscus*), which they use in preference to any other, especially for the hand and foot *puw* 'cables' initially stretched across a watercourse. They use the vine entire for these cables, whereas they split it up into thinner strands for the *porol* 'rib slings'. When bridging rivers in settled areas that are some distance from forest regions where rattan grows, or in other places where it is scarce, bridge-builders/maintainers may supplement it with a combination of the following vines, as second best: *kuwkpuw* jasmine (*Jasminum* sp.), *paerelya* bushkiller (*Cayratia* sp.), *toben* dogbane (*Alyxia* sp.), *maip* waxvine (*Hoya* sp.) and *sebseb* fig (*Ficus* sp.). In addition to these vines, they may also use strips of bark from the *waen* elm (*Trema orientalis*) and *wenet* walnut (*Engelhardia rigida*) for the *porol* 'rib' slings.

The calculated total length of vine in the Pelem vine suspension bridge mentioned in the myth, which spanned a 13-metre-wide Was river gorge, was 715 metres (Chapter 6). The bridge comprised 8 cables (each *c.*13 m long plus 1.3 m in lashing), 24 guys (each *c.*8.5 m long plus 1.3 m in lashing) and 135 slings (each *c.*2.1 m long plus 0.6 m in lashing, comprising a mix of vine and bark strips). A new or well-repaired bridge would have more slings (maybe up to twice as many), and some bridges repaired many times over several years may have considerably more *puw* cables (some of them old and brittle). According to these calculated estimates, the bridge at Paym, which was the longest in the region with a span of 19 metres, requires over 1,000 metres of vine.

Bridge-builders use only the hardest timber available to construct the *pok* 'stages', which must withstand the wet environment adjacent to a river without rotting rapidly. They named as particularly suitable: *pel* Southern beech (*Nothofagus* spp.), *uwk* madder (*Wendlandia* sp.), *waskomb* madder (*Timonius* sp.) and *woliy* coca (*Erithroxylum ecarinatum*).

The time it takes to amass sufficient materials to construct a bridge varies according to its location and distance from sources of suitable vine and timber, and the number of men collecting the materials and the rate at which they work. The collection of raw materials may be particularly time consuming if the builders go to the trouble of gathering considerable quantities of superior *haeluwp* 'rattan' from distant forested regions. The span of the watercourse they intend to bridge is also, of course, a factor, determining the total length of vine necessary. Men often construct bridges at narrow points on rivers because, while dangerous raging white-water-rapid locations, they economize on the amounts of vine that they have to collect. They also minimize the problem of sagging with use (long spans sag lower and may eventually reach to the water, particularly at times of flood flow, when the river may sweep them away) and the problem of excessive swaying movement (which can result, if guys snap, in a bridge twisting over and tipping a person off). Also, it is easier to get the initial vine lines across the river at such locations. As an example of the time to collect materials to construct a suspension bridge, it took six men, with about a two-hour walk to a good source of vine, between two and three weeks of spasmodic working to collect sufficient vine to replace a structure swept away in a flood. Whereas one man, who resided within an hour's walk of forest where vine was plentiful, amassed sufficient materials to construct a new bridge within a week, working constantly because he was keen to connect his house to new gardens cultivated on the other side of the river and a considerable distance from the nearest bridge.

Bridge-builders and users

The men who co-operate to construct a bridge are usually relatives, though friends sometimes join forces. The construction parties coalesce like other groups (Chapters 3 and 7) to pursue an action of mutual benefit to all participants on the basis of unpredictable factors such as propinquity of gardens, location of kin's homes, who happens to be around at the time and feels inclined to help, and so on. Varyingly constituted kin groups were responsible for four out of five observed work-bees on bridges crossing the Was river, two unrelated families co-operated in the fifth to erect a bridge adjacent to their gardens on both banks. Those relatives who co-operated were diversely related, reflecting the

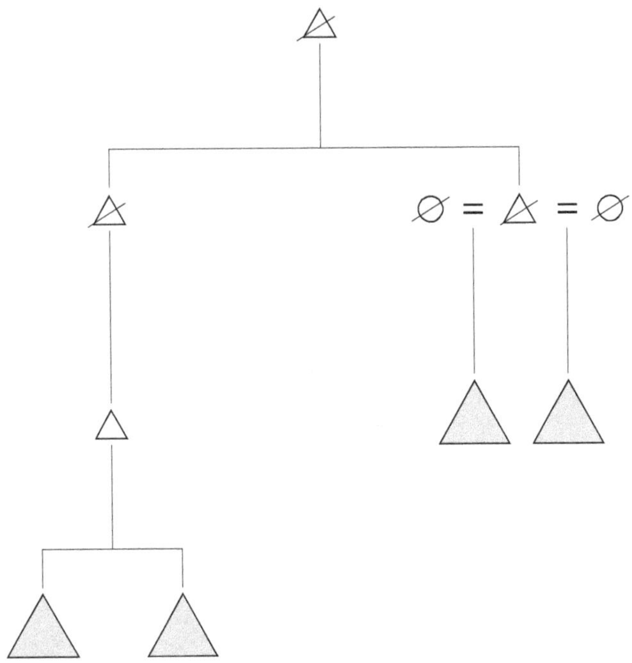

Figure 9.7 Genealogy of relatives who cooperated to repair a bridge.

changeable circumstance that united them. The man who single-handedly collected materials and built a bridge's *pok* 'stages' on his own, for instance, had a close and distant relative of his wife (WB and WFFMFBSSDS) assist him in suspending the vine structure. A somewhat atypical grouping of four agnates came together and co-operated to make extensive repairs to a bridge badly in need of attention (Figure 9.7), which was explicable given their motives, which were to arrange the safe transport across the river of several pigs that they had received as the agnatic party in a mortuary exchange from relatives living on the other side. The relatives who co-operate in erecting suspension bridges usually comprise more randomly constituted groups, such as the work party that came together to construct a new bridge to replace one that was in a dangerously rotten state (Figure 9.8) and the men who united to rebuild a bridge swept away in a flood (Figure 9.9).

Although a bridge is a public utility, it is unusual for more than half a dozen or so men to co-operate in erecting one, and frequently fewer. The builders will be those who urgently require access to both sides of the river; who have, for instance, gardens on both banks. Others are likely to show little interest and not help out, although they may find the bridge convenient when it is completed and use it freely, sometimes to the indignation of those who built it. The man who more-or-less single-handedly erected a bridge, for instance, subsequently considered it his property, complaining about non-family members who used it, though to little effect.

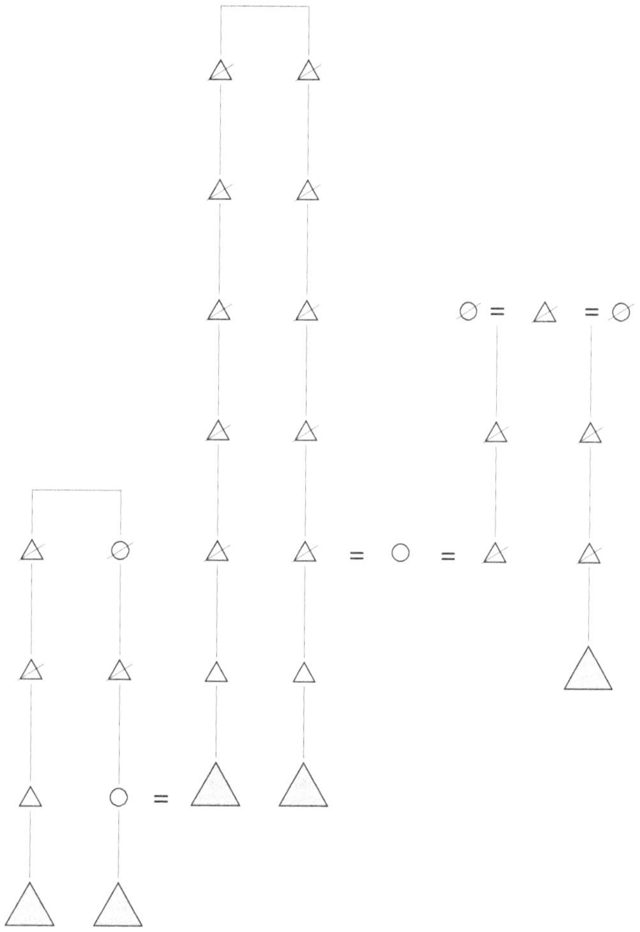

Figure 9.8 Genealogy of relatives who cooperated to construct new bridge to replace dilapidated one.

The egocentricity is further evident in people's attitudes to the materials that feature in the construction of a bridge, individuals remembering what they have contributed and, if the structure is abandoned for any reason, they may turn up to reclaim them, a similar scenario to that of the *sa aenda* longhouses mentioned in the previous chapter. The man with gardens on both banks of the river, for instance, who worked largely alone, initially prevailed on some relatives to help him, and they gathered some necessary materials, but when they found that no one on the opposite bank had any interest in erecting a bridge they left him, taking with them the vine and timber they had collected. The maintenance of a bridge likewise illustrates the individual ethos that pervades Wola socio-political life. It is even more of an individual affair than bridge erection, those attending to a structure that needs some repairs usually working alone. When the solitary bridge-builder subsequently had gardens on only one bank, for example, he lost interest and did nothing to maintain his bridge, which with minimal traffic soon fell into disrepair. It is a graphic manifestation of the 'user-makes' principle that is evident in the manufacture of things generally (Sillitoe 1988a:563), where self-help is a feature of households' political-economic independence, which, as discussed previously (Chapter 7), is central to the stateless order and also figures in the sustainable use of resources.

These attitudes towards infrastructure that serves everyone, one of the few such constructions in the Was valley, reflect an indifference towards community-wide demands and a scarcely developed notion of public service. It further illustrates the individualistic character of the stateless political order, where the mobilization of sizeable groups to co-operate in any enterprise is a challenge (Sillitoe 1979a, 2010a). A related issue is the manner in which those working on a bridge go about the task. They again work without direction. Each knows what they have to do and co-ordinates his actions with the others to achieve their common goal. There is no need for a leader, although someone particularly anxious to erect a bridge may be the driving force behind its construction, rallying one or two less interested relatives to his cause, and a man with experience of 'staking' bridges and adept with *luwsilay* loops may shout, frequently unheeded, directions at those short of experience and less sure of themselves.

While vine bridges have served Highlanders since mythical times, their use poses some risk to users. Brongersma and Venema (1960:106) describe graphically some of the hazards that attend the crossing of suspension bridges from experiences in the Star Mountains of Indonesian Papua.

> It often happens that the weight of the person on the bridge causes some of the slings to break with a loud report; one drops a bit and the bridge starts to sway, the only action one can take is to wait and proceed when the bridge has come to rest again... Not only we, but also our porters were sometimes overcome with fear and stopped in the middle of the bridge; when one looks down one feels that the river is standing still and the bridge is moving. Stopping like this delays the crossing, but it is worse if the bridge starts to sway, for there is a chance the man will fall off.[8]

The Wola are understandably wary of suspension bridges. They are cautious of new ones because men frequently skimp on their rib slings and guys. It is not until some time has elapsed that people cross them with any confidence, once others passing have added vine piecemeal in the interval to give a sturdier structure, and the green materials have stretched and toughened as they dry. Even so, they cross established and well-maintained bridges with care, for a slip may prove fatal and unfortunate accidents

8 I am grateful to Joost Brouwer for helping me translate this passage.

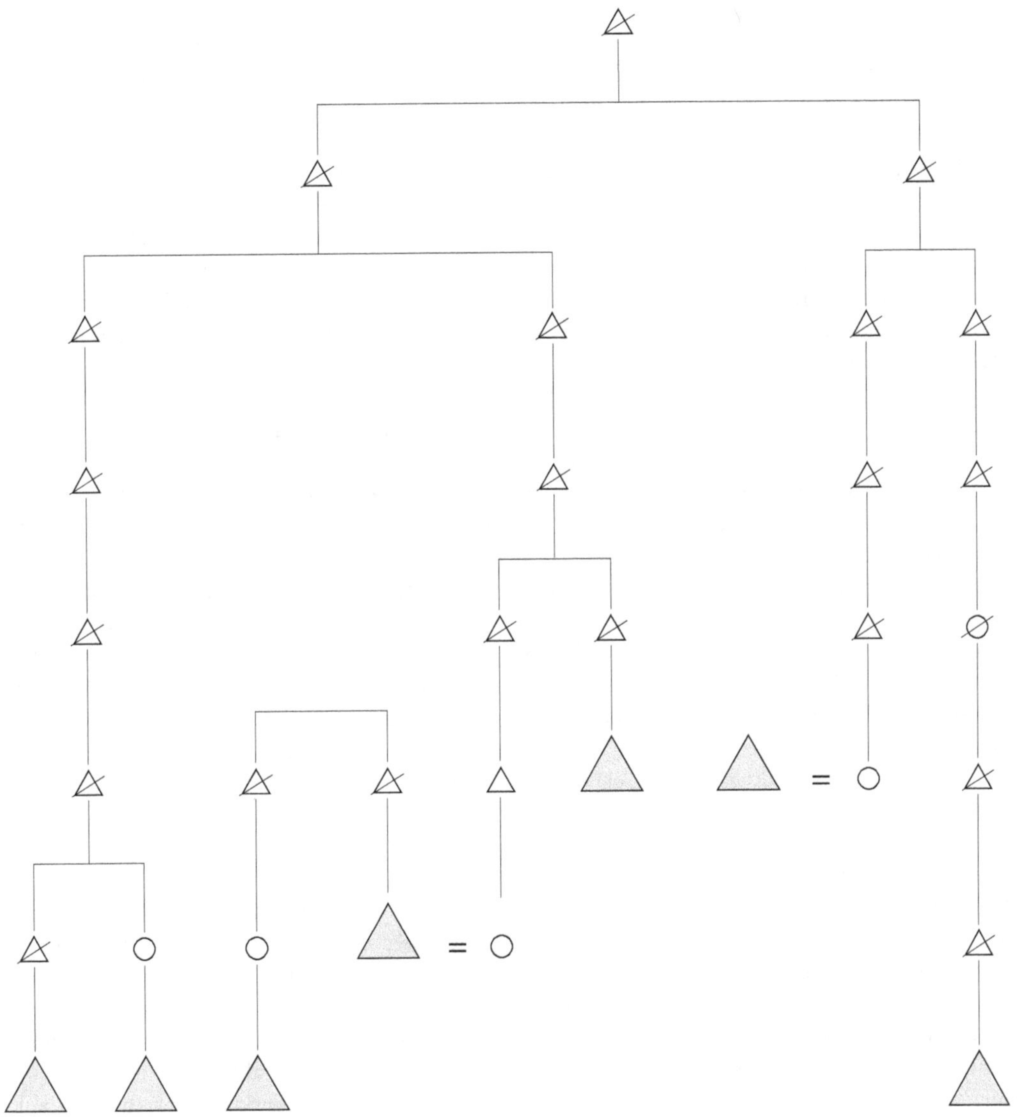

Figure 9.9 Genealogy of relatives who cooperated to rebuild bridge swept away in flood.

sometimes occur. On one tragic occasion, a baby drowned when the net bag in which she was being carried slipped from her mother's head as she stumbled on a wet bridge. They approach with particular care bridges that have fallen into disrepair, with old brittle vine and skimpy slings, as potentially dangerous and liable to collapse. These dangers have prompted Provincial Governments in Highlands regions to replace traditional 'snake's skeleton' bridges with wooden planked structures suspended from steel hawsers; one such was erected to replace the bridge from which the baby fell, upriver from the suspension bridge established mythically with the black snake's assistance. While people appreciate these infrastructure improvements, a problem with them is that, not locally instigated – unlike many innovations in house construction – they are largely beyond their means to maintain when damaged, for instance when rivers flood badly.

Bridging to symbolism?

Knowing something about bridges and rivers, this is an opportune point to explore further the social-constructivist viewpoint, and address the suggestion that the association of a suspension bridge with a snake may offer ingress to symbolic issues – leaving aside the formulaic application of structural, evolutionary or functional interpretations relating the serpent metaphor to some assumed generic concern of humanity. In seeking cues within ethnographic context, we can perhaps trace an association with the belief mentioned in the previous chapter about *towmow* 'spirits' living as snakes underneath hearths and the accompanying rite to blind them by thrusting a trident into a pig's snout held in the ashes of the fireplace (offered to appease the 'snake ghost' residing there, called upon to accept the offering of blood and singed bristles when divination suggested it was responsible for someone's sickness). The idea that snakes may live under hearths possibly relates to the fossorial habits of local snakes, which burrow into soil, leaf litter and dead wood.

People are generally afraid of *sugum* 'small snakes' found in their region, such as the loria forest snake (*Toxicocalamus loriae*), montane keelback (*Tropidonophis statisticus*) and Muller's crowned snake (*Aspidomorphus muelleri*). Fear of snakes relates to beliefs about them being the familiars of ancestor spirits, and apparently has nothing to do with these reptiles being physically dangerous to humans. No one has ever mentioned that snake bites are poisonous and make people sick, and it is possible that snakes occurring at higher altitudes are either non-venomous or not seriously so.[9] Intriguingly, people are less frightened of large snakes, which are not spirit familiars, such as the pythons that occur in lower-altitude forest, including the *burun* 'white-lipped python' (*Leiopython albertisii*) and *humshuwol* (adult) and *glaeshond* (juvenile) 'green pythons' (*Morelia viridis*), which they sometimes catch and eat cooked (Sillitoe 2003:180).[10] The origin of the culturally sanctioned ophiophobia is unknown, lost in the dim past with distant ancestors-cum-snake-spirits, although it could arguably have started through some long-lost contacts with lower regions where deadly snakes occur, the fear lodging deeply in the collective unconscious.

Snake classification is likewise opaque. People do not group these creatures together into a single 'snakes' life-form category, nor a higher taxonomic one equivalent to 'reptiles', with associated identifying characteristics that we might use in excogitating symbolic relationships. The Wola put scary *sugum* 'smaller snakes' in with *elelbiy* 'insects' – which include earthworms that they say give a poisonous bite (Sillitoe 1996:256–7). There is no named life-form class that incorporates the larger pythons (including those found in the Kutubu region), which 'only carry their own names' (Sillitoe 2003:80). This indeterminacy catches ophidian representations generally. There are no cultural contexts that suggest aspects of snake behaviour have metaphorical associations; and no one has ever intimated such – for instance, that snakes occasionally shedding their entire skin might imply renewal. Regarding snakes and their connection with bridges or houses, people do not apparently share a systematic set of socially transmitted ideas, at least that they can articulate, that offer culturally cued opportunities to further symbolic exegesis. When encouraged to discuss myths, such as the snake-bridge story, friends seem puzzled by my quest for metaphoric meaning and associations. Myths are for telling, not for discussion.

While people hold many ideas and beliefs in common, such as 'snake-ghosts' sometimes reside under hearths, they share no overall agreed integrating cosmology. They have individual views, which overlap at some points between some persons and differ at others, and vary in sophistication according to their penchant for confronting life's big questions, as opposed to just getting on with life. It is another instance of the variation that characterizes social life (Chapter 8) and thwarts straightforward normative approaches to behaviour and actors' associated views. At one time, I thought that evangelizing missionaries were responsible for the apparent fragmentation of local supernatural ideas, these falling apart under the Christian offensive that forbade associated activities as evil (Sillitoe and Sillitoe 2009:159–63), and that people were consequently either unwilling to discuss their theology or were embarrassed by its apparent wrongness, it being both intellectually and morally in error. Yet even after several years and knowing some persons well, we do not get far when I try to engage them in reflective discussion. Typical comments are '*sugum taenktay bort na hondowao, kak*' 'I've not seen a snake under a fireplace, I tell you' or '*shumbaen kismesa uwp*' 'our ancestors spoke like this [about such things as snakes under hearths]', implying that they are relaying hearsay beliefs and do not know anything more.[11]

The apparent absence of any cosmological consensus (Lewis 1980), featuring a systematic suite of symbols amenable to interpretation by those with the necessary insight, is an aspect, I now think, of stateless orders. These lack authority figures, by definition, to legislate and arbitrate on such matters, just as they do on taxonomic ones, for instance over the status of snakes (Sillitoe 2003:109–16). While all subscribe to similar ideas that overlap to varying extents, as heirs to the same cultural heritage and speaking the same language, these are potentially subject to variable personal interpretation. The socio-political order affords individuals considerable scope to make their own inferences, so far as they need to, regarding some widely agreed points held in common, such as snake-spirits live under fireplaces. It somewhat reflects Hume's (1896) sceptical philosophy, which undermines the faith we have in the inductive assumptions that inform our knowledge. It is arguably a wise option in view of the contradictory character of human nature. After all, where authorities seek to resolve paradoxical theological questions and manage collective opinion, they spawn endless debates, even conflict (for example, in the Christian tradition that has overwhelmed

9 For example, such snakes as the New Guinea death adder (*Acanthophis* sp.) and New Guinea small-eyed snake (*Micropechis ikaheka*) are possibly absent from the region.

10 Some of the scientific names differ from Sillitoe 2003 following recent taxonomic revisions; identifications follow O'Shea 1996.

11 The tense used in these rejoinders often features evidential markers that signal neither the speaker nor the hearer have any first-hand experience of the matter under discussion, which further relates to Wola-speakers' assessment of the trust they have in any intelligence, the extent of knowing and its tacit aspects (Sillitoe 2010b:20–2).

the Was valley population, should people subscribe to an 'eye for an eye', to which the Wola can readily relate with their revenge ethos, or should they now incline towards 'forgive thy neighbour'?).

This, once more, recommends circumspection in interpreting archaeological remains from stateless orders. Archaeologists regularly turn to symbolic interpretations of unknown or strange things they unearth, such as features of excavated buildings that do not have obvious practical functions, often speculating on ritual associations (e.g. Chapter 4's seminar discussion). This is fair enough with no actors to ask about such things, and only their own imaginations, together with any associated remains, to inform their interpretations. But awareness of how tricky it is to explore the symbolic dimensions of, for instance, built structures, advises caution, for even when we have living people to supply information that relates to such material things out there, it only exists in their heads, if consciously at all, and might be manifest in their behaviour in varyingly understood ways. Furthermore, constructions in the Was valley, even those erected for ritual purposes, indicate that such symbolic associations may be few, suggesting that resorting to symbolism too readily may be inappropriate. On the other hand, it is straightforward in archaeological contexts to distinguish empirical material evidence from metaphysical imaginative interpretations, whereas anthropological enquiries often conflate the two.

If actors are unable or unwilling to account for the significance of many acts and things, which we assume have some emotive symbolic import, and we aspire to make their meaning apparent, the implication is that they exist on an unconscious plane, which returns us to tacit issues, now entirely of the mind, and poses the question of how we can access them. What is the character and validity of any such attempts, as the editors invited me to try regarding cosmological themes? In seeking to identify and interpret symbolism associated with suspension bridges, we might imaginatively explore further possibilities of ophidian imagery beyond their inverted snake skeleton-like structure, and in so doing, turn for inspiration, for instance, to currently popular fractal geometry, which concerns things that display self-similarity and complexity at all scales. The snake myth of bridge construction features elements that repeat in other myths with origin aspects; such as the *ol maerizor injiy* 'humans borne stories' that account for how today's communities came into existence (Sillitoe 2010a:86–7). Somewhat akin to the structural study of myth, we could delve into aforementioned contradictory themes such as the emergence of human sociality in opposition to individual interests. The snake might symbolize individuality, as a solitary creature, but its identification with ancestor spirits signal sociality, particularly when associated with *ol maerizor* origin myths that focus on common ancestry and that locate today's communities on their territories, in the landscape, where we see rivers snaking along valleys and footpaths twisting over mountains. Furthermore, the periodic replacing of vine on bridges, akin to a snake acquiring a new skin with the shed vine rotting and falling away, may serve as a metaphor for change through the generations, where the same structure persists, just as things continue much the same with time spiralling around and not running linearly, and gradual rather than rapid change occurring. Or we might draw, admittedly somewhat ethnocentrically, on the English trope of snaky characters, which ophiophobia arguably intimates, and turn to power issues and the checking of some slippery persons' self-interested behaviour that could compromise others' equal freedom of action. The fractal geometry image catches nicely the focus on equality, all persons repeating more-or-less the same, allowing for gender and age.

The reader can probably come up with further associations by drawing on the Was valley ethnography and snake natural history – such as humans and snakes often interact with water – which is fine, because there is no danger of you distorting the ethnographic evidence, for whatever you come up with is a product of your imagination, informed by your understanding of the evidence. Indeed it is one way that such information about other ways of being-in-the-world may inform future intellectual enquiry. Consequently, it is not that such idealistic (theoretical) approaches to the interpretation of surmised symbols are necessarily wrong – we all have the right to allow our imaginations to run – so much as they may be unverifiable by direct reference to the ethnography, that is, what people do and say. It is doubtful, for instance, that rotten vine on a suspension bridge would remind my friends of snake ecdysis or symbolize to their minds generational change. It is necessary for those of us engaging first-hand with field data and contributing to the ethnographic record, to maintain a clear distinction between evidence and imagination. It returns us to ethno-determinism's aim of anchoring the imagination to actual worldly features, what is going on 'out there' and what actors have to say about it, in addressing externally generated and validated issues, which in anthropological enquiries depend first and foremost on the ethnography. It eschews pure metaphysical idealism, allowing rein to free-floating intellect, which requires faith to enter an imagined world with internally generated and validated 'problems', which postmodern emphasis on the subjectivity of knowledge may appear to licence. In other words, imaginative constructions-cum-deconstructions not 'determined' by ethnography may contribute to social metaphysics but not social science.

The bridge discussion illustrates how ethno-determinism seeks initially to prioritize an inductive approach, albeit acknowledging the inevitability of a deductive aspect, in seeking to further understanding of ethnographic findings on their own terms. The distinction here between inductive and deductive inferences mirrors somewhat that between empiricist and idealist approaches (Chapters 2 and 10). If the aim is to grasp and convey as best we can other ways of being-in-the-world, then we have to try so far as possible to take an inductive stance and view the world anew from that perspective, and struggle not to impose prematurely our deductively informed views. But the challenge is that many aspects of being-in-the-world are unspoken, are even unspeakable, as evident in the foregoing bridge and symbolism digression. All people abide by values and norms, albeit many imbued and accepted tacitly. While they may not apparently openly discuss the implications and frame these as propositional knowledge, these ideological views presumably square in large measure with cultural arrangements and social behaviour. These are accessible to us and we can document them. We have descriptions of customs (such as marriage arrangements), accounts of behavioural sequences (case-method accounts of actual marriages) and perhaps some quantitative data (culled from a sample of marriages regarding issues such as spouse choice etc.). If the actors are only tacitly aware of the implications that we presume pertain, we have to intercede deductively if we are going discuss and analyse these according to the propositional assumptions of our intellectual tradition. We unavoidably draw deductively on our culturally shaped understanding of the world to make sense of the ethnographic data, and deploy what is widely referred to as social theory, though I prefer to talk of social ideology.

Some go further, such as the aforementioned editors, assuming that as the world 'out there' appears on some level the same for all of us, inhabiting the same planet and having the same cerebral capacities (after all cross-cultural interactions are not akin to meeting Martians, for we can all communicate to some extent with one another and convey mutually intelligible ideas), this justifies assuming that deductive inferences apply to all humanity whatever culture, language and historical age, as in structuralism that assumes a certain universal cognitive architecture based on binary processes that often result in contradictions – white versus black – that we struggle to resolve. This approach circumvents the apparent tacit status of many people's understandings and social interactions by claiming that it can pinpoint how for actors these are propositional, and parallel its deductions, in their use of unfamiliar means of communication such as mythical accounts, object designs, ritual sequences and even bridge structures. This affords licence to impose these ideological constructions on others and seek, usually selectively, to find ethnographic evidence to support them. If you believe this – it comes down to faith, as there is no evidence substantiating such exotic communication and indigenous deductions – then the questions posed here regarding access to tacit know-that understanding are baseless. Obviously, I do not believe it. It fails to do justice to the way that different cultures, languages and histories have come up with a multitude of different ways of being-in-the-world and understanding it. It is the reverse of ethno-determinism in imposing an a priori deductive view on the ethnography instead of taking an a posteriori inductive view, and it does not so much solve the tacit-knowledge problem as disguise it by sleight of hand.

According to the inductively inclined ethno-determinist perspective, whatever we deduce should relate to what we induce ethnographically, and we should strive to modify and change our deductions to comply with the ethnographic data, particularly where these challenge the prior understandings that inform our theoretical/ideological deductions, endeavouring not to impose these on the data. The approach, as noted, is consequently eclectic in respect of 'theories', that is it only uses those aspects that fit with the ethnographic evidence, struggling not to manipulate it to what we think we should find, in a bid to substantiate some social ideology qua 'theory'. It undermines the simplistic distinction between descriptive ethnography and analytical anthropology (Chapter 2). It is the implications of these seemingly inescapable deductive interventions that I am exploring here, asking how we can contain them to minimize distortion, or if this is even possible when we seek to convert tacit knowledge into propositional knowledge. The implication is not as some postmodernists suggest that all ethnography is fiction, although some distortion of peoples' views of being-in-the-world seems unavoidable.

It appears that this excursus into ophidian symbolism adds little to our knowledge of suspension bridges or houses, nor indeed of cosmological landscapes or violent waters, but it is germane to this book's argument regarding tacit knowledge, the limits of what we think we know and the problems that attend interpretations that go beyond what actors can consciously recognize (Lewis 1980:17–19). While people may be able to comment on the meanings of some things that we take to be symbols, there are others for which no local exegesis is apparently forthcoming. They may be reluctant to discuss them at all, even seeming perplexed at invitations to interpret such things in words, as frequently occurs in the Was valley. The undertaking of acts in which people use things assumed to have some symbolic significance, with no apparent need to ask why and so on, is a tricky tacit issue. We are wont to assume that humans conceive of things in some way and relate them together. They must have a reason in particular for using objects that otherwise seem to have no purpose. When actors seemingly cannot consciously articulate any meanings, even deny knowing what things symbolize, the reasons are presumably implicit, emotional and unconsciously 'known'.

These unspoken matters apply to most structures on the whole, including houses. We take housing largely for granted, it more often than not 'goes without saying,' as Carsten and Hugh-Jones (1995:3–4) point out, houses being 'so commonplace, so familiar ... that we often hardly seem to notice them.' They stand on the margins of consciousness unless something happens to bring them to our attention, such as a disaster destroying homes. These authors go on to criticize anthropologists for focusing largely on ritual and symbolic issues when they consider houses. They emphasize that 'the house has another side. It is an ordinary group of people concerned with their day-to-day affairs, sharing consumption and living in the shared space of a domestic dwelling ... these everyday activities, carried on without ritual, reflection or fuss ... anthropologists have tended to ignore' (1995:45). They urge them to take more interest in 'everyday life,' which returns us again to the problems discussed here, ordinary life, as we have seen, having significant tacit aspects.

In is apposite in this context that Heidegger (1993 [1954]) draws on the example of a bridge to clarify his thinking about 'dwelling' and 'being-in-the-world.' A bridge unites two river-banks that previously were separate and in so doing creates a new phenomenal relationship in the environment that informs people's experience of the landscape, drawing them tacitly to a certain place on the river to cross it. But how can we progress understanding of these issues in such contexts? If actors do things in these circumstances that have no conscious significance for them, we can hardly expect to bridge ourselves to tacit issues by drawing on what are themselves tacit beliefs and practices – using an object assumed to have some symbolic import pertaining to an interior emotional state, as opposed to using a tool to achieve some exterior physical task – presuming, that is, along with social constructivists, that doing these things has some collective unconscious import, rather than being truly meaningless gestures, or things that had meaning long ago but now forgotten, as some kind of cultural residuals, social changes having made their original meaning redundant.

10

Gender structures and divisions

Another approach to symbolism, as the vine-bridge discussion intimates, is to consider how it relates to the construction and architecture of structures themselves, such as the layout of houses, instead of how symbolism may be read in particular things associated with various structures, such as painted designs applied to some graves. The semiotic view that cultural arrangements (for instance, architectural traditions) generally call for interpretation, is of a different order to addressing specific things that may have symbolic import for actors (for instance, designs on features), whether consciously held, of which they can give an exegesis, or unconsciously held, awaiting the attention perhaps of those with the imagination to uncover their meaning.

A problem for the social-construction approach is that it soon blends ethnographic data with its own, oft times ingenious, interpretations that rely overly on the imagination of commentators. These depend, being hypothetical, on suppositions largely unverifiable by reference to external evidence, and often result in overly structured versions of social life, with actors behaving robot-like rather than as manoeuvring human-beings. While we inevitably impose our ideas and understandings to some extent in seeking to understand other ways of being in the world, as the postmodern critique confirms and as noted previously, the challenge from an ethno-determinist perspective is to distinguish these, so far as feasible, from the ethnography of what actors do and say, which should lead our enquiries and prompt the questions we ask. It is necessary to exercise care in speculating on the significance of things presumed to have symbolic import, such as the design of buildings, particularly when actors themselves are unable to verbalize any meanings. It is questionable to substitute our idealistic views for what we presume is unspoken locally. Such interpretation assumes that researchers can translate tacit intelligence (here encoded in non-verbal cultural practices that communicate thoughts on socio-cultural being) into explicit knowledge, which this book queries. One way forward is to visit the distinction between interpretation and explanation, and to tease out its implications for our understanding of these issues.

House symbolism elsewhere

The architectural-design approach has featured prominently, from the start, in the relatively few anthropological studies of buildings, and illustrates the above problems at different times. According to Souvatzi (2012:176) it comprises two broad strands: 1) 'cultural or ideational', which assumes that the 'built environment' is 'inherently meaningful' and focuses on symbolism and metaphor, drawing on structuralism, linguistics and social psychology; and 2) the 'social', which assumes that built space is 'socially produced' and focuses on socio-spatial action and the 'reproduction of social order', drawing on sociology, geography, architecture and social history.

Until recently, interest in houses focused largely on symbolic and identity issues, extending on previous concerns with the social aspects of household arrangements (Buchli 2013; Lawrence and Low 1990:456–76). The social approaches date back over a century, when they were subject to evolutionary interpretation. For instance, in his last book, Morgan (1881) argued that the domestic architecture and household arrangements of Native Americans reflected the evolution of social organization, notably kinship and property relations.[1] The practice of hospitality, communism[2] and customs relating to food and land in large households, which he discusses in detail,

1 An interesting aside, in view of my book's history (Chapter 1): Morgan originally intended *Houses and House-life of the American Aborigines* to be the final chapter of his book *Ancient Society*, famed for influencing via Marx and Engels the development of communist philosophy, but omitted it because of the manuscript's already large size (the first chapter of *Houses and House-life* is a condensed version of *Ancient Society*'s account of the evolution of society, and the gens, phratry, tribe and confederacy organization of North America that he thought central to understanding the 'houses and house-life of the aborigines').

2 Morgan's view of communism differed significantly from the later Marxist one, as someone who subscribed to the capitalist American ideals of his time. He applied the term to the egalitarian relations of 'primitive society', which he was among the first to glimpse, seeing the significance of the commons approach to property relations.

largely determined the character of 'houses and house-life' in the 'early and middle stages of barbarism'. He posed a general question that continues to be of interest: what does domestic architecture tell us about social organization? But this question has paradoxically led to the ignoring of houses as buildings and the proliferation of sociological vacuities such as 'social relations produce houses' (Rodman's conclusion, Rensel and Rodman 1997:222), which illustrate the sociological-determinism that has dominated since.

The direction taken is illustrated by Littlejohn's (1967) paper, ostensibly on the house among the Temne of Sierra Leone. It scarcely mentions their houses, being an account of people's ideas of space, family, social organization and supernatural beliefs. He is frank about it: 'To bring closer the Temne house and the existence of those who dwell in it I turn now to the household ... [which] can only be understood through a brief description of Temne society ... the core of it is always ... a fragment of a lineage ... I have been considering the Temne house as a structure and set of activities in space' (1967:339, 341, 347). He deals with the actual house, a double-walled structure, in a single early paragraph that cries out for more detail, people apparently occupying the space between the two walls, which is 'partitioned into bedrooms and store rooms' (p. 331–2). While Morgan's cultural-evolution theory is discredited, he at least paid some attention to the actual houses of Native Americans, discussing in particular detail their multiple-roomed structures.[3] Although a sociological focus on relations between occupants that ignores their dwellings may be justifiable in Western contexts, where we have works on architecture and building, it is less so where these are unavailable, and it is certainly questionable from a holistic perspective to omit such information on houses.

More influential, and mystifying, is Lévi-Strauss's (1982, 1987) idea of *sociétés à maison* (house societies), which he advanced to account for certain socio-political arrangements where his 'elementary structures' of kinship (1969) do not apply.[4] There has to be, from a mechanical-solidarity perspective, some other structuring mechanism when kinship-based rules alone are inadequate to social order. Neither his 'house societies' nor his 'elementary structures' of kinship apply to the Was valley. He refers to houses not as buildings but *personnes morales* (jural bodies) that hold 'an estate made up of both material and immaterial wealth, which perpetuates itself through the transmission of its name, its goods and its titles down a real or imaginary line' (1982:174). Initially he came up with the idea to accommodate the *numaym* ('longhouse clan') 'complex kinship' system of the Kwakiutl of the north-west coast of America, who occupied one of Morgan's multiple-roomed structures (Lévi-Strauss 1982:163–87), and subsequently extended it to Austronesia and Africa where he saw 'common social-structural characteristics' (1987:152) underlying complex cognatic orders. A customary feature of 'house societies' is that they are hierarchical and facilitate the organization of relations to maintain rank, with a 'concentration of often considerable wealth or value at stake in the house' (Gillespie 2000:49). The parallels with medieval Europe are striking, where we talk of aristocratic houses (such as the Houses of York, Lancaster, Beaufort, Burgundy and Montbelliard), and where noble families used marriage and inheritance tactically to pass on titles, property, wealth and power, with associated tensions and rivalries within houses (Gillespie 2000:33). Lévi-Strauss sees similar tensions and rivalries in the celebrated *potlatch* competitive-exchange institution of the north-west coast communities.

There is no attention to houses as structures themselves: 'a striking omission ... is any detailed attention to the most obvious feature of houses: their physical characteristics' (Carsten and Hugh-Jones 1995:12). It is a glaring absence when *sociétés à maison* houses often demonstrate rank through grand architecture and spectacular ornamentation. When anthropologists turn their attention from largely sociological issues surrounding housing arrangements to the architecture of buildings, they explore it primarily for its symbolic content, assuming that it metaphorically conveys something about social relations and cosmological ideas (Humphrey 1974; Preston Blier 1987). Houses not only accommodate individual persons, they argue, but also cultural connotations. Buildings have meaning as well as purpose, serving to convey ideas in addition to affording shelter, reaffirming and passing-on culturally shared moral and social values. For some people in South-East Asia, for instance, the traditional stilted longhouse symbolizes the cosmos with heaven, earth and underworld equating, respectively, with the roof space, living floor and stilted underneath (Waterson 1990:93). Some commentators compare building designs with 'texts', and profess they 'read' from them unspoken (even unspeakable) cultural insights, exploring the meanings of being-in-the-world that persons negotiate as they construct living spaces that in turn shape their lives. They may account in this way for features of houses that have no apparent use, which inhabitants take for granted and cannot explain. They may also delve into the definition and creation of domestic spaces as locations that frame social interactions and perceptions; serving, for instance, as symbols of status, signalling wealth and power. These may extend to identity with place and, being an

3 Oliver 2003:7 distinguishes between dwelling as a social process (the act of residing) and physical structure (the building occupied).

4 The 'elementary structures' model speculates that cousin-marriage rules structure the equitable distribution of brides, consequent on exogamy (necessarily instituted from an evolutionary perspective for society beyond the family), which contrasts with 'complex kinship structures'.

aspect of belonging, attachment to other residents (often kin), and possibly feature in the assertion of cultural identity or revival.

A prime example is Griaule's (1975) controversial interpretation of Dogon buildings in Mali as expressions of a complex cosmology, seeing an esoteric anthropomorphic symbolism in the layout of houses and settlements. A house serves as a model of the world, creating a bounded place in universal space; for instance, the house plan represents a man lying on his side having coitus (a narrow passage entry corresponds to the 'sex organ', where liquid from crushed corn equates with semen), and the number of rooms stands for the persons 'issued from the ejected seeds', with the entire oval homestead representing the 'great placenta' from which everything originates. Several interpretations of architecture follow along similar lines, such as Cunningham's (1973) analysis of the design and construction of the Atoni house on Timor, which explores the systematic interconnections between social and cosmological ideas and spatial and living arrangements, with 'the house ... like a model of the cosmos' (1973:234). These associations feature an elaborate series of dualistic oppositions: we see the prominent gender opposition evident in Atoni life, for instance, in the association of males and females with features to right and left, outside and inside the house, respectively. He asserts that in this non-literate culture the construction and layout of the house serves to store and communicate ideas between generations: 'order in building expresses ideas symbolically, and the house depicts them vividly for every individual from birth to death' (1973:204).

Gender associations feature prominently in Bourdieu's (1990 [1970]) interpretation of the Berber house in Algeria, the layout of which symbolizes relations between women and men and mirrors ideas about the world generally, again with a series of the binary oppositions favoured by structuralism (water:fire, night:day, nature:culture). The lower, dark part of the house (adjacent to the stable and water butts) is associated with women and the upper, light part (adjacent to the fire and eastwards facing door) with men. These associations mirror the world outside, the house being 'a microcosm organized by the same oppositions and homologies that order the whole universe ... the house is a world within a world' (1990:277, 282). It symbolizes the sexual division of labour that structures much of Berber life. Inside the house is private space, dark and pre-eminently the female domain of activity, whereas outside is public place, the bright daytime sphere of male occupations. These domestic arrangements structure unspoken daily interaction –subsequently labelled 'habitus' (as discussed in Chapter 3) – serving routinely to socialize persons from a young age into Berber moral and social values.

Structure and semiotics

A similar dualistic interpretation could arguably be made of Was valley residential arrangements, with their men's and women's houses, which returns us to the gender questions with which this book opened. The *enzatenda* 'men's-women's house' described in Chapter 7 captures the arrangements in a single structure. It comprises at the front the bright men's room open to the daylight, facing outwards onto a *howma* 'open-space', where many public events occur, such as marriages and funerals with associated exchange transactions. In opposition, at the rear, is the more private, dark and enclosed women's room together with pigs' stalls that signal the largely unseen female contribution to productive activities that indirectly underpin public activities that are to some extent male centred. The interpretation can be taken further, the residential arrangements symbolizing men's exploitation of women, whose contribution to esteemed communal activities is kept literally and metaphorically in the dark (after Josephides 1983, 1985; Modjeska 1982, 1995; Strathern 1987). The theme is further evident in the ground-plan of the *timp* house. The earth oven in the small enclosure at the rear of the building was used to cook the parts of slaughtered pigs (such as the intestines, stomach and belly-pork strips) that usually go to women, who are excluded from cult activities. According to men, the participation of women and children is prohibited because such rites are dangerous, as they involve interaction with *towmow* 'ancestor spirits' that attack persons and make them sick, even kill them. According to some, if women and children were present they might interfere with the rites, threatening the entire community. It is noteworthy that woman subscribed to the same taboo beliefs and co-operated in arranging the ritual, reflecting collaborative gender relations discussed below (Stewart and Strathern 1999).

While no one has ever suggested that the design of houses or other structures has any such symbolic meaning for them – in the way that churches in England, for instance, have overt symbolic import with a Latin-cross-shaped ground-plan and the altar at the top end of the cross oriented to face east – nonetheless, it is possible to see the gendered accommodation arrangements, which are an aspect of the distinction between female and male domains that is a prominent Wola cultural concern, as symbolizing something about relations between the sexes. Those of a 'constructivist', 'interpretivist' or 'symbologist' persuasion (Geertz 1993; Gell 1998; Munn 1986; Wagner 1981) may even go so far as to think that the gendered layout of houses indicates a 'narrative' about how the Wola 'see' their 'society', that it communicates cultural intelligence, as pointed out earlier. Some commentators claim that people use such symbolic discourse to engage with issues in a way similar to social scientists and, in seeking to establish parity with sociological theories, maintain that they can detect parallels in other cultural milieu – deploying unfamiliar symbolic codes, such as mythical narratives, ritual acts and designs on things (as mentioned in Chapters 3 and 4). In according actors a capacity to represent their social behaviour in such terms, they somehow penetrate tacit awareness, attributing to people explicit – albeit unuttered – knowledge of their social models, for instance the implications of arrangements between the sexes for wider socio-political relations. Such interpretations are controversial in their claims to reveal these unusual intellectual 'discourses', involving the interpretation, by those gifted with the necessary insight, of exotic symbolic codes that somehow – albeit often obfuscatingly and surprisingly – parallel verbalized propositional reason.

Prominent here is structuralism, associated with the work of Lévi-Strauss (1968) among many others, which claims to access deep unconscious structures that are interpreted as humans wrestling with universal conundrums through symbolic play, juggling for instance between contrasting binary oppositions, as in Bourdieu's dark versus light interpretation of a Berber house, extended tongue-in-cheek to a Wola house. The structuralist approach fell out of favour when the dubiousness of its methods became apparent with its cavalier use of ethnographic evidence, to be replaced by other 'isms' such as post-structuralism or social constructivism. The notion of communication remains prominent, albeit with interpretation more circumspect. According to Geertz (1993:14), for instance, 'the aim of anthropology is the enlargement of the universe of human discourse', and he recommends the deployment of a 'semiotic concept of culture' to achieve this end. Social-constructivist approaches seek to further our understanding of how humans constantly create and know their social worlds, acting on their diverse symbolically coded interpretations and understandings to affirm and reproduce their social order. The previously mentioned argument about things serving to corroborate social-identity claims or express aspects of personhood overlaps with such approaches to cultural transmission.

The issue from an ethno-determinist perspective is that my Wola friends have never made comments that intimate such dualistic symbolic connections between their domestic arrangements and gender relations. If any such apprehension exists, it is deeply buried in tacit house-design code, because attempts to encourage people to discuss these arrangements never get far and certainly nowhere near such semiotic interpretations. They are not unusually obtuse, but illustrate a persistent evidential problem with such imaginative conjectures: how can we check their veracity when by definition the symbolic communication, if it occurs at all, is unspoken (unspeakable even) and unconscious? It appears that such idealistic constructivist interpretations of domestic arrangements present us with the same problems of access as other tacit knowledge, including that of positivistic embodied experience that features in the construction of houses.

It seems that to interpret the architecture of Wola dwellings as symbolizing relations between the sexes is akin to seeing the British arrangement of ladies' and gents' toilets as signalling something about gender relations, which if so, as a native I presumably hold as tacit intelligence (just as my Wola friends accept their living arrangements), as it has not occurred to me previously. Regarding this chapter's opening distinction – between the semiotics of overall architectural design and the symbolism of particular features associated with structures – these physical arrangements are of a different order to the symbols put on the doors to indicate which toilet it is appropriate for you to enter depending on your gender. These can carry a significant information load for those with the necessary cultural intelligence, such as the astrological Venus ♀ and Mars ♂ symbols that originate from ancient Mediterranean society.[5] The former signifies a goddess, the moon's daughter, whom we associate with femininity and such traits as fertility, love, beauty, serenity and emotional empathy, and the latter signifies a god, the sun's son, whom we associate with masculinity and such traits as aggression, ego, courage, strength and physical energy. In the West today, the Venusian symbol may be associated with militant feminism and all that implies historically and culturally, and the Martian one with the less prominent 'masculinism movement'.

So what are we to make of Was valley connections between architectural arrangements and gender issues? The ethno-determinist perspective prompts a return to the ethnography.

Gender and houses: interpretation or explanation?

The actions of people following the construction of a house to ensure that it becomes warm and cosy further reflect attitudes about male-female relations. In addition to the practical steps taken lighting fires to warm up and dry out the damp clay floor and green building materials of a 'raw' newly built house, as described in Chapter 7, house owners place an *aend showaip* 'house taboo' on the building, prohibiting any married persons, other than the residents of the house, from approaching it for 'one moon'. This *aend tinjiyay* 'house cover-up' is necessary to prevent anyone who has recently had sexual relations from approaching the house, because if they did so it would not dry out properly but remain chilly and inhospitable. Indeed, from the time they start work on the house, no individual should approach who has recently engaged in coitus, and the builders should remain celibate too. The residents should also abstain from sexual relations for 'one moon' after occupying a new house.

5 Or as in a Tynemouth pub with naval associations, and a landlord with an amusing sense of metaphorical play, a buoy and a gull on respective boys' and girls' loo doors.

Plate 10.1 Cane-grass screen erected in front of new house to protect from sexually contaminated stares.

In order to protect their houses from the gaze of those 'carrying sexual contamination', anxious owners sometimes erect high screens of cut cane grass around their new homes, particularly *enza* 'men's houses' and those adjacent to heavily used footpaths.

This taboo on sexual relations around new houses is one of a range of injunctions that hedge about interactions between women and men (Reithofer 2006:83–99; Sillitoe 1979b). They are of a piece with the design of houses and living arrangements in the Was valley, which signal to even the casual observer something about male-female relations and differentiation between their domains – albeit gradual change has occurred in accommodation arrangements (Sillitoe 2010a:225).[6] One reason that people give for the gendered separation of living arrangements, as noted previously, is men's fear of associating too closely with women, who they believe may make them sick, even fatally ill, particularly if they come into contact with menstrual blood. These beliefs about pollution are a feature of the discrimination between women's and men's ambits – note that the discrimination is a consequence, not necessarily a reason for their existence in a functional sense. The design of houses in the Was valley is not so much a symbolic statement about relations between men and women, as an aspect of the constitution of these relations, necessary to ensure men's well-being (as real as our commitment to sanitation systems for health). When they comment on the separation of domestic space, both men and women talk about protecting the 'health and well-being of men' and ensuring 'warm and dry' versus 'cold and damp' houses, which they take literally to be the consequences of breaking the taboos that surround the building and occupying of houses. The gendered arrangement of accommodation is a reflection of men's genuine fears of being fatally polluted by menstrual blood and the steps taken to reduce the risk. The fears are more than symbolic to the actors, which is not to deny that they convey something about gender relations (Sillitoe 1979b).

It would, of course, be ethnocentric to assume that the state of affairs must be symbolic because we think the fears are actually erroneous. The reason why people came to believe that a non-toxic substance is a pollutant is unknown. We can speculate that sometime in the distant past a man (or men) fell sick and died after some contact with menstrual blood, and that this led to fears

6 This change does not invalidate knowledge about what pertained previously and the political-economic implications, as I argue, particularly as these relate to still extant labour arrangements.

that became widely recognized and passed on down the generations. Once culturally established, we can appreciate how such fears may pass from generation to generation: for instance, I have no first-hand experience of strychnine being poisonous, but I believe that it is, as I have been told so, and I have no wish to experiment on myself to see if it would harm me. Regarding menstrual blood, my experience contradicts Wola beliefs about contact with it. This relates to a methodological dilemma, namely how being unable to subscribe to people's beliefs has implications for participant-observation, explored here as a way possibly to access tacit knowledge (Chapter 11). Several men, concerned for my welfare, tried to persuade me to live separately from my wife, and when we started to build the *enzatenda* house at Honael with its separate male and female accommodation, some of our neighbours concluded that they had convinced me of the danger of us living together and were disappointed when a friend and his family took up residence there and not us – although no one remarked upon it, the pig pens indicated our intentions, particularly in view of my disastrous foray previously into pig keeping (Sillitoe 2003:257–60) and my wife's sensible refusal to be part of that escapade, whose supposed 'quarters' contained the pens. It is a moot point to what extent these differences may distort my understanding of Was valley domestic arrangements, notably their experiential aspects.

If interpretation is a dubious way to further our understanding of such beliefs and associated housing arrangements: what next? There is explanation; notwithstanding that reference to explanation in the social sciences has been unpopular since the latter twentieth century. 'We are concerned with interpretation, not explanation', as a colleague put it to me. Since she drew the distinction to my attention several years ago, I have periodically puzzled over what it means and the implications. Addressing these questions affords an opportunity to deal with criticisms of discussions of gender and housing as overly narrow, such as Souvatzi's (2012:178) about these being 'portrayed as a part of larger dualistic symbolic structures, while extra-domestic spaces and the interlinkages of the house with wider social, economic, and political scales and conditions are much less fully explored'.

Interpretation, as we have seen, involves the translation of people's cultural arrangements to reveal the issues that they think central to their being-in-the-world and their perspective on associated socio-cultural issues. While actors often do not discuss such matters, the assumption is that they encode them culturally in other non-verbal ways that are accessible to those with the necessary translational skills. The problem is that unlike translation from one language to another, which can be tricky enough, there is no way to check on the veracity of any one interpretation against another – we cannot compare the equivalent of one consciously composed string of sounds that have a culturally agreed meaning with another string. When social constructivists claim to access the hidden meanings of things they consider to be symbols, they are claiming to access tacit knowledge, for 'hidden' implies that the import of the symbols are consciously unknown to the actors, who cannot excogitate on their meanings. The problem is that such tacit knowledge, as pointed out, cannot by definition be 'excavated' – i.e. put into words. Indeed it is questionable whether or not such supposed symbolic knowledge exists, as opposed to being an introspective imposition by outsiders.

Explanation, on the other hand, involves the deployment of assumptions and concepts that we think relevant to further understanding of socio-cultural issues, often heavily informed by the *théorie du jour*. It is reductionist in focusing attention on particular concerns thought important to understanding humanity, reducing the complexity of socio-cultural interaction to these issues – such as the evolution of cultural arrangements to further human progress or the functioning of society to ensure survival or order or exploitation or whatever. The boiling down of enquiries to a narrow range of concerns is arguably inevitable given the complexity of human behaviour, as we cannot process all of it, as noted previously regarding holistic aspirations. This approach poses questions that may well be foreign to actors – such as 'what are the implications for this community of having elaborate exchange institutions?' It may assume, for instance, that gendered domestic arrangements have political-economic implications that an outsider can 'see' or hypothesize, of which the actors themselves are again unaware. It advances etic explanations as opposed to questionable, possibly bogus, emic interpretations, which are not truly emic, coming from social scientists who claim to talk about what actors cannot, namely unverbalizable knowledge. Ethno-determinism strives to derive the questions from the evidence of what is heard and seen, being upfront about imposing explanatory frames and the challenge of accessing tacit knowledge.

The explanatory approach seeks to situate what it takes to be symbols within the wider context of associated cultural activities, so as to explicate why actors deploy them in the way they do and the implications for their behaviour and social system, assuming that they do not do things for nothing. According to one school of thought, largely structural-functional in origin, symbolic aspects may contribute, for instance, to the sociological effects of actions that bring communities together, shared beliefs and associated symbols and behaviours acting to promote social cohesion. They help to generate it through what Turner (1969) refers to, in a sophisticated presentation of this position, as a spirit of 'communitas'. In short, they may serve a social-solidarity function. A more fashionable version argues that symbolic activity serves to affirm identity or express aspects of personhood, as in consumer-focused studies of things. In the Was valley, any collective-unconscious import in respect of social solidarity involves not local group identity – an inappropriate way to conceive of the socio-political order – but a more diffuse network reinforcement, which does not necessarily invalidate the hypothesis, although it may in part explain why actors are unaware of such effects with no clearly defined group to focus solidly on. The implicit assumption behind this explanation of symbolic behaviour is that it emerged during human interaction long ago, when ancestors sort of stumbled across it without consciously knowing it – and people have behaved thus ever since.[7]

Nonetheless, while anthropological accounts show how houses may serve as metaphors for society and cosmology, and relate them to the human body in discussing their role in reproducing kinship relations and gender distinctions, they have been 'slow to move … beyond symbolic representation and toward a more actor-oriented approach … the effect tends to be a rather inactive (and timeless) vision of the house as a rather static assemblage of meanings and metaphors' (Souvatzi 2012:178). In addition to overlooking individual tacit agency, they also tend to present arguments as either interpretative or explanatory. The views expressed in the Was valley about living arrangements ensuring well-being draw, for instance, on local ideas about the human body – albeit the focus is not so much on implicit embodied knowledge as explicit bodily knowledge (as evident in Strathern's 1996:65–76 discussion of similar fears, widespread in New Guinea, of menstrual pollution). These architectural features may both constitute people's culturally informed awareness of the body and symbolize aspects of it, thus inviting both explanation and interpretation. The Wola can explain that their gendered residential arrangements keep men's bodies healthy, and they may further interpret the leafy screen erected around newly built houses as a sign to others to avert their eyes and keep away, so as not to risk contamination, rendering them cold and damp. In a similar way, drawing again on this chapter's opening distinction between the symbolism of general architectural layout and specific structural features, we can explain that in England persons who wish to smoke have to do so in shelters outside public premises in order to protect the pulmonary health of others, and we may further interpret the red circle symbol on doors (with a diagonal red line through a squiggle-ended black line depicting a cigarette) as a signal to smokers that they cannot light up.

Social scientists may go further, as pointed out, in explaining or interpreting the often tacit implications of such behaviours and ideas – oft times one assumes flying kites. A widely cited example is Mary Douglas' (1966) purity and danger argument – which parallels Durkheim's and Mauss' functional ideas about the human body and social order (Strathern 1996:9–13) – which posits that pollution beliefs serve to mark out socially important boundaries that persons should not transgress because social disorder is likely to ensue. According to her, we find 'symbolism worked upon the human body. The body is a model which can stand for any bounded system. Its boundaries can represent any boundaries which are threatened or precarious' and it is not possible to interpret beliefs and behaviours that centre on bodily fluids such as menstrual blood 'unless we are prepared to see in the body a symbol of society, and to see the powers and dangers credited to social structure reproduced in small on the human body' (Douglas 1966:115). The fears Wola men have of menstrual blood and associated rites are amenable to interpretation in this vein (Sillitoe 1979b),[8] which shows that incorporation of the individual in the analysis of Was valley social arrangements does not imply eschewal of the classic, collective, sociological focus, or any other that offers insight so long as, from an ethno-determinist perspective, it remains true to the ethnography.

Sexual division of labour

The beliefs and fears that underlie the architectural layout of houses resonate with the sexual division of labour, touched on in previous chapters, which structures many tasks in the Was valley, including house construction. It is possible to see men's fears as reinforcing these labour arrangements, with significant political-economic consequences for connections – or rather disconnections – between production of subsistence needs and acquisition of transactable things, which relate intimately to the acephalous order and the more-or-less equal distribution of power across the community (Sillitoe 2010a).[9] If asked why there is a gendered division of labour in house building, people are likely to point out that it involves the use of axes and that as only men can use these tools they are responsible for the construction. They believe that women can blunt an axe just by handling it (this is one of several taboos that relates to the debilitating effects that women can have on men – Sillitoe 1979b). They may also add that it is a *buriy* 'strong' task, implying that such a heavy undertaking requires men's physical strength.

7 The alternative idea seems improbable, namely that in the distant past some wise ancestors figured it out and overlooked to pass on the explicit knowledge, which became implicit.

8 But not in the way that Douglas (1966:146–8) does, who discusses Meggitt's argument that Mae Enga fears of menstrual pollution result from men marrying women from enemy clans, which does not agree with my understanding of Was valley socio-political arrangements.

9 I have previously depicted these male and female domains in terms of transactors versus producers, although I have subsequently dropped this as too stark, both men and women being involved in transaction and production (Sillitoe 2010a:364).

The Was valley labour arrangements suggest an alternative view of the division of labour to those of sociological studies of the topic such as Émile Durkheim's *On the Division of Labour in Society* that continue to hold sway, and centre on aforementioned solidarity issues (Chapter 2). These arrangements afford an opportunity to follow up on the relevance of sociologically framed assumptions to the Wola political-economic order critiqued in previous chapters. Focusing on the occupational division of labour, sociologists argue that the resulting specialization, increasing the dissimilarities between persons, consequently intensifies their interdependence within increasingly 'complex' societies, what Émile Durkheim (1933) called 'organic solidarity' and subsequent sociologists have elaborated on with further jargon such as Smelser's (1959) 'structural differentiation'. This contrasts with other, 'simple' societies, that feature what Durkheim called 'mechanical solidarity', where social order depends on strong feelings of collective conscience that prompt individuals to subsume their interests to those of the community, which severely punishes those who violate its collective expectations and interests; in contrast to 'organic' societies that supposedly focus on legal rights rather than punishment. The distinction has a strong evolutionary flavour, with mechanical solidarity characterizing 'primitive' societies and organic solidarity 'advanced' societies paralleling Herbert Spencer's (1969) view of social change as a progression socially from homogeneity to heterogeneity and continuing today in modernization views of development that focus, after Smelser, on the integration of increasingly differentiated social arrangements that result as specialization increases.

According to these views, advanced 'organic' societies represent moral progress with stress on values of individuality, equality, liberty and justice. Attitudes and behaviour in the Was valley, where mechanical-like solidarity should pertain, are quite at odds with these notions, up-holding the values of equality and liberty beyond anything imaginable in states that feature occupational specialization. The collective conscience certainly does not smother individual sovereignty, which is not to deny that behaviour is subject to guidance by shared values. When I refer to the freedom of the individual, I am not talking about the philosophers' imagined 'freedom in state of nature', as I point out in Chapter 3, but liberty relative to that experienced in hierarchically organized states. No human being is totally free, in the sense of being free of social constraints. The Wola, like all of us, are born into cultural contexts that inculcate them with values and expected codes of behaviour – they do not make these up for themselves, although they may self-interestedly try to manipulate them.

The expression of political freedom in the Was valley brings to mind Spencer's (1940) advocacy of an individualistic approach to social life and associated laissez-faire policies, whose final evolutionary stage (which predicts a sort of stateless arrangement with the disappearance of central government coercion) ironically intimates the Wola acephalous order – as opposed to the supposedly primitive homogenous society at the bottom of his social evolutionary tree. Intriguingly, coming from the opposite left-wing political pole, Marx's final communist stage reflects a similar acephalous order. And to pile irony on irony, both models give us circular rather than linear evolution. The state and its laws inevitably interfere with liberty, Spencer argued, and are dispensable and would decay, their coercive aspects superseded by voluntary market association. For him liberty was not a question of 'the nature of the government machinery ... [but] the relative paucity of the restraints it imposes' (1940:19). It is necessary he thought, following early liberalism, to overcome central government and its laws that restrict and coerce individual action.

But his 'law of equal freedom' sacrifices equality on the altar of freedom, as the consequences of capitalist-liberal nostrums, such as he advocated, increasingly show today with the ego-centric let-the-market-rip philosophy resulting in an evermore unequal society with a few wealthy and many poor families. The genius of Was valley political-economic arrangements is that they accommodate both liberty and equality; as pointed out in Chapter 8, they achieve what seems impossible in state orders, where these two values confront one another in an either/or relationship – liberty as a core right-wing value and equality a core left-wing one. The valley's residents, on the other hand, are socialized to have the same high regard for both liberty and equality, which predisposes persons to behave in particular ways with certain consciously unacknowledged outcomes that I seek to tease out. According to the sociological view, the uniformity seen in tribal communities reflects 'primitive mechanical solidarity' rooted in the similarity, even homogeneity of individual members of society, while 'advanced organic solidarity' draws on their dissimilarity or heterogeneity. Regarding social consensus in acephalous face-to-face society, the Wola point to egalitarianism as a significant factor rather than anything that intimates individuality-stifling mechanical solidarity, as persons do not expect to differ markedly from one another with the potential for unequal relations, as shown by housing arrangements discussed earlier.

While the egalitarian-value-fostered cultural consensus may promote a certain social solidarity in the sociological sense, this is not because it subsumes individuality under some collective unconscious. Indeed the reverse, the egalitarian ethos promotes individual sovereignty, ensuring the equal distribution of power among all, such that no institution or its office-holders are able to dominate others. Here are egalitarian values that simultaneously accommodate a wide scope for individual freedom of action. They afford considerable latitude for individual variation in behaviour and understandings, with no authority figures to determine what is right and wrong (Sillitoe 2003:111), but the variation is contained by collectively held values and axioms within broadly defined socio-cultural parameters shared by all. They paradoxically allow and yet simultaneously inhibit expression of individuality. The egalitarian values that underpin the same individual freedoms for all are the ones that inhibit expression of individuality in ways that may lead to significant differences between persons coming about with the possibility of some parties wheeler-dealing and securing more power than others, which we may speculate would convert over time into authoritative offices within a hierarchical social system.

This is not to disallow that egalitarian-value-fostered socio-cultural similarity may promote social solidarity in the classic sociological sense – albeit it is necessary to exercise caution with the solidarity hypothesis, given its questionable association with the long-running debate over the big-man led descent-group view of New Guinea Highland's society (Sillitoe 2010a:45–6). But the reverse is equally feasible, namely social homogeneity could undermine solidarity, notably with autarkic households able to supply all their own livelihood wants, their interaction atrophying with all families the same and having no need of others. So what is responsible for 'social solidarity' in the Was valley, if it is not a ruthlessly enforced mechanical process? Dissimilarity is significant here, notably the different lifeways of women and men, as instituted culturally by the sexual-division-of-labour arrangements. Their preoccupation with separate domains paradoxically promotes social interaction. The gendered activities, concerning rights and responsibilities, have a strong moral dimension, with the complementary roles being a significant aspect of relationships. They are arguably identifiable in cultural arrangements more widely, such as in certain ritual activities according to Stewart and Strathern (1999), who propose a 'collaborative model' for female spirit cults. The division of activities along gender lines creates relations of mutual dependency, women and men attending to different but complementary domains, both central to everyday life. They comprise a partnership; no man or woman can properly exist socially alone.[10]

The partnership between women and men extends outwith the family across and beyond the Was valley. The ramifying networked interaction that characterizes Wola social life (Sillitoe 1979a) rests on the establishment of interrelated families through marriage, women marrying men from other families, and their daughters marrying sons of different families again, and so on down the generations with the consequent constant adding of new relations and social interaction as previous ones die off – i.e. the result of exogamy as Tylor (1889) pointed out many years ago. Incidentally, in respect of the gendered structuring of production, there are intriguing – albeit locally unremarked – parallels with female and male roles in reproduction. Women continuously contribute blood and flesh to a foetus growing in the womb, whereas men only do so intermittently during acts of coitus, their semen furthering the growth of bones. Similarly, men undertake productive tasks intermittently, often featuring heavy bursts of labour – as in house building (being involved in the construction of perhaps one or two houses a year) and in the pioneering of new swiddens – tasks that involve the erection of the bone-like frames of houses and fences. Women, on the other hand, contribute continuously to the tasks that fall to them – such as swidden cultivation, undertaking gardening activities daily and netting artefacts such as bags[11] – supplying the food and many of the things their families require to grow and sustain themselves (Sillitoe 2010a:258–329).

Marriage involves protracted bridewealth transactions, which brings us to exchange institutions that are central to social interaction and the obviation of the contradictions apparent in Was valley socio-political life. The way that egalitarian values simultaneously underpin individual freedom for all and inhibit expression of individuality attests the contradictions, thwarting the advent of large differences between persons with potential power differentials, which is a manifestation of the wider individual versus collective interests antinomy. Socio-political exchange institutions, I argue, circumvent the paradox by simultaneously allowing for the sociable expression of collective group-interests (Émile Durkheim's social solidarity) and competitive expression of individual self-interests (Adam Smith's self-love – Chapter 3). Deploying the notion of social solidarity, whether it be thought mechanical, organic, transactional or whatever, questions again the status of our understanding of social interaction when actors themselves do not conceive of the implications of their behaviour in such abstract terms, for instance with respect to social cohesion. Solidarity is a product of their tacitly informed social interactions; it happens without aforethought. In other words, the sexual-division-of-labour arrangements may have political-economic consequences that I, as an outsider, can 'see' or hypothesize, of which the actors themselves are unaware. Similarly, in accounting for what I have heard

10 The consequences of weakening such interdependent relations are evident in Britain currently where marriage partnerships are fragile as a result – not being partnerships at all in the sense that the sexual division of labour promotes partnership – and many families collapse, portrayed in the popular media as a 'breakdown of society' (which is not to condone previously unfair patriarchal arrangements but to question the assumption that the sexual division of labour necessarily implies such relations).

11 In some places people equate women's string bags with the womb and reproduction (MacKenzie 1991), although no one has ever made such a connection directly to me.

and observed during behavioural interactions generally in the Was valley, I address political-economic problems that occur to me, which emerge from my intellectual cultural heritage, while for the Wola associated behaviour is tacit. It is what persons expect to do, for instance in exchange contexts, without pondering issues of liberty and equality, which I underline by borrowing from Adam Smith and referring to 'the invisible hand of exchange' to characterize what occurs during transactional sequences in the Was valley (Sillitoe 2010a:83).

Equal division of power

The division of labour arrangements among the Wola not only challenge orthodox sociological explanations, but also those of economics, such as Adam Smith's *Wealth of Nations* and Karl Marx's *Capital*, whose arguments remain influential. The sexual division of labour effects radically different outcomes to the occupational division of labour, prioritizing other values. The specialized division of labour by occupation, where workers are responsible for different tasks in producing something, may increase productive efficiency and output – as demonstrated by industrialists such as Henry Ford who took Adam Smith's (1993:4) pin-making example to its logical conclusion in establishing the soul-destroying factory production line – and, so it is argued, increase the wealth of nations with market competition ensuring the efficient distribution of resources and products. But as Karl Marx (1974:561–70) pointed out, the resulting capitalist arrangements are amenable to control by a few, who, consolidating their power over others' lives, exploit their labour and enrich themselves at their expense – exemplified in the contrast in England between colonnaded mansions and terraced houses. The sexual division of labour, on the other hand, structures production in such a way that it averts any party gaining such control, promoting unequal relations and curtailing the liberty of the majority, as I have argued previously (Sillitoe 1985, 2010a:365).

Women and men expect to be responsible for different tasks in the Was valley. The household, the basic unit of production, depends on them co-operating to supply its needs. While individuals may sometimes take on the tasks of the opposite sex – in emergencies, for instance, if partners fall sick – they find it awkward and normally adjusted persons do not do so ordinarily. Avoiding one another's activities in this way contributes to people not seeing the potential that extending control over production might afford some persons to increase output, and beyond this discovering the latent power possibilities regarding others' material needs. Every independent household has similar productive capacities, and consequently the same material standing and means-based access to power, so maximizing individual liberty equally for all, after Hobbes' dictum that liberty is power divided among the many, which is in keeping with an egalitarian order where persons do not expect to differ in these respects – as seen for instance in all occupying houses of similar design. The labour arrangements also keep their activities within certain limits, which, together with the absence of any idea of economic growth, have environmental implications regarding the sustainable use of natural resources, as discussed previously (see Chapter 6).

If someone aberrantly decided to increase production with a view to amassing surplus capital, this would mean pushing other household partners to increase their efforts, which they would likely resist; for instance, women who find their spouses too overbearing may leave them and return to their natal kin, who will support them if they agree that the demands made on them are unwarranted (Chapter 3). This can be embarrassing for a man, effectively undermining his household's productive capabilities and cost him something in compensation in negotiating his partner's return. Anyway, the use of surplus as productive capital is a foreign idea; any extra output beyond subsistence requirements finds its way into the exchange domain as transactional capital, where it is stripped of productive associations (Sillitoe 2010a:438–40). A production focus is misplaced, as illustrated by polygynous households, which are often those of *ol howma* 'men of renown'. While they arguably have more female labour at their disposal, they do not use it to increase their productive capital as such (which in part assumes they pioneer and cultivate larger garden areas, albeit women may till and plant areas cleared by other male relatives too, such as those of fathers, brothers and sons – Sillitoe 2010a:187–96). Rather they increase their social capital, for being married to more than one woman increases men's transactional responsibilities and much of the added female labour contributes to meeting these exchange commitments, with wives concerned to see that their husbands deal fairly with their kin. Relations between the sexes make it improbable that a couple would hit upon the idea of increasing their household's production with a view to accumulating a surplus that they might convert into political capital, certainly not extending beyond the customary transactional arena. Even today, when exposed to the capitalist market and experiencing its control over the supply of things necessary to meet material demands, if an eccentric couple did hit upon an Anthony-and-Cleopatra-style bid for power, they would encounter other checks and balances of the acephalous political-economic order, such as the equal access all have to the materials and capital they require to meet their needs locally and the independence of households in supplying their wants. In other words, there would be no one to do business with.

Arguing for a reversal of Marx's portrayal of the implications of the division of labour in capitalist contexts – that the sexual division of labour obviates control of production and the exploitation of many by few – may seem wrongheaded from a feminist perspective, which sees instead an arrangement that exploits women and promotes inequality and dominance, discounting as flawed the identification of egalitarian relations and individual sovereignty as key values (Biersack 2001; Godelier 1986; Jolly 1987; Josephides 1983, 1985; Kelly 1993; MacKenzie 1991; McDowell 1990; Modjeska 1995; A. J. Strathern 1966, 1982, 1987; M. Strathern 1988; Wiessner 2002).[12] But this depends on how you judge what women do compared to what men do (Sillitoe 2010a:330–64), which in capitalist contexts concerns access to money that makes participation in the market possible. The occupational division of labour allows scope for differential ranking of employment, some jobs judged as superior to others and better rewarded; an aspect of hierarchical power relations. Occupations that fall to women often earn less, and domestic work undertaken within one's home, which until recently fell largely to women, brings in no income, which effectively disempowers housewives, having no earnings and control over money, which is central to life in market society. Feminists argue that women should not only have a place in the waged workforce (which many have had for generations), but also that they should do the same work as men for the same pay. This may reduce gender inequalities, viewed from an individualistic perspective, but it reproduces capitalist society's unequal exploitative arrangements; increased numbers of women in wealthy and powerful controlling positions potentially reinforces the divide between rich and poor families, which becomes even larger when there are two high-income earners in a household. The Wola political-economy does not countenance such differentiation in household assets.

It is questionable to seek dominance and exploitation elsewhere, in relations between the sexes; in respect of what women and men do, for instance, under the sexual-division-of-labour code. It is problematic to rank some tasks above others, where we have no income yardstick to measure them by, money being absent in the Was valley until recently and cash featuring minimally today in the local subsistence economy. By what criteria might we identify activities that fall to women as inferior or harder, and argue that men are exploiting their labour in allowing them to be responsible for such tasks? If the local community does not think of activities in this way, it is difficult to avoid the charge of ethnocentricity in doing so, according to what we may judge to be more laborious, less rewarding work. The awkwardness is evident if we reverse things and imagine arguing that women exploit men's labour in the construction of houses. Indeed, the idea that some activities are work and others are not may itself be doubtful (Sillitoe 2010a:351–64). Furthermore, it is untenable to have recourse to the false-consciousness argument – which maintains that people are unable to understand the real nature of their circumstances because they are subject to ideological control by powerful institutional processes that they are unaware have a vested interest in preventing them from seeing things as they are (whereas the more sophisticatedly aware can do so). This relates to tacit awareness of behaviour again. Turning to false consciousness, when local actors' judgements do not match ours, raises starkly the conundrum of outsiders claiming to understand people's lives better than they do themselves, when their thoughts and behaviour are the subject of enquiry. It also affords licence to argue almost any ideological point without need of evidence, which clearly conflicts with ethno-determinism, underscoring the dictum about the need for empirical corroboration that can stand review.

While some men may talk disparagingly about women's occupations, as some women refer disdainfully to what men get up to, this does not indicate exploitation of one sex by the other. Indeed, these attitudes arguably buttress the classless order by discouraging either sex from aspiring to take on or direct what the other does, which could undermine the egalitarian political-economic arrangements, allowing some to seek to control the productive process. It is misleading to adopt a Marxist-like line regarding domestic arrangements: the reverse actually applies in the Was valley. The involvement of women and men with defined domains does not facilitate but instead checks exploitation, discouraging some from trying to outdo and dominate others through production. The relation between the sexes, which is central to the articulation of production with the exchange domain (Sillitoe 2010a:449), features a mutual dependency that counteracts any tendency for some parties to seek to control productive resources at the expense of others, or it even occurring to them. All men recognize that they are similar, as do all women, and while men and women think of themselves as dissimilar – as endorsed by the sexual-division-of-labour arrangements – they mutually depend on one another and enjoy equal everyday liberty within the inevitable constraints of their social order.

The upholding of individual autonomy that is central to the polity applies equally to both sexes. There are curbs that thwart either partner becoming too domineering, which give women, for instance, discreet ways to handle outwardly overbearing male behaviour, such as undue interference in their activities, pushing them to work harder and produce more, which is tantamount to an initial step towards exerting control over the production process. The fears that men have of menstrual pollution, which inform

12 It would be foolhardy, in view of New Guinea's well known cultural variability, to suggest, in collectively citing these authors, that they all hold the same views; they will disagree on many points having worked in different regions at various times, and some have changed their interpretations over time, but they all think women are unequal to men who dominate them, occupying positions of relative social prominence, as seen at public events.

separate housing arrangements, are one such potential lever, albeit more symbolic than actual, as a woman is unlikely to be reckless enough to threaten to pollute a man openly, which would be tantamount to threatening to poison him, but she can make oblique allusions sufficient to warn off a man whose behaviour is unacceptable. There are many examples of females acting freely in ways they choose, sometimes in a manner that is not in the interests of their male partners, such as a wife temporarily withdrawing her labour from her household because she thinks that her husband's behaviour is unreasonable, moving in with her brother's or son's families, for instance, who will support her if they agree with the grounds of her protest – it often costing husbands a compensation payment, as pointed out, to right matters and have their wives, and maybe children, return to them. This freedom of action even affords a wide scope for the few mavericks or eccentrics in a community to behave abnormally, such as engaging without any apparent embarrassment in activities customarily the responsibility of the opposite sex (Sillitoe 2010a:353) – such infrequent exceptions, making others feel uncomfortable, testify to the salience of following the sexual-division-of-labour arrangements.

Incidentally, this discussion of acephalous political-economic arrangements underscores points made about the interpretation of archaeological remains in earlier chapters. While the level of ethnographic detail is unimaginable in prehistoric contexts, it is noteworthy that the sexual division of labour is common across the tribal world and it may pay to look in excavations for evidence of such organization, which occurs in a range of cultures and regions (from the Americas to Africa, Asia and the Pacific). These gendered cultural arrangements comply with cornerstone features of stateless political-economic orders, while other commonly made assumptions, such as those evident in the seminar discussions – that communities have chiefs who wield power, or anomalous objects are evidence of rituals managed by priestly authority figures – do not.

Empiricists versus idealists

After this foray into explanation in contrast to interpretation, it is worth briefly recapping the implications of this time-honoured distinction. The connection between, and weight given to, evidence and imagination in furthering our understanding of issues relates to a schism mentioned in Chapter 2, which extends far back in the history of Western thought, from ancient times, as illustrated in Plato's 'ideal types' versus Aristotle's 'logical reason', up to the current day. On the one hand we have metaphysical arguments, where we are asked to have faith in unverifiable idealism, pure thinking or innate ideas – such as Christianity's 'God', Hegel's 'Absolute Spirit', Kant's *'noumena'* or Heidegger's *'Dasein'*; and on the other hand, empirical arguments, which demand observed evidence from the world 'out there', that we can assess with our senses to back up assertions – such as Locke's *'tabula rasa'*, Hume's 'scepticism', Peirce's 'pragmatism' and Dewey's 'warranted assertions'. The opposed twentieth-century viewpoints labelled 'Continental philosophy' (Schopenhauer, Husserl, Heidegger, Nietzsche, Sartre, Foucault) and 'analytical philosophy' (Frege, Peirce, James, Russell, Wittgenstein, Ayer, Quine) mirror the distinction between these two approaches.

Recent philosophical debates react to the emergence and growing power of natural science; as seen in the impact of systematic enquiries into physical phenomena and startling technological developments. The 'Continental' response to science killing off god, cited in Chapter 2 (for instance in the rubbishing of the Genesis account by the theory of evolution – epitomized in the confrontation between Thomas Huxley and Bishop Wilberforce at the 1860 British Association for the Advancement of Science annual meeting), is to address questions such as: 'why are we here?' and 'what is morality?', and relies heavily on idealism and pure reason in the metaphysical tradition. The 'analytical' response is to model philosophy according to scientific logic and enquires into the character of our understanding, for example focusing on the meaning of words and their relation to empirical reality, and relies heavily on tangible evidence detectable with our senses in the empirical tradition. Among those who discuss the distinction between metaphysical idealists versus empirical realists is Herbert Spencer (1904:3–34) who, seeking to cool down and avoid the heated nineteenth-century debate over evolution's challenge to religious doctrine, distinguished between the science of the knowable, such as palaeontological arguments that deal with demonstrable fossil evidence, and the science of the unknowable, such as religious dogma that demands faith in numina. More recently, Karl Popper (1972:253–92) has distinguished between scientific statements, which are ones that are subject to falsifiability, and non-scientific statements, which are assertions that can never be proved false.

This book draws on aspects of both the phenomenological and positivist approaches to explore the nature of our understanding of the tacitness of being. While ethno-determinism leans towards empirical realism, it does not imply that metaphysics have no place. People's beliefs inform much of their behaviour and ideas, such as those of Was valley residents who believed that their cults, with associated structures and paraphernalia, appeased dangerous and capricious spirit forces (Chapter 8). The extent to which we can achieve a sympathetic understanding of such supernatural beliefs promotes endless debate. They relate to personal emotional states, which I take, after Hume's passions, to be primary phenomena behind our behaviour and understanding of being-in-the-world. Is it necessary to believe, to experience associated emotions to understand them? To what extent are different beliefs accessible and comparable; for instance, when you read that the Wola believe in ancestor spirits, to what extent do these words conjure up a similar image to the Was valley one, and how far does an account of their associated beliefs, about why we fall ill when we do, what it is to die and so on, convey their views? These are purely cognitive images that exist in the mind only, unlike things such as houses. You can sense a Was valley *aenda* 'house' visually from images in this book or encounter one tangibly by dwelling in it if you visit the region, but you will never sense a *towmow* 'spirit'.

But even things 'out there' are not necessarily straightforward to apprehend from a local perspective. The problems of understanding are compounded, for instance, when actors are unable to elaborate on activities associated with things that we may assume to have some symbolic significance, believing them to be efficacious in their own right without need of further excogitation. The target object on the *timp* house illustrates this, which participants just adopted from elsewhere without apparently needing to know its meaning – 'it was what that crowd down the valley did and told us to do when we took the cult over from them, so we did it'. It presumably had some intentional import at its place of origin, but that was lost in its movement. The spontaneous acquiescence is reminiscent of the demeanour of many individuals who comprise English congregations and attend Sunday evensong services, who barely follow the theology of the liturgy and have scant appreciation of the symbolic imagery of church architecture but nonetheless believe that these are significant and emotive. If persons cannot account for aspects of their behaviour and the import of some things, which we take to have some emotional symbolic significance that we seek to unveil, the implication is that we can access unconscious matters, which brings us back to tacit issues.

It is tricky territory. The semiotic-construction approach blurs the line between ethnographic evidence and investigator interpretation in its quest to interpret what people do and say, although it seems initially to comply with the strictures of ethno-determinism, namely that we should take our cue from the ethnographic evidence – as the buzz phrase 'thick description' signals (Geertz 1993), implying adherence to the principle of holism in locating discussion within wider cultural context. Nonetheless, it is often unclear where evidence ends and interpretation begins – thick description becoming opaque extrapolation. In presuming to convert tacit local understandings into formal knowledge it is difficult to know how far the actors are intellectually aware of the issues (which may unavoidably be minimally, according to my understanding of tacit knowledge), and to what extent they are dictated by researchers, who assume to see in actors' behaviour and cultural arrangements what they put into words on their behalf. In short, they impose their own explanations under the guise of interpretations. It often seems that words are put into the mouths of carved masks. It is dubious to presume to speak on behalf of others, in imaginatively interpreting on their behalf aspects of their behaviour and cultural heritage that have no apparent conscious articulable meaning for them, on the assumption that they express these views through non-verbal symbolic 'discourse'. What 'interpretive constructivism' seemingly does is smuggle in the assumptions and concepts that feature in explanation under cover of these being local rather than foreign concerns, again often heavily informed by the *théorie du jour*; such as the binary structure of human cognition, or concerns over personhood and identity, or the implications of gender distinctions.

For all its shortcomings, explanation is straightforward in presenting its assumptions as outsider impositions and not seeking to camouflage them. There is a certain inevitability here, as the postmodern critique makes clear, as we have to draw on what our socio-cultural upbringing (including formal education in subjects such as the social sciences, in institutions such as universities) leads us to think we know in seeking to make sense of socio-cultural arrangements elsewhere. The current focus on 'narrativity' aims to tease out the implications of such rhetorical licence (Carrithers 2005; Meyer and Girke 2012). While there is an unavoidable imposition of our ideas and understandings, ethno-determinism seeks, so far as possible, to distinguish these from what people do and say, allowing these latter to determine what ideas and understandings we draw on and how. Whereas 'symbology' results, as noted, in a mélange, as researchers assume to interpret the otherwise apparently inchoate native view, drawing on their assumed greater conscious insight. In other words, as stressed from Chapter 1 onwards, the problems addressed should emerge out of the ethnography (Sillitoe 2010a:4) and not be imposed by some theoretically defined course, navigating, as Geertz (1993:14) graphically puts it, around the wrecks of sunken theories, whether interpreting cultural communications, of whatever genre, or investigating something else. While the integrity of the ethnography is paramount, this is not to imply that a pure inductive stance and a value-free 'true' record that captures the full complexity of human behaviour and cultural arrangements is possible, for as Popper (1972:46–55) points out, following Hume, there are no objectively secure (or ideology-free) observations; we are all culturally bound and have prior views that inform our understanding of our experiences.

Again, the implication is not that imagination has no part to play; even the most positivistic experimental science relies on leaps of intuition to further understanding. But there is a notable difference between appearing to represent others, in imaginatively interpreting behaviours and things as symbols in ways that they do not do for themselves, and saying that aspects of their cultural being-in-the-world help us to imagine something for ourselves, in the context of our experiences, about humanity and our capacity for social relations and so on. It is one thing, for instance, to claim that recent scientific discussion of fractal geometry helps us better understand what people say and how they behave, deepening our appreciation of humanity's potential regarding socio-cultural arrangements, and quite another to imply that the idea of the 'fractal person' represents their view of the world and their behaviour. The intellectual challenges encountered by ethno-determinism belie any assumptions that it advocates 'mere' ethnographic documentation, and fails, in promoting the centrality of these data, to contribute to anthropology's erudite efforts to further our understanding of humankind. Some versions of 'symbology' invite such criticism more, presenting descriptions, 'thick' or otherwise, of local dialogue and behaviour, and suggesting cultural pointers purportedly to allow readers to interpret for themselves, and eschewing as inappropriate (in seeking to escape postmodern disdain) any further analysis in the context of wider issues.

Social enquiries in the constructivist vein gloss over the significant, and for them awkward, tacit dimensions of human experience. Being-in-the-world comprises more than navigating a culturally coded semiotic system – whether exchanging mutually intelligible sound sequences (words arranged in a language), or socially recognized body gestures (signals such as winks), or using things to send messages (such as clothing to symbolize status) – that awaits excogitation by the culturally aware. The tacit dimensions of behaviour and understandings, which extend to many things including symbols, reveal the limits of interpretation or explanation. Regarding houses in the Was valley, for instance, residents accept the architectural arrangements without comment on the socio-political-economic implications of their gender aspects or connection with sexual-division-of-labour arrangements. While persons share beliefs and agreed codes that inform proper ways to behave, they are often apparently unaware, as we have seen, of the implications and associated regularities in their behaviour and understandings. Although the actors may not be consciously aware of these issues, those trained in social research can apparently interpret/explain their collectively shared tacit knowledge and associated culturally informed interaction. The implication, as pointed out, is that researchers not only know people differently but also know them better than they know themselves, when paradoxically, what they are investigating is of the actors and manifest in their behaviour and what they say.

In this event, how might we, in some measure, access tacit knowing more reliably? This is the question addressed in the following chapters. It comes down to the sensitive use of appropriate field-research methods that feature constant triangulation or checking of ethnographic observations for accuracy and reliability during intensive participation in a community, which in itself is something of an acquired tacit skill (Chapter 11), and assumes competence in the vernacular that comes with an extended stay. It assumes recording of evidence as faithfully as possible, with frank acknowledgment of the foreign, albeit ethnographically determined, intellectual framework used to make sense of it. It may be deeply unfashionable to refer to facts among those currently taken up with the postmodern debate, where subjectivity is the only true reality. But I do not subscribe to the navel-dwelling view that what we and others do and say only exists for us in our minds and each of us may perceive of these differently. The idea that we inhabit imaginary worlds reduces the anthropological endeavour to morally questionable (usually uninvited) involvement in the lives of others to engage in an egotistical exploration of the self. There are objective criteria that exist 'out there', against which we should strive to test any argument's veracity, which is not to say that they are necessarily easy to assess. For instance, if we allow that social mores and beliefs exist as ideas, they comprise notoriously slippery 'facts' such that their evidential value and interpretation is constantly open to challenge, whereas ethnographic data that relate to things, such as house and bridge design and construction, are more securely anchored in the same reality for all of us, which is why I think that it may prove helpful to explore these tricky issues in such material contexts.

11

Passing on knowing

Hands-on participation

It is not only questionable to assume that we can interpret things that we identify as symbols when actors cannot do so, it is also dubious to imply, after social constructivism or interpretivism, that symbols make up anyone's entire culturally mediated reality. Any being-in-the-world comprises more than a semiotic system awaiting interpretation by those with the necessary skills and imagination, whether the symbols have some conscious import, such as the designs painted on some coffins in the Was valley or the bride's behaviour at marriage in handing bridewealth to her kin; or are apparently of unconscious significance, such as some features of cult-houses or ochred pearl-shell crescents. People also engage in culturally shared activities lived without any apparent need of symbolic representation, such as those that attend house construction. When two or three men co-operate together to build a dwelling, for instance, they all know what to do (Chapter 7), and get on with it, without any call for semiotically communicated direction, such as consulting blueprint plans. They engage in little discussion about what they are doing, just proceeding with tasks as they see fit. They converse, of course, but they are more likely to gossip about local goings on than how to proceed with the building construction.

What these two dimensions of being-in-the-world – the symbolic and the pragmatic – have in common, together with much social interaction generally, is that they feature tacit knowing. This is suggestive, with regard to the phenomenological conundrums encountered here. People often cannot express the meaning of things taken to be symbols in words, as discussed in the previous chapters. Whatever their import, if any, it is apparently beyond the verbal, being conveyed by definition via house-architecture code or shield-design code or self-decoration code or whatever (Sillitoe 1980, 1988b). Yet academics who engage in the analysis of symbols think that they can capture what they signify in words. In short, they claim, as pointed out, that they can apprehend tacit knowledge and convert it to explicit knowledge, which is a central challenge addressed in this book. But an appraisal of interpretative approaches to culture questions this assumption, casting doubt on the validity of attempts to interpret symbols for others. Is it feasible to proceed more securely in discussing tacit knowing, and explore the possibility of expressing it explicitly, by considering the other, experiential dimension of culture that relates to pragmatic existence, such as we encounter in construction activities?

Explicit knowledge and formal education

Explicit knowledge, sometimes called theoretical or formal knowledge, is intelligence that we can represent and share consciously using an agreed symbolic code, often language. It entails 'spoken words ... formulae ... maps and graphs, mathematical theory' (Polanyi 1969c:195). It is 'know-what' information or 'know-why' explanation, as opposed to often subconsciously acquired 'know-how' skill and 'know-that' behaviour, and concerns knowledge that holders can co-ordinate, articulate and transmit to others, particularly when formally codified in words and numbers as data, descriptions, principles, specifications and so on (Ryle 1949:25–61). It is an aspect of the materialist viewpoint that there is a world out there beyond our minds to sense and know explicitly. There are various other proposed labels for these two dimensions, for the -what and -how of knowing. For instance, 'iconic' is another term for explicit propositional understanding that depends on language and visual images to convey semiotically patterned knowledge, which contrasts with 'enactive' understanding that involves tacit learning undertaken through participation in an activity such as mastering a skill (Bruner 1966:10–11).

In academia, indeed in our society generally, we prioritize explicit propositional knowledge organized systematically above the intuitive, hands-on, unwritten variety and associated informal practices. A characteristic feature of 'post-industrial society ... is the centrality of *theoretical knowledge* – the primacy of theory over empiricism and the codification of knowledge into abstract systems of symbols' (Bell 1999:20, original italics). The formal education system, from primary through to tertiary levels, privileges the teaching of explicit knowledge and the examining of students' understanding of its representation. The history of polytechnics in Britain, which specialized in vocational education, illustrates the esteeming of the intellectual above the practical, their occupational image and courses undergoing a makeover when granted the status of new universities, as they sought to emulate the academic tradition of the 'elite' old universities. This attitude is even evident in education for applied professions such as engineering, medicine and agriculture, where it is not until you start your apprenticeship in the workforce that you acquire practical skills. On a personal level, the banter that I have encountered with some colleagues when I have mentioned that I am working on a book about house building is, I think, a further reflection of these attitudes. There is likely an element of class consciousness here, as mentioned in Chapter 2, prompting us to downplay technological education as proletarian. Although some educationalists have challenged this hierarchy, we need a better understanding of tacit knowledge to do so effectively, which is a challenge. Its passage between actors is messy and not subject to the regimentation favoured by curricula designers. The Western intellectual tradition that deals in explicit propositional knowledge has problems with such knowing as not amenable, by definition, to standard dialectic analysis featuring premises, evidence and inferences. When we seek to make such knowledge explicit, these procedures fall down and omit things, ultimately missing the point.

We are, it seems, limited in our capacity to deal intellectually with experiential knowledge. Yet, as the postmodern view observes, theoretical knowledge is not entirely explicit, codified and objective, but also involves tacit, intuitive and subjective aspects. Even the natural sciences with their formal laws and mathematical logic depend to some extent on scientists' practical and interpretive skills; there is unavoidably a 'personal coefficient' as Polanyi (1958:17) pointed out some time ago. All knowledge obviously depends on human discrimination. Personal judgement features in the assessment of any propositional representation, inasmuch as we judge the correspondence between our experience of the world and any abstract depiction that purports to represent it and further understanding of it. We use our senses, after Hume (1896), in making such judgements (Gill 2000:44–50; Polanyi 1958:19), which involve being-in-the-world, drawing on both bodily and mental dispositions, in apprehending the evidence and establishing correspondences. While these are culturally informed to an extent, depending on the socially imbued expectations that we grow up with, we need to afford space for individual experience modifying them. We see the consequences of not doing so in the previous collective sociology and 'culture is a system of symbols' views that downplay the individual, and which exacerbate the postmodern impasse. If symbols comprise our entire known cultural world, with language serving as an analogue for reality – after the representational school of philosophy – then our explicit-formal-intellectual way of knowing is not above and beyond things, but wholly comprises such an analogous reality system, and it is difficult to see how we could get outside it to understand anything about another culturally informed way of being-in-the-world.

The view of culture as a system of symbols amenable to interpretation in prose relates to the stress on formal knowledge, reflecting the assumption of Western intellectuals that we can represent everything of interest explicitly in words. The view is arguably ethnocentric, linked to our use of a culturally specific symbolic code comprising twenty-six symbols – the letters of the European alphabet – that seems to have the capacity for infinite expression of ideas. The reproduction of that code in written signs and not only spoken sounds, as I am doing here and we all do when we commit information and thoughts to paper or monitor, exacerbates this tendency perhaps, prompting us to think that we can reduce and reproduce any cultural reality using that literate system of symbols. Those who live in cultures that rely largely on sound codes and spoken language, and depend more overtly on enactive understanding, question these semiotic assumptions. And they remind us of the significance of getting beyond formal didactic to informal experiential learning in accessing tacit knowing and ways.

Tacit knowledge and informal learning

The learning of culturally informed tasks, such as those associated with house construction, reaffirms the point that culture encompasses lived activities that individuals engage in without any apparent need of symbolic representation. Younger persons learn what to do in the Was valley by observing others and mimicking their actions, often with no verbal direction or other symbolic mediation (note children in Plates 7.5 and 7.6). They pick up the knowledge and skills through informal everyday interaction with kin. This features hands-on practice, experience being a significant aspect of learning, which contrasts with formal education passing on explicit knowledge. It is more fun, with a gradual transition from play to work, where children slowly spend increasingly more time on adult pursuits (Sillitoe and Sillitoe 2009:22–4). A boy, for instance, may on occasion assist relatives constructing a house, helping with minor tasks such as fetching materials, holding things and undertaking small jobs, and when an adolescent and sufficiently strong enough, by contributing increasingly to heavy work until he is capable of the full range of adult tasks and takes responsibility for building houses. Both girls and boys learn their roles in the same way, by observing older relatives, playing at grown-up activities and gradually taking on adult tasks. They learn the skills and acquire the knowledge appropriate to their gender: learning reflects the sexual division of labour arrangements that characterize Wola daily life, as discussed in the previous chapter, where men and women undertake different activities. Girls take on adult tasks sooner than boys, accompanying their mothers regularly to gardens, tilling soil for instance and planting sweet potato mounds, starting out with minor contributions, such as carrying a few tubers home in their small net bags, with the load increasing as they grow older. In part this is due, people explain, to many men's tasks, such as building houses and pioneering cultivations, being physically too heavy and dangerous for youngsters to participate in.

In short, there is no formal education: life is an education. Children are unaware that they are learning much of the time, acquiring knowledge and skills by watching and doing in their own time, experimenting until able to undertake tasks. If you ask someone how they know something, or who taught them, usual responses, beyond a blank look, include 'I taught myself' or 'I have always known that' or 'I made it up'. A man questioned about who taught him to 'spear' wooden stakes, for example, may reply that he has always known how to do it, which in a sense, so far as his memory goes, he has, having observed male relatives since infancy and experimented working with them until he mastered the skill. This illustrates how people just pick up knowledge and skills in the course of everyday life, such as how to construct house walls, loop net bags and cook vegetables in earth ovens. It does not preclude individuals asking questions or knowledgeable persons instructing others about something. The everyday words the Wola use for such actions – having no term for teaching as such – relate to its informality. They are *muwalay* 'show', which people use both in showing you something (such as an animal they have caught) and in showing you how to do something (such as how to set an animal trap); and *kakalay* 'tell', which people use both when telling some news (such as so-and-so is searching for a lost pig) and passing on information (such as telling you where you will find a tree suitable for a house ridge-pole).[1] An intriguing consequence of there being no, or scant, intentional teaching is that persons rarely acknowledge that anyone passed the knowledge on to them.

The way persons take the credit themselves for what they know relates again to the prominence of individual sovereignty in the Was valley. It parallels the 'user-makes' principle that features in house construction and artefact manufacture generally (Chapter 7, Sillitoe 1988a:563–71), where persons expect to be responsible for the manufacture of things that their households need and do not rely on others. It ensures that they have control of supplying their wants, rather than other parties, who could use any such capacity to exert power over them, in the Marxist sense as discussed in Chapters 7 and 10, by threatening to withhold necessities. That individuals consider themselves autodidacts, who know what they know through their own efforts, further negates the possibility of authority figures. They discover knowledge for themselves, being self-taught for all intents and purposes, even inventing, to their minds, some of what they know, and relying heavily on personal experience in doing so. It underlines the cultural relativity of the idea that culture is a body of shared knowledge and ways, much of it symbolic, which persons pass on. The absence of any notion of formal education underscores further that no one recognizes anyone else as an authority on any matter. No person has the right to decide or tell others what they might know, or has authority to know things as experts that others of their gender do not know. (The exception is persons who knew certain incantations and ritual procedures used to access the spirit world, although this knowledge was structured in such a way as to prevent holders using it to exert any power over others, who could secure the services of other persons or acquire the knowledge for themselves for a modest exchange of wealth –Sillitoe and Sillitoe 2009: chapter 8.)

The 'informal' experiential approach to learning in the Was valley is central to education in many parts of the world where learning involves doing and reflecting on direct experience, achieving meaning through practice. A report on First Nations education in Canada catches its flavour well: 'The first principle of Aboriginal learning is a preference for experiential knowledge. Indigenous pedagogy values a person's ability to learn independently by observing, listening, and participating with a minimum of intervention or instruction. This pattern of direct learning by seeing and doing, without asking questions, makes Aborginal children diverse learners.' (Battiste 2002:15). The experiential approach focuses on individuals learning by doing, where information content and tasks undertaken – such as in building a house – inform the processes of knowledge transmission and acquisition (Coy 1989:1; Lave 1982:186). It is an ancient, arguably primordial approach to learning, sometimes called 'informal education' and 'auto-didacticism'. All humans learn through experience to some extent, from infanthood onwards. It is arguably the most effective approach to learning, being deeply rooted in our way of knowing what we think we know about being-in-the-world. As the Chinese saying attributed to Confucius over two millennia ago puts it: 'Tell me, and I shall forget. Show me, and I may remember. Involve me, and I will understand.'

The formal and informal distinction has featured for many years in education, with the work of Jerome Bruner (1960), who distinguished between 'analytical' and 'intuitive' thinking, widely cited for stressing experiential and environmental factors in education, and arguing that school curricula should aim to foster intuitive 'grasping' of understanding, which is to reach 'plausible but tentative formulations without going through the analytical steps

1 Other words with the same meanings are *buwalay* 'show' and *kanjay* 'tell' (*kanjay* is used when telling persons directly and *kakalay* for telling them indirectly).

by which such formulations would be found to be valid or invalid' (1960:13). The intuitive focus continues in Donald Schön's (1983) 'reflective practitioner' formulation. Formal schooling recognizes such learning as 'experiential education', arranging programmes to organize and control it through structured activities. It has long featured in vocational-professional career training. It seeks to facilitate students' direct subjective experiences as a more natural way to gain knowledge. Learning by direct participation in, and reflection on, life's everyday events and experiences promotes deep learning, it is argued, as it is more consistent with how we normally learn and develop. Such education may feature learning by yourself or 'self-teaching'. It contrasts with 'didactic education' in which teachers pass on explicit knowledge to learners through prescribed approaches to study, designed to achieve specified learning outcomes.

The centrality of experience in the learning process has led to the emergence of 'experiential learning theory' in education circles. It challenges the dominant cognitive and behavioural learning theories that overlook the part subjective experience plays in learning. It builds on the experiential aspects of Dewey's pragmatic philosophy, Lewin's social psychology and Piaget's cognitive-developmental epistemology (Kolb 1984; Kolb, Boyatzis and Mainemelis 2000). Implicit learning (Berry and Dienes 1993) is a closely related idea; it applies where individuals are not consciously aware of having learned something. According to experiential learning theory, 'knowledge is created through the transformation of experience' and features a 'combination of grasping and transforming experience' (Kolb 1984:41). The 'grasping' of experience features two processes: 'concrete experience' and 'abstract conceptualization'; and the 'transforming' of experience involves two processes too: 'active experimentation' and 'reflective observation'. 'Concrete experience' features observations and prompts reflections. These are incorporated into, and used to refine, abstract concepts from which new inferences are drawn, and further actions ensue that are tested against new experiences. According to the theory there are two kinds of learners: the 'doers' and the 'watchers'. In 'grasping' information, the former rely more on their senses and tangible experience of the world, while the latter depend on abstract analysis and symbolic representation. In 'transforming' experience, the former incline to doing things, that is they prefer active experimentation, while the latter watch others, that is they opt for reflective observation. Both approaches may feature in learning, although often not simultaneously: when constructing a building, for instance, it is not possible to manoeuvre a crane ('concrete experience') and consult the operating manual ('abstract conceptualization') at the same time. According to the theory, we develop a preferred 'learning style' due to our environment, past life experiences and (unspecified) hereditary factors. The 'learning style' in the Was valley inclines markedly towards the 'doer' approach.

The acquiring of tacit knowing may feature implicit learning,[2] as Polanyi's (1966:7–8) reference to 'subception' indicates. It concerns how we acquire, or 'dwell in', such new knowledge. The process is subject to heated debate in psychology. Experiments to probe subliminal learning (Reber 1993:17)[3] suggest that individuals possess knowledge that they are unaware they have learned (Baumard 1999:54–8). An intriguing question is how individuals learn to exercise such unconscious skills, when in a sense, due to their ambiguity, they do not know. It involves a 'process of unconscious trial and error by which *we feel our way* to success and may continue to improve on our success without specifiably knowing how we do it' (Polanyi (1958:62, original italics). The example of learning to use a stone axe given in Chapter 2, to illustrate the discussion of subsidiary awareness, demonstrates the process. Initially, the young men helping to build the house were puzzled when stone axes did not handle like steel ones. What their older relatives told them did not make much sense at first. They also focused their awareness on the axe and not the task at hand, which made them clumsy. But gradually they became familiar with the new rhythm and focused less on the axe in their hands and more on the task. When they had interiorized the knowledge, it became tacit and subsidiary to the task in which they used the tool, such that the skill existed in relation to the task on which they now focused their attention fully. In achieving competence, they became unaware of how they achieved it when I questioned them about it. They 'just knew' how to handle an axe. This slipping into unconsciousness is 'accompanied by a newly acquired consciousness of the experiences in question, on the operational plane. It is misleading, therefore, to describe this as the mere result of repetition; it is a structural change achieved by a repeated mental effort aiming at the instrumentalization of certain things and actions in the service of some purpose' (Polanyi 1958:62).

The transmission of such knowledge requires exposure to it in action. It occurs through practice under the guidance of a knowing person, as in 'situational learning', the principle of which is that we only effectively gain such knowledge through 'real life' involvement in activities where it features; that is by practical experience and not

2 For reviews of implicit learning. see Berry and Dienes 1993; Jimenez 1997; Marescaux 1997; and Stadler and Frensch 1998.

3 A common experiment involves showing subjects sequences of letters governed by some generative rules – such as DEFKLM and JKLPQR – and others that are not, and asking them to say which sequences are 'correct'. Subjects are often able to identify the 'correct' sequences, without being able to specify how (i.e., verbalize the rules), having become subconsciously aware of them (Reber 1967).

by formal education. Indeed, any attempt to reduce such knowing to a formal teachable format will at best distort and at worst destroy it. The problematic nature of the relationship with conventional education is evident in the professions (Sternberg 1999) where the acquisition of vocational competence involves practice, and verbal instruction has little significant effect on performance; albeit instruction improves persons' ability subsequently to answer formally posed questions. It queries the relationship between attempts to talk about skills, as in this book; and knowing them, which is a central problem that it addresses. A review of how such knowledge is passed on may help us better understand the relationship between talking about tacit knowing in theory, as here in the context of New Guinea Highlands house construction, and doing such knowing in practice The concern for task mastery, and consideration of related matters such as material properties and social arrangements, takes us beyond formal content transmission to informal learning practices, focusing on, as A.J. Strathern pithily puts it, 'action and performance, on doing rather than being, or the being that resides in doing, that issues from and is expressed only in doing' (1996:202).

Apprenticeship learning

It is through apprenticeships that we pass on much skilled knowledge, and it is on these arrangements that the formal education system has focused in seeking to extend understanding of such knowledge transmission. The apprenticeship approach concerns learning through integration into the practices that comprise an occupation. It is among the oldest and most effective ways for experts to pass on both explicit and tacit knowledge to novices, and was prominent from the Middle Ages until recent industrial times.[4] As Brown, Collins and Duguid (1989:453) point out: 'Before schools appeared, apprenticeship was the most common means of learning and was used to transmit the knowledge required for expert practice in fields from painting and sculpting to medicine and law.'

Apprenticeship is necessary to acquire the 'personal knowing' (Polanyi 1958:49), that aspect of tacit understanding that we cannot fully specify. Such training, as Coy (1989:1–2) points out, 'is personal, hands on, and experiential' and features 'where there is more to performing the role at hand than reading a description of its content can communicate', for aspects of it 'can only be experienced.' In his discussion of apprenticeship in the acquisition of building skills, Trevor Marchand (2001, 2009) stresses this aspect, where the 'learning process involves little or no verbal communication, the apprentice must rely on his/her eyes, ears, and sense of touch' (Marchand 2001:138). While mimicking activities is prominent, this learning does not preclude some overt communication, for when we go beyond simple tasks that an individual may imitate by observing someone doing them (such as carrying a pile of thatching material), to complex tasks that demand complicated cognitive processes (such as calculating the dimensions of a house plan), there is a need for some explanation, as we cannot tell from observation what is going on in the actor's mind.

Apprenticeship involves a learner within a context of practice (Pratt 1998), working side-by-side with acknowledged experts to learn tasks in hands-on activities. Apprentices master the necessary skills by learning to perform under the guidance of such experienced practitioners, who help them make the subliminal connections, for it is 'By watching the master and emulating his efforts in the presence of his example, the apprentice unconsciously picks up the rules of the art, including those which are not explicitly known to the master himself.' (Polanyi 1958:53). The learners encounter various real workplace problems and learn how to solve them. They gradually become more capable at performing tasks and integrating skills, tackling increasingly challenging assignments. Thus apprenticeship fosters independent practice and encourages students to master domain-related principles in various contexts (Barab and Hay 2001:72) such that it provides the scaffolding not only for understanding and performing tasks but also for solving problems and dealing with unusual and difficult situations (Collins, Brown, and Newman 1989). The skills learned build up into sequences of increasing complexity until practitioners become masters who can cope with various and novel demands.

There is a notable difference between apprenticeship as training into specialized work structured according to a hierarchical occupational division of labour and the acquisition of skills under an egalitarian sexual division of labour regime (Chapter 10), which we need to bear in mind when considering what studies of formal apprenticeship can teach us about tacit knowing. For instance, under an occupational regime, Marchand (2001:136) tells us that '[q]uestions are rarely posed, as they would likely be construed as a challenge to the Master's authority' (although given the 'nature of this form of knowledge, it cannot be "talked about" anyway'), and apprenticeship 'not only serves to reproduce knowledge about a craft, but also acts as a valve to regulate the numbers permitted access

4 Some speculate that the apprentice way of learning is ancient – the acquisition of skills and tacit knowledge featuring 'mimetic' behaviour that involves a deeper and more primitive part of the brain – and that it contributed during the pre-human 'speciation' phase of hominin evolution to the astonishingly rapid growth of the brain, notably the neocortex, by promoting certain aspects of memory, for instance those regarding suitability and use of certain materials (materiality perspective), while simultaneously favouring certain social arrangements (cultural evolution perspective) that also contributed to brain expansion (Boivin 2004:69; Donald 1991; Ingold 2000:358; Renfrew 2007:112–14).

to this realm of specialized knowledge'.[5] Elsewhere he describes hierarchies, featuring labourers up to master builders, that equate somewhat with social and ethnic divisions, and guilds of masons to which master builders control access, all of which contrasts with Wola learning arrangements that are an informal part of everyday life and pass on knowledge of all tasks to everyone equally, subject to relevant gender ascriptions. Nonetheless, there are parallels in the acquisition of tacit knowing, with an apprentice finding 'knowledge is sedimented in his bodily performance ... [such] know-how does not necessitate a consciously aware form of reflection' (Marchand 2001:24), acquired as it is through mimicry and practice.

There are various models of apprenticeship learning, some more straightforward than others. According to one (Pratt 1998), there are three aspects: 'activity', which concerns practical participation (including decision-making, with experience); 'sociality', which concerns student interaction within an interrelated web of relations (both with teachers and fellow learners); and 'authenticity', which concerns the connection between craftsmen and users of their products (involving an awareness of customers' views). Another model (Brandt *et al.* 1993) focuses on different stages of activity: firstly, novices discover what works by successfully using skills to solve problems under expert guidance; secondly, they have sufficient practical and theoretical knowledge to recognize and tackle standard problems semi-independently; and thirdly, they progress beyond basic skills to perform at a level approaching that of those fully competent in the field. According to this model, participants find that the apprenticeship 'learning experience expands their awareness of the factors that should be considered, helps them organize and pay attention to their thought processes while handling difficult tasks, problems ... previously ignored or regarded as unimportant'. Yet another model (Hansman 2001) elaborates further on the series of phases an apprentice passes through, progressing from novice to journeyman and finally to master craftsperson. The first phase is 'modelling', where learners observe whole activities without being aware of the finer points of tasks, seeing the wider practice without being able to specify its aspects. The second is 'approximating', where apprentices start to mimic with guidance the teacher's actions, who shares 'tricks of the trade' (Hansman 2001:47); in this phase novices reflect on what and why they plan to do things and how they as would-be craftspersons compare with the master. The third phase is 'fading', where the teacher's assistance decreases, and the apprentices, still under the watchful eye of the master, start working increasingly independently as their capabilities increase. The fourth is 'self-directed learning', where apprentices graduate to be a journeyman able to work alone on mastered tasks, learning more as they go (from novel problems, mistakes etc.), turning when need be to expert assistance. The fifth and final phase is 'generalizing', where craftspersons are able to apply their skills in various contexts and solve new problems confidently, their ability developed to its full potential.

The 'cognitive apprenticeship' approach – featuring 'anchored instruction', 'learning communities' and 'in-situ assessment' – seeks to delve deeper into the mechanisms of apprentice-style learning, using 'situated learning theory' (Brandt *et al.* 1993; Collins 1991; Collins *et al.* 1989; Collins *et al.* 1991; Dewey 1933; Rogoff 1990; Tisdale 2001). In this learning arrangement, students participate in activities and interact socially within 'communities of practice' centred on knowledgeable persons from whom the less experienced learn (Lave and Wenger 1991; McLellan 1994). Communities of practice may overlap, such that a master in one may be a novice in another. This approach to education engages learners in the 'real world' of the workplace, involving goal-oriented activities and solving everyday problems, while affording access to guidance and featuring built-in reinforcement (Kerka 1997). The cognitive apprenticeship approach actively promotes independent exploration, whereas in traditional apprenticeships it occurs more covertly as novices acquire mastery of the skills. It features six stages.[6] Knowing comprises doing, its rationale is

5 After discussing the term apprenticeship, Marchand (2001:136, 149) concludes that the Arabic circumlocution suggests that it is appropriate, albeit his 'study is firmly rooted in Western philosophical discourses about the mind which were entirely unfamiliar to my Yemeni co-workers' (Marchand 2001:26).

6 The stages of the cognitive apprenticeship model are as follows. Firstly, 'modelling', this involves demonstrating a process and giving reasons for it (Collins 1991). The instructor serves as 'cognitive role' model, both demonstrating and explaining certain actions – modelling expert performance – so that learners can follow their thinking processes (Shunk 2000). The modelling requires teachers both to demonstrate physically how they tackle different tasks and to discuss their thoughts, showing how they cope with difficulties and encouraging novices to proceed the same way. This think-aloud approach helps learners to acquire a conceptual model while observing an activity, making explicit the experts' approach to problem-solving (Collins *et al.*1989; Collins, 1991). According to situated learning theory, it also gives students confidence to see experts puzzling over difficulties (Shunk 2000). Secondly, 'coaching' is 'the process of overseeing the student's learning' and it is 'the thread running through the entire apprenticeship experience' (Collins *et al.* 1991). It involves knowledgeable persons giving assistance to learners to accomplish tasks, until they can do them more or less unaided. The process may begin with helping them decide on an appropriate approach and finish with feedback on their performance. There are various coaching strategies, including giving hints, offering verbal and non-verbal support, commenting on problem diagnosis, discussing difficulties encountered by

subordinate to skill. It is situated performance, rather than propositional competence. Knowledge exists within the routines of the community of practice, not as abstract propositions, and learning consequently involves mastering the routines and integration of complex procedures. It entails 'legitimate peripheral participation' by novices (Lave and Wenger 1991); that is, those who hang around doing minor tasks, gradually taking on responsibility for more significant ones. It is legitimate, as all parties accept such newcomers into the margins of the community of practice; it is peripheral, as they undertake the minor jobs; and participation features centrally as they learn by doing. All of which is reminiscent of learning in the Was valley. Otherwise, these attempts to specify approaches to apprenticeship systematically, contrast markedly with the region's idiosyncratic practices. They reflect our centralized culture's urge to assert explicitly and organize formally the educational process so as to control and direct it, which depends to some extent on literate communication. Such attempts to manage learning differ starkly from the uncentralized Wola approach, where knowledge passes on implicitly through informal interaction that is unregulated, with no party in control, which complies with oral arrangements. While we might expect such laissez-faire arrangements to promote continuous innovation and wide variation in house designs, the egalitarian ethic, as we have seen (Chapter 8), counters any such tendencies.

Culture as doing

While the steps these models identify are somewhat mechanistic and fail to capture the fluidity and quirkiness of skill acquisition (Chapter 7), as is usual with attempts to formalize any tacit activity into propositional terms, they do offer some insights into the learning process. They recognize that learning is dynamic and recursive, with students repeatedly revisiting and pondering on their activities, and not a serial process featuring the one-off absorption of subject matter. They allow that task-based knowledge is not acquired language-like via 'linear-sentential' propositions (Bloch 1991:184, 192; Marchand

learners, evaluating task execution and final outcomes, and working with novices to correct their weaknesses. In short, coaching involves 'doing whatever it takes to assist learners in their learning, from start until finish' (Brill *et al.* 2001). Thirdly, 'scaffolding', this also features masters assisting apprentices to achieve goals. It is an aspect of coaching and receives considerable attention in the cognitive-apprenticeship literature. It refers to the framework supporting novices when they work at levels beyond their current abilities. It involves a 'cooperative problem-solving effort by teachers and students in which the express intention is for the students to assume as much of the task on their own as possible, as soon as possible' (Collins *et al.* 1989). It entails two simultaneous processes: support of students by experts and its gradual withdrawal as they can perform tasks. The scaffolding idea draws on the 'zone of proximal development' concept in educational theory, which is the difference between the 'actual developmental level as determined by independent problem-solving and the level of potential development as determined through problem-solving under adult guidance or in collaboration with more capable peers' (Vygotsky 1978:86). When a novice is in the 'zone', such support can give necessary encouragement and confidence to achieve a task. Scaffolding also reduces learners' frustration, knowing that they are not alone in trying to master a task (Brill *et al.* 2001). Help should be readily available, as they may give up if left struggling alone to make progress for too long. And the support should be just sufficient to help them achieve tasks beyond their ability and develop the skills necessary. Fourthly, 'articulation' involves verbalizing aspects of the apprenticeship experience, which, featuring tacit knowledge, is limited by definition. By discussing practices, novices may refine and expand their understanding. A variety of strategies may facilitate this, such as discussions, presentations and problem-solving. It may involve apprentices talking about both activities and understandings, bringing issues into the open to clarify them, or separating aspects of what they know from their skills to perceive these more effectively (McLellan 1994). Fifthly, 'reflection' is closely associated with articulation, both often occurring simultaneously. According to the cognitive-apprenticeship approach, it is an important, and often neglected, aspect of instruction. It involves standing back from a task and thinking about it, and should occur regularly. Reflection fosters insights and promotes revision of ideas. The aim is to look back at key points during an activity and consider what has been done with a view to understanding it better and improving on performance. The master actively promotes and guides reflection, for instance posing questions about recent activities and encouraging students to ask questions during the learning process. Sixthly, 'exploration', where learners increasingly undertake activities independently of teachers and engage in solving problems alone. It is modelled on 'discovery learning' and 'inquiry-based' approaches where students obtain new knowledge for themselves and develop problem-solving skills that lead them to investigate and make their own discoveries (Bruner 1960; Shunk 2000). This stage promotes learner autonomy in setting achievable goals, formulating problems, gathering necessary information and testing questions (Collins 1991; Collins *et al.* 1989). They experience what it is to be an expert because they are increasingly thinking and performing like one. Initially the mentor may direct and guide exploratory activities, posing problems that require creative modification of previously learned techniques, encouraging students to test hypotheses, methods and approaches after the way experts solve problems (Collins 1991). The goal is to use what they have learned to find and solve their own problems. While experts may initially encourage exploration, they will increasingly withdraw and encourage student autonomy (Rogoff 1990); again this features 'fading' through gradual reduction in 'coaching' and removal of 'scaffolding' support as novices progressively manage on their own, and are encouraged to set achievable goals and tackle solvable problems.

2003:42). They illuminate how practical knowledge depends on sequences of cognitive 'connections' that develop while undertaking habitual, culturally informed routines (Bloch 1991:193–4). Furthermore, the knowledge gained depends not only on the structuring of the learning process but also on the way persons cognitively organize what they come to know. A Wola house-builder can rapidly decide, for example, if a location is a good house site according to slope, drainage, aspect, rocky outcrops and so on. Such a fast assessment precludes going through a series of language-like propositions to gauge suitability. Instead, an individual compares a mental image that he has of 'a good house site' with what his senses tell him about the place in question. He may in the process modify his mental model, in view of that particular place's potential as a house site. The psychological theory of 'connectionism' illustrates the process, which maintains that knowledge acquisition involves the instantiation of 'best exemplar' mental images of things and the cumulative linking of these via cognitive networks, which we subsequently draw on in undertaking tasks (Bloch 1991:185).[7]

Cultural factors inform the structuring of connections in considerable part. This underlines how culture partly exists as shared unspoken understandings acquired during participation in the same customary tasks. It is through undertaking the same routines that community members constantly reaffirm and review their shared approach to the world and how to conduct themselves in it. Experiential approaches play down the extent to which such 'partially shared understandings of the world' (Frake, 1969:29; Bloch 1991:183) shape the outcome of skilled learning processes. They suggest that any skill acquired knowledge is equally available to all persons who are willing to learn it (Ingold 2000:167), shaped by shared prototypical comprehension of the world that organizes understanding during learning (Quinn and Holland 1987). This may apply biologically but not culturally. It is difficult otherwise to comprehend how it is that persons organize, understand and apply knowledge differently in various socio-cultural contexts. The cultural dimensions of knowledge transmission and acquisition ineluctably augment the experiential aspects, informing the cognitive processes that order understanding of the world. They feature, broadly speaking, culturally shared and socially structured habitual ways that actors need to be party to, to operate effectively in a community (Bloch 1991).

It is necessary to consider the consequent shaping of the knowledge-acquisition process – not only by the physical and material aspects of any task, but also the social and cultural contexts in which learning occurs – as featuring socio-culturally mediated cognition (Ellen 2006:3; Lave 1982). Numerous ethnographic examples illustrate the process. In West Africa, for example, the Gonja weavers of northern Ghana have an apprenticeship system organized around extended family groups under elder experts, linked together in co-operative networks (Goody 1982). The arrangement is a loose confederation of producers and their apprentices, acknowledging obligations to knowledgeable and experienced elders. A boy starts from the position of bobbin-winder and weaving apprentice, and progresses to become a weaver in his own right, then a master weaver and finally a cloth trader and garment tailor supervising the work of adult relatives and their youth apprentices,[8] who are subject to his authority and guidance. The Vai and Gola tailors of Liberia similarly pass on their tailoring 'culture of practice' to the next generation with little formal instruction, or verbal communication of any sort, between master tailors and their apprentices (Lave 1982:182). The apprentices not only learn a trade, acquiring the skills that lead from apprentice to master, but also how to engage in buying and selling to support their family. They purchase their own cloth, cut and sew it under the eye of a master whose work they in turn observe closely, to produce garments of increasing complexity, which reflect the wearers' social status in the community. Involving wider social categories, Liberian tailoring illustrates the need to understand the broad cultural context informing the learning process, in addition to the various raw materials and the physical and mental complexity of various tasks involved with skill acquisition. In short, both of these African examples show how we have to understand knowing and learning as socio-culturally situated activities.

A notable difference between the acquisition of skilled knowledge in these West African societies and in the Was valley is that they feature formal apprenticeship arrangements similar to those in Europe, where experts, even craft-like guilds, control access to the knowledge. The comparison, as mentioned above, starkly highlights again

7 The interlinking of knowledge domains is illustrated, for instance, in the basketry practices of the Nuaulu of Seram, where individuals' understanding of their environment and the properties of raw materials such as rattan and dye plants, their social position and access to techniques and designs possessed by kin, their weaving ability and awareness of the practical uses that different baskets serve, coalesce in configuring the acquirement of basketry skills (Ellen 2009:260–9).

8 The Gonja weave their cloth in long strips, a bobbin boy under a weaver's direction first measuring out the thread, winding it back and forth between iron pegs. The blue cloth is renowned, and dyeing is a skilled operation involving the fermentation of indigo leaves in open pits, in which hanks of thread are submerged. Weavers skein the dyed thread on to their simple looms, the attachment of the warp threads in the correct pattern to obtain the desired design being a skilled operation; they wind the weft thread onto bobbins. The elders sew some of the woven cloth into garments, and they manage sales, which requires experience and acumen, and they purchase thread for further weaving.

the significance of socio-political context to any skilled knowledge tradition. Although among the Wola some persons may know things that others do not know, there is no specialized occupational division of labour (as discussed in Chapter 10) and any individual can master any skill they wish by watching someone, probably a kinsperson, engaged in an activity, and maybe discussing it with them. There are no structures controlling knowledge and training, which is in keeping with an acephalous polity where power is diffused equally throughout the community, and capital control is beyond the reach of any party. The absence of experts may allow for more idiosyncratic individual variety, with no authorities to determine standards, albeit practice remains within culturally defined customary limits. The value put on equality features here, inhibiting variation beyond the point that intimates the emergence of unequal relations and power differentials within communities (Chapter 8).

It is necessary to balance the individual experiential approach with the collective intellectual one. The former focuses on the part physical and material contexts play in the knowledge transmission and acquisition process, searches for the cognitive underpinnings of how individuals acquire skill and understanding (Bloch 1991; Marchand 2003), and assumes that learning is underpinned by reciprocal engagement with environmental 'affordances' (Ingold, 1992:42, 2007; Knappett 2004) during materially engaged activities. The latter focuses on the socio-cultural schemes that order acquired knowledge and inform decision-making, and presumes that we arrange what we know into coherent shared cognitive representations to guide thought and action (Ellen, 2006:9; Quinn and Holland 1987) that are communicable in part and shared to some extent. These are, again, aspects of the classic individual versus society antinomy of social philosophy that I think helps us appreciate the Wola political-economy (see Chapters 2 and 3; Sillitoe 1979a, 2010a). It is necessary to deploy the two approaches in combination and conceive of knowledge transmission and use as a complex outcome that depends on how any environment presents itself to an individual and what opportunities it affords for action, and the actor's understandings of these issues as informed by culturally shared knowledge about the world that structures organization and use of acquired intelligence.

The experiential approach of apprenticeship learning shows how partial it is to consider the passing on of knowledge as solely the transfer of culturally determined and symbolically transmitted understandings between minds with their subsequent deployment impacting on the outside world, or that such understandings and how to act on them are open to all individuals in the same way. Although sharing the same natural environment, material circumstances and social arrangements shapes learning similarly for all culture bearers, differing experiences and aptitudes lead to different outcomes. Transmission involves taking on an integrated repertoire of skills and knowledge, thereby acquiring the 'ability to perform with a technology' (Stone 2007:73; cf. Richards 1993:67), rather than the transfer of discrete bits of knowledge between generations. The distinction is reminiscent of that between learning to apply particular pieces of knowledge mechanically versus understanding the underlying general principles creatively, such that one can innovate and not just reproduce. It is a distinction currently overlooked in much formal Western education with its regimented lists of 'course objectives', 'learning outcomes' etc.

From evolution...

The newly skilled may innovate creatively on tradition according to their individual talents and modify future practice, introducing fresh ideas into their cultural heritage. Extending on the embodied-knowing image, Ingold (2000:5) underscores the possibility of innovation and modification by referring to the passage of practical knowledge through socialization and practice as it being 'regrown' in each generation. Learners are active co-authors rather than passive recipients of knowledge: they do not just receive knowledge but actively 'regenerate' it for themselves through their engagement with their cultural and material inheritance (Ingold 1996:41, 2000:167). The scope for individual variation is the seed of change, underlying the mutations that occur in knowledge during transmission. The cultural approach to knowledge acquisition, which focuses on how the individual brain organizes its understanding of the world according to socially shared ideas about 'the general order of things', has consequently to accommodate potential change occurring continually at both the individual and collective levels. It has to cope with learning as an ongoing process that involves the ordering of continuously acquired and modified knowledge according to cultural frameworks that are themselves subject to continuous review. This returns us to cultural evolution (Chapter 4), which addresses the processes by which these cognitive blueprints come into existence and are passed on and modified during both childhood socialization (Quinn and Holland 1987:12) and a lifetime's experiences (Ellen 2006:29).

The evolutionary approach to changes in technology and material assemblages addresses knowledge transmission issues at the collective population level, complementing the practical individual and community linguistic focuses of the experiential apprenticeship and cultural cognitive approaches, respectively. It has a long history, dating from Victorian times (Penniman 1965; Stocking 1987), albeit passing out of favour for most of the twentieth century, though now enjoying a renaissance. It relates directly to archaeological interests in seeking to understand how material assemblages and related technologies change, often over long time spans (see Chapter 4). The approach is Darwinian at root, seeing cultural change as a phylogenetic process involving 'descent with modification' of shared knowledge and practices (Darwin 1859:171), drawing parallels, for instance, with evolutionary biology and population genetics, arguing that differences result from a cultural 'founder effect' (Lycett 2009:2). Current research seeks to understand the patterns and speed of change by investigating modifications according to one of three dominant routes of cultural transmission – either parent-to-offspring ('vertical'), between generations but non-parental ('horizontal-oblique') or between peers of the same generation ('horizontal-contagious') – using cladistic analysis, for instance, to analyse trends (Tehrani and Collard 2009:293). Less change is anticipated where 'vertical transmission' predominates, which features direct inheritance (or descent with restricted scope for diffusion) that limits the spread of innovations. On the other hand, innovations are expected to spread rapidly across and between populations where 'horizontal-oblique' and 'horizontal-contagious' transmissions predominate.[9]

The evolutionary approach, seeking to monitor and correlate technical and social change with shifts in knowledge transfer, pays little attention to the processes that feature in the transmission and acquisition of knowledge and skills. If anything, it views these as the passing on of discrete blocks of knowledge to receptive agents. Yet individuals learn to make things not so much by acquiring a specific 'cognitive model' from the generation above, but more, as discussed, through a learning process with a 'structured bias' that accommodates individual variation (Ellen 2009:271–2). The approach also overlooks how different contexts may impact on transmission, and the changes occurring in what people know. And it is largely limited to giving broad-scale accounts of change, focusing on knowledge and skill transmission over considerable time periods. While it is possible to correlate the passing on of knowledge and skills with some social and generational changes, and further our understanding of gross patterns of transformation, this cultural-evolution model is limited in considering transmission with respect to issues identified as relevant beforehand: the relationship between knowledge-holder and knowledge-recipient. The experiential approach, focusing on the processes by which people acquire and share knowledge and skills (e.g. Ingold 2000; Lave 1982), takes us beyond the view that cultural intelligence is simply 'passed along' from one mind to another between generations, to the connection between transmission context, arrangements and extent of change. It suggests that the cultural circumstances in which knowledge is passed on will influence capacity to innovate, impinging on rate and nature of change more than does the mode of transfer according to the relationship between teachers and learners (Buechler 1989:42, 47; Coy 1989:1; Singleton 1989:16–17).

This is apparent in the Was valley, where egalitarian socio-political values inhibit innovation, as pointed out in Chapter 8. Although the acephalous context may impede inventiveness, over time the ancestors of today's residents obviously evolved or developed today's style of architecture, with individual ingenuity presumably serving as the driver of changes. The prominence of tacit knowledge in house building further reinforces the conservative tendency, people just getting on with it, following customary practice, and not consciously thinking about construction design. It also suggests how today's architecture came into being. It seems unlikely that one or several persons thought consciously about house design at some time in the past and came up with today's building plan. The myth about a snake inspiring the design of vine bridges underscores the point (see Chapter 9). Rather, the design came into being through the generations as people modified – probably in small degrees overall – on then current practice. They did so most likely following tacit inspiration; that is, on the spur of the moment, while in the act of building: 'I'll try this way of doing it.' We have no idea what the prototype shelter looked like millennia ago, and there is little point in engaging in speculation on the point unless archaeologists find excavatable evidence (such as discussed in Chapter 4). We might assume that the format originated on the coast – among those distant ancestors who according to the archaeological evidence were the first humans to venture into the New Guinea highlands some tens of millennia ago (Sillitoe 2003:222–37) – and was adapted to the chillier montane environment.

The response of persons today to questions about why they construct houses the way they do supports the tacit interpretation of their evolution. When asked, for instance, why their houses have two *bez* 'ridge-poles' and not one, usual replies, after some puzzlement at such a question, are along the familiar lines of *shumbaen bismiyuwp* 'the ancestors did it like that' or it is *naun shumbaen maen* 'our ancestors' custom', or phrases to that effect. The Wola, as pointed out, often have recourse to such responses about following the ways of their ancestors when asked questions for which they have no other answer, which presumably do not occur to them (Sillitoe 2010a:473). It is possible to speculate that at some time in the past, someone building a house after a recent *dorpind* 'sky fall' event – catastrophic fallout from a volcanic eruption (Blong 1982) – decided while the event was fresh in the memory, to construct a

9 Among other such models of cultural transmission, Hewlett and Cavalli-Sforza (1986:924) argue that 'many-to-one transmission', which features several persons focusing instruction on a 'recipient', results in uniformity, whereas the reverse, 'one-to-many transmission' generates rapid change. It is a strange formulation (e.g. in our society, formal education features both modes) and not as popular as the vertical and horizontal transmission formula.

roof with twin ridge-poles to give a considerably stronger structure, as discussed in Chapter 7, which could better withstand the weight of pyroclastic fallout. Accounts of *dorpind* (Sillitoe 1996:132–3) refer to roof structures and may be interpreted as having a similar moral to the three little pigs' story about building sturdy houses so the wolf could not huff and puff them down. Such a roof structure would also be better able to withstand earthquake tremors, as noted, with the double-ridge arrangement contributing to the flexibility of the framework and making it less likely to collapse when shaken by earth movements. The twin ridge-pole innovation consequently caught on and is now customary building practice.

The tacit acceptance of knowledge passed on down the generations is of a piece with the existence of a single overall house design across the region (with some minor variations such as numbers of rooms, small differences in wall construction etc. – many of which relate to pragmatic considerations about the availability of suitable building materials). While this seems, at first glance, to be at odds with the degree of liberty that the acephalous socio-political arrangements afford individuals – with no authority to prevent experimentation, we might expect people to try all manner of different structures – the tacit approach to house construction, along with the value put on equality as discussed in Chapter 8, inhibits any such conscious inventiveness beyond occasional inspired innovations. If these prove particularly useful, they will likely catch on, with everyone who wants to, free to adopt them. Such changes, by definition, do not reflect emerging power differences between families, as being beyond the restricted use of any single party. The primeval circumstances speculated to have prompted house design are analogous to those that probably surrounded the emergence and subsequent evolution of today's social order, which did not come from the deliberations of some committee of the wise, but rather emerged tacitly during social interaction over the generations (Chapter 4). Beyond mythical accounts involving the adventures of humans and animals (Sillitoe 1979a:43–5, 2010a:86–7), people respond to questions about the origins and reasons for their social arrangements by resorting again to 'we are following our ancestors' customs' kinds of answer. In seeking to go beyond these 'it's tradition' responses, which social scientists consider inadequate, we find ourselves back again with interpretations-cum-explanations that presume to understand the forces behind the origin of socio-cultural interaction and behaviour better than the actors themselves.

The use of social ideology qua theory to access tacit issues inaccessible to people is reminiscent of the biologists' application of evolutionary theory to other animals. Whatever knowledge creatures such as tortoises and finches have, it is certainly tacit, if not largely instinctive – that is biological in origin, not learned from fellow like-creatures. Evolutionary theory supplies a coherent explanation of such behaviour and accounts for otherwise inexplicable evidence, such as the fossil record (to the embarrassment of scriptural accounts, contributing to the undermining, as mentioned in Chapter 2, of theological philosophy). But unlike other animals, human beings can teach one another things and think about their existence and actions, like social philosophers, and their culturally apprehended value-loaded thoughts inform their behaviour. They consequently intervene consciously in their environment, both natural and social, and seek to manipulate it, leading to highly complex interactions, as discussed in Chapter 4, that stymie any straightforward application of evolutionary theory to human existence. These complex interactions also query the natural-selection premise that reproductive success and the passing on of genetic heritage is behind everything. Many 'successful' persons in contemporary society, for instance, prioritize their careers and decide to have no children, passing on no genes at all. In this event, cultural evolutionists seek to identify a natural selection proxy, which is often the ability to intervene technologically in the world and modify the environment. It is the view that informs much economic development discourse. The squat, thatched and smoky houses of the Was valley are less evolutionarily advanced, from this standpoint, than the concrete, steel-and-glass, high-rise constructions of European cities. But the advanced credentials of high-rise structures are considerably less secure from the perspective of their environmental impact, sustainability or detrimental effect on social relations and interactions. What is a 'fit' arrangement in the absence of any survival-of-the-fittest mechanism in cultural contexts? Many families would prefer a humble thatched cottage to a high-rise apartment, and ironically enough, few of the architects responsible for such apartment blocks live in them.

...to diffusion

There are further problems for cultural evolution. It only partially helps us understand technological change, even on its own terms. It runs into the same problems as its nineteenth-century forerunner with respect to the simultaneous effects of diffusion, as 'Galton's Problem'[10]

10 The eponymous problem concerns the statistical issue called autocorrelation today, where comparisons are made using data that are not statistically independent. It is named after the Victorian polymath Francis Galton following his criticism of Tylor's comparative use of information from a range of past and present cultures to support his argument about the evolution of marriage and descent institutions as indicated by correlations with social complexity. Galton pointed out that the cross-cultural inferences were spurious, as similarities between cultures could be the result not only of evolutionary processes but also borrowing or common descent, and without controlling for the latter two one cannot draw valid conclusions about the former.

pointed out, in confounding both processes. While the 'horizontal-oblique' and 'horizontal-contagious' modes of transmission arguably encompass diffusion, the focus on evolutionary change within particular populations limits consideration of diffusional processes. It is difficult anyway in the modern world to sustain the implied view of cultures as discrete entities. The processes of diffusion are glossed today as globalization, a new name for a not particularly new idea (diffusion suggesting a dated and not overly successful theory). The voluminous globalization literature discusses the socio-cultural changes resulting from global communications and trade, which are responsible for considerable change in small-scale societies, including the disruption of local knowledge transmission.

One view of such change is that it involves the erosion and non-transfer of culturally shared knowledge and understanding of the world, even leading to the loss of tradition. As Polanyi (1958:53) notes, with respect to generational relations, 'A society which wants to preserve a fund of personal knowledge must submit to tradition,' pointing out that 'an art which has fallen into disuse for the period of a generation is altogether lost'. If certain tasks are undertaken less frequently, fewer persons may share related knowledge such that some of it becomes threatened, impoverishing the technology.[11] Knowledge mutations and changes in associated practices relate to concerns over loss of cultural identity, which is a theme in vernacular architecture studies, discussed in Chapter 1, that seek to document and protect building traditions from globalizing forces. While concerns about cultural identity are not yet apparent in the Was region, the intrusion of the outside world into the valley in the mid-twentieth century has brought globalizing forces with a vengeance, disrupting the tacit acquisition of knowledge passed on conservatively down the generations (which accorded with the Wola view of time, mentioned in Chapter 3, depicted as a tight spiral, with little change occurring within living memory). Subsequently, rapid diffusive change has occurred in a couple of generations. The *mbolin* 'white's' style house (Chapter 8), seen here and there along the valley, is one of the results and among the first to come to a visitor's notice, even though many families continue to prefer their squat traditional houses.

Evidence that younger generations in many small-scale communities have a diminished awareness of their cultural heritage compared to their forbears, following exposure to intensifying globalizing forces, is further stimulating interest in the processes by which knowledge and skills are acquired and transmitted. The phylogenetic evolutionary approach, focusing on social change processes, offers some, albeit limited, understanding of associated disruption of knowledge transmission between generations. However, with its population level focus, it affords little insight into the ways individuals acquire and share knowledge. It is possible, for instance, that the apparent connection between interrupted skill transmission and altered parent-child relationships is a figment of research that concentrates wholly on family-centred learning (Tehrani and Collard, 2009:289, 295). According to some, state-sponsored education and settlement is more responsible for the absence or partial transmission of knowledge and skills, and the consequent loss of cultural heritage-cum-capital in contemporary small-scale communities (Ohmagari and Berkes 1997; Zent 2001). Such forced-pace diffusion has occurred when those from one region have conquered those in another, as when Europeans explored and colonized much of the world, setting out to transform communities, such as those of New Guinea, to their ways or, as they put it, to 'civilize them'.[12] The efforts of missionaries to convert 'pagans' to Christianity is a prominent example, their establishment of formal schooling, disruptive of traditional ways, being one of their major successes, in their view, along with saving souls. The process continues today largely under the banner of development, and is particularly evident where agencies fund education programmes of various kinds. The expressed aim of these activities is to change peoples' ways of being-in-the-world. Not all such change is forced on people, many seek education voluntarily. They may want to know more, for instance, about the changing world and its formidable technology, or the capitalist system and how better to access its alluring consumer goods.

Sometimes education leads to problems similar to those caused by colonial interference. Ironically, some of these educational problems are the result of ignorance. The indigenous-rights movement argues forcefully that it is necessary to know about the cultures and histories of communities where agencies presume to intervene with education programmes. They need to ask more searching questions about the relevance of the Western style of education and avoid some of its more crass and harmful assumptions. In Papua New Guinea, for instance, it results in disaffected school leavers unable to find

11 The basketry tradition of the Nuaulu, mentioned previously, is a contemporary illustration, with individuals ceasing to pass on certain less popular designs today, together with knowledge of associated materials, such as that of the dye plant *Lawsonia inermis* that has no other use and is at risk of cultural loss (Ellen 2009:261–5).

12 The consequences have proved dire for some peoples. A cruel example is the 'stolen generation' of Australian Aboriginals, where children were taken from their parents and put into boarding schools with the expressed aim of breaking the link with their natal cultures and promoting their assimilation into White Australian society. The intellectual and emotional damage inflicted on those so treated amounts to a crime against humanity.

appropriate employment, who swell the ranks of so-called *raskol* (violent criminal) gangs that menace the nation's fragile social order and contribute to the impression of a failed state. It would be preferable to empower local people to take responsibility for their education, helping them to shape any arrangements to best meet their cultural expectations and aspirations in a globalizing world (Sillitoe 2015:17–21). This view underpins the long-running advocacy of vernacular education that challenges the idea that conventional European style schooling alone gives a 'proper' education (Nichol 2011); pointing out that this is only one approach and in the context of the overall history of humankind, a strange, even bizarre one.

Furthermore, encouraging people to draw upon their own pedagogic traditions and values may help promote the continuance of currently dwarfed minority ways of seeing and being-in-the-world. Language loss epitomizes threats to these ways. In New Guinea, for instance, an island renowned for the variety of its languages, there is growing concern, with the spread of Pidgin lingua franca, at their disappearance, together with associated encoded cultural knowledge. This globalizing erosion of the many different knowledge traditions and accumulated wisdom that have evolved over millennia is analogous to loss of species, in impoverishing humanity's cultural heritage. It is not just a matter of aesthetics, as the emergent idea of bio-cultural diversity shows, which builds on the realization that environmental conservation has to involve those living in any region because their culturally informed activities have invariably contributed to the natural environment seen today. The advocates of resilience theory and institutional economics (Berkes *et al.* 2003:20–1; Dolsak and Ostrom 2003) draw attention to this issue, highlighting the cost of the loss of knowledge between generations, in large part tacit, which promotes environmentally sustainable lifeways. There is something to learn from these other cultural traditions where they represent sustainable environmental adaptations (Chapter 6), lessons that the capitalist order urgently needs to heed with the planet under serious threat from its industrial activities. It is possible that not only how but also what is taught in Western style education may literally prove to be a dead-end for humanity.

This suggests that we need to pay close attention to these other ways of being-in-the-world and passing on associated lore, and to reflect on the possibility that Western approaches to education may not necessarily be the best and certainly do not have all the answers. It is a further reason for bothering with a study such as this one (Chapter 1), of construction activities in the New Guinea mountains. The above critique advocates reform of our overly examined and bureaucratic system of formal education with its learning outcomes, teaching strategy documents, lists of key skills and so on (Amit 2000; Shore and Wright 1999), which make teaching and learning such an unpleasant experience, particularly when viewed from the perspective of the stress learners experience, with the system intent on marking out successes and failures rather than passing on wisdom to all. This is an aspect of hierarchical social arrangements that feature an occupational division of labour, where education serves to sort persons according to career prospects, notably allowing the advantaged class to keep control of highly salaried, privileged positions (the primary purpose of British private education), which again contrasts markedly with an egalitarian order and sexual division of labour such as that of the Was valley. Other ways of learning can perhaps help us to put some humanity back into education, which, in the broad sense of passing on accumulated cultural knowledge, is a defining characteristic of what it is to be human. They may also help us recognize the contribution of tacit understanding and pragmatic experience to what we know, beyond the explicit knowledge that formal education largely deals in.

In view of the promotion of formal education in Papua New Guinea and downplaying of traditional approaches to education (Nichol 2011), it is intriguing to consider the benefits that educationalists argue the cognitive-apprenticeship strategy, which mirrors local pedagogic approaches, can bring to learning (Brill *et al.* 2001). It offers ways to tackle some of the problems facing current formal education by promoting positive learning environments and encouraging pupils through guided involvement, and using partnership learning to address disciplinary and authority issues. The advantages include authenticity (novices engaging directly with real tasks and expert practice through situated learning); motivation (students actively participate in meaningful tasks and make their own discoveries, which gives a sense of ownership); retention (learners better retain knowledge gained in such a way and can draw on it to solve novel problems); and higher-order reasoning (with expert guidance and support, students explore new ideas and make discoveries using their own judgment). The apprenticeship approach, as outlined earlier, highlights the centrality of doing in the acquisition of knowledge and the situational, context-dependent character of learning. It bridges the gap between theoretical and practical knowledge – whereas predominantly academic education downplays the latter – by embedding learning in activities that promote mastery of skills and subject knowledge. It is particularly appropriate in addressing the conundrums of a postmodern age that features the decentring of knowledge, where individual learners are enmeshed in particular networks of relations. The apprenticeship approach acknowledges that learning features an element of enculturation, not always readily transferred between contexts, knowledge being embedded in local contexts and languages rather than based on universal laws. It complies with increasing recognition of the contribution of intuition, pragmatic experience and tacit knowledge to what we think we know.

Participant-observation as apprenticeship

The comments about alternative approaches to education, familiar to those with any knowledge of anthropology, apply to the discipline not only in the formal context of the classroom, teaching about anthropological issues and communicating something about other ways of being-in-the-world; but also in the informal one of the field, learning about socio-cultural issues and accommodating so far as possible to other ways of knowing about the world. Participant-observation, referred to at several junctures in previous chapters, features prominently in the field context. The fieldworker using this approach seeks to study a community by participating in activities, so far as possible, which assumes that some knowing persons (key informants in yesterday's terms, collaborating mentors in today's) teach us as novices (perhaps even idiots, when levels of ignorance approach mine on first entering the Was valley) about their way of life (Spradley 1980). We seek to achieve familiarity with people through close involvement in their lives over an extended time period, by living among them (to the extent practicable). The maxim is learn to 'do as the locals do' and establish rapport with the community; that is, 'talk the talk' and 'walk the walk' (DeWalt *et al.* 1998). This allows us, when we have people's trust and are a familiar presence, to see goings on, enquire into issues in depth and check the reliability of information by triangulating between individuals at different times and in various contexts. Fieldwork involves a range of methods in addition to observing and participating in daily life, including interviews, surveys and biographical enquiries; while these are largely qualitative they may also include, as in this study, collection of quantitative data. But participant-observation is central.

The practice of participant-observation, particularly the participation aspect necessary to engage with tacit knowledge, is intriguingly somewhat akin, given the foregoing discussion, to entering into an apprenticeship (Lave 2011). Researching building skills, Marchand makes the connection, championing participant-observation-cum-apprenticeship 'within anthropology's canon of standard practice' for enabling 'academics to acquire some level of first-hand experience, and possibly "expertise", in the practices that they theorize and write about exposed viscerally to the learning environments and livelihoods of fellow workers ... able to interact more competently within the multiple mediums and nuanced forms of communication that are employed in the transmission of skills' (Marchand 2010:S10). But there is something disingenuous in intimating progression from menial labourer to top apprentice in a few months of fieldwork, when it takes local Malians several years to master mud-brick production, wall and ceiling construction, and the sculpting of rooftop crenulations. Elsewhere Marchand notes that he 'was not privy to the masons' secrets ... Only apprentices who work closely with their masters ... over a period of many years' learn everything (2009:9), but he overlooks investigating the implications of such constraints on participant-observed knowledge. They are significant in my experience.

It was through participation that I became aware of the skill entailed in house-wall erection. This illustrates how taking part helps gain an inkling of some tacit issues, and also the limitations of the approach. Wall construction involves the driving in of substantial wooden stakes around a marked-out house plan (as related in Chapter 7), an activity in which I took part after acquiring some thick industrial gloves to protect my tender hands from splinters and grazes. These suggest a feeble attempt at participation, which is probably a fair reflection of what many of us achieve, for in truth my participation was admittedly fairly limited. My friends watched my endeavours closely, with some amusement on occasion, feeling responsible for seeing that that I did not injure myself as a soft-handed academic, and they never left me to work alone. The erection of the walls of a house sounds and looks easy enough, involving the *aezay* 'spearing' of pointed stakes into the ground, which men do by holding them with both hands and driving them into the soil. It takes several thrusts to get a stake firmly established, and men wiggle the stake to and fro each time to loosen it and pull it from the ground for the next thrust, placing it in exactly the same place. It is necessary to drive in the stakes in a straight line with their tops all about the same height above the ground. Sometimes a man sights along the wall line and gives directions while another drives in stakes, as described previously; other times an individual working alone stops periodically and stands back to sight along the wall for straightness. The driving in of the stakes is a mechanical operation, the movement of arms and hands subject to subsidiary attention. The house-builders focus their attention on seeing and achieving a true line, and placing the stakes firmly in the soil. It was trying my hand at stake 'spearing' that taught me this, for I was unable to achieve a true line with my concentration focused on the stake in my hands and the shock to my body when it entered the soil. The tacit skills involved in wall erection do not cease there. The depth to which men need to drive in stakes varies with soil type, as does the 'spearing' technique. In a *tiyptiyp* subsoil with high volcanic ash content they drive stakes in deeper than in a *hundbiy* clay subsoil occurring on limestone, and deeper still if driving stakes into a *bombraysha* subsoil high in organic matter (Sillitoe 1996:276–89). You can not only sense when you have 'planted' a stake right with your eyes, watching its alignment, but also by the 'feel of the stake', which relates to its firmness, such as resistance to sideways movement, and the solid sound it makes as it enters the soil. I never managed to see, feel or hear a stake properly.

The tacit knowledge drawn on in house construction extends beyond practical building tasks. When out with men collecting raw materials, it seems that they know where to go in the forest or nearby cane grassland to find what they need without need to search for it. They just go to the stand with the saplings they need or the tree with particular bark or whatever. In a sense they do not think about it. The mention of a place name, as recounted in Chapter 6, can conjure up an image of certain available resources. Even when they ask relatives if they can help locate something, they usually just give a place name, sufficient to convey the information. Again, I cannot pretend to have achieved this level of geographical knowledge of the Was valley, even though I have walked through much of the region and visited many of the places men go to collect building materials. In order to participate meaningfully, that is to have the required tacit awareness of resources to know where to go to collect what, it would be necessary to get on top of the Was valley system of place names that convey information about availability of suitable natural resources, who has rights to use those found where and so on. In short, I shall need to know the geography like the residents know it.

The degree to which participant-observation gives accurate insight into others' lives varies, but whatever we achieve, it is illusory to think – as the postmodern critique points out – that we can fully understand actors' perceptions of their cultural reality, however much we engage in and experience activities with them (Geertz 1984; Rosaldo 1986). Critiques of the approach raise various problems. The fieldworker's presence, for instance, is likely to influence participants' behaviour somewhat. The researcher's personal interests and views will inform what is picked up as relevant and noteworthy and what is not.[13] They will also invariably influence the interpretation and evaluation of data, as discussed in the previous chapters. On the positive side, we may more readily apprehend the tacit dimensions of social behaviour in a foreign community because we do not take its norms for granted like persons who are brought up in the culture and accept these without question as normal. While these customary ways implicitly inform native-born behaviour, to an outsider they may seem strange; for instance, the lengths to which the Wola go to avoid stepping over others and their possessions is striking and relates to men's fear of women's poisoning powers, which leads on to gender relations, as this book discusses at various junctures. This is to argue, paradoxically, that to participate 'fully' and know like a native is limiting, and relates to the comment made several times about social researchers assuming to observe and know more than actors. Whatever we observe and participate in, our account will anyway be partial and biased. A reverse danger is 'going native' in striving for an in-depth understanding of a community, although this is improbable in the New Guinea Highlands, even if feasible. Any account resulting from someone truly 'going bush,' which is unlikely (after all, it assumes the fieldworker suffers amnesia and forgets the conventions of academic research or the reverse, that the natives have such an academic tradition), would presumably be difficult to evaluate by any disciplinary criteria.

Participant-observation, as the term suggests, involves both observing and participating in a community. An intriguing issue is how we join in and stand back at the same time, that is experience activities as a participant and analyse what is going on as an observer. We are either a participating observer or an observing participant, as it is difficult to be both simultaneously. The distinction implies that we either watch people and talk to them with notebook in hand recording what we see and what they say while occasionally trying our hand at the activity, or we put the notebook to one side and join in fully, asking them to teach us with a view to subsequently recalling our newly gained knowledge and experience. While not mutually exclusive, it is difficult simultaneously to participate and observe in any measure. When some friends were building a house adjacent to ours, for instance, I joined in with various tasks, such as driving in wall stakes, and gained some insight into the work involved and what they achieve in constructing houses entirely from locally available materials. But when I documented in detail the construction of the house at Honael, as detailed in Chapter 7, I carried a notebook, camera and clock, which precluded me taking part to any significant extent.

While participant-observation is not about both participating and observing at the same time but engaging back-and-forth in both during fieldwork – which mirrors formally in some measure what we do informally in everyday life moving to-and-fro between practical engagement and intellectual detachment – it suggests that we act both as subjective participant and objective observer, in striving to

13 A recent critique, in the context of tacit knowledge enquiries, comes from Ingold, although it is difficult to know what to make of it. Regarding ethnographic research, he asserts that the 'real problem... lies not in the alleged contradiction between participation and observation, which is a chimera, but in the disconnection of the art of description from observational practice' (2011b:224), because we leave the field where we observe and go to the library to describe what we have seen. This comment on participant-observation seems to overlook that most of us use a notebook to take down details of what we observe and experience in the field, not relying on memory recall of these when settled in the library. Anyone who has worked in New Guinea, for instance, will know that it is not possible to remember, even follow, the details of many complex exchange sequences without taking on-the-spot notes. It is this resort to notebook recording, after all, that makes participation problematic, because it inhibits full engagement in events, to the extent that it is possible anyway.

understand what being-in-the-world means for those we live with. On the one hand, we seek to engage closely with a community, with all that involves personally, while on the other striving somehow to stand back and record what we observe without allowing our personal selves to intrude. It implies that we achieve an improbable combination of a subjective and objective perspective. The postmodern critique undermines this aspiration, as pointed out at several junctures already, casting doubt on the validity of our observations in comparison to actors' understandings, asserting that we can never really know other communities, beyond our own subjective experiences of them. Whatever I do, I cannot achieve the viewpoint of my Highlander friends, let alone tell others about it. And there is no objective stance, however I conduct my observations, given the cultural relativity of my knowledge.

The intention, in conceding that participant-observation is largely an aspiration, is not to disparage the method. Those working in this way have unquestionably furthered our understanding of humanity's diversity. Maybe I am biased, having used the method during my Was valley research, but it has contributed significantly to the discipline's bedrock ethnographic legacy, without which, however defective, it could not 'construct or deconstruct' its theoretical edifices. It is a key method in ethno-determinism, opening up unanticipated cultural doors, maybe revealing unknown unknowns, when used in an open-minded fashion. It is a sound approach to working with people who lack the means to leave a record of their lifeways, particularly those undergoing rapid change – as increasingly is the case, with the aforementioned loss of languages and associated wisdom and knowledge that enrich our understanding of ways to be-in-the-world.

The methodological implications of participant-observation have been puzzled over for many years, with the shortcomings informing contemporary shifts in research practice. The move towards collaborative and participatory approaches (Sillitoe 2012b, 2015), for instance, addresses the postmodern dilemma to some extent, in seeking to meet demands for equal relationships with host communities by enabling the co-formulation and undertaking of research. In striving to incorporate people meaningfully into the research process as partners not informants, it includes those who, as socialized members of communities, know them as natives. While considerable challenges of translation may remain, the arrangement diminishes considerably the subjective-outsider-distortion effect. Participant-observation continues to feature in such contemporary participatory research, albeit as only one method among several. There is, after all, no substitute for outsider researchers taking part in local insider activities to advance meaningful local participation, let alone gain understanding of them.

The problem of interest here is not to what extent we can participate in or observe activities but what we can claim to do with what we learn, beyond doing them. The assumption is that the intelligence we gain through participant-observation, which we acquire by joining in various goings-on, is not only knowable to us to some extent but also that after serving our field apprenticeship, we can communicate whatever it is that we think we know to others. But the previous discussion of tacit issues questions this conviction, intimating the limits of participant-observation, and new participatory methods come to that, in accessing such knowing and making it known. After all, the apprenticeship and related educational models may, in attempting to formalize tacit learning, give us some idea of how we acquire skilled knowledge but they do not help us understand what it is to have it, once we have mastered it. In brief, to what extent is it possible formally to convey what we learn observing and participating in activities such as building houses?

12

Knowing the tacit

The propositional approach to knowledge may be unable to represent some aspects of tacit awareness, nonetheless what kind of an explicit handle can we get, if any, on know-how, such as a craftsperson subconsciously acquires and employs, or know-that intelligence that we all routinely use in everyday social life? In view of the personal nature of the transmission of such knowing – as apprenticeships exemplify in enabling novices to acquire and develop the skills of a trade through hands-on participation – we can anticipate some challenges in capturing it in explicit terms. These problems run through this book as it struggles to deal with the tacit dimensions of construction and other knowledge in the New Guinea Highlands.

Indeed, it is arguable that tacit and explicit competencies differ to such an extent that it is confusing to refer to both as knowledge, suggesting an improbable similarity (Gourlay 2002:12). While there is a large philosophical literature on what constitutes knowledge – for example, the 'externalist' theories relating to causality, reliability and counterfactual issues, and 'internalist' theories, such as foundationalism, coherentism and subjective awareness (Honderich 1995:447) – it is not necessary, even if space permitted, to engage in a review of associated tangled epistemological debates. The key issue of interest here is that knowledge, as the philosophers put it, comprises 'justified true belief' (Nonaka and Takeuchi 1995:21, after Plato) or 'true warranted belief' (Gourlay 2002, after Klein 1998).[1] These definitions accord with the empiricists' theory of knowledge, that of philosophers such as Locke, Mill, Berkeley, Hume, Peirce and Ayer, as discussed in Chapter 2 in the context of the ethno-determinist perspective. Induction has a prominent role in this view, with secure knowledge founded on true belief derived from experience (Russell 1948), where our senses feature (after Hume 1896), all this in turn relating to the veracity of facts, those fragments of information that are a significant component of knowledge, which we assess as either true or false, and comprise the elements from which we construct our views and theories of life and the world.

It is difficult to see how we can apply this definition of knowledge to tacit understanding. If it depends on bodily awareness, how can we consciously know what we embody is true? It is only proven in action. Unable to express such comprehension in words, it is not possible to produce reasons to warrant belief and justify its truth. We can only do this for explicit knowledge – albeit according to the sceptical philosophical tradition we can never be sure of achieving such true belief because we cannot rely totally on what we induce to be true from previous experience (while the sun may come up again tomorrow, one day it might not – indeed according to astronomers, aeons from now, it will not do so). The conundrum of whether experiential knowing comprises knowledge is another artefact of Western academia's exclusive focus on dialogic propositional understanding, which is at the heart of our struggle to deal with tacit awareness. It is evident in discussions of tacit issues from Polanyi onwards, who often referred to 'knowing' instead of 'knowledge', shifting the focus to a process rather than an entity (Blackler 1995; Gourlay 2002). But whether we talk of an activity ('knowing') or a thing ('knowledge'), we still face the problem of conveying what is tacitly known using our explicit, wordy and symbolic code, as Marchand makes clear with respect to Yemeni minaret builders, where 'requests for explanations of knowledge … resulted in incoherent or incomplete responses as they were not accustomed to verbalising their "knowing that" … In summary, the builder cannot explain "how" he knows.' (2001:176, 220).

Managing knowledge conversion

Some management experts claim to have cracked the problem. The centrality of tacit knowledge to many company activities has prompted business management to take an interest in making it explicit, in an attempt to increase efficiency and hence profits. Such knowing comprises expertise within a company that is not formally

1 Assessing the probable truth or reliability of knowledge imparted by speakers is a feature of the Wola language, which allows that knowledge is often partial or based on unreliable evidence, as signalled by incorporating 'evidential' markers in verb constructions (Sillitoe 2010b).

recognized, for instance in training manuals, but is necessary for its effective running. The amorphous quality of this knowledge poses challenges for conventional management practice, namely how to access what is not on the record and encourage staff to communicate about things they do not normally discuss. In this respect, organizational management distinguishes between knowledge 'stocks', which are the explicit corporate capital such as databases, files and manuals found in offices, and knowledge 'flows', which are the movement, sharing and dissemination of information around an organization. The organizational approach highlights that much of a firm's knowledge circulates tacitly among employees, who acquire skills through informal on-the-job training, and that such knowledge is held in and communicated via unofficial social networks that hierarchical management structures cannot tap into and control, and may inadvertently subvert if they try to do so.

Some management specialists, distinguishing 'technical' from 'cognitive' tacit knowledge, talk of tacit knowing as the property of both individuals, manifest in action-based skills and learned through doing, and groups, manifest in shared practices and unspoken understandings among members (Baumard 1999:30–3; Choo 1998:117–19). While the tacit-knowing property of individuals and groups is not the same, both feature transmission of subliminal knowledge. Those who have the requisite experience, in whom tacit knowing resides, may find it challenging to formalize what they know in such a way that they can share it with others who lack their experience. The challenge according to some business-management experts is to convert such tacit knowing into explicit 'stock' so that its 'flow' can be managed around the company. While it may be difficult to articulate, they claim to have ways of doing so. They have come up with the theory of 'organizational knowledge creation', which they maintain can codify tacit knowledge and convert it into explicit knowledge.

Organizational knowledge creation seeks to 'amplify' workers' tacit knowledge, promoting its deeper understanding to create 'new knowledge'. The aim is to explain 'organizational creativity, change, and innovation ... the ultimate outcome of organizational knowledge creation is product and process innovations ... defining new problems and creating, exploring, and experimenting with new solutions' (Nonaka and von Krogh 2009:644, 646). It sets out to explain 'how new ideas come forth in innovation, not only how individuals tap into rich practices and acquire the tacit knowledge' (Nonaka and von Krogh 2009:645). According to the theory, interaction between individuals with diverse backgrounds promotes innovation (Brown and Duguid 1991; Roberts 2006) and 'practitioners "intuitively" recognize when the "rules" should be discarded or reformed in light of emergent circumstances', albeit they may not be able to 'reflect on the "rules" of performance while they are engaged in the practice' (Nonaka and von Krogh 2009:645). In respect of inspiration and new insights, Scharmer (2000) distinguishes between 'not-yet-embodied tacit knowledge' and extant 'embodied tacit knowledge', the former dependent on leaps of imagination that a person is unable to explain *ex post facto*. It is the sort of 'impossible to pin down' awareness that gives companies a competitive advantage and is hence the Holy Grail of business studies that seek to elucidate how interactions within corporations might stimulate such innovative advances, without necessarily seeking to convert it into explicit propositional knowledge. It is intriguing to note that the reason for seeking to capture tacit knowing is to promote innovation and change, which contrasts starkly to the conservative forces traditionally present in small-scale social orders such as that of the Was valley where the reverse pertains (Chapter 8), ensuring a largely unchanging way life, until recently anyway. The acceptance of the ineffability of tacit knowing by such populations may be, as we shall see, a more realistic view of being-in-the-world.

Organizational knowledge-creation theory uses the idea of a tacit-to-explicit continuum to track how organizations create new knowledge through a 'process of knowledge conversion', the two types of knowledge apparently '"oscillating" to mutually enhance each other' (Nonaka and von Krogh 2009:642, 643). The goal is to advance on 'the somewhat static view of "knowledge assets" ... by explaining how tacit and explicit forms of knowledge interact to create new knowledge' (Nonaka and von Krogh 2009:638). The assumption is that 'knowledge is created and expanded through social interaction between tacit knowledge and explicit knowledge', and this interaction is christened 'knowledge conversion' (Nonaka and Takeuchi 1995:61). One scheme for promoting the process features four modes of conversion (Nonaka and Takeuchi 1995:63–9). One of these, the 'externalization' mode, involves the conversion of tacit knowledge to explicit knowledge by using analogies, metaphors and models to articulate it, albeit with opaque results. It parallels what anthropologists do in interpreting tacit social behaviour and unverbalized cultural symbols. The other three modes are 'combination' (explicit to explicit, using formal education and training to pass on codified knowledge systematically), 'internalization' (explicit to tacit, internalizing knowledge that is first verbalized, by learning through doing) and 'socialization' (tacit to tacit, acquiring skills through observation and practice, as in an apprentice learning from a master).

These modes compare with the stages of the 'levels of learning approaches', such as Bateson (1973:250–79) formulated in asking why people improve at doing things with practice, even fairly meaningless tasks. He proposed four kinds of learning arranged hierarchically, each higher level incorporating the lower ones. They are i) experiencing something directly through the senses (a nettle stings);

ii) generalizing about such experiences (avoid all nettles); iii) contextualizing such generalizations and extracting implicit rules[2] (though generally avoid nettles, may enter nettle patch to help someone); and iv) contextualizing such knowledge existentially (what taking a risk with nettles signals about me). The move between the first level and the next two adumbrates the distinction between tacit and explicit learning, and according to Bateson, understanding the nature of the shift is one of the major challenges facing contemporary educational theory.

Other approaches to organizational knowledge creation maintain that they can capture tacit knowledge by 'task analysis' involving structured interviewing of experts and/or observation of them, and by action learning that consciously replicates the master-to-apprentice craft-learning process of observation and practice. By recording these activities in fine detail (as in Chapter 7's account of house building) it is apparently possible to turn tacit into explicit knowledge. A widely cited example of such knowledge conversion concerns the development of an automatic bread-baking machine by a Japanese company. The traditional Japanese engagement with the tacit dimensions of knowledge, as exemplified in *koan* riddles, may account for why some of Japan's firms sought early on to make it explicit in business contexts. Some claim that their efforts to do so underpin the success of Japanese companies. In order to perfect the dough-kneading process, the company apprenticed an employee to a master baker to learn the skill, which involved stretching and twisting the dough in a particular way. The employee subsequently conveyed the motion to design engineers, who replicated it in the bread-baking machine (Nonaka and von Krogh 2009:638, 645; Nonaka and Takeuchi 1995:103–6; Ribeiro and Collins 2007). The person learned the skill of bread making

> through socialization, observing and imitating the head baker, rather than through reading memos or manuals. She then translated the kneading skill into explicit knowledge. The knowledge was externalized by creating the concept of "twisting stretch". In addition, she externalized this knowledge by expressing the movements required for the kneading propeller, using phrases like "more slowly" or "more strongly".
>
> (Nonaka and Takeuchi, 1995:105)

In certain contexts it is possible, it is argued, to open in this way the 'black box' of internalized subsidiary awareness that features while concentrating or focusing attention on actions elsewhere (cf. Bateson, 1979:43). This is what management studies claim to do in 'converting' tacit to explicit knowledge. They assume that 'knowing how' (Ryle 1949:27–32) to do something, such as knead bread properly, is only waiting to be put into words or rules by an actor reflecting sufficiently on what he or she is doing, whereas when it exists only in use it remains unexpressed in rules. It seems that knowledge-creation theory assumes that while certain practical understanding may only be achieved through apprenticeship, it is not qualitatively different in content to knowledge acquired 'through reading memos and manuals', as it can be reduced to similar articulated rules, and even be reproduced by machines, following appropriate 'conversion' methods; that is, it can be 'translated' into explicit knowledge.

Anthropological and archaeological conversions

According to another 'cultural model' approach, task-based knowing is not of a different order to knowledge expressible in language (Quinn and Holland 1987:3–5). Knowledge that informs understanding of processes, of how to perform tasks, draws on the same prototypical 'image-schema' as linguistically expressed knowledge of the world that depends on basic relationships between entities and concepts (Quinn and Holland 1987:25; cf. Marchand, 2003:33). Culture features in the selection of metaphors considered appropriate for structuring the prototypical schema that inform understanding and action (Quinn and Holland 1987:30). This approach, incorporating culture in the know-how acquisition process – where it inevitably features as we do not invent associated activities for ourselves anew – seeks to understand the relationship between collective representations, individual cognition and task capabilities.

Furthermore, tacit understanding may include 'ostensive knowledge' (Collins 2001b:73), that is, intelligence conveyed with gestures, although not communicable in words, which demands personal contact to pass on. Human behaviour involves a range of non-verbal 'paralinguistic' signals such as nods, shrugs, arm gestures, facial expressions and eye movements that complement and enhance communication, as is evident from infancy onwards, when such signing features centrally in interaction between adults and young children. We commonly communicate using both non-linguistic and linguistic means interwoven together in everyday interaction (Geertz 1993:6–12). They occur simultaneously and one is not necessarily more primitive than the other, as Bateson (1968:614) argues 'paralanguage has blossomed side by side with the evolution of verbal language'; albeit kinesics is not well adapted to trading in ideas that are central to our stock of explicit knowledge. Actors may pick up on non-linguistic signals subliminally and these implicitly inform their behaviour and understanding. Such knowing requires participation to follow the gestures that feature in any activity, as they are only made in the context of action, another pointer to the relevance of participant-observation, discussed in the previous chapter. Nonetheless, it is conceivable that some

2 Sometimes called 'deutero-learning'.

such communication could be made explicit, in view of the aptitude of humans to invent words for activities and things when they recognize them. The combination of explicit and tacit dimensions in the same activity, where 'mutually complementary', affords an opportunity to access the latter (Nonaka and von Krogh 2009:640). This notably occurs when we express explicit knowledge in utterances that draw by definition on tacit knowledge of language. The knowledge-creation approach particularly concerns activities that we consciously concentrate on while learning them, but which become increasingly tacit as we acquire the capacity to undertake them; as evident, for instance, in Artificial Grammar Learning,[3] which 'starts as slow and consciously modifiable, but with repetition, it gradually becomes automated' (Pothos 2007:230). It assumes that while we may not reflect on such knowledge once acquired, we may nonetheless consciously access it if prompted to do so.

Some anthropologists find this linguistically attuned cultural focus conducive to their interests. For example, Marchand (2001, 2003), who, in exploring the conversion of tacit knowing into explicit knowledge in his enquiries into Yemeni building practices, turns to Fodor's (1981) 'language of thought' approach to mental states and representations, and his subsequent, though now dated, modularity of mind thesis.[4] After noting the challenges faced by 'an inquiry limited to registering and analysing propositional statements ... as the traditional builders were not proficient at verbally explaining *how* they know', he proposes to adopt 'a modular approach to *thinking about thinking*. Establishing an analytic framework which considers not only propositional concepts in language, but sensorimotor representations in performance' (Marchand 2001:156 – original italics). But it is unclear how the 'modular approach' overcomes the problems of converting tacit into explicit knowledge (even if recent neuroscientific research (Zimmer 2014) had not undermined the approach's credibility).[5] Nonetheless, Marchand claims to show that the 'language-biased window can be expanded by demonstrating how a careful consideration of the skill-related performances inculcated during the course of a training programme reveals and distinguishes different types of cognitive operations ... in particular those associated with object descriptions and spatial relations' (2001:168).[6] In so doing, he maintains that it is possible to convert 'non-linguistic knowledge to ... explicit discourse', albeit he adds the contradictory rider that the 'knowledge itself changes character in the process' (Marchand 2001:176, citing Bloch 1998:7). The 'non-propositional nature of the builder's expertise' means that its translation into 'another domain, namely a propositional one encoded by language' involves a 'substantial loss of specificity and content' (2001:244), which appears to contradict claims to convert tacit knowing into 'explicit discourse'.

Other attempts to convert tacit knowing into explicit knowledge in anthropological contexts that focus specifically on building traditions are likewise contradictory. In a recent collection of anthro-architectural essays, Ingold (2013), for instance, claims to refute Polanyi's assertion that tacit knowing is beyond words. The refutation involves distinguishing between what it is to know and to tell: 'we can tell of what we know through practice and experience, precisely because telling is itself a modality of performance ... [such that] what is not explicated may still find expression in spoken or written words' (Ingold 2013:109), and we thus mysteriously have explication of the 'not explicated'. The claim to capture the tacit in formal terms involves the conflation of two meanings for the English verb to 'tell', which are a) telling or speaking something in words, such as telling someone a story about an event; and b) telling or sensing something is likely to happen, such as my Wola friends can tell by the way an axe goes in how timber is likely to split. Apparently, 'both in the verbal relating of stories and in the coupling of sensory awareness with

3 Artificial Grammar Learning is a method used in cognitive psychology to research language learning by investigating peoples' ability to learn made-up languages.

4 The modularity thesis maintains that the mind comprises innate neural modules or relatively independent structures that have distinct evolved 'domain specific' functions relating to particular perceptual and linguistic processes, which in turn connect to a 'central processing' part of the mind that relates together their various activities and contents about things in the world (Fodor 1983). But as with phrenology, its precursor, experimental evidence questions this view. Current neuroscientific research (Zimmer 2014) indicates that the 'domain specificity' and associated 'encapsulation' (insulation) assumptions are erroneous, showing that mental functions involve large overlapping networks of subsystems, that when individuals perform different tasks similar brain areas are active and that the same brain injuries in different subjects cause many varying mental deficiencies.

5 Similar opacity regarding the relationship between tacit and explicit knowledge characterizes Marchand's later work on Malian building traditions, where he 'takes a broad encompassing perspective on what constitutes knowledge, one that crucially moves beyond propositional forms of knowledge expressed in text and the spoken word to include, at the very least, the skill-based practices and performances enacted by the socialized and enculturated body of the craftsman' (2009:8), albeit he confusedly uses the written word to do so.

6 Intriguingly, given criticism of Descartes' mind and body dualism in the cognitive sciences (Chapters 2 and 5), his doctrine of innate ideas apparently helps overcome the tacit to explicit barrier, for a 'Neocartesian psychology maintains that mental structure is mirrored in the organization of the propositional content of mental states, and that analysis of this organization will explicate the innate mental structure' (Marchand 2001:168–9).

material variations – telling is a practice of correspondence', the suggested conjunction implying that a skilled person can tell a story that conveys what they can tell in action, which is inferred as tacit (Ingold 2013:111).

> These two senses of telling are closely related ... because for those who listen ... the telling of stories is an education of attention. Through it, things are pointed out to novices, so that they can discover for themselves.
>
> [In this way] novices grow into the knowledge of their predecessors through a process that could best be described as one of "guided rediscovery", rather than receiving it ready-made ... To tell, in short, is not to explicate the world, obviating the need for would-be practitioners to inquire for themselves.
> (Ingold 2013:110)

This surely amounts to an affirmation of Polanyi's argument,[7] 'guided rediscovery' equating with apprenticeship-like acquisition of tacit knowing for oneself, which cannot be handed over 'ready-made'. The substitution of 'stories' for 'specifications' does not make the tacit explicit. It arguably conveys little about house building at all, the implication that you could know what it is to build a house by listening to stories about others' experiences, likely told with 'ostensive' gestures, seems improbable. It amounts to a convoluted affirmation of the reflexive approach, discussed further below, advocating subjective accounts of what it is like to do something, rather than what actually to do, that is, how to experience it and so know it tacitly yourself. In order to achieve this latter level of understanding it is necessary to engage in house-building tasks, which presupposes knowledge of the specifics of house construction.[8]

A recent attempt in an archaeological context to deal with specifications and show how a craftperson's 'movements and techniques can be translated into ... text' (Høgseth 2013:96) proposes – perhaps appropriately enough – using a system of virtual hieroglyphics. It draws on a notation system used in dance choreography[9] because 'To succeed in translating movements and techniques into an academic format, it is necessary to invent and develop a notation system that helps analyse details related to the craftsperson, the tool, and the motions.' Consequently, the method needs to use 'signs to focus on and capture the dynamics in the craftsperson's movements and the relations between rhythm, speed, force, technique, etc.; it also points to how the body is tied to materials and the use of tools' (2013:100–1).[10] While the approach seeks to further our understanding of tacit issues from an archaeological perspective, as I propose in Chapter 4, it unfortunately encounters the same problems as other attempts to make 'non-discursive knowledge discursive' (2013:104), requiring us to master a strange new notation to boot in supposedly converting embodied know-how to words. The use of notation to show steps in certain crafts, such as knitting and crocheting, is a common long-standing practice (Jørgensen 2012); a forerunner of the *chaîne opératoire* approach (see Chapter 2). But these notation systems assume that you are going to follow them in practice, acquire the skill necessary to make something and then dispense with them. It is disingenuous to suggest that they overcome 'the difficulty encountered in verbally articulating craft knowledge that has become "second nature" to the practitioner', and that such 'notation is used ... to bypass these problems' (Makovicky 2010:S96). They are aids to acquiring tacit knowing, not formal substitutes for it.

The focus on 'technical-rationality' in conversion-to-explicit-knowledge approaches may stunt practical understanding and action, as Schön (1983) argues in distinguishing between 'reflection-in-action' and 'reflection-on-action'. He advances an epistemology of practice 'in which the knowledge inherent in practice is be understood as artful doing' (Usher *et al.* 1997:143). Central to this reflective practice and learning approach is that actors have two theories of action (Argyris and Schön 1974). There is one implicit in what they do, 'theories-in-use'; and a second they explicitly adopt in telling others about their actions, 'espoused theory'. The discrimination is between action and attribution, between the way we execute and then review our actions. Reflection-in-action involves 'thinking on your feet'. It draws on previous experience and understandings, and relies on intuition in determining actions as a situation unfolds. We have to puzzle things through as every circumstance is unique, which implies

7 As Ingold further confirms later in the same chapter, telling us that 'Concentrated in skilled hands are capacities of movement and feeling that have been developed through life histories of past practice' (2013:115), and describing a practical class with students making string that 'offered an experience of tactility' (2013:120).

8 According to Ingold (2013:110) a 'complete specification would offer no guidance', providing 'information *about* the specified, about the materials to be used, about parts and their dimensions, about movements to be made', unlike stories apparently that 'issue *from* moving bodies and vital materials, in the telling'. This appears to contradict Ingold's phenomenological position in which he argues for a focus on the 'becoming' of things in human hands, which implies specified movements with materials; it is certainly at odds with this book.

9 The Sutton Movement Writing system.

10 In an ethnographic context, Farnell (1994) discusses the use of another dance movement notation, 'Labanotation', to record the bodily motions involved in embodied acts, illustrating the method with North American sign language (which may be tacit for experienced users, though they can nonetheless give verbal definitions of their actions).

not following established textbook ideas and techniques closely but innovating as necessary. We rapidly reflect on and generate an understanding of the new phenomenon drawing on previous ideas (Schön 1983:68). Reflection-on-action occurs later, after the event. It is to go over actions subsequently and expand on ideas about activities and practice. The notion of repertoire features here, with practitioners amassing information, examples and images of actions from experiential knowledge to draw upon later (Dewey 1933:123). Persons make use of such experiences and understandings that come from 'reflection-on-action' during future 'in action' contexts. They see any new situation as both similar to and different from one in their repertoire, which serves as a 'sort of' familiar precedent, without necessarily specifying in what way similar or different (Schön 1983:138). It serves as a basis for further thought and action. We may not have a complete grasp of things before we act, but drawing on our repertoire, previous experience and current frame of reference, we usually respond appropriately and manage. This view of practical acts as unique engagements informed by previous experience of working with similar materials recalls the recent anthropological and archaeological materiality-cum-affordances-of-materials line of argument and the 'flow'-focused critique of *chaîne opératoire* step-by-step approaches to technical activities (Chapters 2 and 7).

Critiques of knowledge conversion

While the 'reflection-in-action' and 'reflection-on-action' distinction appears to have validity, the relationship between them is unclear, as again they seem to be inextricably entwined in reality. Persons can describe how they think in undertaking some actions, using a repertoire of ideas and images (albeit when in the 'zone' and time is short, opportunity for reflection may be limited). But for activities that draw on tacit knowing, conscious reflection is limited for both 'in' and 'on' contexts; with 'in'-engaged practitioners not reflecting focally on their physical acts but doing, and 'on'-engaged ones being unable subsequently to reflect verbally on these actions. While verbal and non-verbal signals may feature in achieving 'tacit knowing' (Polanyi 1969a), it is wrong-headed to argue that such knowing, when 'we attend to something ... can be interpreted without loss into a sign-process or semiotic framework' (Gourlay 2002:18). As Geertz comments in his review of Levi-Strauss's work: 'One can become conscious of one's grammatical categories by reading linguistic treatises just as one can become conscious of one's cultural categories by reading ethnological ones. But, as acts, both speaking and behaving are spontaneous performances fed from underground springs.' (Geertz 1993:354). It is from these underground springs that tacit behaviour flows, which normative-rule focused social science largely overlooks. Tacit knowing involves more than signalling of whatever sort, it centrally involves doing.

While non-verbal 'ostensive' signals may feature in achieving 'tacit knowing', if it is possible to reflect on these signals and translate them into words, that is, bring them into discursive conscious focus, we are dealing not with tacit but implicit knowledge. They comprise different sorts of knowing (Collins 2001b:72–3; Janik 1988:54–8), there being a subtle difference between implicit and tacit categories that goes beyond neither being readily made explicit. Implicit knowledge may not yet be formally understood or expressed according to any intellectual principles or theory, but may be in the future; whereas it is not possible similarly to codify tacit knowledge on current evidence. In short, the distinction is between what is not said but could be if prompted, and what cannot be said ever. Regarding the construction of house walls in the Was valley, for instance, youngsters come to realize that they have to drive stakes in firmly and aligned, by seeing what others do and hearing comments about getting the stakes in right. It is subliminally obvious that a true wall line is necessary to ensure a stable structure without anyone pointing it out explicitly, but if an ignorant foreigner asks house builders why they go to such lengths they could likely formulate this implicitly held explanation.

When making this differentiation, implicit and codified knowledge are not always clearly distinguishable (Polanyi 1958). Implicit knowledge falls in the grey area between wholly tacit and wholly explicit knowledge, which comprise poles on a continuum (Howells 1996; Nonaka and von Krogh 2009). In this grey area 'the tacit cooperates with the explicit, the personal with the formal' (Polanyi 1958:87) and we have a certain limited capability, perhaps, to make some knowledge found towards the tacit pole explicit in some circumstances. When management experts claim to convert tacit knowing to explicit knowledge they are referring to such implicit knowledge that is potentially amenable to overt symbolic representation. According to Baumard (1999:2, 59), we may not want to convey verbally what we know implicitly, whereas with tacit knowing we cannot do so. Regarding implicit knowledge not yet put into words, this may be due to 1) power issues such as concealing trade secrets, intentionally or unintentionally, 2) unrecognized knowledge that no one has tried to document such as certain craft know-how, or 3) widely held and unquestioned presuppositions such as what constitutes normal behaviour. There are no insurmountable obstacles to making such knowledge explicit, once recognized, unlike things inexpressible in words; that is, those 'aspects of human experience which are *wholly* knowable self-reflectively ... but by their very nature are incapable of *precise* articulation' (Janik 1988:54, original italics). This truly tacit knowledge may relate 1) to the senses, such as the characteristic smoky smell of a Wola house, which we can only gain by experiencing the sensation, or 2) to practices that involve 'open-textured rule-following', such that efforts to make them explicit lead to a 'rules-regress' impasse (Collins 2001a; Johannessen 1988).

According to one commentator, 'the appropriation of "tacit knowledge" in management studies' is a 'great misunderstanding', because viewing it as 'knowledge not-yet-articulated – knowledge awaiting for its "translation" or "conversion" into explicit knowledge ... ignores the essential ineffability of tacit knowledge' (Tsoukas 2003:424). It is an impoverished view of 'practical knowledge'. The apprentice baker, for instance, acquired more than 'technical knowledge' (Oakeshott 1991:12–15), without realizing it. While the bread-making machine episode illustrates how a mechanical appliance may duplicate human skill, it does not make the tacit knowledge explicit. Transfer to other humans still involves the traditional apprenticeship process. While a simulated procedure that reproduces a behavioural sequence that for a human involves tacit knowing may suggest that such a task performance does not of necessity demand tacit knowledge, it does not imply that tacit knowing is unnecessary for the successful performance of the task by human beings. The apprentice baker could not express the tacit or 'ineffable' (Polanyi 1962:87–95) aspects of what she learned in a series of propositions, even though it was evident in the loaves she baked. In order to 'convert' her newly acquired bread-making skill into mechanical motion, she had to become focally aware of her subsidiary kneading knowledge. In other words, her attention was no longer focused generically on making bread, but specifically on thinking about bread kneading, which was not the same thing. Indeed, turning subsidiary awareness to focal attention, and trying to specify the particulars of a skill that are 'logically unspecifiable' (Polanyi 1962:56), would disrupt the undertaking of the activity, reducing the artisan to a clumsy apprentice again (see Chapter 2).

Subsidiary awareness cannot be specified on its own, as it has to be viewed in association with the focus to which it attends (Chapter 2). Subsidiary knowledge is 'not known in itself but is known in terms of something focally known, to the quality of which it contributes; and to this extent it is unspecifiable ... expert[s] ... know many more things than they can tell, knowing them only in practice, as instrumental particulars ... the pondering of a judgement in terms of such particulars is an ineffable process of thought' (Polanyi 1958:88). When separated from focus activity and examined independently, subsidiary knowledge loses its *raison d'être* and changes its nature. While we can turn our attention to subsidiary awareness, we cannot do so in the context of actions where it features in a subsidiary capacity. Indeed focusing on such awareness after completion of the activity that defined its existence – which derives by definition from connection to that activity – is to see it in a context that differs significantly from the original locus of action. A new set of subsidiary particulars will inform this new context of action, and so we find ourselves in an infinite regress. This relates to Wittgenstein's idea of the 'open-textured character of rule-following behaviour' (Janik 1988:56–8), namely that it is not possible to specify the rules of any action fully, because to explain these involves further rules, which in turn require more rules etc. In the final analysis, rule-following depends on doing, as it is necessary to learn how to undertake something before one can appreciate the import of following the rules to do it.

Some have sought to draw on the authority of Wittgenstein's – admittedly obscure – later philosophy to argue, after Polanyi, that it is futile to seek to reduce tacit to explicit knowledge (Gill 2000; Janik 1992; Johannessen 1992). Practice is central to its acquisition and paying attention to this dimension brings up the role of intuition and hunches. The consequent amorphous quality of tacit knowledge challenges conventional research practices, posing considerable methodological problems. It is arguably beyond conventional enquiry and documentation, as the lone statement that concludes Wittgenstein's controversial *Tractatus* intimates: 'Whereof one cannot speak, thereof one must be silent.'[11] (1922:75). Although on another occasion he was more optimistic, noting that 'the very things that are most obvious can become the most difficult to understand. What has to be overcome is not difficulty of the intellect but of the will.' (Klagge and Nordmann 1993:161). And more enigmatically, 'Something that we know when no one asks us, but no longer know when we are supposed to give an account of it, is something that we need to *remind* ourselves of.' (Wittgenstein 1953:no. 89, original italics). The distinction between 'saying' and 'showing' perhaps hints at the way we might steel the will or remind ourselves. We can say something in a proposition, such as 'Wola houses are squat', and while we can say something similarly propositional of the phrase itself, such as 'it is a short phrase', it can only show that houses are squat (Kenny 2005:48). If we are unable to say something, we can try to show it. And this is what this book seeks to do.

While it may be impossible to 'convert' skilful knowing with its ineffable personal insights into explicit knowledge, this does not imply that we cannot talk about practical skills – such as the baker's 'twisting stretch' – otherwise I could not write this book. It is not that any activity that involves some technical 'knowing-how' precludes a certain cognizable 'knowing-that' (Dretske 1991:116). Any such activity likely involves knowing where and in what order to do things to achieve a task, including what tools and materials to use in performing it.[12] And such

11 *'Wovon man nicht sprechen kann, darüber muss man schweigen'* (Wittgenstein 1921:75).

12 And vice versa, interpretation of explicit knowledge may require tacit knowledge (Polanyi 1966:7).

'knowing-that' may be largely tacit, or at least implicit, as not usually reflected on consciously or put into words. Yet, falling within the implicit 'grey area', it does not comprise of corporeal experiences beyond the ambit of reasoned discussion. It may become evident during the acquisition of skills when intentionally accessed and articulated during instruction of an apprentice in correct procedures, body movements and so forth. It is possible in this way to reflect on practical activities, during and after their performance, and draw attention to certain aspects and issues – as evident in instructions such as 'try doing it this way', 'compare this with that', 'use this material' – so long as we realize that we are not capturing the entire skilled repertoire explicitly, but only those 'technical knowledge' aspects that we can articulate as directions and propositions (*contra* Ambrosini and Bowman 2001; Eraut 2000). While the novice may need such declarations of knowledge to acquire skills, they are unnecessary for an expert, who has achieved a tacit understanding through practice that renders articulation redundant. Consequently, some aspects of tacit task-related knowing – although by no means all – are similar to our everyday dispositions to make assumptions, and in certain circumstances persons can become aware of them. They are implicit in that they are usually unconscious but can be accessed, for example under questioning or on tackling an unforeseen problem.

Practitioners and, more particularly, apprentices may have their attention drawn to unremarked aspects of activities and, in reviewing these, gain insights and see new connections. The use of language to instruct and interrogate actions may bring forth novel understandings because those 'aspects of things that are most important for us are hidden because of their simplicity and familiarity. (One is unable to notice something—because it is always before one's eyes.)' (Wittgenstein 1953:no.129). Such a procedure may bring things to light that we already know but, as Wittgenstein points out, do not know that we know, of which we need to 'remind ourselves'. It was in this fashion, by having design engineers 'remind' her – by drawing her attention to what she had unconsciously taken on from the master baker – that the apprentice from the bread-machine company explicitly formulated the 'twisting stretch' concept. It is in this manner that we can talk about skilful knowing, which, although ineffable in the final analysis, we may 'remind' ourselves about and notice previously unremarked features and so deepen our understanding. The implication is that we need to pay attention to fine ethnographic detail, as 'ethno-determinism' advocates, to capture as much as possible of what may be overlooked, if we intend to further understanding of tacit knowledge. When we review our understanding in this way, we may give 'prominence to distinctions which our ordinary forms of language easily makes us overlook' (Wittgenstein 1953:no.132). While this is a commonplace way of gaining an 'explicit understanding of our everyday, practical activities … [using] the self-same methods for drawing *our* attention to how people draw each other's attention to things' (Shotter and Katz 1996:230 – original italics), it has significant implications for our ability to understand and document such knowledge.

When tacitness is absolute and articulation is impossible, we have 'ineffable' knowledge. In this event, it can only be passed on 'by example from master to apprentice. This restricts the range of diffusion to that of personal contacts.' (Polanyi 1958). Even knowledge-creation theory has to admit that truly tacit knowledge is inaccessible to 'conversion', for at the 'extreme end of the continuum, tacit knowledge is not accessible through consciousness … [being] tied to our physiology and sensory and motor functioning … [and] our history of physical movement in the world' (Nonaka and von Krogh 2009:641, 642). New knowledge of the kind sought by management studies through organizational creation theory does not result from 'converting' tacit to explicit knowledge – which strictly speaking is not possible as experiential knowledge is only manifest in action (von Krogh and Roos 1996) – but emerges through largely unmanaged social interactions where persons 'remind' each other about implicit aspects of what they do that may have escaped their attention previously, prompting them to explore novel connections. 'We cannot operationalise tacit knowledge but we can find new ways of talking, fresh forms of interacting and novel ways of distinguishing and connecting… New knowledge comes about not when the tacit becomes explicit, but when our skilled performance – our praxis – is punctuated in new ways through social interaction.' (Tsoukas 2003:426). Management experts who accept this view argue that tacit know-how can 'flow' within organizations without any elusive 'conversion'. Companies need to focus on creating environments that facilitate its passage and learning, not bogus conversion to explicit knowledge. The 'situated learning' approach and associated 'communities of practice' discussed in the previous chapter are relevant here (Lave and Wenger 1991; Wenger 1998). These comprise informal associations that emerge spontaneously in the workplace as persons cooperate to achieve tasks. They are consequently beyond explicit managerial control and need to remain so to be effective in creating and transferring knowledge in organizations; albeit companies can facilitate the communication of such knowledge by encouraging creative interaction, such as employing participatory methodologies that allow innovative visual and practical approaches. These informal arrangements mirror the way in which people customarily pass on knowledge in the Was valley through personal interaction, largely with kin. It is likely that high-tech companies may learn something from the way they do so, such face-to-face social orders being subtly complex in respect of informal learning and knowledge corroboration.

The supposed 'conversion' of tacit knowledge to explicit knowledge using analogies, metaphors and models produces opaque results at best, and scarcely advances our efforts to get on top of experiential knowledge. While one may rue the 'division between propositional and embodied learning ... that, ironically, paints practice theorists into a corner when observing skill education' (Downey 2010:S26), it appears currently to be a tight corner, from which it may be inappropriate to think of extricating ourselves. It is arguably distorting to portray embodied knowledge in opposition to explicit knowledge in this way, for 'Tacit and explicit knowledge are not the two ends of a continuum but the two sides of the same coin: even the most explicit kind of knowledge is underlain by tacit knowledge.' (Tsoukas 2003:425). It is paradoxical that any explicit understanding features tacit knowing, as in our grasp of the symbols used to convey such intelligence, where 'explicit knowledge must rely on being tacitly understood' (Polanyi 1969b:144). This contradiction captures this book's struggle to convey the experiential knowing that informs construction activities in the Was valley. While it is sometimes possible to know 'subsidiary elements' and relations between them explicitly, and such knowing may further understanding in other ways (such as complex linguistic rules of syntax go beyond the average speaker's subsidiary awareness of language), this ability to 'paraphrase the cognitive content of an integration' does not overcome the point that 'the sensory quality which conveys this content cannot be made explicit' (Polanyi 1968:32). Although we may strive to describe activities and behaviour fully, as attempted here for Highlands' house construction, including dependence on subsidiary features, the resulting description, as situated cognition studies point out (Clancey 1997; Gourlay 2001), is obviously not the same as the activity. How feasible is it explicitly to access 'the sensory quality' of doing and being any further, if at all?

Making sense of the senses

There are broadly speaking two fundamentally different 'realist' ways to make sense of sensory experience; that is, to conceive of the link between an external object and its internal representation. While they broadly confirm Hume's (1896) assertions about the senses, they take us way beyond anything that occurred to him regarding the 'passions'. According to those of a dualist persuasion, inexplicable mental operations go on within us such that we cannot reduce our interpretation of sensory experiences merely to physical processes. An inference is not due to mechanistic cerebral activity but is a conscious action that occurs over and above a sensory input. On the other hand, according to those of a behaviourist persuasion, we can explain the entire process physically; that is, reduce sensory experiences to physiological accounts. The brain is simply a bodily organ that processes sensory data and coordinates the things we do correspondingly.

The Wola entertain both points of view without any apparent intellectual problem. On the one hand, they distinguish between mind and body, as mentioned in Chapters 2 and 5, in the Cartesian dualistic fashion of some Western philosophy. They see the material and spiritual as two aspects of reality, recognizing that everyone has a tangible *tokor* 'body' animated by an intangible *wezow* 'life-force', which is also one's 'shadow'. All living persons have a *wezow* by definition, which at death departs the *tokor* to become a *towmow* 'ancestor spirit' that more often than not is malevolent towards its living kin, visiting sickness on them. While there may be no empirical evidence scientifically speaking for the existence of such spirits, the fact that people fall ill and die is sufficient proof for those who believe in them. When missionaries, who also subscribe to the mind-body distinction, told people about their beliefs in the soul, they readily understood and converted to Christianity (transferring with minimal proselytizing and so evading *towmow* ancestor attacks, among other reasons – albeit persons continue to fall ill and pass away).

On the other hand, the Wola talk of the *konay* 'intellect' that resides in the *hibiy-aumuw* 'liver-heart' region of the chest, which, although it is not visible if you cut open the body, suggests a physiological view. It relates to the *konem* 'thoughts', which are the *konay* in action so to speak, processing reflections, opinions, decisions and so on. What you know *beray* 'sits' in the *konay*, in your chest organs; in contrast to our skull-located brain ontology. The organization of thoughts about particular issues involves acts of *nendsay* 'cognition'. The Wola talk of knowing something in terms of the senses, as *henday* 'seeing' or *pangay* 'hearing' it. They say that the *konay* belongs to individuals and anticipate that their *konem* about issues may differ. A frequently heard phrase when persons act in a certain way is *obun konem* 'it is her/his idea'. While their *konay* informs their actions and calculations, individuals do not act alone. Others influence and shape their *konay*, initially as mentors when growing up and subsequently through ongoing social interaction, with kinsfolk particularly prominent. Individual experiences mould one's *konay* too, it growing and changing throughout life. After discussing similar ideas of Melpa speakers (who distinguish between *noman* 'mind' and *min* 'soul'),[13] A. Strathern (1996:43) comments that they are 'the reverse of Descartes's' formulation for they 'link rather than separate mind and body' and consequently 'their dualistic concepts [are] closer in operation to the holistic ideas of the phenomenologists'.

The dualist perspective argues that it is not our brains that are conscious but our noumenal selves. We are physical beings that have minds that make us enigmatically

13 Several languages in New Guinea make similar distinctions, as Strathern (1996:76–84) illustrates in reviewing several of them.

more than the sum of our parts. When we have a sensory experience, 'see' something for instance, it does not simply register as a brute physical fact in the brain, for we consciously 'interpret' it. The occurrence of illusions demonstrates this, it is argued, where sense data originating from something may prompt quite different interpretations. The related 'indirect realist' approach accepts that physical objects in the external world are responsible for our sense experiences. It seeks to account for our ability to recognize, classify and discuss things according to such sensory information without recourse to Platonist universals (abstract ideal templates of phenomena that have an eternal otherworldly quality, which are heavily criticized regarding epistemological access). A device deployed by philosophers in debates about the mind, to represent the subjective property of such conscious experience, is the quale, which comprises sensory experiences internal to the brain and 'is directly intuited, given ... it is purely subjective' (Lewis 1956:121). According to Dennett (1988), qualia have four properties: 1) ineffable (incommunicable, only accessible by direct experience); 2) intrinsic (irreducible, whatever the connection to other things); 3) private (methodical interpersonal comparisons not possible); and 4) directly or immediately apprehensible in consciousness (experiencing a quale is to know it). Focusing on the link between external object and internal representation, qualia concern the phenomenal character of experience, notably what is happening with the 'intentional gaze', and extend on the phenomenological view of reality that assumes that we only ever have direct access to sense experience and not the external world.[14]

It is difficult to demonstrate the existence of qualia, these not being amenable to communication, and in arguing for them philosophers resort to hypothetical thought experiments, such as the oft cited 'Mary's room' scenario (Jackson 1982). It proposes that Mary has lived from birth in a monochrome room where everything is black and white. She knows every fact of physics about other peoples' colour experiences, but it is only when she leaves the room that she learns what it is to see colours.[15] In other words, she acquires the quale that equates, for instance, to experiencing seeing green, when she sees green. While we can imagine a Meri[16] from the Was valley isolated in a room, a more feasible scenario would be trying to explain to her what it is like to drink a Burgundy wine, giving her all the organoleptic facts about its taste (which would make more sense than the physics of light). But she will only know the taste when she sips the wine and acquires the relevant quale. Consequently, qualia are 'perceptual experiences, which no amount of purely physical information includes' (Jackson 1982:273) and they have real properties, as a person who has access to a quale differs from someone who does not. Philosophers use such arguments to refute 'physicalism' in which all things, including the mind, comprise only of physical properties explainable by the laws of nature. Although Meri may possess all the physical information about a flavour 'out' in the world, she does not comprehend the experience of tasting wine 'in' her head, hence there is a non-physical aspect to our taste knowledge. A starker illustration would be conveying the experience of drunkenness to people who drink only water, as my Wola friends did when I first knew them.

A favourite family amusement when we share a bottle of wine is to describe the flavour to one another -- such as 'suggests fresh cassis and ripe blackberries with intense aromas of red blossom, enticingly spicy with classic truffle notes combined with hints of citrus' – and see who is nearest to what the label tells us. We invariably disagree, but presumably experience the same wine quale taste. The 'anthropology of the senses' or 'sensory anthropology' assumes that we can know what others taste and smell (or sense through their ears, eyes and skin) because if we drink what they drink we experience the same sensations, these being biologically universal to our species (Pink 2009:73–6). It is a questionable position; for instance, people who have only ever drunk water, such as many of my Was valley friends, seem to have finer-tuned taste buds able to distinguish different water sources; whereas persons such as myself have the edge taken off their taste sense by drinking wine, coffee, beer and so on. Even if we allow duplication of experience, we still face the problem of conveying in words the actual taste sensation of eating cassowary meat or pandan nuts to those who have not eaten these things; beyond an autobiographical (reflexive) account of our gustatory experiences consuming them, after 'sensory anthropology', together with a discussion of associated social responses and local comments.

The qualia idea has been subject to considerable and tortuous philosophical criticism (Ludlow, Nagasawa and Stoljar 2004; Tallis 2011; Wright 2008), embracing that intriguing character the 'philosophical zombie', a hypothetical being identical to a human except without

14 The quale concept resonates with the philosophy of Bishop Berkeley, who, observing that we are only aware of material things in the world by virtue of mental representations and not directly (all we have are the inner images and no way to get outside our heads and see that they really match the way things are out there), went on to deny the existence of an external world underlying our inner experience because all that exists for us are the ideas in our minds.

15 Arguably, Mary's inability to learn to see green is due to the inadequacy of language to convey experiences, albeit this admission, in the absence of proposing a different way of communicating, does nothing to advance a written review such as this one!

16 The name Meri is the New Guinea Pidgin version of Mary; the word *meri* also means female.

conscious experience.[17] The notion of qualia straddles, philosophically, the physical and metaphysical; they are undetectable, prompting questions such as 'do they exist or not beyond the imagination?' According to some critics it is difficult to establish a connection with real entities employing qualia, as virtual constructs they obfuscate the link behind a mask of pre-conceptual awareness. They run into the classic mind-body dualism problem of how a non-physical entity such as a quale connects with the physical world. For 'Brain is Brain, and Mind is Mind, and we don't know how the twain meet' (Koestler 1967:203). How can something psychic – quale, conscious, soul – cause effects in the material world, such as trigger neurons to fire in the brain (probably resulting in some corporeal action), when this requires physical energy to occur? A further criticism of the 'Meri style' reasoning is that it unrealistically separates 'know-that' knowledge from 'know-how'.[18] Learning communicates 'know-that' phenomenal information, whereas the experiences to which qualia refer, transmit 'know-how' tacit abilities (Lewis 2004). In everyday life the two occur largely simultaneously, we gain information (e.g. that a wall comprises woven bamboo) and ability (to recognize images of such walls), whereas Meri only acquires 'know-that' intelligence from what we tell her, until she drinks and gains the 'know-how' ability to recognize the wine-taste sensation, which involves not further information but experience.

The information and ability distinction is starker if, instead of something cerebral such as colour or taste recognition, we imagine Meri mastering a physical skill, such as netting string designs (Sillitoe 1988a:242–60). This returns us from a shaky imaginary world to the solid real one, which residents of the Was valley would appreciate, much of whose knowledge involves 'know-how' skills. The implications of the distinction for knowing apply even to apparently straightforward tasks such as felling trees. It is possible, for instance, to show someone, such as an ignorant anthropologist, that felling a tree involves making two cuts on opposite sides of the *iysh tay* 'base of the trunk': one higher and shallower, called the *tib*, and the other lower and deeper, called the *maend*; this second cut is intended to control the direction in which the tree will topple. Yet no amount of 'know-that' information, about gravitational field and vector forces for instance, can prepare you to judge the direction of fall of a tree, which depends on several interrelated points, such as crown shape; the direction, however slight, the tree is leaning in; the wind's strength and direction; the density and heaviness of the timber; and the surrounding vegetation, such as vines growing between neighbouring trees that can 'pull' the crown in a certain way. The only way that you can gain the skill to judge how a tree is likely to topple is to fell many trees and arguably acquire the necessary repertoire of associated qualia or build up the requisite experiential knowledge. The way in which other, variable situational factors contribute to the outcome of the tree falling further illustrate the notion of things having agency (see Chapter 5) in addition to the human actor, who relies on bodily experience. A tree-feller cannot know for certain how any timber will respond to his axe blows and Bateson, who sees the mind as analogous to an ecosystem, points out the agentive implications, where

> Each stroke of the axe is modified or corrected, according to the shape of the cut face of the tree left by the previous stroke. This self-corrective (i.e., mental) process is brought about by a total system, tree-eyes-brain-muscles-axe-stroke-tree; and it is this total system that has the characteristics of immanent mind ... But this is *not* how the average Occidental sees the event sequence of tree felling. He says "I cut down the tree" and he even believes that there is a delimited agent, the "self", which performed a delimited 'purposive' action upon a delimited object.
>
> (Bateson 1973:318)

Neuroscience and making sense of experience

The veracity of the qualia concept is largely beside the point so far as this book goes, which can accept that there is 'a wholly non-conceptual component of perceptual experience' (Lowe 1996:101), representation of which presents challenges. Nonetheless, is it possible to get a more explicit hold on such experience? Some think that advances in neuroscience, furthering our understanding of how the brain functions, may offer some scope. Indeed, it is possible to identify neurological properties of qualia (Ramachandran and Hirstein 1997), which include: 1) invariable neural representation (green is always sensed as green, we do not sometimes see it as red and others as yellow etc.); 2) being subject to interpretation (a tree may be seen as a source of building materials, a place of shade, a roost for birds, a reminder of a pleasant memory, a symbol of strength etc.); 3. involve short-term memory (for interpretation to occur, which implies involvement of neurological circuits that generate attention, otherwise we have only a reflex action). But this does not overcome the problem of how to connect such non-physical, arguably imaginary entities to the physical real world that we experience through our senses. The behaviourist approach avoids this impasse by arguing that it is all physical, and occurs with no need for any numinous intervention by

17 If you hit such a zombie with a stinging nettle it would not feel any pain, although it would behave as if it did so!

18 Also, arguably, if Mary knows everything physical about the experience of seeing colour, she effectively has omniscient intelligence, which she can use to know the sensory experience of seeing green (Dennett 1988).

a non-physical soul-like entity. According to this line of argument sensory experiences are products of physiological activity within the brain, which is just a bodily organ like any other, such as the heart or liver.

While some philosophers may critique this empirical approach to understanding embodied knowledge as too positivistic, even flawed, most scientists accept the behavioural argument. From the Enlightenment onwards, when thinkers such as Descartes, Bacon and Galileo promoted experimentation and exploration of the world, scientists have amassed empirical evidence-based data that fundamentally challenge metaphysical explanations of being, as pointed out previously (Chapter 2), such that today most scientists would consider it unexceptionable to assume that knowing and thinking are bodily functions that occur largely in the head. The field of neuroscience is advancing our understating of the physiological mechanisms, burgeoning recently with technological developments in neuroimaging (see Chapter 5), such as magnetic resonance imaging (MRI) and positron emission tomography (PET), which allow scientists to study the anatomy of subjects' brains non-invasively, and to scan their cerebral activity as they undertake different tasks, by physically tracking blood flow or oxygen consumption in different parts of the cerebrum (Rilling 2008:8–11).

Yet neuroscience has had scant impact on anthropology (Brown and Seligman 2009:31), as a recent overview comments 'In spite of important discoveries over the last three decades ... brain science scarcely touched anthropology, except in its biological wing... Continued insulation from brain sciences appears untenable in the long run, potentially increasing our field's marginalisation, especially as neuropsychology and imaging turn to questions of cultural diversity.' (Downey 2012:243). Regardless of the lack of interest, the neologism neuroanthropology has emerged recently to label anthropology's limited engagement with neuroscience (Downey 2012; Rilling 2008). It is an example of anthropology's apparent capacity to incorporate other subjects within its boundaries by adding anthro- and ethno- prefixes or suffixes to other disciplines' names.[19] This is both a strength in facilitating interdisciplinary engagement and a weakness in suggesting a discipline that cursorily covers everything to do with humanity and nothing in specialist depth. It is a contradiction to which my ethno-determinism contribution to the neologism thicket plays. Furthermore, the coming together of the biological and sociological sides of anthropology in neuroscientific enquiries, to look into tacit knowledge for instance, promises to further the rapprochement of these two wings of the discipline, together with archaeology (as seen earlier in human ecology, and increasingly promoted by broad-based anthropology departments, such as ours in Durham), in a way that avoids the problems referred to in previous chapters that attend attempts to 'rehabilitate' evolution.

The hesitancy of some branches of the social sciences to engage with neuroscience is partly attributable to suspicion of its more sweeping assertions. The idea that neuroscience can explain how humans tick, which Tallis (2011) dubs 'neuromania', is arguably a reductionist fallacy, reminiscent of bumps-on-the-head phrenology. There are intriguing parallels with the 'ecological' approach to materiality (see Chapter 6), in that both ultimately reduce understanding of what is going on to molecular interactions. In neuroscience these involve ions such as Ca^{2+} causing electrical activity across axon neuron membranes, triggering the release of neurotransmitters such as glutamate into synaptic gaps, which bind to receptors in adjacent dendrite neurons generating an action potential that stimulates further connections (Rilling 2008:7–8; Tallis 2011:16–19). It is this cerebral physico-chemical activity that comprises thought.

While certain localized (or modularized) parts of the brain 'light up' when scanned, this does not explain behaviour or emotions, it only indicates what parts are active; for instance, 'PET images do not explain *how* neuronal activity contributes to mental behaviour' (Stout 2005:280 – original italics), they only indicate *where* the activity occurs in the head, such scans showing blood flow or oxygen consumption in different parts of the brain alone. They do not show what areas of higher activity are doing, nor do they rule out the possibility that other areas of lower activity may contribute in some way. Just knowing what area of the brain is active when doing something – such as felling a tree – is only a step towards advancing understanding of human cognition. Indeed the same areas repeatedly feature in a range of activities and are credited with underlying entirely different functions, which prompts Tallis (2011:82) to draw an amusing analogy with a 'crooked estate agent' who lets 'out the same bit of real estate simultaneously to different clients'. Even if scans did exactly match up particular areas of brain activity with particular thoughts and actions, what would this signify? Neuroscience suggests a melange of answers ranging from brain activity coinciding with, causing or comprising psychological states, which suggests that nerve impulses sparking through the brain are somehow able to appear to us as things other than what they are.

There is no consensus over what comprises 'internal' representations and how neurons, which individually represent nothing, realize these cognitive patterns, namely our thoughts, through the activation of complex neural networks that extend throughout the body and via it to

19 Other social sciences that have the same urge to incorporate new fields follow the same strategy, such as archaeology that has recently coined the neologism 'neuroarchaeology' (Malafouris 2009), as discussed in Chapter 5, to label enquiries that draw on brain science.

the material world beyond, as pointed out by the 'extended mind' hypothesis (Chapter 5). It seems that cerebral electro-chemical activity overlays two appearances: on the one hand, perceiving something via the senses, such as seeing a green leaf, and on the other, producing a matching brain state. The mystery is how we get from the physical firings of synapses to perception and ultimately to self-awareness, without conjuring up a supervising consciousness, which is the thing we want to explain. The 'inside' appearance is only available to the person with the brain, and neuro-cognitivists arguably do the same as believers in immaterial mentality to accommodate it: they smuggle into the head not one mind but many minds (or, according to Tallis 2011:188–9, 'homunculi' tiny observers), in the guise of countless neuronal circuits.

The relationship between the brain and the mind is ambiguous. Although the mind, which we think of as supplying our representations and understandings, derives somehow from cerebral functioning, it comprises an abstract entity undetectable in the physical activity of neurons. The quale seeks to fill the gap, although as an idea itself it does not circumvent the mind's abstract existence, which prompts some to eschew reference to the mind at all, as too poorly defined to contribute to constructive debate, and even reminiscent of a dualistic spirit force (Gosden 2010:40). The network of billions of neurons in the human brain produce what complexity theory recognizes as emergent properties, that is synergies way beyond the activity of any single neuron, resulting in the mystery of the mind or consciousness; which leads to neuroscience trying to reduce our awareness to the electro-chemical impulses of our grey matter. The complexity increases by many orders of magnitude given the trillions of cognitive connections built and maintained through culture that link humans together in ways that transcend biology, going way beyond the organic functioning of our individual 'en-brained' bodies, making it fatuous to reduce human awareness to brain circuitry alone.

Robotic knowing

While caution is in order, it would be perverse to dismiss the neuroscientific approach to skilful knowledge without some further consideration. The 'causal theory of perception' argues that a path exists between an externality and perception. Retinal images of a thing, for instance, connect to neural activity in the brain. While the visual input that we perceive when light rays hit the retina is our evidence of real entities, it is essentially hypothetical, a virtual view. The activity in the brain, involving complex neurological pathways starting with the sense organs, is of a quite different order to the signal arriving from the external world. The light waves hitting the eyes do not propagate to the cerebrum's occipital lobe. While sensory-induced activity in the brain may produce a real-world simulation, it does not convey knowledge about an entity's properties. Raw sensations are 'non-epistemic', that is evidence lacking information, until the mind conceptualizes them, its response informed by memory and featuring previously established neural circuits. We interpret the neuro-physiological firing that sense experiences stimulate in the brain according to previous understandings that are continually updated through a process of complex neural feedback. We learn how to perceive things and their properties, and such perception involves continual adaptation, always open to revision with experience of the real and correction by other persons. A child in the Was valley comes across wood from an early age and senses that it is a hard substance that can sometimes cause pain, for instance that as hot embers it burns, and remembers this (as a quale, if you wish), tagged as fearful and to be avoided. Subsequently, she refines this view of wood with further experiences, learning that not all wood, such as a mother's digging stick, hurts to touch, until perceptions match the real world and behaviour produces results that fit desired outcomes. When the child acquires language, she will revise some of her understandings further according to what others say about phenomena such as wood (Harman 1990). This approach to knowing concedes that all of our understandings of the real are necessarily tentative, according to Hume's (1896) sceptical argument.

When we pay no attention to sensory stimuli, that is, they are not the focus of current conscious interest, they effectively become invisible and explicitly knowledge-less, which is the basis of tacit understanding. It is arguable that pointing out that such knowing is tacit avoids addressing the question of what mental operations occur in activities that feature such knowledge: 'How does one produce behaviours of that kind?' (Fodor 1981:74). This class of mental operations is open, some argue, to analysis using the so-called 'language of thought' approach, which maintains that the brain operates in a way analogous to a computer program, with a match-up between syntactic and semantic processes. It proposes that correlation of the mapping of neural impulses inside the brain with perceptions of goings-on outside in the world is reminiscent of how a computer binary-code programme generates an image, although the code is not an image in itself but an otherwise incomprehensible array of numbers and letters. This line of argument relates to artificial intelligence, where research in the latter part of the twentieth century has fuelled discussions about the possibility of making all knowledge explicit (Goranzon and Florin 1990; Goranzon and Josefson 1988). It arguably allows us to address the question of how 'unspecified' the processes are that underpin tacit knowing.

According to the artificial-intelligence approach, if we can express in language the code that allows a computer to simulate human actions, we can infer that it represents the cause of those acts: 'If machines and organisms can produce behaviours of the same type and if descriptions of machine computations in terms of rules, instructions, etc., that they employ are true descriptions of the etiology of their output, then the principle that licenses inferences from like causes to like effects must license us to infer that the tacit knowledge of organisms is represented by the programs of the machines that simulate their behaviour' (Fodor 1981:78). The inference that a computer-robot can exactly replicate human behaviour is itself a tall order, but even if we assume that this is possible, it does not permit us to infer from 'like causes to like effects' that the same thing is going on internally in both. It brings to mind criticisms of the bread-making machine mimicking the baker. The inference that computer-generated acts that simulate human actions replicate the same internal processes is questionable, for what goes on in a computer is likely to be completely different to human cognitive processes (Dreyfus and Dreyfus 1986). Different internal arrangements can produce the same results (this even applies at the mechanical level – for instance, a film camera functions quite differently to a digital one).

The technologies now available to neuroscience allow us to map the neural impulses in the brain and correlate with perceptions of goings-on out there, such as our encounters with wooden objects. It is conceivable that in the future we shall know exactly the 'structural isomorphism' between outside-world objects and inside-head impulses, that is, synapse by synapse and cell by cell what neurological activity in the brain relates to seeing the colour green, tasting red wine or knowing how a tree will topple, and that spark of enchained cerebral activity might be identified as a quale that has a transient almost non-physical existence. But it is difficult to see what such brain scanning may contribute to our understanding of tacit issues other than locating the parts of the cerebrum that 'light up' during actions that feature such knowing and skills. While mapping neural pathways as they fire electrical pulses in response to sense stimuli may be analogous to diagramming the binary switching that occurs in a microchip during particular computer operations (after the way that digital code produces particular letters in word processing or symbols in mathematical calculations), the ability to locate and map the neural pathways does not get us far in understanding what it is to have tacit knowledge and acquire skills for ourselves (no more than knowing the letters of the alphabet code allows us to know what it is to write a poem or mathematical symbols what it is to solve a differential equation). The 'Chinese room' scenario of Searle (1981) neatly illustrates the point, showing how a computer just manipulates symbols without understanding their significance.[20]

The computer analogy falls down because machines have no capacity to think independently, a central issue in artificial intelligence and robotics, which for some comes down to them having no 'spirit' or 'life', or 'ghost' in Ryle's (1949:15–16) memorable phrase.[21] This returns us to the dualist perspective. Even so, the behaviourist view is equally valid, for although no machine currently comes anywhere near simulating the incredible complexity of the neural networks that comprise the human brain, which result in synergies way beyond the sum of their individual parts (with a capacity to generate and manipulate qualia, if you will), it is nonetheless arguable that the functioning of the human brain ultimately depends on biochemical processes with no ethereal input; that 'all mental events can be reduced to the operation of the "automatic telephone exchange" in the brain' (Koestler 1967:202). Whether you subscribe philosophically or scientifically to the embodied position, it seems that cognitive science, and notably neurobiology, offers a way forwards – blatantly empirical – to further our understanding, particularly of tacit knowing, without necessarily embarking on the goose chase of converting it to explicit propositional knowledge. But there is currently no prospect that the objective approach of science may account for the subjective aspect of experience (Nagel 1974:436), and so overcome the problems that a study such as this faces engaging with tacit skilled knowing, and we have to resort it seems to less thoroughgoing ways of conveying 'know-how'.

The reflexive way out

A response to the conundrums that surround the apparent inaccessibility of embodied knowing, explicitly speaking, is to take phenomenological ontology at face value (notably existential versions that inform postmodernism, which maintain that objectivity is an illusion and subjectivity the only true reality) and accept that we only ever have direct access to our sense experiences and not the external world. The only 'true' perspective, divorced from the existential immediacy of reality in this way, depends on individual experience. If all that any of us can know is Plato's play of shadows then that is all that we can report on. It has led to the reflexive approach in the social sciences that assumes that if we can only ever have entrée to our own subjective view of the world, then we can only legitimately claim to know and report on our personal experiences. The reflexive approach is in part a rejoinder to the postmodern critique

20 In this thought experiment you imagine yourself locked in a room into which cards are posted with Chinese characters on them. You have another pile of cards with Chinese characters on them and a book that matches the cards you receive with those you have, and you post the matching card out of the room. To those outside the room it looks as if you are answering questions about a story in Chinese, even though you know no Chinese and are simply manipulating what are to you meaningless characters.

21 Subsequently popularized by Koestler 1967.

that ethnographic writing comprises an unequal discourse, the observer constructing a view of the observed, even appearing to know more about behaviour than actors know themselves by assimilating it into an imposed all-encompassing social-theory narrative (representing a dubious 'objectification' of the 'Other').

A recent example of such a reflexive account of practical experience is Erin O'Connor's attempt (2005, 2006, 2007, n.d.) to 'get at the tacit understandings, experiences and skills of handicraft' (2005:183) by enrolling at a New York art studio to learn the craft of glassblowing, 'to evaluate how glassblowing is read by the glassblower ... to reflect upon the ebb and flow of sensations, techniques, and modes of consciousness' (2005:185). It is a graphic account of the participant aspects of participant-observation. She describes how she learned to use the tools such as blowpipe and furnace, and to handle the molten material such as 'gathering' and 'twirling' glass, which she describes as a move 'from cognitive readings of practice towards corporeal readings' (2005:183). She points out that 'To reveal and elucidate these tacit processes, the ethnographer must access these tacit understandings through immersing herself in the practice itself ... a myriad of subtleties of the art begin to emerge ... through practicing and reflexively considering that practice.' (n.d.:8). She relates how she acquired practical knowledge, developing a tacit understanding of the relationship between body and skill through her dynamic engagement with the materials, where she found the various tools 'through techniques actually becoming a part of her' (2005:188); a graphic account of subsidiary awareness. She attempts to communicate the role of 'material imagination', albeit she maintains that she can scarcely convey with words or images how glassblowers design and make things, for 'In glassblowing, the material imagination is not one of light, of visual observation of reflective forms, but rather one of heat. To get at the heat and therefore to come to know this imaginative mode, the ethnographer must be *in situ*.' (n.d.:10). She describes her physical experiences: 'Blowing glass is hot – a shirt clings to the torso, hair becomes wet, rivets on jeans singe the skin atop the pelvic bone, and moistness sheathes the entire body.' (n.d.:10).

There is something limited intellectually, dubious even, about recourse to such a reflexive approach. It is skilled professionals – such as builders, glassblowers, blacksmiths or whoever – who have spent careers working, acquiring and honing associated tacit knowledge in their fields, who might give us authentic reflexive accounts, not researcher amateurs, who drop in for a while and try their hand at tasks and then go away and write about their vicarious experience of other's experience. There are many examples of craftspersons writing about their work through the ages, as related in Chapter 2, which remind us that while phenomenological postmodern attention to the subjective dimensions of experiential knowing is trendy, it is not new. One of the many accounts that seek to convey something about craft work is George Sturt's (1923) account of the wheelwrights' trade, [22] mentioned in Chapter 2 and cited approvingly several times in *Made in Niugini* to illustrate points about tacit knowing and technical change. The book's idiosyncratic style illustrates the difficulty of capturing something of skilful knowing,[23] as in the following, which is part of Sturt's account of sharpening a pit-saw:

> I got a little 'cut' into the saw. 'She' felt right, when I passed my hand gently along the teeth ... one felt a sort of bite from the newly filed teeth. And she looked right too, when I squinted along to see if the teeth stood out evenly in their two rows, this side and that. It is true, in the act of sawing afterwards an occasional jump of the saw almost jarred one's arms out and probably the top-sawyer swore a little ... a facetious wheelwright took to calling one of the saws 'Old Raspberry', in allusion to her scratching character. He also hinted that the teeth were 'all uncles and aunts,' they were so uneven.
>
> (Sturt 1923:61)

The distance, culturally speaking, between practitioners and researchers is an issue. While a New Yorker may readily cotton on to issues raised by some of the city's glassblowers, the gap between myself and New Guinea house-builders makes my experiences much less convincing in conveying aspects of their skilled being.[24] It is questionable, particularly in such cross-cultural contexts, whether an outsider's

22 As a personal aside, in reflexive vein, Sturt's book appeals to me, as he was a Surrey man from Farnham, and on my mother's side I am from a Surrey family – Sturt writing of a time familiar to my maternal grandfather as a lad – and my forebears were also craftsmen such as journeymen carpenters, plumbers and decorators. I subsequently learned that the uncompromising critic F.R. Leavis praised the book for its freshness and honesty.

23 Another autobiographic account of the same genre is Rose's (1937) book of his life as a Buckinghamshire carpenter in the late nineteenth century (cited by Marchand 2012:261). Comparing his account with the reflexive description by an anthropologist of sawing a wooden plank in two (Ingold 2011b:51–3), illustrates the gulf between craftspersons and academics when considering such skilled activity.

24 As Nagel comments in his droll 'what's life like for a bat' musings: 'There is a sense in which phenomenological facts are perfectly objective: one person can know or say of another what the quality of the other's experience is. They are subjective, however, in the sense that even this objective ascription of experience is possible only for someone sufficiently similar to the object of ascription to be able to adopt his point of view—to understand the ascription in the first person as well as in the third, so to speak. The more different from oneself the other experiencer is, the less success one can expect with this enterprise.' (1974:442).

somewhat contrived experiences of an activity – leaving aside for now the issue of whether or not you can find the words accurately to convey your experiences – match closely in any way those of habitual actors or professionals, who may themselves be reluctant or unable to express their experiences in words. The reflexive approach is a doubtful way to plumb the intellectual challenges of conveying tacit experience, beyond writers frankly presenting their experiences as 'my time with the Wola or whoever' travelogue-cum-tourist accounts, which while a popular genre since Victorian times do not seriously engage with the epistemological issues nor do justice to such peoples skilled heritage.

These challenges relate to the widely acknowledged limits of participant-observation. My attempts to participate in Wola house building, as mentioned in Chapter 7 for instance, were a pale reflection of experiential reality. In order to avoid nasty splinters, I wore industrial gloves when undertaking tasks with rough wood, such as attempting to drive in wall stakes. Otherwise, I should never approach anything near tacit awareness, focusing all the time on the 'present at hand' task, being concerned not to hurt myself. The alternative would be to spend some time engaged in heavy manual labour sufficiently to callus and toughen up my hands (although splinter injuries might still occur, our Wola friends regularly coming to us for first-aid treatment of such wounds). It is an aspect of how the materials worked on may affect the worker (see Chapter 6), and reminds us that it is not only the worker who has an effect on the materials (Julien and Rosselin 2002). While it is conceivable that I could harden my palms enough after an appropriate period of physical work (maybe aided with a suitable lotion such as methylated spirit), the time it would take for me to approach my friends' competence with an axe would be considerably longer, if I could ever achieve it. Their experience of various house-construction tasks illustrates Mauss' (1935) 'bodily techniques' idea (Ingold 2001; Warnier 1999) that considers 'the ways in which men, from society to society, traditionally know how to deploy their bodies'[25] in various activities, which, in postulating that these are culturally specific, queries the possibility of properly acquiring a 'cultural anatomy' beyond that into which one is socialized, let alone communicating the experience to others in words. In short, while I could write an amusing Bill Bryson style account of my somewhat incompetent building experiences, such a reflexive exercise would convey little about Wola tacit knowledge.

Furthermore, subjective accounts give us little idea about how actually to do things, such as construct a house, a lapse discussed in Chapter 2 with respect to vernacular-architecture studies. They focus instead on the writer's personal experiences of undertaking various tasks, acquiring associated tacit knowledge to various, usually ill-defined, extents; we rarely know how far they progress beyond complete novice. After reading O'Connor's colourful accounts, you have no more knowledge about how practically to blow glass than when you started – for this you need to study appropriate manuals and serve an apprenticeship. All that you come away with is some appreciation of what glassblowing was like for the author and what the experience might be like for you. This is further apparent in Marchand's (2001, 2009) reflexive-cum-case-history studies of Yemeni minaret building and Malian mud-brick house construction, which as he notes, focus 'not on analyses of the material artefact but on its producers and the processes of their production' (Marchand 2009:12) – exploring, for instance, masons' associations, their social networks and changing fortunes. Although he includes some technical information on the construction of minarets and houses, his books fall far short of giving full accounts; for instance, I had to nip into a neighbourhood minaret, reading the Yemeni volume while in Doha, to satisfy my curiosity about the construction of the spiral stairway that is integral to the structure.

The reflexive response to the challenge of accessing tacit skilled knowing, comprising authors' accounts of their experiences of activities, hardly addresses the problems of dealing with tacit knowledge in explicit terms, which is what documentation and discussion in academic contexts implies.[26] It evades them altogether. We have to engage with more than autobiography if we are going to tackle the intellectual challenges of resolving tacit with explicit knowing. These reflexive short-comings are evident to Marchand in his forays into building construction, and he seeks to compensate for them by setting accounts of his experiences as a labourer in social and historical context, largely via the tried-and-tested case-study approach describing the daily life and social interaction of actors, while weaving his 'own apprenticing experiences into the narrative' (Marchand 2009:12). As he puts it:

> I had access to their concepts either through what they told me or through what I could observe ... I was able to conceptually share some of their everyday imagery by ... regularly engaging with them in work and social

25 My translation of 'les façons dont les hommes, société par société, d'une façon traditionnelle, savent se servir de leur corps' (Mauss 1935:271).

26 Other accounts that seek to get at embodied knowing likewise founder regarding the conundrum of matching tacit awareness with the explicit propositions that a written record presupposes. The 'cultural phenomenology' of Csordas (1994), for example, focuses on 'somatic modes of attention' and points out that bodily awareness is 'culturally constructed' (as when ritual healers see visions that inform them about the healing process) with the subjective experience remaining a mystery.

> relations [in other words, undertaking participant-observation] … my ensuing representation of the builder's teaching-learning process is predominantly an account of my recorded observations of my fellow workers, overlaid with my own, mainly image and sensorimotor-based, concepts.
>
> (Marchand 2001:157–8)

Although, as he comments at several junctures, the latter tacit 'concepts' were not 'propositional' and expressible in language, nonetheless he attempts to convey something about them using words.

13
Doing it to know it

While reflexive accounts of one's experience undertaking tasks – such as glassblowing – may illuminate the experience of doing such things somewhat, the insight gained is limited if words can only catch a shadow of what one tacitly knows. Regarding house-building in England, for example, no amount of reading others' subjective reflections will enable you to know what it is to plaster a ceiling or build a brick wall. You have to do it to know it, which presupposes a sound knowledge of how to do it – that is, the technical aspects of the activity, so that you can engage in it and know it. In his study of Yemeni builders, Marchand makes much the same inference, being 'forced to conclude that an attempt to discuss or explain their specific skill-related knowledge would be counterproductive ... the practical experience that constitutes the knowing how eludes being "fixed" by language ... ideally, any form of performative knowledge enjoins a performative demonstration' (2003:42–3). Likewise, if you wish to acquire the tacit knowing that informs house construction in the Was valley, which this study cannot convey, you have to build a Wola-style house. You will learn something different from the propositional account of house building this book offers, which while academically informative is experientially deficient, given that engagement in construction is necessary to knowing how to build a house like a native. This is not a reaffirmation of the reflexive approach, which comprises a second-hand account of another's subjective experience of doing something. It is about you, the reader, having the first-hand experience to know it.

A postgraduate student highlighted the import of these problems for me, as a craftsperson firmly of the opinion that if you want to acquire skilled knowledge, you have to acquire the skill; no amount of words can compensate. She studied basketry crafts in South East Asia (Davy-Ball 2009), and as someone trained in making fine jewellery, she approached the research in the same way, by literally apprenticing herself to some basket makers and learning how to weave various basket designs, producing her own baskets and returning with boxes full of woven objects of various sizes and patterns, which for her demonstrated the knowledge she had acquired. In her thesis she gave a marvellously detailed account of basket making, in the tradition of craft manuals, but she found it hard to engage in theoretical discussion of what she had learned, as for her it was part of her skill repertoire and any persons wanting to access what she had learned would have to use her thorough account and persevere at making baskets. Only when they could make them would they know what she did in some measure. Finally, I managed to persuade her to engage in some discussion of various material-culture theories – symbolism and design, identity issues and consumption, participatory museology, and so on – to satisfy the examiners, but for her it was an arid intellectual exercise that missed the richness of skilled knowledge.

The implications of engaging with tacit knowledge give a novel spin to the participation entailed by the participant-observation approach, discussed previously in relation to apprenticeship. They affirm the approach's advocacy of participation to understand other ways; albeit discussions of the method pay scant attention to the point that many of these ways are tacit and hence pose particular problems of access. While I engaged in participant-observation, I 'spoiled' the authenticity of the learning process, like many fieldworkers, by engaging in 'thin participation', paying more attention to observation, documenting in detail and consciously reflecting on what we did in building houses. It seems that 'thick participation', with some attendant observation, is necessary to achieve authentic tacit knowledge, which implies working on the construction of not only one house but several, until you have internalized the experiential knowing and can engage unconsciously in the work; that is, until you can just do it.

The standpoint that knowledge involves personal experience puts an intriguing slant on arguments about overcoming the dualistic subject-object perspective that some argue characterizes Western thought since the Enlightenment (e.g. Foucault 1973; Rorty 1991), with the emergence of the scientific method that sees the world comprising objects 'out there' that the subject enquirer observes and analyses. In order to achieve the prior, undifferentiated view ascribed to others, and connect with the world, being in it rather than on it, we have to participate as subjects, engaging, rather than separating ourselves and objectifying 'out there'. This recalls attempts to overcome the dualistic Cartesian mind-body perspective, discussed earlier (Chapters 2 and 5). The argument that, in order to know it, we should not stand back as objective observers of the world echoes again the tensions and contradictions inherent in the participant-observation method, which assumes that we can be both subjective and objective at the same time. Even more problematic is the implication that to achieve such an undifferentiated and authentic view of house building in the Was valley it is necessary to go to the Highlands and participate in life there, learning to construct houses by building houses, which will strike most readers as impractical to say the least, if not crazy. It brings to mind Wittgenstein's comment in the penultimate paragraph of the *Tractatus Logico-Philosophicus* (1992), that anyone who understands that book will see that its propositions are senseless.

Neuro-imaging to know

Why would any reader remotely contemplate building a pokey and smoky Wola house that few can probably imagine living in comfortably?[1] Yet if you really want to know about Was valley housing and construction practices, including the requisite tacit-knowledge perspectives, it seems that you need such experience – you have to build and occupy a house, which seems ridiculous. We seem to be at the sort of impasse beloved of postmodernists. Have we reached the limits of what we can aspire to know? Perhaps we can imagine neuroscience extending the limits in the future, as discussed in the previous chapter. If what we think of as our mind (or at least our cognitive capacity for reasoning, remembering and knowing) is the result of chemical-electrical reactions in the brain (and not some spirit entity that occupies the body and may depart from it, notably at death), it is conceivable that we shall be able precisely to map such cerebral activity and so know what it is, neurally, to know anything.

We are aware that the complex tacit skills that feature in human tool use involve cerebral activity that integrates sensory-motor adaptations that regulate unconscious perceptions together with cognitive processes that provide contextual information and conceptual knowledge, such as properties of things and task requirements, as Chapter 5 recounts with respect to stone-tool-making experiments. In a neuroimaging experiment, for instance, seeing and naming tools activated the left dorsal premotor cortex, while referring to tool uses (e.g. sawing) not only increased activity there but also caused firing elsewhere in the posterior of the frontal lobe associated with coordinating complex body movements, as occur when using tools (Grafton *et al.* 1997). This confirms that cognitive motor information contributes to our mental images of things such as tools, our understanding of which depends not only on appearance but also use. Indeed, there is growing neuroimaging evidence that 'motor systems are activated by images of objects or words associated with particular action sequences. For example, when passively viewing a hammer, motor areas associated with grasping objects is activated.' (Brown and Seligman 2009:37). The brain activity occurs primarily in the left parietal cortex, which instead of functioning as a straightforward store of skill memories, dynamically integrates multiple associated representations computed via a distributed network of largely left- and some right-lateralized neuronal circuits, resulting in an appropriate routine for the task in hand. The evidence from 'apraxic patients and functional neuroimaging studies converge on the hypothesis that the human left cerebral hemisphere is asymmetrically involved in the processes', which result in 'the formation of an internal praxis representation that can be used to guide contextually appropriate actions' (Frey 2009:28).

While the neural networks that comprise the human brain are, as noted, very complex, it is possible that we may someday be able to understand and reproduce them, matching up outside world things and inside head representations; that is, to trace neuron by neuron what synaptic activity in the brain relates to recognition and engagement with particular things. We may also be able to replicate that spark of neurological activity exactly in another brain, such that persons can visualize the same green colour, savour identical wine tastes or anticipate how trees will fall (assuming identical neural circuitry). In this event, we could communicate tacit knowledge between individuals by stimulating their brains appropriately to fire the necessary neural pathways, and they could experience and understand anything, such as building a Was valley house. If we reproduce in a person's brain the cerebral electrical activity experienced by someone engaged in a skilful activity that depends on tacit knowing, it seems that she or he may come to 'know' vicariously the experiential knowledge without actually undertaking the activity. In this scenario, if sufficient electrodes could be attached to a Wola man's head, and all the cerebral activity recorded as he engaged in some act (such as driving in the stakes for the wall of a house or aligning the ridge-poles or rafters) could be reproduced exactly in my head, then I should perceive what he knows, including its tacit aspects.

1 Even when I had the opportunity to occupy such a house, I did so only occasionally for a few days at a time, notably when visiting elsewhere, albeit I was in the Was valley to experience and learn about living there. I occupied a *mbolin aend* 'white's house' because an *aend henenj* 'traditional house' was too cramped and dark not only for personal comfort but also to undertake research efficiently, notably organizing field notes and maintaining equipment properly (Sillitoe and Sillitoe 2009:171–5); yet another illustration of my limited participation.

In other words, we may extend what we know through a virtual experience. But this seems paradoxical, that we should be able to gain experiential knowledge without physically engaging in the activity that comprises the experience. And maybe this is no more than science fiction, because in addition to having the brain waves we should presumably need to have the associated physiological sensations, the muscle activity and sensory experiences to truly access the tacit knowing, especially as 'many skills take a long time to acquire because they actually necessitate physiological change: stronger muscles, greater flexibility, more acute perceptual-motor ability, and slowly developed, incrementally learned patterns of behaviour' (Downey 2010:S33; also Ingold 2000:375). It seems that we are a long way from such artificial reproduction of tacit knowledge, if we ever shall be able to do so, and until then the only way to 'know' how to blow glass, plaster a wall or throw a pot is to find a glassblower, plasterer or potter who is willing to take you on as an apprentice and instruct you in acquiring the skill and associated tacit knowledge.

Furthermore, this cerebral transfer scenario assumes that the architecture of all human brains is identical, with the same neural synaptic pathways, when we know that they vary between individuals, just like our physiognomy. The neurological plasticity of the human brain, mentioned previously (Chapter 5), indicates the extent of the variation, where training can affect 'physiological change in the brain, nervous system, bones, joints, sensory organs ... altering the organic architecture of the subject' (Downey 2010:S27). Psychological research shows that our perceptual systems adapt and improve when consistently deployed on certain tasks; dramatic evidence of such neuroplasticity is the reconfiguration of the brains of persons suffering apraxia to compensate for cerebral damage (Pascual-Leone *et al.* 2005). In respect of tacit knowing, this variation in brain circuitry also likely correlates with differences in ability and skilfulness, responsible for the disparities between those who are naturally dextrous and those who are 'all thumbs'. Archaeologists, as discussed earlier, have seized on the point that a key aspect of the 'recursive relationship between people and the world is the plasticity of both brains and objects' to support their materiality argument about human cognition, in which 'brains help make new objects, which in turn help create new brains' (Gosden 2009:109). Brain scans show that routinely engaging in particular activities results in the differential development of parts of the brain, in their relative shape and size. In this event, my suggestion that neuroscience might stimulate my brain to 'truly' know house building from a Wola perspective is a vain hope, as the relevant areas of my brain are poorly developed, lacking necessary expansion through engagement in associated activities.

All may not be lost, however, for the necessary cerebral circuitry could arguably be developed, the fundamental structure of our brains being the same, having not changed since the Pleistocene according to evolutionary psychologists, and brain plasticity allowing for reconfiguration of neural connections. The computer – which would play a central role in such cognitive transference – serves as an analogy: if we load several identical desktop computers with different combinations of software, they will be capable of different things, but we can further programme them with the missing software if necessary to do the same things. But there is probably a limit to this with our brains, which differ fundamentally from computer chips. While we may all be born with brains that are biologically identical regarding their potential, they develop particular cerebral circuitry from birth onwards depending on experiences. Socialization in our early years plays a significant part here. It seems that once many of the circuits have developed they become relatively fixed by adulthood, or sooner. This explains why gifted gymnasts, ballerinas, violinists, golfers and so on, who depend on particularly sophisticated embodied knowledge, often start on their careers early in life, whether due to some natural urge-cum-aptitude or parental pressure-cum-guidance.

The prominence of juvenile experiences is notable in language learning, such that it is difficult to acquire the skills of a native speaker after childhood; for a European to master a tonal language, for instance, or even hear some tones, let alone reproduce them. Language acquisition is a key aspect of social enculturation, which from infancy inculcates values, expectations and responsibilities about how to behave as a socialized person in your community. It is culturally specific and informs all activities. Indeed cultural context complicates things further, as there is growing evidence that it too moulds the brain's plasticity to some extent: 'Crosscultural functional neuroimaging studies reveal differences in brain function across cultures, and neurodevelopmental studies demonstrate that experience can shape both brain structure and function, providing a mechanism by which people of different cultures learn to act and think differently.' (Rilling 2008:27). Neurological research has recently revealed differences in neural processes for a range of activities. Socio-cultural practices, values and beliefs affect brain functioning, from perception and memory to emotion and social cognition, arguably with implications for a range of human behaviours (Downey 2012:245; Park and Huang 2010). A widely cited example is the different ways in which Asians and Europeans view faces and scenes, with the former focusing on context and relative size and the latter on object and absolute size, which has fed into dubious reductive arguments about holistic versus analytic perceptions reflecting collectivist versus individualist value orientations (Chua *et al.* 2005; Nisbett and Miyamoto 2005). This generalizing about mental differences between persons from different cultures is potentially dangerous, as mentioned in Chapter 5, as it smacks of racial stereotyping. It challenges the 'psychic unity' of humankind, an assumption central to anthropology since Boas (1911) propounded the idea to oppose racist views (Downey 2012:252). While cerebral disparities may be due to cognitive habits and not to differences in intellectual potential, we are treading a fine line in pointing out that 'brains can be grown into different configurations, not because we are, at birth, fundamentally and irreducibly different, but because we become "en-brained" from our own distinctive combination of genes, epigenetic influences, environmental factors, experiences, learning, and even understandings of ourselves' (Downey 2012:252). These 'en-brained' views further indicate a need for anthropology to engage with neuroscience to head off hare-brained arguments, by informing such enquiries about culture and encouraging necessary subtlety in research that looks into the human brain and cultural issues (Brown and Seligman 2009).

It seems questionable that one could ever faithfully reconstruct any activity, such as house building, without reproducing all the other features of Was valley life, and learning the activity in their informal way along with others that comprise their cultural universe. Consequently, the suggestion that to appreciate the tacit knowing that informs house building in the Was valley you need to build a house, does not imply following a blueprint such as this book supplies, which would be inadequate, but rather living as a Wola speaker and learning not only how to construct a house but also the Was way of being-in-the-world to appreciate residents' socio-culturally mediated intentionality. If we concede that 'real' native-like understanding implies taking on the full gamut of Was valley cultural experience, we hit another dead end, because as the postmodern critique points out, and as discussed earlier in appraising participation-observation, it is not possible to enter fully into another culturally socialized world, but only partially. One can never learn to build a house in the same way as my friends, who start as children to acquire the skills and knowledge, largely through watching and mimicking relatives rather than by formal observation and enquiries (Chapter 11). Any 'cerebral–transfer' procedure will not solve the dilemmas of cultural relativity because we cannot undo our brain's circuitry, that is return to infancy and be re-enculturated. This amounts to an empirical scientific 'reductionist' affirmation of the ideological postmodern 'rationalist' position that it is wrongheaded to think of knowledge as culturally neutral, disembodied, objective and universal.

To know or not to know

A further complication is that tacit understandings also feature prominently with regard to appreciation of wider cultural contexts. Looked at another way, neuroscience may help further understanding of such culturally informed tacit know-that social behaviour and its propagation, which 'practice theory' identifies and then puts into the black box of the unconscious. From this perspective, a promising way to integrate 'neurological research and cultural theory is to treat culture, not as information, but rather as skill acquisition … Reconceptualizing enculturation as 'enskilment' links to robust findings on neural architecture and functioning.' (Downey 2012:249). Progress to date has been slow, with considerable controversy. The argument over 'mirror neurons' illustrates the dissension. These are observed in the pre-motor and parietal cortices of macaques, and fire when the monkeys perform an object-orientated action or see another animal do so (Rizzolatti and Craighero 2004). Some argue that the existence of such neurons furthers the learning of skills and behaviour through imitation, and may also relate to the physiological basis of culture and language, indicating how we imbue socially inculcated values and behaviours, for 'Mirror neurons seem to facilitate social cognition by helping individuals to understand others' actions.' (Brown and Seligman 2009:36). But they have yet to be observed in human brains (Tallis 2011:190), although some think that recent neuroimaging data indicate the existence of similar networks associated with imitation between persons (Downey 2010:S29).

Whatever the simulation or mirroring mechanism, the embodied approach to cognition, which advocates attention to experience-elicited states in constituting knowledge rather than abstract symbolism, arguably addresses the roots of 'automatic' social attitudes and actions, addressing head-on their sensorimotor and emotional aspects. It is consequently arguable that 'An experimental approach in cultural neuroscience has the potential to further illuminate the neural substrates and cognitive and emotional implications of such forms of social embodiment.' (Brown and Seligman 2009:37). This approach to socio-cultural issues also focuses attention on the individual, with a 'fundamental shift in the location of culture's generator – away from an ideational (or ambiguous) structure – and a shift in scale – to the level of the individual's nervous system … [which is] a powerful "person-centred" refocusing of our theoretical energies' (Downey 2012:247). It fits with my long-running advocacy of focusing attention on the individual in social life (see Chapter 2), which has been a central issue for me in understanding Was valley socio-political-economic arrangements.

The marked experiential dimensions of social being – where much of our culturally situated behaviour is more akin to involuntary eye twitches than to voluntary knowing winks (Geertz 1993:6–7, after Ryle 1971:480–1) – have further implications for knowing life in the Was valley. Again, you have to live it to know it. In order to understand what social being comprises for native persons, it would be necessary to internalize their cultural expectations and responsibilities and gain acceptance in the community such that you can behave like a fully socialized member without reflecting on latent social implications, on structure and process, institutions and agency. The degree to which this is possible for a foreigner inculcated into another cultural tradition is a moot point. Even if you achieved this level of understanding, while you might approximate a native person you would no longer be an ethnographer, engaging in anthropological enquiries – you would be participating entirely on local convention-complying terms and no longer observing on foreign note-jotting ones! However limited the extent to which we are able to engage with another cultural way of being-in-the-world, the need to strive for it underlines, as the 'ethno-determinist' approach stresses, the importance of the long-time anthropological maxim of holism (Chapter 1), often subscribed to more in word than deed (Bubandt and Otto 2010).

The implications of seeking to understand other socio-cultural ways of being-in-the-world are further complicated by their experiential dimensions. Convincingly accessing tacitly subscribed to know-that cultural norms presents particular challenges, because these concern transient social behaviour informed by abstract ideology or Hume's 'secondary reflective impressions'. The Scottish Enlightenment philosopher established that we can only know lively impressions or sensory experiences with certainty, and even directly connected ideas are copies subject to error, going on to argue that what we take for warranted beliefs are not founded on reason (matching impressions or physical experience to ideas or mental processes leads us astray) and so indicating the dubiety of induction (subsequently built on by Popper in his falsifiability view of science). It is suggested that a way to gain some understanding of such elusive know-that knowing, which disappears immediately after any act in which it features, is to investigate know-how tacit skills that relate to concrete things. This returns us to the seminar argument (Chapter 4) and the part experimental archaeology might play in furthering our understanding of such 'enskilment'. But further reflection questions its feasibility, this approach running into the contradiction that in order to understand know-how activities, it is necessary, as pointed out, to comprehend the wider know-that informed cultural context, of which know-how activities are purportedly going to further understanding. This may be awkward for anthropology, but it seems doubly so for archaeology with no prospect of trying to join in activities and experience socio-cultural life, however imperfectly, in long-gone communities.

If we assume that you could become as proficient as a Was valley resident in house building – a questionable assumption, as we have established, for you already carry around culturally determined understandings of the world and what it is to be in it; it also supposes that you have the requisite manual dexterity – how will you pass on what you know, and further interpret or explain it according to the canons of our intellectual tradition? We confront another paradox, for if you seek to understand the subjective character of such experience, this is 'fully comprehensible only from one point of view ... [and] any shift to greater objectivity – that is, less attachment to a specific viewpoint – does not take us nearer to the real nature of the phenomenon: it takes us farther away from it' (Nagel 1974:445). You might try to get across something of the subjective experience in a reflexive account (Chapter 12), your success depending on how adept your prose at conveying such a skilled undertaking, but no matter how 'thick' your description, some of the activity will be beyond words, involving Wolacentric cultural issues and tacit awareness that elude such documentation. It implies that limited reflexivity is the only possibility and any theorizing on the back of it will take us even further away from the subjective reality of construction activity. Furthermore, someone reading such an account will still have no idea, as pointed out, about how to build a house. Even if you follow Chapter 7's record of house construction manual-like as a series of directions, and manage to master the skills to build a house, so understanding what house construction is for a Was valley resident, you will not be able to communicate what you have learnt any better than a native, further confirming the rather contrary nature of participation-observation. In truth, you are unlikely to achieve much more than I have in passing on such knowledge, although you will know more than you will from just reading this book, having apprentice-like experience of building tasks.

A deal of what humans know and do is plainly not subject to conversion into discourse, which poses problems for those of us who work in literate academic traditions. The challenge of using the written word to capture tacit knowing is perennial (Keane 2005; Marchand 2003:31). We all have 'to confront the same issues of representation ... no academic writer can escape these issues', and finding 'a way of moving beyond ... [the] stifling division between academic study and observational practice' has so far proved intractable, and mystical exhortations 'to re-embed the practices of reading and writing, and above all of thinking, within our observational engagements with the world around us' are unlikely to give much traction (Ingold 2007:35, 36). Embodied experience, comprising a significant part of human cultural life, has no symbolic content per se, and we cannot represent it according to any such code. This endorses the postmodern point mentioned above, that we cannot access another's being-in-the-world except partially, and underscores the critique that even if possible, we could not corroborate the understanding with others without them also having the experience; both parties, for instance, would need to embody knowledge of house building (know how to build like a Wola individual, not using this book as a manual). Virtual experience may offer a way to access such knowing in the future, overcoming current problems of communication, but as noted there may be insurmountable problems here too. Neuroscience currently has no such capacity to pass unspoken intelligence between brains and we have to fall back on less comprehensive explicit ways of conveying 'know-how' skill knowledge. It is questionable if phenomenal experience will ever be scientifically reducible to brain architecture and whether artificial reproduction of tacit knowing can occur. Arguably, it is unlikely that it will be possible to reproduce the associated individually variable, culturally nuanced, cerebral synergies that are way beyond the sum of their individual synaptic parts. Anyway, for many of us, there must somehow be more to knowing than this, which brings us back to the ethereal spirit and indeterminate qualia and metaphysical speculation, which sceptical empiricism seeks to get beyond.

Specifying doing to know

It is difficult to avoid a certain sense of disappointment in completing yet another book that seeks to further understanding of life in the Was valley, in view of its off-the-wall (or build-your-own-wall) conclusions and litany of shortcomings and paradoxical dead ends. It appears that we can go no further, after concluding that to understand Wola house construction you have to build a house. In short, that it is only through experience that you can really know about it. The situation is reminiscent of Russell's (1946:685) comment on Hume that 'he developed to its logical conclusion the empirical philosophy of Locke and Berkeley, and by making it self-consistent made it incredible. He represents, in a certain sense, a dead end: in his direction it is impossible to go further … I cannot but hope that something less sceptical than Hume's system may be discoverable.'[2]

The postmodern critique elaborates on some of the problems that attend giving precedence to ideas, which involve explicit knowledge and propositional understanding. But is it correct to conclude that other modalities of knowing are not amenable to intellectual enquiry and that accounts such as this one are irredeemably flawed, if not a self-deceived waste of time? While we have to admit to the limitations debated in postmodernism – of which some early ethnographers were also aware – it is feasible surely to capture some aspects of tacit knowing using a range of approaches. Even Plato suggests that we can get out of the allegorical cave and see beyond its shadow play, or updating the metaphor, that we are not necessarily confined entirely to our own subjective cinemas. While phenomenological discussions about the subjectivity of direct experience and tacit consciousness point out considerable dilemmas, it is arguably possible, after Malinowski nearly a century ago, to 'make directly for the question of practical means to overcome some of the difficulties involved' where people may 'feel or think or experience certain psychological states in association with the performance of customary acts [but] … are not able to formulate these states, to put them into words,' which he rightly considered 'the real Gordian knot … of social psychology' (1922:22).

Recognizing the challenges that tacit knowing present to explicit reasoning does not rule out a rigorous approach that breaks down activities into manageable steps for exposition, which may facilitate further analysis and understanding in a propositional manner. The 'rules-for-action' may only be 'unspecifiable' (Polanyi 1958:55) in the sense that a skilled actor does not need to be able consciously to state them in order to perform actions. This is what Polanyi's discussion of tacit knowledge suggests: 'the aim of a skilful performance is achieved by the observance of a set of rules which are not known as such to the person following them' (1958:49). Indeed, it is arguable that the efficacy of the 'rules' may depend in part on them remaining beyond the actor's ken, as conveyed in the notion of subsidiary awareness. Many physical aptitudes show this: effectively plastering a ceiling, for instance, is only possible by not thinking about hand and eye coordination. It is necessary that the acquisition of such subliminal skills is unconscious, such that the body that mimics the actions of another person 'knows' something more than the mind that otherwise directs the body.

While a skilled person may not need to know the rules or precepts explicitly, they nonetheless exist; for even rules 'not known as such' are rules.[3] In this event it should be possible to reproduce tacit knowing to some extent, without going through an apprenticeship, by way of a thorough account that somewhat reveals the 'hidden' rules. The propositional approach of standard academic accounts such as this one may inevitably overlook aspects of tacit knowing, but even so it is possible to capture something, albeit only partially, through comprehensive ethnography-led enquiries. The transfer of tacit awareness depends on the imitation of physical gestures, and we can seek to document these in writing, drawing, photography, film and computer simulation, as here. An observant person can describe many physical actions in words; such as, 'grasp a hammer by the handle end with thumb aligned along it, swing firmly in an arc to connect with the nail,' and so on. It is through such detailed accounts of what persons do in deploying

2 It is arguable that Russell's pessimism is misplaced, because if humanity found a way beyond Hume's scepticism the consequences could be dire, lacking the wisdom to handle the resulting knowledge of human behaviour, which could be used equally for bad as good ends (another paradox that illustrates the paralogism of arguments about dispensing with dualistic thought): the stuff of nightmares that feed dystopian literature.

3 I understand 'rules' to encompass replicated accustomed acts that necessarily comply with natural laws.

embodied knowledge, of the sort that ethno-determinism advocates, that we come as near as it is currently feasible to communicating such knowing, given the limitations of doing so outside engaging in an apprenticeship. In his study of Yemeni minaret building, Marchand (2003:46, 48), who claims to have served an apprenticeship, reaches a similar conclusion, 'knowledge as manifest in their practice … must be considered as the direct subject of anthropological enquiry… [which] calls for good-old-fashioned fieldwork with an expanded critical notion of what "intelligence" is … the most important contribution … is in our ethnographic accounts.'

A manual-like account of tasks, or the *chaîne opératoire* associated with an activity – whatever its shortcomings (see Chapters 2 and 7) – will allow others who want to know more to attempt to reproduce them, gaining some awareness of the nature of the actions and further insights into related experiential knowing (albeit limited, as pointed out, as not learned by observing a skilled knowledge bearer and lacking related cultural insights, being unable – as in a virtual simulation – to reproduce the full culturally contextualized embodied experience). Furthermore, visual images –whether in motion in film or video, or in static illustrations, such as diagrams and photographs – of skilled knowledge-bearers undertaking tasks, like those associated with house building, can capture some of what cannot be put into words (see the short film 'House at Honael' that complements this study).[4] The use of drawings to transmit information, which may include notation (as in Chapter 7), is common practice among craftspersons and has long featured in material culture studies too, albeit few anthropologists have any use for this powerful means of representation beyond recording the occasional genealogy.[5] The distinction between 'knowing' (an activity) and 'knowledge' (a thing) points to the value of exploring such other means of communication to supplement and extend on the conventional written approach. The alternative is to give a reflexive and postmodern-aware account, which, as pointed out, may give you some idea of another's subjective awareness of the experience of engaging in tasks that involve tacit knowing, but no matter how self-critical, unless technically detailed to boot, will scarcely allow you to achieve an awareness of what it comprises for yourself.

The ethnographically engaged approach is admittedly limited in making the truly tacit aspects of behaviour known. Any 'rules-for-action' that we can reproduce explicitly are implicit, not tacit, according to distinctions made earlier. They are only 'unspecifiable' in the modest sense that a skilled actor is not consciously aware of them in performing actions. But other precepts are literally 'unspecifiable' (Polanyi 1958:55) in the absolute sense that they are beyond specification, as the master/apprentice relationship implies (Chapter 12) and the opaqueness of the acquisition of associated skills shows. Consequently, it is argued that 'These hidden rules can be assimilated only by a person who surrenders himself … uncritically to the imitation of another.' (Polanyi 1958:53). The many different local beers brewed in England illustrate the point. In Masham, a North Yorkshire village, for instance, they brew 'Old Peculiar', a sweet dark beer. If we concede that some physical differences between this beer and others accounts for its distinct qualities, then a comprehensive study of Old Peculiar brewery procedures, together with physico-chemical research into the properties of the ingredients used, should allow us to specify exactly the formula to produce the beer anywhere by anyone. But this appears currently not to be possible. When the giant Scottish and Newcastle brewing company acquired Theakston, the brewers of Old Peculiar, and attempted to brew the beer in its Tyneside industrial plant – with extensive technical expertise – it could not reproduce it. The beer demands local Masham water and the old brewer's olfactorily linked skills to produce it. Even if, ultimately, all differences have a physical explanation, we do not currently have the means (tests, instruments or whatever) to reveal these, and in their absence tacit skills, involving embodied passage directly from person-to-person, offer the only way to convey the knowledge. Which returns us to our conclusion that you have to do it to know it. While a considerable amount of ink has been spilled discussing embodied knowledge, few have yet faced up to the point that the only way you can really learn it is by doing it. The ethno-determinist approach at least gives a sufficient ethnographic blueprint to build a house, for instance, or to imagine oneself doing so, and intimates the implications.

Ethnography and anthropology

Indisputably, we cannot achieve a full and true understanding of the native point of view, as postmodernism argues. The so-called emic perspective – that of the native – is aspirational only, even though what we purport to apprehend comprises what people say and do. Any insider native view, including our own, comprises 'impressions' in Hume's terms, or common sensory experiences of phenomenal things in the world that inform 'simple ideas',

---

4 See www.community.dur.ac.uk/paul.sillitoe/wola/houses. This assumes that people give permission for such film documentation. There are also various methodological issues regarding film, such as the way the person holding the camera decides how to represent what is going on, which may distort local views too (Heider 2006:10; Ruby 2000:151; Stasch 2011b).

5 'It is extraordinary that in all the debate about 'writing culture', the assumption has always been that the graphic part of ethnography is writing and not drawing … Given that by all accounts, drawing is an immensely powerful tool of observation … why has it been all but forgotten in anthropology.' (Ingold 2011b:222).

which in turn through aggregation and reflection comprise culturally informed 'complex ideas'. The latter, socio-culturally shaped sense-informed products of the mind, may allow purely imaginary 'secondary' complex constructs, which Hume (1896:3) illustrates by describing how his impressions and thoughts about gold and pavements allow him to go on to imagine a city paved with gold. We are obliged to adopt an etic perspective – that of the outsider researcher – in formulating our, in turn, culturally cum disciplinarily informed understanding of what people say and do, founded on what we understand of their impressions and ideas about being-in-the-world, yet framed somewhat by ours. While any intellectual perspective necessarily comprises 'complex ideas', ours involves doubly complex ideas, as it seeks to include native complex ideas, which we can only understand second-hand, for we cannot entirely experience how others relate socio-culturally to their sensory-impression-informed simple ideas, in building up their complex comprehension of the world. Yet paradoxically, as pointed out several times, we may appear to know more about them than they know themselves when we study them. Consequently, you, reader, are experiencing a complex of complex ideas. The upshot is that our ideas about what people say and do, whether interpretation or explanation, are products of the imagination. It is necessary to relate our confection of complex ideas constantly to the ethnographic evidence – the central point of ethno-determinism – which is our shared connection to the world out there and sensory awareness of it, or risk drifting off into an entirely fictional abstract one. This is to heed Hume's (1902:165 – original italics) sceptical conclusion:

> When we run over libraries, persuaded of these principles, what havoc must we make? If we take in our hand any volume; of divinity or school metaphysics, for instance; let us ask, *Does it contain any abstract reasoning concerning quantity or number?* No. *Does it contain any experimental reasoning concerning matter of fact and existence?* No. Commit it then to the flames: For it can contain nothing but sophistry and illusion.

Regarding the experiential dimensions of knowing, ethnography supplies evidence of knowledge as it manifests itself in behaviour in the world 'out there'. While we may use such evidence to explore embodied knowledge further, we only perceive the tip of the iceberg, using language to make known what we think we know, for we soon reach the limits of dialogue, as we have seen, when it comes to such knowledge, which relates directly to 'impressions'. But without sufficient ethnographic evidence, we scarcely perceive even the tip, and can only plumb the depths of imaginary icebergs. Those who think that ethnography is mere description, or worse boring, and who, instead of allowing it, ethno deterministically, to drive enquires, cherry pick facts to fit their theoretical points, deal in abstractions that verge on culturally relative – ethnocentric even – illusions. Ethnography introduces you to another way of being-in-the-world and, short of engaging in participant-observation yourself, it is as near as you can get to knowing it. While immersing yourself in the ethnography can only get you so far in appreciating another perspective, without engaging with ethnography you cannot pretend to have made any attempt. So, if you are browsing this book without studying the detailed account of house construction (I doubt anyone will try to build a house and really experience associated tacit knowledge), you are only going to gain a superficial understanding of the Was valley view of building.

Yet anthropology is more than ethnography (see Chapters 2 and 9), which is the thrust, I think, of such apparently opposed statements as 'ethnography is Anthropology, or it is nothing' (Sahlins 2002:12) and 'anthropology is not ethnography' (Ingold 2008).[6] It is necessarily so, for we inevitably impose some structure on any ethnographic account in writing it, which implies some process of author-imposed ratiocination, whether interpretation or explanation. In other words, ethno-determinism acknowledges that we inevitably draw on theory-cum-ideology for intellectual guidance, but it argues that we strive to let the ethnography guide enquiries, not allow academic concerns overly dictate their direction. And it advocates drawing on any appropriate mix of such theories-cum-ideologies that might help illuminate the ethnography, including sociological and environmental ones. While this mongrel attitude has irritated some critics,[7] such a pluralistic approach is now acceptable, with postmodernism advocating hybridity, pastiche and multi-vocality. This book extends metaphorically on building construction to advocate handyman- or *bricoleur*-like[8] advancement of ethnographically driven understanding that makes use intellectually of whatever comes to hand in any particular cultural context.

---

6 There is arguably also a turf war aspect to declaring ethnography the proper preserve of anthropology in the context of several social sciences currently claiming to use ethnographic approaches.

7 They have variously charged me with being naive and more recently unnuanced, even atheoretical (a meaningless charge, as noted in Chapter 1); or more kindly, 'idiosyncratic' and refusing 'to toe the orthodox line' (Golub 2011:826). In the same review, Alex Golub talks about 'wishing for a more clearly articulated theoretical basis' (2011:825) to my work, which suggests that I have successfully escaped any identifiable theoretical preoccupation – a central aim of ethno-determinism – though hopefully this discussion makes my intellectual position clearer, albeit the nuances and paradoxes surrounding tacit issues may continue to lend an inescapable opacity.

8 To modify Lévi-Strauss's (1972:16–33) term.

Ethnography, as noted, is not self-contained and inevitably involves comparison, which gives it an objective, even a universal aspect, in drawing on cross-cultural understandings of other communities to further those of the place where we subjectively participate and observe. It seems, working in the Western intellectual tradition of empirical scientific enquiry, we are inevitably trapped within a hermeneutic circle in which the subject-object distinctions pertain, and no amount of philosophical hand-wringing is going to get us out of it (Chapter 1). Unless, that is, we prioritize the abstract over the concrete, overlooking the ethnography, and opting for some idealistic approach that allows for imaginative play and Alice-in-Wolaland fiction. But no amount of metaphysics ever built a house. If we wish to convey knowledge about 'impressions' that we sense 'out there' and focus our 'intentionality' on, around which we build our 'ideas' and shared perceptions – for instance about how to build a house – we need to be an observer and document what is going on; that is, see the house as an object and depict its construction. But we also have to participate, as we have seen, to gain an inkling of related tacit knowledge.

While we may think that what we sense and know in observing and participating, and subsequently write about, refers to the same concrete reality for us all, we struggle to represent what we intuit may be the actors' embodied experience of it. In the same way as we use the idea of 'culture' to make the behaviour of others intelligible to us, so we may use the idea of 'embodiment' to make their experiences familiar to us. Attention to the physical person and awareness of bodily circumstances are prominent concerns in the Was valley and relate to interest in well-being; for example, having glossy bouncy hair and glistening firm skin are signals of health. These link to gender issues, as discussed, such as having separate housing and living arrangements, which intimate men's fears of women making them sick (Reithofer 2006:83–99; Sillitoe 1979b; Sillitoe and Sillitoe 2009:77–80, 109–11). But we need to beware of eliding people's emic focus on bodily experiences and their attention to their being with, for instance, phenomenological philosophy (Anderson 2011:235), as though they represent an alternative symbolic 'discourse' addressing the same issues. In the same way as we cannot assume that our use of the idea of 'culture' captures the interests and concerns of the actors we seek to understand, without some unambiguous ethnographic evidence to that effect. The etic perspective, whether formulated as interpretation or explanation, risks becoming imaginary without supporting ethnographic evidence for the emic connections. But some ambiguity seems unavoidable, with aspects of the local perspective unspoken, indeed unspeakable. This returns us to the tacit conundrum and the problems of engaging with undeclared understanding, as noted by Malinowski (1922:11) nearly a century ago, 'natives obey the forces … of the tribal code, but they do not comprehend them'. Such actors are unable to see outside their 'customary ways' let alone being able to interpret or explain them, and may even act as if unaware of them.

Social scientists think that they can penetrate the unspoken tacit dimensions without necessarily acquiring the embodied knowledge (which if they did acquire would presumably be tacit, not explicit, for them too – i.e. beyond consciousness and discussion), which leads them to appear to know, as noted, more about what social actors are doing than the actors themselves. Following Hume's sceptical argument, this problem of the relation between tacit 'impressions' that pertain to concrete existence and 'ideas' expressed using symbols is irresolvable using any explicit intellectual approach devised to date, for you can only really know, as argued, through experience. Some may ask, 'Who cares?' After all, a zoologist studying rabbit behaviour does not expect to understand it from the rabbit's view, investigating behaviour from an ethological perspective. But humans, as reasoning beings, consciously intervene in the world to change it to serve their ends, in ways well beyond any other creatures, and they are self-aware of some aspects of their behaviour (quite different to rabbits we assume, although we cannot talk to them about it, which is the point). In short, culture extends human capacities way beyond biological endowment and demands an ethnographic perspective. Consequently, we have to access people's views somehow to understand their behaviour. We cannot approach it rabbit-like, restricting our research to the equivalent of observing instinctive aspects of behaviour, because these are inevitably mixed up with ratiocinate activity, informed by the collective experience of previous generations. If we have learned anything from the dualistic 'nature versus nurture' tail-chasing debate, it is that lesson. If we are going to approach these issues from a biological direction, it will be via neurological science, as discussed, to track what is going on in the brain when engaged in certain behaviours.

Arguably, the above conundrums apply to all human interaction, whether the actors belong to a different community or the same one, because as individuals we sense things differently, our sensory organs varying in what they pick up, and we further perceive these signals differently having diverse learning experiences and personal histories that inform our impressions and ideas and criteria for knowing. This is, after all, the source of disagreements and arguments, as engaged in here. It relates to the issue of individual variability, which complicates matters further, and is particularly evident in the Was valley with the degree of sovereignty extended to individuals by the acephalous polity (Chapter 2). The upshot is that none of us can entirely know the complex of ideas held by others, even close kin. Yet, while we all differ in the ideas that we hold, those held by members of the same community will overlap to considerable, if varying, extents. And there is sufficient overlap of understanding and interpretation between us as individuals to allow us to think that our sensory experiences and associated ideas do match up, and behave as if we are sensing and knowing exactly the same things 'out there', such that we can share and inform each other's perceptions about them. The overlap allows us to assume that some of our perceptions of things are universal, and even talk about truth and objectivity. But the variation is much greater when persons from different cultural traditions who speak diverse languages interact, with knowledge being a culturally informed cooperative construction sustained by interacting individuals. Accommodating such cross-cultural variation is a perennial anthropological challenge.

Let's stop hurtling around our hermeneutic circle as if we may find a way out. A certain privileging of our intellectual position is inevitable if we are going to make any sense of what people are up to in terms familiar to us, which we can understand. After all, I argue in English, which inevitably transmogrifies meanings expressed in the Wola language, based on culturally mediated different experiences and diverse categories. Acknowledging the limits of our enquiries may authorize us to engage in interpretations and explanations of behaviour. Nonetheless, it is important that we give what ethnographic evidence we can – albeit always inevitably tainted by our intellectual preoccupations that determine what we seek to understand – and distinguish it as clearly as we can from our ruminations, as the ethno-determinist approach argues. Attention to the factual world we sense 'out there' is not as hapless as suggested by some postmodern social thinkers, who presumably occupy an imaginary world.

Imposing order on the Was valley

The moral seems to be: know thy limits. We cannot know how it is to be in the world from another perspective, particularly one informed by a foreign culture – that is, achieve the same conscious awareness of being, let alone tacit knowing. While this reaffirms the postmodern perspective, it does not imply that the only option is reflexive accounts of our subjective experiences, which is to eschew any objective enquiry and forgo analytical rigour. We can legitimately seek to answer questions and develop hypotheses that derive from ethnography, and address these from our culture's intellectual perspective. Among those that have occurred to me in the Was valley are asking how people maintain a sustainable farming system that allows them to cultivate the same land for many years under a nominally shifting regime, which I address by investigating, so far as possible, local natural-resource management practices and knowledge, and relating these to the principles of environmental and agricultural science (Sillitoe 1996). Another issue that concerns me is the extent to which our category of the 'economy' can further understanding of material and social relations, which I tackle by explicating, again so far as I can, local ideas and behaviour, and contrasting these with the assumptions of economic science (Sillitoe 2010a); an endeavour continued here in investigating the construction of fixed capital assets. Another question central to my enquiries is how the dense population of the Was valley contrives social order, living in a stateless polity that prioritizes individual political sovereignty and egalitarian relations, which I address by enquiring into local views and behaviour, again to the extent possible, and comparing with our sociological and political suppositions (Sillitoe 1979a).

Regarding this last question about stateless arrangements, socio-political exchange institutions feature prominently in the regulation of the acephalous order, as previous chapters point out (see also Sillitoe 1979a). These are integral to arrangements that keep power labile and afford individuals a degree of freedom of action that it is hard for those of us familiar with centralized government to conceive; as mentioned in discussing how individual persons are responsible for house construction (Chapter 7). Equality is a significant aspect of the political order too. There is some disagreement about egalitarianism being a feature of Highlanders' social life, particularly with respect to gender relations, which some interpret as unequal. The sexual division of labour, for instance, is seen as an arrangement that allows men to exploit women (Chapter 10). This inequality is seen in housing arrangements too; or is it? Women may occupy single rooms only but they infrequently have to share their space with other adults (except occasionally widowed mothers or unmarried sisters). Some may share their quarters with pigs, but the animals' body heat can warm houses considerably of a chilly night. It depends what you decide to prioritize, which underlines the importance of ethnography, listening to what people have to say and seeing how they behave. While women and men acknowledge different roles and responsibilities, this is not evidence that they think of their relations as unequal or that they feature exploitation (Sillitoe 2010a:330–64).

Though exchange is a ubiquitous feature of Was life, and people have a large vocabulary for various transactions (Chapter 3), and though these events are a popular topic of conversation, particularly among men, my Wola friends are not apparently able to discuss their socio-political implications regarding individual sovereignty and equality. They are able to express the values that I take as central – with comments regarding freedom of individual action, such as a person's behaviour is *obun konem* 'her/his will/wishes' or *mbinyon turiymon* 'their pleasure', and regarding equality that *hombuwn bombortuwbiy* 'all are the same' or *sem menden taengat inj* 'no family is different [from another]'. But the question that I ask about the part exchange plays in their stateless society does not apparently occur to them. They live their acephalous order and take it tacitly for granted, as they are socialized to do. They are too busy interacting socially according to values inculcated from birth onwards to reflect on their political-economic implications. Furthermore, it is unrealistic to suppose that people might ask questions that presuppose a comparative political perspective when, until fairly recently, they had no experience of the institutions that those of us familiar with state government associate with, and think necessary to, social order. What familiarity they have of state arrangements comes from perplexing Australian colonial authority, which often seemed arbitrary, even capricious, in the way it dealt with persons, and following national independence, what amounts today to a failed state. Consequently, my attempts to understand their political-economic order, which contrast it with ideas drawn from experiences of life in a state context, inevitably appear to know it differently to them. In doing so, for instance, I use the term 'political-economy', although there is no analogous word, and hence consciously expressed concept, in the Wola language. In imposing it on the collective behaviour that I have observed, and comments that I have heard actors make about it, I consequently run into the contradiction, as pointed out several times, of appearing to know more about the implications of their social interaction than the actors do themselves, when it is their perceptions and behaviour that are of interest.

Social science deploys abstractions such as economy and society largely to explore philosophically, not explain scientifically, the import of collective human interaction. While writers seek not to reify these abstract 'complex ideas' and attribute to them a concrete existence, they do so indirectly in according them the power to create social solidarity, manage conflict, give meaning to life, effect symbolic communication, direct leadership, organize production, ensure distribution of resources and so on. In so doing they may intimate in an unavoidably tautologous way that it is for such reasons that these concepts exist, exerting their ethereal force.. The values, expectations and responsibilities that persons hold in common and pass on between generations, which inform the behaviour of those socialized with them, evolved (and continue to evolve/ change) in response to the tacit emotional drives of individuals interacting together and behaving in concert. They order collective living, and the co-operation central to human success, without, as argued, anyone reasoning logically about the implications for 'society'. We can only speculate, as pointed out in Chapter 4, on the processes that occurred as our proto-human ancestors started to realize long ago that the synergies achievable through co-operation were considerably greater than acting on one's own, the play-off of individual against group interests becoming increasingly significant as collaborating groups of individuals interacted and negotiated values and norms that they came to share and agreed should guide their collective behaviour. According to the unequal-gender-relations standpoint, men dominated the process and manoeuvred to subjugate women, in a way recalling the Victorian social evolutionists' patriarchal horde idea. In this view, the esteemed exchange institutions of New Guinea evolved as forums for men to compete for social standing, while excluding women, even exploiting them, their labour supplying transacted wealth, which is, I argue, to misrepresent the complexity of exchange behaviours and source of transactable things.

In exploring social interaction in the Was valley, I consequently and inevitably deploy English language categories and associated social-philosophical assumptions. But cognizant of the strictures of empirical philosophy about the implications of being in the concrete world, and prioritizing ethnographic evidence, I nonetheless aspire to some science-like rigour (without implying inappropriately seeking to prove or falsify hypotheses experimentally). The tensions in striving to accommodate both positivistic and idealistic perspectives are seen in the long-running 'is anthropology a science or humanity' debate (Carrithers 1992:146–55), which is ultimately irreconcilable for reasons discussed here. The tacit dimensions of behaviour starkly reveal the limits of these aspirations and the problems facing the social sciences. While we should constantly relate abstractions such as economy or society back to the ethnography, to ensure that the assumptions these 'complex ideas' introduce correlate with what people say and do (or risk coming up with a purely imaginary symbolic interpretation that misrepresents their being-in-the-world), the ethnographic evidence can prove elusive, it being tacitly within the actors. If I ask my Wola friends why they engage in exchange activities, and I prod them beyond their stock *shumbaen bismiyuwp* '[because] our ancestors did so' answer, they come up with comments such as *imbiy na wiy* '[I'll have] no name [if I don't take part]' or *dimb ora na bayaemiy* '[only] very inadequate persons do not do so' or *mbinyom niy kogit homobtuw* 'they [who enter into transactions with me] would be angry with me'. These attitudes again challenge the reciprocation thesis that dominates discussion of exchange in Melanesia, drawing on Mauss (1970), which maintains that persons engage in transactions and meet obligations because some ethereal-cum-moral, even spirit-like force adheres to the things that they give to one another and obliges a return (Chapter 3), for no one has ever intimated the existence of such a force to me.[9]

While the reference above to persons being angry arguably foreshadows a certain awareness of social-order

9 The plethora of terms for exchange activities further indicates, as pointed out in Chapter 3, the inadequacy of 'gift', the popular gloss for reciprocal transactions associated with this viewpoint.

issues – i.e. disputes may occur if you do not meet your transactional obligations – I have never been able to engage anyone in further abstract discussion of these issues. The responses to the question 'why do persons participate in exchanges' refer to personal emotional states or, after Hume (1896), 'passions' and associated human desires, such as a need for the regard of others, to be liked and have self-respect, even a certain admired standing, if ambitious, and the social sway that may come with it. The phenomenological turn invites us to focus our attention on experience of such feelings, although it is not popular to argue that the emotions underpin human behaviour, a stance that was severely frowned on until relatively recently. On one occasion, I recall Lucy Mair, who could be a fierce critic, crossing out the word 'feel' in a sentence that I had written, 'they ~~feel~~ obliged', and writing in the margin, 'people do not feel, they think', which though it confused me at the time, taught me something valuable, as evident in my recall of the comment decades later as I puzzle over the implications of explicit and tacit knowledge informing human behaviour. Various emotions motivate persons in the Was valley to engage in transactions, to judge from my friends' comments, albeit collective values and norms, by which persons appraise behaviour, inform their emotional responses; that is, expectations and responsibilities inherited and shaped down the generations, which culture 'carries', and that give some predictability and order to 'impression'-based social interaction. If Lucy were here today, I should explain that I seek to pay attention both to 'passions' and 'ideas', as they jointly feature in human behaviour in any cultural context, but that the 'passions' are prior, more primeval. The 'cultural programming' of our responses to emotional states, and probable modification of them from their 'natural' state, probably started long before modern humans emerged 100,000 or more years ago.

It is possible that the technologies increasingly available to neuroscience, as discussed above in respect of 'know-how' tacit knowing, may allow mapping of cerebral activity associated with different emotional states, for instance feelings of affection or aversion, and associated behavioural sequences. They promise scientific engagement with these issues, featuring robust evidence by which to assess hypotheses about social behaviour. But we are far from understanding the 'passions' and tacit 'know-that' awareness, or their contribution to social interaction. Indeed it is questionable, as pointed out, if such understanding is feasible, as it implies turning emotional awareness into rational intelligence, which will presumably transmogrify it into something quite different. Just as the prospect of turning tacit knowing into propositional knowledge, if possible at all, threatens to transform it into something else. This conundrum is at the centre of studies, such this one, that deal in conscious understanding, which by definition cannot accommodate other ways of knowing without seeking to change and misrepresent them. Furthermore, individuals vary regarding their experience of emotions, depending on personal history, which includes dealings with kin and extent of parental affection, state of relationships and interaction with other persons, and so on. The gradient of feelings that can attend any particular emotion make comprehensive understanding of their contribution to social interaction improbable given the complexities of previous experiences, both individual and collective, together with differing socio-culturally conditioned responses and expectations, themselves the sedimented product of generations of previous emotionally informed social behaviour, and subject to change in light of subsequent generations' experiences.

Variation is a ubiquitous aspect of social life, as noted, in view of individuals' differing personal histories and emotional responses, together with diverse community circumstances. Regarding gender relations, for instance, individual women's and men's opinions about these vary, depending on previous and current relations, together with personality. This points up the fundamentality of the individual perspective, which I argue is necessary to comprehend social behaviour in the Was valley. My variable 'Melanesian individual' perspective takes issue with the sociological 'moral collective' approach that proposes normative obligations to 'clans' are central to ordering Highlander social life (Sillitoe 2010a:39–44). The sociological approach jumps straight to 'complex ideas', whereas a focus on 'simple ideas' that relate directly to the 'impressions' that our senses detect and 'passions' that we feel, which contribute to 'complex idea' building blocks, reveals the danger that the resulting disconnection with reality 'out there' poses in allowing the sociological imagination to invent a symbolic cultural world for the unaware, or at best tacitly aware, natives. It is understandable, perhaps, that the normative sociological focus came to dominate, because in jumping straight to the etic perspective it avoids the intellectual problems that dog attempts to access the emic view, much of it tacit. The associated problems emerged long ago, as seen in a Western Pacific context, for instance, in Malinowski's (1922:176) comments about the implications of Trobriander 'customs' being 'beyond the comprehension of the natives'. Nor did he shy away from the implications of unconscious drives shaping observed behaviour, arguing for a focus on 'primitive psychology', which was unfortunate given the considerable problems encountered regarding the relevance, cross-culturally, of many Western psychological assumptions (particularly the dominant Freudian ones of his time). These strengthened the sociologists' contention that a sociological, overwhelmingly Durkheimian, approach is the more coherent in the study of small-scale community life, as evident in his protagonist Radcliffe-Brown's insistence on 'comparative sociology'.[10]

Divisions of labour

While we inevitably start from our etic perspective or ontological viewpoint, the ethno-determinist perspective advocates that we hold it as loosely as possible. It challenges approaches that arguably aggravate our predisposition to arrive in a community already socially enculturated to expect certain things and understand being-in-the-world a particular way (by, for example, subjecting students to an academic education, filling their heads with social theories that are elaborate, even esoteric, intellectual perspectives). In deploying associated notions of society, economy, gender or whatever, researchers may search for aspects of related assumptions that they think identifiable in the emic viewpoint, such as in peoples' comments and behaviour, maybe expressed in cryptic symbols that require interpretation, used as evidence to support etic views. In other words, we arguably invent 'their' culture for 'them' (Wagner 1981:4). In doing so, we should strive to allow the emic ethnography to guide our enquiries so far as we can and not impose inappropriate ideas; such as assuming local residential clusters in the New Guinea Highlands necessarily display group solidarity, when close observation of behaviour questions it.

The holding of our unavoidable etic perspective 'loosely' is itself a tacitly acquired skill. It is not that we participate as best we can just to get an inkling of others' tacit knowing, but also to hone the tacit skills necessary to work ethno-deterministically and develop our own pliant, implicit way of being. This relates to the aforementioned apprenticeship qualities of participant-observation (Chapter 11; Lave 2011). This way of working requires experience and practice to achieve the state of mind that allows us to hold our socio-cultural conditioning sufficiently freely to let those with another socio-cultural perspective genuinely (so far as possible, anyway) to determine the direction of enquiries. While it is not viable to slip our own socio-cultural anchor, it is possible with familiarity and the will, to allow considerable slack in the chain, which is as much as we can expect to achieve, given the unavoidable strictures of postmodernism. While our participation is inevitably limited, and possibly distorted, we can sail along as if in another community to some extent, such that its members contribute significantly to setting the course of investigations, albeit we cannot dispense with our intellectual compass in pondering where such participation takes us. This implies that laying out a methodological chart for persons to follow in working ethno-deterministically – that is, putting it down in words, as in a methods manual – is unfeasible, because as with acquiring any tacit skill, it demands practice not words; furthermore, the encounter will differ significantly according to socio-cultural context, so we shall not all have the same experiences. (This methods issue relates to the foregoing point about the distortions courted in filling students' heads with theories prior to the participatory experience.) Those who have the aptitude can adopt the free-fall mind necessary, but in my experience those who constantly seek guidance about what to do in methodology handbooks struggle. A project involving me and a person with an excellent background in geological science, for instance, which was to research local community and mining company relations in Papua New Guinea, faltered because of his experimental mindset, expecting me to tell him what to do next and direct him to appropriate texts (after laboratory manuals) even when he was in the field on the other side of the world.

The tacit approach to field research may strike some as too opaque, the nuances of its theoretical-cum-ideological view too ambiguous. But it suggests a way to tackle the question of how to get a handle on tacit knowing generally, namely that we need to see the problem from a new perspective in which we suspend our usual intellectual expectations. In short, an approach that taps into the intuitive, creative and emotional dimensions of experiential knowledge, as opposed to trying to understand it according to the dialectical, rational, evidential canons of explicit knowledge. An exemplar is the Japanese intellectual tradition that accepts we may not readily perceive and express knowledge, and which seeks to convey it by having students come to terms with *koan* riddles, mentioned in the previous chapter in the context of claims that a Japanese company's development of a bread-making machine demonstrates the conversion of tacit into explicit knowledge (Nonaka and Krogh 2009; Ribeiro and Collins 2007). But this seems to amount, again, to arguing that the only way really to understand embodied knowledge is to learn the activities in which it features and puzzle over whether there is anything more we can do with such knowing other than do it.

The variation noted in individual experiences of doing things further complicates engagement with tacitness. It is possible that there is a gender dimension here, which gives an intriguing spin to accessing embodied knowing issues. The implications are interesting with respect to participant-observation, suggesting that women and men may come to know things somewhat differently. Some feminist

10 If we take the perspective of currently popular rhetorical analysis, which looks for possible covert factors (Carrithers 2005), it is likely that European class consciousness played a part too (cross-cultural psychology continuing as a prominent strand in American anthropology), the academic establishment siding with one of its own – the superlatively self-confident, Cambridge educated, opinionated, suave Englishman with aristocratic pretensions against the brash, Jagiellonian educated, pushy, self-glorifying egocentric Polish minor noble.

scholars focus on subjectivity, relatedness and intuition to ask if women's and men's embodied knowledge differs, proposing, for instance, that certain intuitive faculties may be linked to gender. After suggesting that 'there are ways of thinking that we don't yet know about … ways which traditional intellect denies, decries, or is unable to grasp', Adrienne Rich (1977:284) goes on to ask 'whether women cannot begin, at last, to think through the body, to connect what has been so cruelly disorganized – our great mental capacities, hardly used; our highly developed tactile sense; our genius for close observation; our complicated, pain-enduring, multi-pleasured physicality'. It is feasible that gender results in women and men attending to the world via their bodies in different ways (Gallop 1988), a tendency possibly reinforced during socialization, albeit we need to treat such a suggestion with care, so as not to appear to endorse possible gender associations with questioned dualistic oppositions seen in the mind-body debate – contrasting feminine with masculine, nature with culture, and intuition with reason – when feminism has been at pains to 'deconstruct' associations with such a Cartesian-cum-structuralist worldview. If women do apprehend things somewhat differently, female researchers may more readily intuit some aspects of tacit knowing, notably those associated with know-that social relations and interaction, although whether they will be any more able explicitly to communicate what they access is a moot point.

The Wola would appreciate the view that tacit knowing features gendered differences, as it complies with their sexual division of labour arrangements (Chapter 10). It is self-evident to them that such labour arrangements result in women and men having different sets of skills and associated embodied knowledge. In house building, for example, many tasks involve the use of axes, both in preparing materials and constructing buildings. The axe is exclusively a male tool (Sillitoe 1988a:43–55), which women should not use, or even touch, because, as noted, if they do so they may blunt the cutting edge. Any gendered activities, as these sexual-division-of-labour arrangements highlight, have further intriguing methodological implications, complicating the suggested generic gender differences in accessing embodied knowing. They suggest that male and female fieldworkers potentially experience different skills when engaging, as appropriate to their gender, in participant-observation. Even if their participation distorts native conventions, by trying to imitate the behaviour of the opposite sex, they arguably experience things differently. This variable gender dimension, added to the cultural and individual aspects, further complicates the task of convincingly accessing and passing on tacit knowing in any sure replicable way, let alone communicating what it is explicitly.

Some may ask, to return to an opening question: 'Why bother?' Why trouble with this phenomenological mind-wringing, when arguably the time is past that we might assume to go and conduct fieldwork in other communities with a view to furthering our understanding of human socio-cultural arrangements? It seems arrogant in the twenty-first century to presume to undertake such research, when communities can, and should, undertake it for themselves. This is certainly the attitude of some communities, such as in Amazonia, which feel that they have been poorly served (or even misrepresented, or exploited and abused) by anthropological research. And this augurs change, with calls for collaborative and engaged research with indigenous colleagues (Sillitoe 2012b, 2015), if they agree to us joining them. The implication is that we should all now focus on 'anthropology at home', as the only politically acceptable way to proceed, and maybe on occasion facilitate research collaboratively with communities elsewhere, if they wish to avail themselves of any co-operation.[11] A problem with 'anthropology at home' is that while we may tacitly 'know' the embodied aspects of our culture, as members of it, these may be even more inaccessible to us explicitly than to a foreigner, because we are conditioned to accept so much of our socially informed way of being-in-the-world. It is difficult to stand back and take a dispassionate view; so, for instance, someone who has certain religious convictions is unlikely to consider them critically from an agnostic social-science perspective (as illustrated in Chapter 2's mention of a believer's blind faith in the face of scientific findings). Though this trips over the paradoxical point that such a dispassionate stance is alien to that of the actors, which participant-observation seeks to access (in addition to aforementioned assumptions of greater knowledgeability). In short, 'anthropology at home', devoid of the cross-cultural aspect that is central to anthropology, is ethnographically limited sociology. Wherever it comes from, we still need ethnography from beyond home to further comparative understanding of humanity.

In my defence, as an apparently old-school anthropologist, if I had not engaged in the research programme that this book is part of, much knowledge of Was valley culture and ways before the arrival of the outside world would now be lost (Sillitoe 2010a:468–9). Whenever cultural formations disappear, we lose forever unique examples of ways that humans have come up with to be in the world (as when languages become extinct and we lose

---

11 Less common currently, people elsewhere may develop a tradition of anthropological enquiry and give us a dose of our own nosey medicine by conducting field research in Europe. While a Wola ethnography of England is currently an improbable prospect, a question that may occur to such a researcher – mirroring my stateless order question – is how we tolerate centralized state control over our lives and lack of personal freedom; albeit any person who undertakes such an endeavour will presumably be educated in European cultural-academic ways and no longer able to see the world from a stateless perspective.

associated knowledge), which is not only a loss in itself, but also diminishes the fund of evidence that may further comparative understanding of humanity. When I first went to live in the Was valley there was scant awareness in the region of anthropological research,[12] let alone persons to undertake it. The idea was strange. It took months to convince people that I was there to understand their way of life. They thought it extraordinary that someone from the technologically overwhelming world beyond their region should have any interest in their ways, which they thought obvious anyway, reflecting the assumption of tacit-knowledge bearers that there was nothing to know. And currently none of my friends show any interest in engaging in the sort of ethnographic work that I have undertaken, although they seem grateful to have a record of their *shumbaen* 'ancestors'' ways.

These comments relate to the social changes that have occurred and are occurring, often glossed as globalization. While many communities in the Highlands and beyond are increasingly aware of the implications of research and want to ensure that use of any knowledge originating with them is fair, this is only one aspect of such change. (In Papua New Guinea currently, it is communities with companies exploiting mineral resources on their territories that are largely interested in research, to establish their land rights and ensure their entitlement to ensuing royalties.) The challenges of change are considerably more. In addition to the cultural, gender and other epistemological barriers that we have to contend with in dealing with tacit knowing, we face a moving target to boot. We are inevitably running to catch up, without the prospect of ever doing so at current rates of accelerating change. It is reminiscent of the predicament of the ancient Greek philosopher Cratylus (a follower of Heraclitus), who found communication impossible on the grounds that change (Heraclitus' flux) is constant and so it is impossible to speak of anything, as nothing is fixed and knowable and so expressible in words – as a consequence, according to Aristotle, he abandoned speech and limited his communication to gestures (Allan 1954:271–2).[13] Sometimes, when dizzy with intellectual tail-chasing, I glimpse the wisdom of his mute actions. Yet academic expectations to capture Was valley culturally mediated experience explicitly in words have proved too compelling to resort to Trappist vows (just yet anyway), regardless of the considerable hurdles that I have encountered, some apparently insurmountable, which bring to mind the following lines of Tennyson's *Ulysses* poem:

> 'Yet all experience is an arch wherethro'
> Gleams that untravell'd world whose margin fades
> For ever and forever when I move...
> And this gray spirit yearning in desire
> To follow knowledge like a sinking star,
> Beyond the utmost bound of human thought'.

The changes in building styles and architecture in the Was valley, with *mbolin aenda* 'whites' houses' and the demise of cults with their associated structures, are aspects of the socio-cultural transformations. Such rapid change compounds the challenge of comprehending the experiential aspects of life. If these changes result in the demise of the traditional way of construction, which seems probable in view of what has happened elsewhere (Coiffier 1990; Rensel and Rodman 1997; Ruff and Ruff 1990), it will be necessary to know how to build a house not only to advance any future enquires into traditional house construction, but also into the associated tacit knowing, regardless of possibly insurmountable limits. If change results in the disuse of traditional tools and related technology, the loss of associated embodied knowledge is potentially irreversible. It becomes extinct. With regard to stone tools, for instance, no one living in the Was valley today has first-hand experience of using these tools routinely in undertaking everyday tasks. The knowledge has disappeared in a couple of generations, for a significant part of it was tacit, learned by informal hands-on education, through observing and mimicking skilled persons (see Chapter 2), and with their passing the associated know-how is perhaps lost forever. If neurological advances result in possible communication of embodied knowledge to any extent using virtual-world technology, and New Guinea Highlanders have stopped constructing structures such as discussed in this book (just as they have stopped making many artefacts documented in *Made in Niugini*), the only way to experience and 'know' and pass on something of such capabilities, with no extant knowledge-bearers, will be to replicate the construction activity (leaving aside the question as to whether it would ever be possible to reconstruct activities faithfully without reproducing and situating them holistically within wider Highlander cultural life). Such a circumstance is reminiscent of experimental archaeology (Chapter 4), only with detailed ethnographic information to guide activities. It may further advance archaeological understanding, featuring technologies of the sort used by long-ago populations responsible for the remains of buildings found in prehistoric excavations. Any such advances, if feasible, will require detailed information about the process and skills – that is, a step-by-step guide

12 Just before the outbreak of the Second World War, F.E. Williams (the Papuan Government Anthropologist) spent a few weeks in the neighbouring Ak valley, and while he enquired into peoples' lives more than other government officers on that patrol (Williams 1939), they considered him to be one of them.

13 Today, he would probably be diagnosed with 'selective mutism', which is an anxiety disorder in which a person, who although quite able to speak, is unable to do so other than with a few close persons such as parents and siblings.

Plate 13.1 Over to you: what you have to build to know.

to follow – which implies rigorous ethno-determinist-like research.

Why should the building of smoky squat houses interest anyone, especially if the builders discontinue doing it? Beyond pure ethnographic curiosity and awareness of humanity's cultural aptitude and diversity, any such interest seems improbable, until put in the context of the loss of skills, which returns us to division of labour issues again, not sexual this time, but occupational. Since the eighteenth century, when manufacturers became aware of the increases in production achievable with labour specialization (Chapter 10), measured in terms of volume of goods produced – such as pins turned out, to cite Adam Smith's (1993) classic example – we have witnessed the progressive loss of embodied craft skills that are no longer passed on in practice. The advance of mechanized technology has furthered the occupational division of labour, each arguably stimulating the other, with robots today increasingly replacing humans, which in turn is driving attempts to make tacit knowing explicit and reproduce associated skills, as in the bread-making machine. The change from 'hand-tool to the machine, is not from the simple to the complex, but is rather tantamount to the withdrawal of the producer, in person, from the centre to the periphery of the productive process' (Ingold 2000:298). Skilled persons, previously common, disappear with ever increasing mechanization of production, with a few pockets of dedicated craftspersons keeping some tacit skills alive.

The nineteenth-century Arts and Crafts Movement decried the trend early on, but could scarcely halt the relentless profit-driven process involving labour specialization, which has arguably made goods available to people on a scale unthinkable previously, increasing material standards of living greatly but at considerable cost in respect of monotonous and stressful working conditions, with humans reduced to cog-like machine operatives. In his censorious but influential book *The Seven Lamps of Architecture*, Ruskin (1849:194, 196), for instance, rails against his contemporaries:

> we are, perhaps, hardly enough in the habit of inquiring, with respect to any particular form of luxury or any customary appliance of life, whether the kind of employment it gives to the operative or the dependant be as healthy and fitting an employment as we might otherwise provide for him. It is not enough to find men absolute subsistence; we should think of the manner of life which our demands necessitate; and endeavour, as far as may be, to make all our needs such as may, in the supply of them, raise, as well as feed, the poor.

The manufactured goods that arrived when the outside world entered the valleys of the New Guinea Highlands much impressed people, and they have turned into avid consumers, who currently show no signs of ruing possible threats to cultural identity and everyday life in taking up associated foreign ways, which experience elsewhere suggests is likely at some time, as discussed in Chapter 1 with respect to interests in vernacular architecture. If, and when, that time comes, the heirs to Highlander tradition may find ethnographic accounts valuable cultural resources.

While they may be latecomers to consumerism, the residents of the Was valley are keen to catch up, contributing significantly to the rapid pace of change. A noteworthy aspect is the erosion of their previously independent ways, as has occurred before in other places where 'the consumer revolution implied the "withdrawal of the producer", a successive deconstruction of everyday skills through the consumption of the readymade ... the process in which people, in the transition from preindustrial production to consumption of technologies, in many ways lost control over their everyday technologies' (Flitsch 2008:269). The increasing dependence on outside imported goods, starting with steel tools and gradually extending to other things such as manufactured building materials, presages the end of the Was autonomous 'user-makes' organization of production, which is an integral aspect of egalitarian political-economic arrangements. It is these changes that may herald the collapse of associated acephalous ways, as a local hierarchy emerges with some parties controlling access to the global market and imports, rather than the imposition of the nation-state. If associated occupational labour arrangements come to structure everyday life, as opposed to the traditional sexual division of labour, people may come to experience exploitative employment as they increasingly enter the capitalist market to acquire its manufactured goods.

Furthermore, while the loss of knowledge and embodied skills, which accompanies such social and material change with the passing of local traditions, may otherwise seem inconsequential, these are an integral part of a locally well-adapted way of being-in-the-world built on the trial-and-error wisdom of generations. The suppliers of the new imports look less well adapted. Although occupational specialization and mechanization may mark technological progress, the environmental costs are proving high and may turn out in the long run to be unsustainable, if not catastrophic. We have something to learn from more sustainable technologies such as those of the Was valley (Chapter 6), however improbable at first sight, or how uncomfortable their houses may seem to us. But humanity looks set to lose such sustainable ways of being-in-the-world with the demise of knowledge and skills integral to them. The embodied standpoint underscores the implications in arguing that the only way really to understand and appreciate the tacit aspects of associated experiential knowledge, of Was valley house construction for instance, is to do it, to build a house. Accounts such as this can give only a glimpse of such knowing, which we can only really comprehend in the doing. The labour implications now concern you, the reader. In short, go construct and dwell in a Wola house to acquire the necessary tacit skills and truly know about house building.[14] It is a conclusion that will doubtless seem far-fetched – though a future experimental archaeologist may actually take on the challenge – yet it is where sceptical phenomenological assumptions take us.

14 PS, the Health and Safety Office at my institution advises that I point out that working with an axe can be dangerous, particularly if you are inexperienced.

APPENDIX

Glossary of Wola construction terms

aend bez ma taebtuw	to put *bez* ridge-pole in place
aend bez tonga mbotuw	to tie green ridge-pole between two stakes to keep it straight before use.
aend bubtuw/aend bay	to build a house
aend buwk tongaebtuw	to lash *buk* pandan-bark sheet in position
aend buwk tor baebtuw	to put *buk* in place, and stake and lash
aend dem	log along bottom of *aend hwimb*
aend dol aesobtuw	to drive in wall stakes
aend dol or *dol*	house wall
aend gaimb payobtuw	to cut/collect cane-grass thatch
aend hobol kobtuw	to dry house out (with fires)
aend homb lubtuw	to thatch horizontally along ridge
aend hwimb	bark lining/facing
aend hwimb paeraebtuw	to fix *shwimb* bark lining on *waenuw* wall of *enza*
aend iysh haerubtuw	fires lit to dry out new houses
aend kor bubtuw	to make fire-screen
aend kundiyubtuw	to dismantle a house with a view to rebuilding
aend maenjet mubtuw or *aend tomb bubtuw*	to lay out house plan with saplings and vines
aend pinj	ridge-pole supports
aend pinj aesobtuw	to drive in ridge-pole supports
aend showaip	house embargo (*tambu*)
aend suw tombobtuw	to prepare a house site
aend dombep	wooden framework for fireplace
aend tinjiyubtuw	to thatch house
aend tominj	thin saplings lashed on semicircular rear of roof
aend wem sawobtuw	to cook meal for helpers in earth oven (*mumu*)
aend wil	porch and sitting room
aenda maenjet mubtuw/ aend tomb bubtuw	to mark out plan of new house
aendashor koma	aerial roots of *aendashor/ paym* pandan
aenda munk	house yard/site
aendashor shor waenuw hwimb tongaebtuw	fastening *aendashor/paym* leaves in position on wall.
aenk	pandanus
aenk el	old pandanus
aenk tor	thatching gutter of pandan bark
aenk tor bubtuw	to construct *aenk* pandan-bark gutter for thatching
aenk tor bubtuw iysh aesobtuw	driving in stakes to hold *aenk tor* gutter in place
aenk/pundin lubtuw	to beat *pundin* or *aenk* pandan trunk to loosen bark
baerob	vine binding along top of *maenpat*
bez or *aend bez*	ridge-pole
bez tongaebtuw	to lash *polpen* on *bez*
bez kaembobtuw	to cut off end of *bez*
bokor lubtuw	chopping tree trunk transversely in two
bor mor homeabtuw	to make axe cuts along pandan trunk in preparing *buk* sheets
buk (or *aend buwk*)	bark sheet (from *pundin* or *aenk* pandans) for house wall
buk boraelubtuw	to jam *buk* in position with *pongow* stakes prior to lashing
buk iy muw tuwaebtuw	pulling pith from inside of *aenk* trunk
buk lubtuw	to beat *aenk* trunk with clubs to make bark pliable and soften pith
buk poror lubtuw	to lift *buk* in place on wall
daengbiyp	logs laid along bottom edge of leaves
daengbiyp bubtuw	laying logs along leaves to secure in place
dem (or *dombep*)	sill log

dem (or *dombep*) *lubtuw*	to put *dombep* logs in place (for fireplace, outside drainage and pig stalls)
deriyau	door
deriyau hae komb waerubtuw	to make slats for door
deriyau haekomb	door slats
deriyau haekomb bobtuw	to close slats on door
deriyau waerubtuw	to cut out door opening
dombep (or *dem*)	sill log
dul	wooden wedge
duwn hondobtuw	to sight along stakes for straightness
em gəmb	small garden near house
enzatenda	house with male and female rooms
gaimb	cane-grass (*pitpit*)
gaimb haeray	burning off dry cane-grass and other vegetation during site preparation
gaimb payobtuw or *aend gaimb payobtuw*	to cut cane-grass /collect thatch
gaimb tay aezay	levering out cane-grass roots
gamonk	beading/batten
gamonk dekel bubtuw	to split *gamonk* saplings in half for beading
gamonk maen kobtuw	to temporarily lash *gamonk* in place
gamonk paym shor howubtuw	tiling using tucked-under-leaf method
gamonk paym shor maerubtuw	tiling using folded-pandan-leaf method
gamonk tongaebtuw	to lash *gamonk* beading permanently in place
gəmb	digging stick
gənk bubtuw	splitting tree trunk longitudinally in two
haeluwp dekel bubtuw	to split rattan vine
haeluwp ya	rattan vine
hedpong	sapling that goes from *nunung* wall to the crutch of *polpen* at the ridge
hedpong haerubtuw	to put *hedpong* in place at *nunung*
hiypaip aesobtuw	to drive in short stakes of front wall
hiytpaip	short stakes in low front wall of *wil aend*
holol	wooden club
homb / humb	roof crest/ridge
hwimb	any wall surface covered with *paym* pandan leaves (e.g. *waenuw hwimb* or *maenpat hwimb*)
hwimb pongow	sticks used to secure *paym* leaves
iyport lor	row of lashing on inside of house wall
iyb hedla	eaves
iyb hedla horaebtuw	to lash saplings (4 or 5 rows) round the bottom of *polpen* over eaves
iyb gən	drainage way
iyb gən mubtuw	to dig drainage way round house
iysh aesobtuw	driving in stakes
iysh humbiy howubtu	to put bark sheets round eaves prior to thatching
iysh tongaruwk	lashing gutter stakes to eaves
jiy bubtuw	cutting a half trunk longitudinally into quarters
ka	lathes
ka bubtuw (or *tongaebtuw*)	to lash on *ka* lathes
kat tor	pig-stall roof
koma	aerial root of *paym/aendashor* pandan
komberlay	front of house (near *aend wil*)
kor	stretchers lashed onto *polpen* between outside wall and ridge
kor tongaebtuw	to lash the *kor*
kuwngis	sapling that fits in roof crotch
kuwngis tongaebtuw	to lash *kuwngis* in crotch
maembow pes	eave wadding
maembow pes bubtuw	to stuff moss or leaves into gaps round eaves
maenpat	longer stakes on low front wall of *aend wil*
maenpat aesobtuw	to drive in longer stakes on side of front low wall
mem tay (or *taenk tay*)	fireplace
momberael	stones surrounding a fireplace
momberael ubtuw	to line edge of fireplace with stones
momenda	house where men and women sleep in same room
nay wem	*mumu* (earth-oven) meal for thatchers
ndul bubtuw	to weave along top of *hiytpaip gamonk*
nunung	bell end of roof on *tenda*
ordbay munubtuw	to break sapling in two to make *hedpong* from *kungis*

paym shor	leaves of *paym* or *aendashor* pandan
paym shor torbaebtuw	to put *paym* leaves on *waenuw* wall
pendet bay	rolling stakes down slope
pinj or aend pinj	ridge-pole supports
polpen (or *aend polpen*)	roof rafters/stays
polpen haerubtuw	to place *polpen* in place
polpen ndibdiy kaembobtuw	to cut ends off *polpen* along ridge
polpen tongaebtuw	to lash *polpen* at eaves
pongow ma haebtuw	to jam *pongow* sticks in place on dividing wall
pongow or *iysh pongow*	jamming sticks/stakes
pongow tongaetuw	to lash jamming sticks/stakes in place (together with *gamonk*)
pongow waerubtuw	to make jamming sticks
pongow waeruw ma haebtuw	to make and drive in *pongow* sticks
pəgaenda	shelter used in daytime
puw hondobtuw	cutting *aenk* trunk open lengthwise
sal	sprigs of casuarina leaves hung at *aend wil* entrance
sal aengaybayobtuw	to bend saplings in place under *sal* fringe to hold it up
sal gamonk	front-eave arch sapling
sal gamonk tongaebtuw	to lash *gamonk* in place round front of house
sal sobtuw	to put on *sal* fringe – push in
senz kobobtuw	to collect/pull up grass thatch
sezel bombobtuw	to drive in *sezel* cane
shonon	eave batten
shonon gamonk	thin beading on upper part of wall between *wil aend* and sleeping room
showmay kuwl	pig stalls
suw tombobtuw	to dig out a house site
tabok	lashing on *gamonk* above door (figure-of-eight weave)
tabok lubtuw	to weave figure-of-eight binding on *gamonk* over doorways
taembok	bamboo
taenk tay (or *mem tay*)	fireplace
tay gamonk	bottom *gamonk* on outside wall (permanently lashed)
tenda	women's house/women's room
togol	short stake
togol haebtuw	to drive in short stakes to hold door slats
tongok lubtuw	splitting *jiy* lengths into stake-sized pieces
tubael	room
tuwmainj tongaebtuw	to lash *tuwmainj* across top of *waenuw* wall
tuwmainj tongaebtuw	to lash *tuwmainj* in bell end
tuwmaink or *siymaenk*	sapling hoop on roof
uwpuluwk (or *puluwk*) *aenda*	men's sleeping room
waendiyubtuw	to sharpen stakes
waenuw	wall or wall stakes
waenuw aesobtuw	to drive in stakes of interior dividing walls
waenuw waerubtuwwaenuw lor payobuw/waenuw tombobtuw	to level off site for house so even and ready for building (i.e. final levelling)
waenday payobtuw	to sharpen *polpen* meeting on *hedpong*
wenet humbiy	bark from *iysh wenet* tree
wenet humbiy howobtuw	laying *wenet* bark along eaves
ya	vine, bark strips

REFERENCES

Abel C. 2000. *Architecture and Identity: Responses to Cultural and Technological Change*. Oxford: Elsevier/Architectural Press.

Adams, F. and Aizawa, K. 2009. Why the mind is still in the head. In P. Robbins and M. Aydede (eds.) *The Cambridge Handbook of Situated Cognition* (pp. 78–95). Cambridge: Cambridge University Press.

Agricola, G. 1950 [1556]. *De Re Metallica* (trans. H.C. and L.H. Hoover). New York: Dover Pubs.

Aizawa, K. 2010. The coupling-constitution fallacy revisited. *Cognitive Systems Research* 11:332–42.

Alberti, L.B. 1988 [1452]. *De Re Aedificatoria. On the Art of Building in Ten Books* (trans. J. Rykwert, R. Tavernor and N. Leach). Cambridge, MA: MIT Press.

Alexander, C. 1979. *The Timeless Way of Building*. Oxford: Oxford University Press.

Alexander, C., S. Ishikawa and M. Silverstein 1977. *A Pattern Language: Towns, Buildings, Construction*. Oxford: Oxford University Press

Allan, D.J. 1954. The problem of Cratylus. *The American Journal of Philology* 75(3):271–87.

Allen, B.J. 2001. Boserup and Brookfield and the association between population density and agricultural intensity in Papua New Guinea. *Asia Pacific Viewpoint* 42(2/3):237–54.

Allen, B.J. and R. Crittenden 1987. Degradation and a pre-capitalist political economy: the case of the New Guinea highlands. In P. Blaikie and H. Brookfield (eds.) *Land Degradation and Society* (pp. 143–56). London: Methuen.

Allport, G. 1937. *Personality: A Psychological Interpretation*. New York: Holt, Rinehart, and Winston.

Ambrosini, V. and C. Bowman 2001. Tacit knowledge: some suggestions for operationalization. *Journal of Management Studies* 38(6):811–29.

Amit, V. 2000. The university as panopticon: moral claims and attacks on academic freedom. In M. Strathern (ed.) *Audit Cultures: Anthropological Studies in Accountability, Ethics and the Academy* (pp. 215–35). London: Routledge.

Anderson, A. 2011. *Landscapes of Relations and Belonging: Body, Place and Politics in Wogeo, Papua New Guinea*. Oxford: Berghahn.

Andrews, P. A. 1999. *Felt Tents and Pavilions: The Nomadic Tradition and its Interaction with Princely Tentage* (2 vols). London: Melisende (Kölner Ethnologische Mitteilungen, Sonderband).

Arboleda, G. 2006. What is Vernacular Architecture? www.vernaculararchitecture.com

Argyris, C. and D. Schön 1974. *Theory in Practice: Increasing Professional Effectiveness*. San Francisco: Jossey-Bass.

Aristotle 2012. *Aristotle's Nicomachean Ethics* (trans. R.C. Bartlett and S.D. Collins). Chicago: University of Chicago Press.

Ashby, M.F. and K. Johnson 2010. *Materials and Design: The Art and Science of Material Selection in Product Design*. Oxford: Butterworth-Heinemann.

Asquith, L. and M. Vellinga (eds.) 2006. *Vernacular architecture in the Twenty-first Century: Theory, Education and Practice*. London: Taylor and Francis.

Audouze, F. 1999. New advances in French prehistory. *Antiquity* 73:167–75.

Barab, S.A., and K.E. Hay 2001. Doing science at the elbows of experts: issues related to the science apprenticeship camp. *Journal of Research in Science Teaching* 38(1):70–102.

Barad K. 2003. Posthumanist performativity: toward an understanding of how matter comes to matter. *Signs* 28:801–31.

Barck, K. 1994. Materiality, materialism, performance. In H.U. Gumbrecht and K.L. Pfeiffer (eds.) *Materialities of Communication* (pp. 258–72). Stanford: Stanford University Press.

Barnes, J. 2005. *Arthur and George*. London: Jonathan Cape.

Barnes, J.A. 1962. African models in the New Guinea Highlands *Man* 62:5–9.

Bateson, G. 1968. Redundancy and coding. In T.A. Sebeok (ed.) *Animal Communication: Techniques of Study and Results of Research* (pp. 614–26). Bloomington: Indiana University Press.

——— 1972. *Steps to an Ecology of Mind: Collected Essays in Anthropology, Psychiatry, Evolution and Epistemology*. London: Paladin.

——— 1979. *Mind and Nature*. Glasgow: Fontana.

Battiste, M. 2002. Indigenous Knowledge and Pedagogy in First Nations Education: A Literature Review with Recommendations. Report for the National Working Group on Education and the Minister of Indian Affairs, Ottawa. www.afn.ca/uploads/files/education/24._2002_oct_marie_battiste_indigenousknowledgeandpedagogy_lit_review_for_min_working_group.pdf

Baumard, P. 1999. *Tacit Knowledge in Organizations*. London: Sage

Bay, J-H. and B.-L. Ong (eds.) 2006. *Tropical Sustainable Architecture: Social and Environmental Dimensions*. Oxford: Architectural Press.

Bayliss-Smith, T. and J. Golson 1999. The meaning of ditches: deconstructing the social landscapes of New Guinea, Kuk, Phase 4. In C. Gosden and J. Hather (eds.) *The Prehistory of Food: Appetites for Change* (pp. 199–231). London: Routledge.

Beaune, S.A. de 2013. De la beauté du geste technique en préhistoire. *Gradhiva* 17:26–49.

Beckmann, J. 1777. *Anleitung zur Technologie*. Gottingen: Verlag der Mittwe Bandenhoef.

Behrmann,W. 1950–1 Die Versammlungshäuser (Kutlhäuser) am Sepik in Neu-Guinea. *Die Erde* 2:305–27.

Bell, D. 1999. *The Coming of Post-industrial Society: A Venture in Social Forecasting.* New York: Basic Books.

Berkes, F., J. Colding and C. Folke 2003. Introduction. In F. Berkes, J. Colding and C. Folke (eds.) *Navigating Social-ecological Systems: Building Resilience for Complexity and Change* (pp. 1–29). Cambridge: Cambridge University Press.

Berlin, B. 1992. *Ethnobiological Classification: Principles of Categorization of Plants and Animals in Traditional Societies.* Princeton: Princeton University Press.

Berliner, N. 2003. *Yin Yu Tang. The Architecture and Daily Life of a Chinese House.* Boston: Tuttle.

Berndt, R.M. and P. Lawrence (eds.) 1973. *Politics in New Guinea.* Seattle: Washington University Press.

Berry, D.C. and Z. Dienes 1993. *Implicit Learning: Theoretical and Empirical Issues.* Hove: Lawrence Erlbaum Associates.

Besson, J. 1999 [1578]. *Theatrum Instrumentorum et Machinarum.* Washington, DC: The Dibner Library of the History of Science and Technology Smithsonian Institution Libraries Digital Edition.

Bidet, A. 2007. Le corps, le rythme et l'esthétique sociale chez André Leroi-Gourhan. *Techniques and Culture* 48–9(1):5–38.

Biersack, A. 2001. Reproducing inequality: the gender politics of males cults in Melanesia and Amazonia. In T. Gregor and D. Tuzin (eds.) *Gender in Amazonia and Melanesia: An Exploration of the Comparative Method* (pp. 69–90). Berkeley, CA: University of California Press.

Binford, L.R. 1962. Archaeology as anthropology. *American Antiquity* 28:217–25.

——— 1978. *Nunamiut Ethnoarchaeology.* New York: Academic Press.

——— 1983. *In Pursuit of the Past.* New York: Thames and Hudson.

Bird A. 2007. Perceptions of epigenetics. *Nature* 447(7143):396–8.

Bird-David, N. 1999 "Animism" revisited: personhood, environment, and relational epistemology. *Current Anthropology* 40(S1):567–91.

Biringuccio, V. 1942 [1540]. *Pirotechnia* (trans. C.S. Smith and M.T. Gnudi). New York: American Institute of Mining and Metallurgical Engineers.

Blackler, F. 1995. Knowledge, knowledge work and organizations: an overview and interpretation. *Organization Studies.* 16(6):1021–46.

Bloch, M. 1991. Language, anthropology and cognitive science. *Man* (NS) 26:183–98.

——— 1998. *How We Think They Think: Anthropological Approaches to Cognition, Memory, Literacy.* Oxford: Westview.

——— 2006. *L'anthropologie cognitive a l'epreuve du terrain.* Paris: Fayard.

Blong, R.J. 1982. *The Time of Darkness: Local Legends and Volcanic Reality in Papua New Guinea.* Canberra: Australian National University Press.

Boas, F. 1911. *The Mind of Primitive Man.* New York: Macmillan.

Bøgh, A. 1996. Abundance and growth of rattans in Khao Chong National Park, Thailand *Forest Ecology and Management* 84(1–3):71–80.

Boivin, N. 2004. Mind over matter? Collapsing the mind-matter dichotomy in material culture studies. In E. DeMarrais, C. Gosden and C. Renfrew (eds.) *Rethinking Materiality: The Engagement of Mind with the Material World* (pp. 63–71). Cambridge: McDonald Institute Monographs.

——— 2008. *Material Cultures, Material Minds: The Impact of Things on Human Thought, Society, and Evolution.* Cambridge: Cambridge University Press.

Boserup, E. 1965. *The Conditions of Agricultural Growth: The Economics of Agrarian Change under Population Pressure.* London: Allen and Unwin.

Bourdier, J.-P. and N. Al Sayyad (eds.) 1989. *Dwellings, Settlements, and Tradition: Cross-Cultural Perspectives.* Lanham: University Press of America.

Bourdieu, P. 1970. La maison kabyle ou le monde renversé. In J. Pouillon, and P. Maranda (eds.) *Échanges et communications. mélanges offertes à C. Lévi-Strauss à l'occasion de son 60è anniversaire* (pp. 739–58). The Hague: Mouton.

——— 1977. *Outline of a Theory of Practice.* Cambridge: Cambridge University Press.

Bowden, R. 1990. The architecture and art of Kwoma ceremonial houses. In N. Lutkehaus, C. Kaufmann, W.E. Mitchell, D. Newton, L. Osmundsen and M. Schuster (eds.) *Sepik Heritage: Tradition and Change in Papua New Guinea* (pp. 480–90). Bathurst: Crawford House Press.

Boyd, D.J. 1984. The production and management of pigs: husbandry options and demographic patterns in an Eastern Highlands herd. *Oceania* 55:27–49.

Bradley, B.A. 1975. Lithic reduction sequences: a glossary and discussion. In E.H. Swanson (ed.) *Lithic Technology: Making and Using Stone Tools* (pp. 5–14). The Hague: Mouton.

Bradley, R. and M. Edmonds 1993. *Interpreting the Axe Trade: Production and Exchange in Neolithic Britain.* Cambridge: Cambridge University Press.

Brandt, B.L., J.A. Farmer Jr. and A. Buckmaster 1993. Cognitive apprenticeship approach to helping adults learn. *New Directions for Adult and Continuing Education* 59:69–78.

Brill, J., B. Kim and C. Galloway 2001. Cognitive apprenticeships as an instructional model. In M. Orey (ed.), *Emerging Perspectives on Learning, Teaching, and Technology.* epltt.coe.uga.edu

Broadbent, S.G. and D.M. Broadbent 2004. Traditional vs modern construction practices. Geneva: Shelter Centre, Information and Research for Reconstruction sheltercentre.org/sites/default/files/IREC_TraditionalVsModernConstructionPractices.pdf

Brongersma, L.D. and G.F. Venema 1960. *Het witte hart van Nieuw-Guinea.* Amsterdam: Scheltens and Giltay.

Brookfield, H.C. 2001. Intensification and alternative approaches to agricultural change. *Asia Pacific Viewpoint* 42(2/3):181–92.

Brookfield, H.C. and P. Brown 1963. *Struggle for Land: Agriculture and Group Territories among the Chimbu of the New Guinea Highlands.* Melbourne: Oxford University Press.

Brown, B. 2004. Thing theory. In B. Brown (ed.) *Things* (pp. 1–22). Chicago: University of Chicago Press.

Brown, D. 2009. *Māori Architecture – from fale to wharenui and beyond.* North Shore: Raupo Penguin.

Brown, J.S., A. Collins and S. Duguid 1989. Situated cognition and the culture of learning. *Educational Researcher* 18(1):32–42.

Brown, J.S. and P. Duguid 1991. Organizational learning and communities-of-practice: toward a unified view of working, learning, and innovation. *Organization Science* 2(1):40–57.

Brown, P. 1973. *The Chimbu: A Study of Change in the New Guinea Highlands.* London: Routledge and Kegan Paul.

Brown, P. and Brookfield, H.C. 1967. Chimbu settlement and residence: a study of patterns, trends and idiosyncracy. *Pacific Viewpoint* 8:119–51.

Brown, R.A. and R. Seligman 2009. Anthropology and cultural neuroscience: creating productive intersections in parallel fields. In J.Y. Chiao (ed.) *Cultural Neuroscience: Cultural Influences on Brain Function* (Progress in Brain Research 178):31–42.

Browne, D. 1974. *The House Building Book.* New York: McGraw-Hill.

Bruner, J S. 1960. *The Process of Education.* Cambridge, MA: Harvard University Press.

——— 1966. *Toward a Theory of Instruction.* Cambridge, MA: Belknap Press

Brunskill, R.W. 2000 [1971]. *Vernacular Architecture: An Illustrated Handbook* (4th edn.). London: Faber and Faber.

——— 2006 [1981]. *Traditional Buildings of Britain: An Introduction to Vernacular Architecture and its Revival.* London: Cassell.

Bubandt, N. and T. Otto 2010. Anthropology and the predicaments of holism. In T. Otto, and N. Bubandt (eds.) *Experiments in Holism: Theory and Practice in Contemporary Anthropology* (pp. 1–15). Chichester: Wiley-Blackwell.

Buchli, V. 2013. *An Anthropology of Architecture.* London: Bloomsbury.

Buechler, H. 1989. Apprenticeship and transmission of knowledge in La Paz, Bolivia. In M.W. Coy (ed.) *Apprenticeship: From Theory to Method and Back Again* (pp. 31–50). New York: SUNY Press.

Buller, D.J. and V.G. Hardcastle 2000. Evolutionary psychology, meet developmental neurobiology: against promiscuous modularity. *Brain and Mind* 1:307–25.

——— 2005. Modularity. In D.J. Buller (ed.) *Adapting Minds: Evolutionary Psychology and the Persistent Quest for Human Nature* (pp. 127–201). Cambridge, MA: MIT Press.

Bulmer, R.N.H. 1976. Selectivity in hunting and disposal of animal bone by the Kalam of the New Guinea Highlands. In G. Sieveking, I. Longworth and K. Wilson (eds.) *Problems in Economic and Social Archaeology* (pp. 169–86). London: Duckworth.

Campbell, S. 2002. *The Art of Kula.* Oxford: Berg.

Candea, M. (ed.) 2010. *The Social after Gabriel Tarde: Debates and Assessments.* London: Routledge.

Carrithers, M. 1992. *Why Humans have Cultures: Explaining Anthropology and Social Diversity.* Oxford: Oxford University Press.

——— 2005. Why anthropologists should study rhetoric. *Journal of the Royal Anthropological Institute* 11(3):577–83.

Carsten, J. and S. Hugh-Jones 1995. Introduction. In J. Carsten, and S. Hugh-Jones (eds.) *About the House: Lévi-Strauss and Beyond* (pp. 1–46). Cambridge: Cambridge University Press.

Carter, T. and E.C. Cromley 2005. *Invitation to Vernacular Architecture: A Guide to the Study of Ordinary Buildings and Landscapes.* Knoxville: University of Tennessee Press.

Chalmers, D.J. 2011. Foreword. In A. Clark *Supersizing the Mind: Embodiment, Action and Cognitive Extension* (pp. ix–xvi). Oxford: Oxford University Press.

Chappell, J.M.A. 1966. Stone axe factories in the highlands of east New Guinea, with a note on linguistic boundaries and the axe quarries. *Proceedings of the Prehistoric Society* 32:96–121.

Chemero, A. 2003. An outline of a theory of affordances. *Ecological Psychology* 15(2):181–95.

Childe, V.G. 1946. Archaeology and anthropology. *Southwestern Journal of Anthropology* 2:243–51.

Choo, C.W. 1998. *The Knowing Organization: How Organizations use Information to Construct Meaning, Create Knowledge, and Make Decisions.* New York: Oxford University Press.

Chua, H.E., J.E. Boland and R.E. Nisbett 2005. Cultural variation in eye movements during scene perception. *Proceedings of the National Academy of Sciences* 102(35):12629–33.

Chudley, R. and R. Greeno 2010. *Building Construction Handbook* (8th edn.). Amsterdam: Elsevier.

Clancey, W.I. 1997. The conceptual nature of knowledge, situations and activity. In P. Feltovich, R., Hoffman and K. Ford (eds.) *Human and Machine Expertise in Context* (pp. 247–91). Menlo Park: AAAI Press.

Clark, A. 1997. *Being There: Putting Brain Body and World Together Again.* Cambridge, MA: MIT Press.

——— 2003. *Natural Born Cyborgs: Minds, Technologies and the Future of Human Intelligence.* Oxford: Oxford University Press.

——— 2010. Material surrogacy and the supernatural: reflections on the role of artefacts in 'off-line' cognition. In L. Malafouris and C. Renfrew (eds.) *The Cognitive Life of Things: Recasting the Boundaries of the Mind* (pp. 23–8). Cambridge: McDonald Institute for Archaeological Research.

Clark, A. and D.J. Chalmers 1998. The extended mind. *Analysis* 58:7–19.

Clarke, D.L. 1968. *Analytical Archaeology.* London: Methuen.

Clarke, W.C. 1971. *Place and People: An Ecology of a New Guinea Community.* Berkeley: University of California Press.

Cleary, G.J. and C.R.B. Blackburn 1968. Air pollution in native huts in the Highlands of New Guinea. *Archives of Environmental Health* 17(5):785–94.

Clifton-Taylor, A. 1987. *The Pattern of English Building.* London: Faber and Faber.

Clune, J., J.-B. Mouret and H. Lipson 2013. The evolutionary origins of modularity. *Proceedings of the Royal Society* 280(1755):20122863. DOI: 10.1098/rspb.2012.2863.

Coiffier, C. 1983 Maison des hommes: maison masque ou maison masquée? In F. Lupu (ed.) *Océanie, le masque au long cours* (pp. 195–202). Rennes: Ouest France Université.

——— 1990. Sepik river architecture: changes in cultural traditions. In N. Lutkehaus, C. Kaufmann, W.E. Mitchell, D. Newton, L. Osmundsen and M. Schuster (eds.) *Sepik Heritage: Tradition and Change in Papua New Guinea* (pp. 491–500). Bathurst: Crawford House Press.

Coles, A. and P. Jackson 2007. *Windtower.* London: Stacey International.

Collins, A. 1991. Cognitive apprenticeship and instructional technology. In L. Idol and B.F. Jones (eds.) *Educational Values and Cognitive Instruction: Implication for Reform* (pp. 121–38). Hillsdale, NJ: Lawrence Erlbaum Associates.

Collins, A., J.S. Brown and A. Holum 1991. Cognitive apprenticeship: making thinking visible. *American Educator* 15(3):6–11, 38–46.

Collins, A., J.S. Brown and S.E. Newman 1989. Cognitive apprenticeship: teaching the craft of reading, writing, and mathematics. In L.B. Resnick (ed.) *Knowing, Learning, and Instruction: Essays in Honor of Robert Glaser* (pp. 453–94). Hillsdale, NJ: Lawrence Erlbaum Associates.

Collins, H.M. 1974. The TEA set: tacit knowledge and scientific networks. *Science Studies* 4:165–86.

——— 2001a What is tacit knowledge? In T.R. Schatzki, C.K. Knorr and E. von Savigny (eds.) *The Practice Turn in Contemporary Theory* (pp. 107–19). London: Routledge.

——— 2001b Tacit knowledge, trust, and the Q of sapphire. *Social Studies of Science* 31(1):71–85.

Conkey, M.W. 2006. Style, design, and function. In C. Tilley, W. Keane, S. Kuechler, M. Rowlands and P. Spyer (eds.) *Handbook of Material Culture* (pp. 355–72). London: Sage.

Connolly, B. and R. Anderson 1987. *First Contact: New Guinea's Highlanders Encounter the Outside World.* New York: Viking.

Conroy, E. 2011. Masdar Tower is a big experiment. *The National UAE* (6 March): www.thenational.ae/news/uae-news/masdar-tower-is-a-big-experiment

Corsellis, T and A. Vitale (eds.) 2005. *Transitional Settlement: Displaced Populations.* Oxford: Oxfam.

Coudart, A. 1992. Sur l'analogie ethnographique et l'ethnoarchéologie et sur l'histoire des rapports entre archéologie et ethnologie. In J. Garanger (ed.) *La Préhistoire dans le Monde* (pp. 248–63). Paris : Puf.

——— 1994. Maisons néolithiques, maisons de Nouvelle-Guinée. L'ethnologie comparée du choix social et technique. In B. Latour and P. Lemonnier (eds.) *De la Préhistoire aux Missiles Balistiques. L'intelligence Sociale des Techniques* (pp. 228–52). Paris: La découverte.

Coudart, A. and S. Cleuziou 2005. Du rapport entre l'ethnologie et l'archéologie : il n'est, dans le domaine des sciences, aucune discipline supérieure ou inférieure aux autres. *Les Nouvelles de l'Archéologie* 99:48–55.

Coy, M.W. 1989. From theory. In M.W. Coy (ed.) *Apprenticeship: From Theory to Method and Back Again* (pp. 1–11). Albany: State University of New York Press.

Craig, B. 1988. Relic and trophy arrays as art among the Mountain Ok, Central New Guinea. In L. Hanson and A. Hanson (eds.) *Art and Identity in Oceania* (pp. 196–210). Bathurst: Robert Brown.

Crawford, A.L. 1981. *Aida: Life and Ceremony of the Gogodala.* Bathurst: Robert Brown and Associates with the National Cultural Council of Papua New Guinea.

Crittenden, R. 1982. Sustenance, Seasonality and Social Cycles on the Nembi Plateau, Papua New Guinea. Ph.D. thesis, Australian National University, Canberra.

Crouch, D.P. and J.G. Johnson 2001. *Traditions in Architecture: Africa, America, Asia, and Oceania.* New York: Oxford University.

CSIRO. 1965 *Lands of the Wabag-Tari area, Papua New Guinea.* Melbourne: CSIRO Land Research Series No. 15.

Csordas, T.J. 1994. *The Sacred Self: A Cultural Phenomenology of Charismatic Healing.* Berkeley: University of California Press.

Cunningham, C.E. 1973. Order in the Atoni house. In R. Needham (ed.) *Right and Left: Essays on Dual Symbolic Classification* (pp. 204–38). Chicago: University of Chicago Press.

Dant, T. 2005. *Materiality and Society.* Maidenhead: Open University Press.

Darlington, C.D. 1969. *The Evolution of Man and Society.* London: George Allen and Unwin.

Darwin, C. 1859. *The Origin of Species.* London: John Murray.

David, N. and C. Kramer 2001. *Ethnoarchaeology in Action.* Cambridge: Cambridge University Press.

Davidson, C.H., C. Johnson, G. Lizarralde, N. Dikmen and A. Sliwinski 2006. Truths and myths about community participation in post-disaster housing projects. *Habitat International* 30(1):100–15.

Davy-Ball, M.D. 2009. *'Betek, tali ngan atap' 'Knots, string and blades': Production and use of organic utility objects by the Orang Ulu of Sarawak.* Doctoral thesis, Durham University.

l'Orme, P. 1981 [1567]. *Architecture de Philibert de L'Orme, conseiller et aumosnier ordinaire du roy, et abbé de saint Serge lez-Angers: oeuvre entière contenant onze livres.* Bruxelles: Pierre Mardaga éditeur.

DeCristoforo, R.J. and A. 2007. *Housebuilding: A Do-it-yourself Guide* (rev. edn.). New York: Sterling.

Deleuze G, and F. Guattari 2004. *A Thousand Plateaus: Vol. 2: Capitalism and Schizophrenia.* London: Continuum.

DeMarrais E., C. Gosden and C. Renfrew 2004. Introduction. In E. DeMarrais, C. Gosden and C. Renfrew (eds.) *Rethinking Materiality: The Engagement of Mind with the Material World* (pp. 1–7). Cambridge: McDonald Institute Monographs.

Dennett, D. 1988. Quining qualia. In A. Marcel and E. Bisiach (eds.) *Consciousness in Modern Science* (pp. 42–77). Oxford: Oxford University Press.

Derrida, J. 1992. *Given Time: 1. Counterfeit Money* (trans. P. Kamuf). Chicago: Chicago University Press.

Descartes, R. 2008. [1637] *Discourse on the Method* (trans. J. Veitch). New York: Cosimo.

DeWalt, K.M., B.R. DeWalt and C.B. Wayland 1998. Participant observation. In H.R. Bernard (ed.) *Handbook of Methods in Cultural Anthropology* (pp. 259–99). Walnut Creek, CA: AltaMira Press.

Dewar, R.E. 1984. Environmental productivity, population regulation, and carrying capacity. *American Anthropologist* 86(3):601–14.

Dewey, J. 1933. *How We Think*. New York: D.C. Heath.

Di Paolo, E.A. 2005. Autopoiesis, adaptivity, teleology, agency. *Phenomenology and the Cognitive Sciences* 4(4):429–52.

Dias, B.G and K.J. Ressler 2013. Parental olfactory experience influences behavior and neural structure in subsequent generations. *Nature Neuroscience* (doi:10.1038/nn.3594).

Dolsak, N. and E. Ostrom 2003. The challenges of the commons. In N. Dolsak and E. Ostrom (eds.) *The Commons in the New Millennium: Challenges and Adaptation* (pp. 3–35). Cambridge, MA: MIT Press.

Donald, M. 1991. *Origins of the Modern Mind*. Cambridge, MA: Harvard University Press.

Dotov, D.G., L. Nie and M.M. de Wit 2012. Understanding affordances: history and contemporary development of Gibson's central concept. *AVANT* 3(2):28–39.

Douglas, M. 1966. *Purity and Danger: An Analysis of the Concepts of Pollution and Taboo*. London: Routledge and Kegan Paul.

Downey, G. 2010. 'Practice without theory': a neuroanthropological perspective on embodied learning. In T.H.J. Marchand (ed.) *Making Knowledge. Journal of the Royal Anthropological Institute* 16(Issue Supplement):S22–S40.

——— 2012 Neuroanthropology. In R. Fardon, O. Harris, T.H.J. Marchand, M. Nuttall, C. Shore, V. Strang and R.A. Wilson (eds.) *The SAGE Handbook of Social Anthropology* vol. 2 (pp. 243–60). London: Sage.

Dretske, F. 1991. *Explaining Behavior: Reasons in a World of Causes*. Cambridge, MA: MIT Press.

Dreyfus, H. 2002. Intelligence without representation: Merleau-Ponty's critique of mental representation. *Phenomenology and the Cognitive Sciences* 1(4):367–83.

Dreyfus, H. and S. Dreyfus 1986. *Mind over Machine: The Power of Human Intuition and Expertise in the Era of the Computer*. New York: Free Press.

Duly, C. 1979. *The Houses of Mankind*. London: Thames and Hudson.

Durkheim, É. 1933 [1893]. *On the Division of Labour in Society* (trans. G. Simpson). London: Macmillan.

——— 1964 [1895]. *ThE RULEs of Sociological Method*. London: Routledge.

Edwards, P.J. and P.J.Grubb 1977. Studies of mineral cycling in a montane rain forest in New Guinea: I. The distribution of organic matter in the vegetation and soil. *Journal of Ecology* 65:943–69.

——— 1982 Studies of mineral cycling in a montane rain forest in New Guinea: IV. Soil characteristics and the division of mineral elements between the vegetation and soil. *Journal of Ecology* 70:649–66.

Elleh, N. 1997. *African Architecture: Evolution and Transformation*. New York: McGraw-Hill.

Ellen, R.F. 1982. *Environment, subsistence and system: the ecology of small-scale social formations*. Cambridge: Cambridge University Press.

——— 1988. Fetishism. *Man* 23(2):213–35.

——— 2006. *The Categorical Impulse*. Oxford: Berghahn.

——— 2009. A modular approach to understanding the transmission of technical knowledge: Nuaulu basket making from Seram, Eastern Indonesia. *Journal of Material Culture* 14:243–77.

——— 2011. The eolith debate, evolutionist anthropology and the Oxford connection between 1880 and 1940. *History and Anthropology* 22(3):277–306.

Eraut, M. 2000. Non-formal learning and tacit knowledge in professional work. *British Journal of Educational Psychology* 70:113–36.

Evans, I., L. Smiley and M.G. Smith 2002. *The Hand-sculpted House: A Practical and Philosophical Guide to Building a Cob Cottage*. White River Junction (Vermont): Chelsea Green Pub Co.

Fahlander, F. 2004. Archaeology and anthropology: brothers in arms? – on analogies in 21st century archaeology. In F. Fahlander and T. Oestigaard (eds.) *Material Culture and Other Things: Post-disciplinary Studies in the 21st Century* (pp. 157–80). Göteborg: Gotarc Series C no 61.

——— 2008. Differences that matter: materialities, material culture and social practice. In H. Glørstad and L. Hedeager (eds.) *Six Essays on the Materiality of Society and Culture* (pp. 127–54). Lindome: Bricoleur Press.

Fahlander, F. and T. Oestigaard (eds.) 2004 *Material Culture and other Things. Post-Disciplinary Studies in the 21st Century*. Göteborg: Gotarc Series C no 61.

Farnell, B.M. 1994. Ethno-graphics and the moving body. *Man* 29(4):929–74.

——— 2000. Getting out of the habitus: an alternative model of dynamically embodied social action. *Journal of the Royal Anthropological Institute* 6:397–418.

Fathy, H. 1986. *Natural Energy and Vernacular Architecture: Principles and Examples with Reference to Hot Arid Climates* (ed. W. Shearer and A. Ahmed Sultan). Chicago: University of Chicago Press for the United Nations University.

Feacham, R.G.A. 1973. The Raiapu Enga pig herd. *Mankind* 9(1):25–31.

Fergusson, J. 1865. *A History of Architecture in all Countries*. London: John Murray.

Fewster, K. 2001. The responsibilities of ethnoarchaeologists. In M. Pluciennik (ed.) *The Responsibilities of Archaeologists: Archaeology and Ethics* (pp. 65–73). British Archaeological Reports International Series 981. Oxford: Archaeopress

Firth, R. 1927. The Maori hill-fort. *Antiquity* 1:66–78.

——— 1951. *Elements of Social Organisation*. London: Watts.

Flitsch, M. 2008. Knowledge, embodiment, skill and risk: anthropological perspectives on women's everyday technologies in rural Northern China. *East Asian Science Technology and Society* 2(2):265–88.

Fodor, J.A. 1981. *Representations: Philosophical Essays on the Foundations of Cognitive Science*. Cambridge, MA: MIT Press.

——— 1983. *The Modularity of Mind*. Cambridge, MA: MIT Press.

Forge, A. 1966 Art and environment in the Sepik. *Proceedings of the Royal Anthropological Institute of Great Britain and Ireland for 1965*, pp. 23–31.

Foucault, M. 1973. *Birth of the Clinic: An Archaeology of Medical Perception* (trans. A.M. Sheridan). London: Routledge.

Fowler, C. 2004. *The Archaeology of Personhood: An Anthropological Approach*. London: Routledge.

Frake, C.O. 1969. The ethnographic study of cognitive systems. In S.A. Tyler (ed.) *Cognitive Anthropology* (pp. 28–41). New York: Holt, Rinehart and Winston.

Frankel, S. 1986. *The Huli Response to Illness*. Cambridge: Cambridge University Press.

Franz, J. and E. Papyrakis 2011. Online calculators of ecological footprint: do they promote or dissuade sustainable behaviour? *Sustainable Development* 19(6):391–401.

Frey, S.H. 2009. Tool use, communicative gesture and cerebral asymmetries in the modern human brain. In C. Renfrew, C. Frith and L. Malafouris (eds.) *The Sapient Mind: Archaeology Meets Neuroscience* (pp. 21–31). Oxford: Oxford University Press.

Gallop, J. 1988. *Thinking Through the Body*. New York: Columbia University Press.

Garrow, D. and T. Yarrow 2010. Introduction: archaeological anthropology. In D. Garrow and T. Yarrow (eds.) *Archaeology and Anthropology: Understanding Similarity, Exploring Difference* (pp. 1–12). Oxford: Oxbow Books.

Gazzaniga, M.S., R.B. Ivry and G.R. Mangun 2009. *Cognitive Neuroscience: The Biology of the Mind*. New York: W.W. Norton and Company.

Geertz, C. 1984. From the native's point of view: on the nature of anthropological understanding. In R.A. Shwedder and R. LeVine (eds.) *Culture Theory: Essays on Mind, Self, and Emotion* (pp. 123–36). New York: Cambridge University Press.

——— 1993. *The Interpretation of Cultures*. London: Fontana Books.

Gell, A. 1998. *Art and Agency: An Anthropological Theory*. Oxford: Oxford University Press.

Gellner, E. 1985. *Relativism and the social sciences*. Cambridge: Cambridge University Press.

Gerrans P. 2005. Tacit knowledge, rule following and Pierre Bourdieu's philosophy of social science. *Anthropological Theory* 5(1):53–74.

Gibbs, R. 2006. *Embodiment and Cognitive Science*. Cambridge: Cambridge University Press.

Gibson, J.J. 1977. The theory of affordances. In R. Shaw and J. Bransford (eds.) *Perceiving, Acting and Knowing*. Hillsdale, NJ: Erlbaum.

——— 1979. *The Ecological Approach to Visual Perception*. Boston: Houghton Mifflin.

Gill, J.H. 2000. *The Tacit Mode: Michael Polanyi's Postmodern Philosophy*. New York: State University of New York.

Gillespie, S.D. 2000. Lévi-Strauss maison and société à maisons. In R.A. Joyce and S.D. Gillespie (eds.) *Beyond Kinship: Social and Material Reproduction in House Societies* (pp. 22–52). Philadelphia: University of Pennsylvania Press.

Gillespie, S.D., R.A. Joyce and D.L. Nichols 2003. Archaeology is anthropology. In S.D. Gillespie. and D.L. Nichols (eds.) *Archaeology is Anthropology* (pp. 155–69). Arlington: American Anthropological Association.

Givoni, B. 1969. *Man, Climate and Architecture*. London: Elsevier.

Glaser, B.G. and A.L. Strauss, 1968. *The Discovery of Grounded Theory: Strategies for Qualitative Research*. London: Weidenfeld and Nicolson.

Glassie, H. 2000. *Vernacular Architecture*. Bloomington: Indiana University Press.

Glørstad, H. and L. Hedeager 2008. On the materiality of society and culture. In H. Glørstad and L. Hedeager (eds.) *Six Essays on the Materiality of Society and Culture* (pp. 9–32). Lindome: Bricoleur Press.

Godelier, M. 1986. *The Making of Great Men*. Cambridge: Cambridge University Press.

——— 1999. *The Enigma of the Gift*. Cambridge: Polity Press.

Godelier, M. and J. Garanger 1973. Outils de pierre, outils d'acier chez les Baruya de Nouvelle Guinée *L'Homme* 13:187–220 (in English: Stone tools and steel tools among the Baruya of New Guinea *Social Science Information* 18:663–78).

Godelier, M. and M. Strathern (eds.) 1991. *Big Men and Great Men: Personifications of Power in Melanesia*. Cambridge: Cambridge University Press.

Golub, A. 2011. Review of *From Land to Mouth*. *Pacific Affairs* 84(4):825–6.

——— 2014. *Leviathans at the Gold Mine: Creating Indigenous and Corporate Actors in Papua New Guinea*. Durham: Duke University Press.

Goody, E.N. (ed.) 1982. *From Craft to Industry: The Ethnography of Proto-industrial Cloth Production*. Cambridge: Cambridge University Press.

Goody, J. (ed.) 1958 *The Developmental Cycle in Domestic Groups* (Cambridge Papers in Social Anthropology 1). Cambridge: Cambridge University Press.

Goranzon, B. and M. Florin (eds.) 1990. *Artificial Intelligence, Culture and Language: On Education and Work* (pp. 45–55). London: Springer-Verlag.

Goranzon, B. and L. Josefson (eds.) 1988. *Knowledge, Skill and Artificial Intelligence*. Berlin: Springer-Verlag

Gosden, C. 1999. *Archaeology and Anthropology: A Changing Relationship*. London: Routledge.

——— 2009. Social ontologies. In C. Renfrew, C. Frith and L. Malafouris (eds.) *The Sapient Mind: Archaeology meets Neuroscience* (pp. 105–17). Oxford: Oxford University Press.

——— 2010. The death of the mind. In L. Malafouris and C. Renfrew *The Cognitive Life of Things: Recasting the Boundaries of the Mind* (pp. 39–46). Cambridge: McDonald Institute for Archaeological Research.

Gould, R. 1980. *Living Archaeology*. Cambridge: Cambridge University Press.

Gourlay, S.N. 2001. Situated cognition and knowledge: a contribution towards conceptual clarity for knowledge management. In S. Tsuchiya and J.-P. Barthes (eds.) *Proceedings of the Eighth International Symposium on the Management of Corporate and Industrial Knowledge, October 22–24* (pp. 29–43). Compiegne, France.

——— 2002. Tacit knowledge, tacit knowing or behaving? Kingston Business School: myweb.tiscali.co.uk/sngourlay/PDFs/Gourlay%202002%20tacit%20knowledge.pdf

Grafton, S.T., L. Fadriga, M.A. Arbib and G. Risolatti 1997. Premotor cortex activation during observation and naming of familiar tools. *Neuroimage* 6:231–6.

Graves-Brown, P.M. (ed.) 2000. Introduction. In P.M. Graves-Brown (ed.) *Matter, Materiality, and Modern Culture* (pp. 1–9). Routledge, London.

Gregory, C. 1982. *Gifts and Commodities*. London: Academic Press.

Griaule, M. 1975 [1948]. *Conversations with Ogotemmeli: An Introduction to Dogon Religious Ideas*. Oxford: Oxford University Press.

Griffin, D.R. *2001 Reenchantment Without Supernaturalism: A Process Philosophy of Religion*. Ithaca, NY: Cornell University Press.

Gropius, W. 1965 [1935]. *The New Architecture and the Bauhaus*. Cambridge, MA: MIT Press.

Guidoni, E. 1978. *Primitive Architecture*. New York: Abrams.

Haddon, A.C. 1912. The houses of New Guinea. In *Festskrift tillegnad E. Westermarck: i anledning av hans femtioårsdag, den 20 november 1912* (pp. 17–58). Helsingfors: J. Simelii Arvinjars Boktryikeriaktiebolag.

Hall, R. de Zouche (ed.) 1972. *Bibliography on Vernacular Architecture*. Newton Abbot: David and Charles (for the Vernacular Architecture Group).

Hallpike, C.R. 1977. *Bloodshed and Vengeance in the Papuan Mountains*. Oxford: Clarendon Press.

Hampton, O.W. 1999. *Culture of Stone; Sacred and Profane Uses of Stone among the Dani*. College Station: Texas A. and M. University Press.

Hansman, C.A. 2001. Context-based adult learning. *New Directions for Adult and Continuing Education* 89:43–51.

Hanson, L., B.J. Allen and R.M. Bourke 2001. Mapping land resource vulnerability in the Highlands of PNG. In R.M. Bourke, M.G. Allen and J.G. Salisbury (eds.) *Food Security for Papua New Guinea* (pp. 284–90). Canberra: ACIAR Proceedings No. 99.

Hardy, K. and P. Sillitoe 2003. Material perspectives: stone tool use and material culture in Papua New Guinea. *Internet Archaeology* 14. intarch.ac.uk/journal/issue14

Harman, G. 1990 The intrinsic quality of experience. In J. Tomberlin (ed.) *Philosophical Perspectives 4: Action Theory and the Philosophy of Mind*. Atascadero, CA: Ridgeview Pub. Co..

——— 2010. Technology, objects and things in Heidegger. *Cambridge Journal of Economics* 34:17–25.

Harré, R. 2002: Material objects in social worlds. *Theory, Culture and Society* 19(5/6):23–33.

Harvey, G. 2005. *Animism: Respecting the Living World*. London: Hurst and Co..

Haun, L. 2008. *How to Build a House*. Taunton: Habitat for Humanity.

Hauser-Schäublin, B. 1990. In the swamps and on the hills: traditional settlement patterns and house structures in the Middle Sepik. In N. Lutkehaus, C. Kaufmann, W.E. Mitchell, D. Newton, L. Osmundsen and M. Schuster (eds.) *Sepik Heritage: Tradition and Change in Papua New Guinea* (pp. 470–9). Bathurst: Crawford House Press.

Hawker, R.W. 2008. *Traditional Architecture of the Arabian Gulf: Building on Desert Tides*. Ashurst: WIT Press.

Heath, K.W. 2009. *Vernacular Architecture and Regional Design: Cultural Process and Environmental Response*. London: Architectural Press.

Heidegger, M. 1962 [1927]. *Being and Time* (trans. J. Macquarrie and E. Robinson). New York: Harper and Row.

——— 1971. *Poetry, Language, Thought* (trans. A Hofstadter). New York: Harper and Row.

——— 1993 [1954]. Building dwelling thinking. In D.F. Krell (ed.) *Basic Writings: From Being and Time (1927) to The Task of Thinking (1964)* (pp. 343–64). New York: Harper Collins.

Heider, K.G. 1970. *The Dugum Dani: A Papuan Culture in the Highlands of West New Guinea*. Chicago: Aldine.

——— 2006. *Ethnographic Film* (rev. edn.). Austin: University of Texas Press.

Henare, A., M. Holbraad and S. Wastell 2007. Introduction: thinking through things. In A. Henare, M. Holbraad and S. Wastell (eds.) *Thinking through Things: Theorising Artefacts Ethnographically* (pp. 1–31). London: Routledge.

Herdt, G.H. (ed.) 1982. *Rituals of Manhood: Male Initiation in Papua New Guinea*. Berkeley: University of California Press.

Hewlett, B.S. and L.L. Cavalli-Sforza 1986. Cultural transmission among Aka Pygmies. *American Anthropologist* 88(4):922–34.

Hide, R.L. 1981. Aspects of Pig Production and Use in Colonial Sinasina, Papua New Guinea. Ph.D. dissertation, University of Columbia.

Hides, J. 1936. *Papuan Wonderland*. London: Blackie and Son.

Hipsley, E.H. and N.E. Kirk 1965. *Studies of Dietary Intake and the Expenditure of Energy by New Guineans*. Noumea: South Pacific Commission (Technical Paper 147).

Hockings, J. 1989. *Traditional Architecture in the Gilbert Islands: A Cultural Perspective*. St Lucia: University of Queensland Press.

Hodder, I. 1982a. *Symbols in Action*. Cambridge: Cambridge University Press.

——— 1982b. *The Present Past*. London: Batsford.

——— 2011. Human-thing entanglement: towards an integrated archaeological perspective. *Journal of the Royal Anthropological Institute* 17(1):154–77.

——— 2012. Archaeology and anthropology: the state of the relationship. In D. Shankland (ed.) *Archaeology and Anthropology: Past, Present and Future* (pp. 65–76). London: Berg (Association of Social Anthropologists Monograph 48).

Høgseth, H.B. 2013. The language of craftsmanship. In M.L.S. Sørensen and K. Rebay-Salisbury (eds.) *Embodied Knowledge: Historical Perspectives on Belief and Technology* (pp. 95–105). Oxford: Oxbow Books.

Holbraad, M. 2011. *Can the Thing Speak?* Open Anthropology Cooperative Press (Working Papers No. 7). www.openanthropcoop.net/press.

Holy, L. 1979. The segmentary lineage structure and its existential status. In L. Holy (ed.) *Segmentary Lineage Systems Reconsidered* (pp. 1–23). Belfast: The Queen's University Papers in Social Anthropology, Vol. 4.

Holy, L. and M. Stuchlik (eds.) 1981. *The Structure of Folk Models*. London: Academic Press.

——— 1983. *Actions, Norms and Representations: Foundations of Anthropological Enquiry*. Cambridge: Cambridge University Press.

Honderich, T. 1990. *The Consequences of Determinism*. Oxford: Oxford University Press.

——— (ed.) 1995. *The Oxford Companion to Philosophy*. Oxford: Oxford University Press.

Howarth, C. and P. Swadling 2009. Prehistoric stone mortar from Papua New Guinea. *Artonview (National Gallery of Australia)* 59(3):28–9.

Howells, J. 1996. Tacit knowledge, innovation and technology transfer. *Technology Analysis and Strategic Management* 8:91–106.

Hume, D. 1896 [1739]. *A Treatise of Human Nature (ed. L.A. Selby-Bigge)*, 3 vols. Oxford: Clarendon Press.

——— 1985 [1748]. Of the original contract. In E.F. Miller (ed.) *Essays Moral, Political and Literary* (pp. 467–81). Indianapolis: Liberty Press.

——— 1902 [1777]. *Enquiries Concerning the Human Understanding and Concerning the Principles of Morals (ed. L.A. Selby-Bigge*, 2nd edn.). Oxford: Clarendon Press.

Humphrey, C. 1974. Inside a Mongolian tent. *New Society* (31 Oct.):273–5.

——— 1988. No place like home in anthropology: the neglect of architecture. *Anthropology Today* 4(1):16–18.

Humphrey, G. 1923. *The Story of Man's Mind*. Boston: Small, Maynard and Co..

Husserl, E. 1982 (1913). *Ideas Pertaining to a Pure Phenomenology and to a Phenomenological Philosophy. First Book: General Introduction to a Pure Phenomenology* (trans. F. Kersten). The Hague: Nijhoff.

Hutchins, E. 2005. Material anchors for conceptual blends. *Journal of Pragmatics* 37:1555–77.

Hyndman, D. 1991. The Kam Basin homeland of the Wopkaimin: a sense of place. In A. Pawley (ed.) *Man and a Half: Essays in Pacific Anthropology and Ethnobiology in honour of Ralph Bulmer* (pp. 256–65). Auckland: Auckland University Press (Polynesian Society Memoir 48).

Ingold, T. 1992. Culture and the perception of the environment. In. D. Parkin and E. Croll (eds.) *Bush Base Forest Farm: Culture, Environment and Development* (pp. 39–56). London: Routledge.

——— 1996. The optimal forager and economic man. In P. Descola and G. Palsson (eds.) *Nature and Society: Anthropological Perspectives* (pp. 25–44). London: Routledge.

——— 2000. *The Perception of the Environment. Essays in Livelihood, Dwelling and Skill*. London: Routledge.

——— 2001. Beyond art and technology: the anthropology of skill. In M.B. Schiffer (ed.) *Anthropological Perspectives on Technology* (pp. 17–31). Albuquerque: University of New Mexico Press.

——— 2007. Materials against materiality. *Archaeological Dialogues* 14(1):1–16, 31–8.

——— 2008. Anthropology is *not* ethnography. *Proceedings of the British Academy* 154:69–92.

——— 2009. The wedge and the knot: hammering and stitching the face of nature. In S. Bergmann, P.M. Scott, M. Jansdottir and H Bedford-Strohm (eds.) *Nature, Space and the Sacred: Transdisciplinary Perspectives* (pp. 147–161). Aldershot: Ashgate.

——— 2010. The textility of making. *Cambridge Journal of Economics* 34(1):91–102.

——— 2011a. Introduction. In T. Ingold (ed.) *Redrawing Anthropology: Materials, Movements, Lines* (pp. 1–20). Farnham: Ashgate.

——— 2011b. *Being Alive: Essays on Movement, Knowledge and Description*. London: Routledge.

——— 2012. Toward an ecology of materials. *Annual Review of Anthropology* 41:427–42.

——— 2013. *Making: Anthropology, Archaeology, Art and Architecture*. London: Routledge.

Jackson, F. 1982. Epiphenomenal qualia. *Philosophical Quarterly* 32:127–36.

Jacob, P. 2003. Intentionality. *Stanford Encyclopedia of Philosophy*: plato.stanford.edu/entries/intentionality.

Janik, A. 1988. Tacit knowledge, working life and scientific method. In B. Goranzon and I. Josefson (eds.) *Knowledge, Skill and Artificial Intelligence* (pp. 53–63). Berlin: Springer-Verlag.

——— 1992. Why is Wittgenstein important? In B. Goranzon and M. Florin (eds.) *Skill and Education* (pp. 33–40). Berlin: Springer-Verlag.

Jimenez, L. 1997. Implicit learning: conceptual and methodological issues. *Psychologica Belgica* 37(1/2):9–28.

Johannessen, K.S. 1988. Rule following and tacit knowledge. *AI and Society* 2:287–301.

——— 1992. Rule-following, intransitive understanding and tacit knowledge. An investigation of the Wittgensteinian concept of practice as regards tacit knowing. In B. Goranzon and M. Florin (eds.) *Skill and education: Reflection and Experience* (pp. 41–61). Berlin: Springer-Verlag.

Johnson, M. 1987 *The body in the Mind: The Bodily Basis of Meaning, Imagination, and Reason*. Chicago: University of Chicago Press.

Jolly, M. 1987. The chimera of equality in Melanesia. *Mankind* 17:168–83.

Jones A. 2004. Archaeometry and materiality: materials-based analysis in theory and practice. *Archaeometry* 46(3):327–38.

Jones, K.S. 2003. What is an affordance? *Ecological Psychology* 15(2):107–14.

Jørgensen, L.B. 2012. Writing craftsmanship? Vocabularies and notation systems in the transmission of craft knowledge. In W. Wendrich (ed.) *Archaeology and Apprenticeship. Body Knowledge, Identity and Communities of Practice* (pp. 240–54). Tucson, AZ: University of Arizona Press.

Josephides, L. 1983. Equal but different? The ontology of gender among Kewa. *Oceania* 53(3):291–307.

——— 1985. *The Production of Inequality: Gender and Exchange among the Kewa*. London: Tavistock.

Joyce, R.A. 2012. Life with things: archaeology and materiality. In D. Shankland (ed.) *Archaeology and Anthropology: Past, Present and Future* (pp. 119–32). London: Berg (Association of Social Anthropologists Monograph 48).

Jowitt, G. and Shaw, P. 1999. *Pacific Island Style*. London: Thames and Hudson.

Julien, M.-P., and C. Rosselin 2002. C'est en laquant qu'on devient laqueur. De l'efficacité du geste à l'action sur soi. *Techniques et culture* 40:107–24.

Just, R. 2004. Methodological individualism and sociological reductionism. *Social Analysis* 48(3):186–91.

——— 2005. In defence of rules: Pierre Bourdieu en Grèce. *Journal of Mediterranean Studies* 15(1):1–24.

Kant, I. 2001 [1800]. *Lectures on Ethics* (trans. P. Heath). Cambridge: Cambridge University Press.

Keane, W. 2005. Signs are not the garb of meaning. On the social analysis of material things. In D. Miller (ed.) *Materiality* (pp. 182–205). Durham, NC: Duke University Press.

Kelly, R.C. 1977. *Etoro Social Structure: A Study in Structural Contradiction*. Ann Arbor : University of Michigan Press.

——— 1993. *Constructing Inequality: The Fabrication of a Hierarchy of Virtue Among the Etoro*. Ann Arbor: University of Michigan Press.

Kembol, L., N. Koleala, J. Kutal, P. Kutapae, T. Pyakalya, A. Ruhan and F. Yakopya Pakene 1976. *Enga Housing and Enga Tradition*. Boroko: Institute of Papua New Guinea Studies (Occasional Paper 4).

Kenny, A. 2005. *Wittgenstein*. Williston, VT: Wiley-Blackwell.

Kent, S. (ed.) 1990. *Domestic Architecture and the Use of Space: An Interdisciplinary Cross-cultural Study*. Cambridge: Cambridge University Press (New Directions in Archaeology).

Kerka, S. 1997. Constructivism, workplace learning, and vocational education. *ERIC Digest* 181. Columbus: ERIC Clearinghouse on Adult, Career, and Vocational Education.

King, G. 1998. *The Traditional Architecture of Saudi Arabia*. London: I.B. Tauris.

Kingston, S. 2003. Form, attention and a southern New Ireland life cycle. *The Journal of the Royal Anthropological Institute* 9(4):681–708.

Klagge, J.C. and A. Nordmann (eds.) 1993. *Ludwig Wittgenstein: Philosophical Occasions 1912–1951*. Indianapolis: Hackett Pub. Co..

Klein, P.D. 1998. Knowledge, concept of. In E. Craig (ed.) *Routledge Encyclopaedia of Philosophy* (pp. 266–76). London: Routledge.

Knapp, R.G. 1990. *The Chinese House: Craft, Symbol, and the Folk Tradition*. Hong Kong: Oxford University Press.

Knappett, C. 2004. The affordances of things: a post-Gibsonian perspective on the relationality of mind and matter. In E. DeMarrais, C. Gosden and C. Renfrew (eds.) *Rethinking Materiality: The Engagement of Mind with the Material World* (pp. 43–51). Cambridge: McDonald Institute Monographs.

——— 2005. *Thinking Through Material Culture: An Interdisciplinary Perspective*. Philadelphia: University of Pennsylvania Press.

——— 2007. Materials with materiality? *Archaeological Dialogues* 14(1):20–3.

——— 2011. Networks of objects, meshworks of things. In T. Ingold (ed.) *Redrawing Anthropology: Materials, Movements, Lines* (pp. 45–63). Farnham: Ashgate.

Knorr-Cetina. K. 1999. *Epistemic Cultures: How the Sciences Make Knowledge*. Cambridge, MA: Harvard University Press.

Koestler, A. 1967. *The Ghost in the Machine*. London: Hutchinson.

Kolb, D. 1984. *Experiential Learning: Experience as the Source of Learning and Development*. London: Prentice-Hall.

Kolb, D.A., R.E. Boyatzis and C. Mainemelis 2000. Experiential learning theory: previous research and new directions. In R.J. Sternberg and L.F. Zhang (eds.), *Perspectives on Cognitive, Learning, and Thinking Styles*. Mahwah, NJ: Lawrence Erlbaum.

Kramer, C. (ed.) 1979. *Ethnoarchaeology*. New York: Columbia University Press.

Lakoff, G.P. and M. Johnson 1999. *Philosophy in the flesh: the embodied mind and its challenge to western thought*. New York: Basic Books.

Landtman, G. 1920. Papuan magic in the building of houses. *Acta Academiae Aboensis Humaniora* I(5):3–28.

Langness, L.L. 1964. Some problems in the conceptualization of Highland social structures in New Guinea: the Central Highlands. *American Anthropologist* 66:162–82 (special publication, ed. J.B. Watson).

Latour, B. 1991. Technology is society made durable. *The Sociological Review Monograph* 38:103–31.

——— 1992. Where are the missing masses? the sociology of a few mundane artifacts. In W.E. Bijker and J. Law (eds.) *Shaping Technology/Building Society: Studies in Sociotechnical Change* (pp. 225–58). Cambridge, MA: MIT Press.

——— 1993. *Nous n'avons jamais été modernes*. Cambridge, MA: Harvard University Press.
——— 1999. *Pandora's Hope: Essays on the Reality of Science Studies*. Cambridge, MA: Harvard University Press.
——— 2002. Gabriel Tarde and the end of the social. In P. Joyce (ed.) *The Social in Question: New Bearings in History and the Social Sciences* (pp. 117–32). London: Routledge.
——— 2005 *Reassembling the Social : An Introduction to Actor-Network-Theory.* Oxford: Clarendon Press.
Latour, B. and S. Woolgar 1986. *Laboratory Life: The Construction of Scientific Facts*. Princeton, NJ: Princeton University Press.
Laugier, M.A. 1753 [1985]. *Essai sur architecture*. Santa Monica: Hennessey and Ingalls.
Lave, J. 1982. A comparative approach to educational forms and learning processes. *Anthropology and Education Quarterly* 13(2):181–7.
——— 2011. *Apprenticeship in critical Ethnographic Practice*. Chicago: University of Chicago Press.
Lave, J. and E. Wenger 1991. *Situated Learning: Legitimate Peripheral Participation*. Cambridge: University of Cambridge Press.
Lawrence, D.I. and S.M. Low 1990. The built environment and spatial form. *Annual Review of Anthropology* 19:453–505.
Lawrence, R.J. 1983. Interpretation of vernacular architecture, a critical review. *Vernacular Architecture* 14:19–28.
Le Corbusier, 1985 [1923]. *Vers une architecture. Towards a New Architecture*. New York: Dover Publications.
Leach, E.R. 1984. Glimpses of the unmentionable in the history of British anthropology. *Annual Review of Anthropology* 13:1–23.
Lederman, R. 1986. *What Gifts Engender: Social Relations and Politics in Mendi, Highland Papua New Guinea*. Cambridge: Cambridge University Press.
——— 1990. Big men large and small? Towards a comparative perspective. *Ethnology* 29:3–15.
Lemonnier, P. 1976. La description des chaines opératoires: contribution a l'étude des systèmes techniques. *Techniques et Culture* 1:100–5.
——— 1983. L'étude des systèmes techniques, une urgence en technologie culturelle. *Techniques et Culture* 3:11–26.
——— 1986. The study of material culture today: toward an anthropology of technical systems. *Journal of Anthropological Archaeology* 5(2):147–86.
——— 1992. *Elements for an Anthropology of Technology*. Ann Arbor: Museum of Anthropology, University of Michigan (Anthropological Paper 88).
——— 2012. Pourquoi les maisons? Parce que les barrières, tiens! *Les Nouvelles de l'Archéologie* 127:49–53.
Leonardi, P.M., B.A. Nardi and J. Kallinikos 2012. *Materiality and Organizing: Social Interaction in a Technological World*. Oxford: Oxford University Press.
Leroi-Gourhan, A. 1968. L'expérience ethnologique. In J. Poirier (ed.) *Ethnologie générale* (pp. 1816–25). Paris: Gallimard.
——— 1993 (1964). *Gesture and Speech* (trans. A. Berger). Cambridge, MA: MIT Press.
Lévi-Strauss, C. 1968 *Structural Anthropology* (trans by C. Jacobsen and B.G. Schoepf). London: Allen and Lane.
——— 1969. *The Elementary Structures of Kinship*. (trans. J.H. Bell, J.R. von Sturmer and R. Needham). London: Eyre and Spottiswoode.
——— 1972. *The Savage Mind*. London: Weidenfeld and Nicolson.
——— 1982. *The Way of the Masks* (trans. S. Modelski). Seattle: University of Washington Press.
——— 1987. *Anthropology and Myth: Lectures, 1951–1982* (trans. R. Willis). Oxford: Basil Blackwell.
Lévy-Bruhl, L. 1926. *How natives think*. London: George Allen and Unwin.
Lewis, C.I. 1956. *Mind and the World-Order: Outline of a Theory of Knowledge*. New York: Dover Publications.
Lewis, D. 2004. What experience teaches. In P. Ludlow, Y. Nagasawa and D. Stoljar (eds.) *There's Something about Mary* (pp. 77–103). Cambridge, MA: MIT Press.
Lewis, G. 1980. *Day of Shining Red: An Essay on Understanding Ritual*. Cambridge: Cambridge University Press.
Littlejohn, J. 1967 [1961]. The Temne house. In J. Middleton (ed.) *Myth and Cosmos: Readings in Mythology and Symbolism* (pp. 331–47). Garden City, NY: The Natural History Press.
Longacre, W.A. and J.M. Skibo (eds.) 1994. *Kalinga Ethnoarchaeology: Expanding Archaeological Method and Theory.* Washington, DC: Smithsonian Institution Press.
Loupis, G., R. Ruff and W. Ruff 1982. Notes on architecture of the New Guinea Highlands. Lae: Department of Architecture and Building, PNG University of Technology.
Lowe, E.J. 1996. *Subjects of Experience*. Cambridge: Cambridge University Press.
Ludlow, P., Y. Nagasawa and D. Stoljar (eds.) 2004. *There's Something About Mary*. Cambridge, MA: MIT Press.
Lycett, S.J. 2009. Understanding ancient hominin dispersals using artefactual data: a phylogeographic analysis of Acheulan handaxes. *PLoS One:* 4, 10, e7404. doi:10.1371/journal.pone.0007404.
Macaulay, D. 1979. *Motel of the Mysteries*. Boston: Houghton Mifflin.
——— 1980. Motel of the mysteries. *Reader's Digest* (April):52–9.
Macintyre, M. 1989. Better homes and gardens. In M. Jolly and M. Macintyre (eds.) *Family and Gender in the Pacific: Domestic Contradictions and the Colonial Impact* (pp. 156–69). Melbourne: Cambridge University Press.
MacKenzie, M.A. 1991. *Androgynous Objects: String Bags and Gender in Central New Guinea*. Amsterdam: Harwood Academic.
Makovicky, N. 2010. 'Something to talk about': notation and knowledge-making among Central Slovak lace-makers. In T.H.J. Marchand (ed.) *Making Knowledge. Journal of the Royal Anthropological Institute* 16(Issue Supplement):S80–S99.

Malafouris, L. 2004. The cognitive basis of material engagement: where brain, body and culture conflate. In E. DeMarrais, C. Gosden and C. Renfrew (eds.) *Rethinking Materiality: The Engagement of Mind with the Material World* (pp. 53–62). Cambridge: McDonald Institute Monographs.

——— 2009. 'Neuroarchaeology': exploring the links between neural and cultural plasticity. *Progress in Brain Research* 178:251–9.

——— 2013. *How Things Shape the Mind: A Theory of Material Engagement.* Cambridge, MA: MIT Press.

Malafouris, L. and C. Renfrew 2010. The cognitive life of things: archaeology, material engagement and the extended mind. In L. Malafouris and C. Renfrew (eds.) *The Cognitive Life of Things: Recasting the Boundaries of the Mind* (pp. 1–12). Cambridge: McDonald Institute for Archaeological Research.

Malinowski, B. 1922. *Argonauts of the Western Pacific.* London: Routledge and Kegan Paul.

——— 1935. *Coral gardens and their magic* (2 vols.). London: George Allen and Unwin.

——— 1960. *A Scientific Theory of Culture and Other Essays.* New York: Galaxy.

Marchand, T.H.J. 2001 *Minaret Building and Apprenticeship in Yemen.* Richmond: Curzon.

——— 2003. A possible explanation for the lack of explanation; Or, 'why the master builder can't explain what he knows': introducing informational atomism against a 'definitional' definition of concepts. In J. Pottier, A. Bicker and P. Sillitoe (eds.) *Negotiating Local Knowledge: Power and Identity in Development* (pp. 30–50). London: Pluto Press.

——— 2009. *The Masons of Djenné.* Bloomington: Indiana University Press.

——— 2010. Making knowledge: explorations of the indissoluble relation between minds, bodies, and environment. In T.H.J. Marchand (ed.) *Making Knowledge. Journal of the Royal Anthropological Institute* 16 (Issue Supplement):S1–S21.

——— 2012. Knowledge in hand: explorations of brain, hand and tool. In R. Fardon, O. Harris, T.H.J. Marchand, M. Nuttall, C. Shore, V. Strang and R.A. Wilson (eds.) *The SAGE Handbook of Social Anthropology*, vol. 2 (pp. 261–72). London: Sage.

Marcscaux, P.-J. 1997. Can dynamic control task knowledge be communicated? *Psychologica Belgica,* 37(J/2):51–68.

Marglin, S. 1996. Farmers, seedsmen, and scientists: system of agriculture and system of knowledge. In F. Apffel-Marglin and S. Marglin (eds.) *Decolonizing Knowledge: from Development to Dialogue.* Oxford: Clarendon Press.

Marx, K. 1974 [1867]. *Capital: A Critical Analysis of Capitalist Production* (ed. F. Engels, trans. S. Moore and E. Aveling). London: Lawrence and Wishart.

Master, K.M. 1974. Air pollution in New Guinea: cause of chronic pulmonary disease among Stone-Age natives in the Highlands. *Journal of the American Medical Association* 228(13):1653–55.

Mauss, M. 1935. Les techniques du corps. *Journal de Psychologie Normale et Pathologique,* 32(3–4):271–93

——— 1970. *The Gift: Forms and Functions of Exchange in Archaic Societies* (trans. I. Cunnison). London: Cohen and West.

Mawe, T. 1982. *Mendi Culture and Tradition: A Recent Survey.* Waigani: National Museum and Art Gallery of Papua New Guinea.

May, J. 2010. *Buildings Without Architects: A Global Guide to Everyday Architecture.* New York: Rizzoli.

May, T. 2003. When is a Deleuzian becoming? *Continental Philosophy Review* 36:139–53.

McAlpine, J.R., G. Keig and R. Falls 1983. *Climate of Papua New Guinea.* Canberra: CSIRO and Australian National University Press.

McDowell, N. 1990. Competitive inequality in Melanesia: an exploratory essay. *Journal of the Polynesian Society* 99:179–204.

McLellan, H. 1994. Situated learning: continuing the conversation. *Educational Technology* 34(10):7–8.

McLeod, M.D. 1993. Collecting for the British Museum. *Quaderni Poro* 8:13–111.

Mead, M. 1939. Native languages as field-work tools. *American Anthropologist* 41(2):189–205.

Meggitt, M.J. 1957. House building among the Mae Enga, Western Highlands, Territory of New Guinea. *Oceania* 27(3):161–76.

——— 1977. *Blood is Their Argument: Warfare Among the Mae Enga Tribesmen of the New Guinea Highlands.* Palo Alto: Mayfield Publishing Company.

Merleau-Ponty, M. 1962 [1945]. *Phenomenology of Perception* (trans. C. Smith). London: Routledge and Kegan Paul.

——— 1968 [1964]. *The Visible and the Invisible* (trans. A. Lingis). Evanston: Northwestern University Press.

Meskell, L. (ed.) 2005. *Archaeologies of Materiality.* Malden, MA: Blackwell.

Meyer, C. and F. Girke (eds.) 2012. *The Rhetorical Emergence of Culture.* Oxford: Berghahn.

Miller, D. (ed.). 2001. *Home Possessions: Material Culture Behind Closed Doors.* Oxford: Berg.

——— 2005. Materiality: an introduction. In D Miller (ed.) *Materiality* (pp. 1–50). Durham, NC: Duke University Press.

——— 2010. *Stuff.* Cambridge: Polity Press.

Milner-Gulland, E.J. and R. Mace 1998. *Conservation of Biological Resources.* Oxford: Blackwell.

Modjeska, C.N. 1982. Production and inequality: perspectives from central New Guinea. In A.J. Strathern (ed.) *Inequality in New Guinea Highlands Societies* (pp. 50–108). Cambridge: Cambridge University Press.

——— 1995. Rethinking women's exploitation: the Duna case and material basis of big man systems. In A. Biersack (ed.) *Papuan Borderlands: Huli, Duna, and Ipili Perspectives on the Papua New Guinea Highlands* (pp. 265–85). Ann Arbor: Michigan University Press.

Moore, H. 1982. The interpretation of spatial patterning in settlement residues. In I. Hodder (ed.) *Symbolic and Structural Archaeology.* (pp. 74–9). Cambridge: Cambridge University Press.

Morgan, L.H. 1965 [1881]. *House and House Life of the American Aborigines*. In *Contributions to North American Ethnology*, vol. IV. Chicago: University of Chicago Press.

Munn, N. 1986. *The Fame of Gawa: A Symbolic Study of Value Transformation in a Massim (Papua New Guinea) Society*. Durham: Duke University Press.

Murphy, R.F. 1972. *The Dialectics of Social Life*. London: George Allen and Unwin.

Nabokov, P. and R. Easton 1989. *Native American Architecture*. Oxford: Oxford University Press.

Nagel, T. 1974. What is it like to be a bat? *Philosophical Review* 83(4):435–50.

Nas, P.J.M. (ed.) 2006. *The Past in the Present: Architecture in Indonesia*. Rotterdam: NAi Publishers.

Nichol, R. 2011. *Growing Up Indigenous: Developing Effective Pedagogy for Education and Development*. Rotterdam: Sense Pubs..

Nisbett, R.E. and Y. Miyamoto 2005. The influence of culture: holistic versus analytic perception. *Trends in Cognitive Science* 9(10):467–73.

Noble, A.G. 2007. *Traditional Buildings: A Global Survey of Structural Forms and Cultural Functions*. London: I.B. Tauris.

Noë, A. 2004. *Action in Perception*. Cambridge, MA: MIT Press.

Nonaka, I. 1994. A dynamic theory of organizational knowledge creation. *Organization Science* 5(1):14–37.

Nonaka, I. and H. Takeuchi 1995. *The Knowledge Creating Company: How Japanese Companies Create the Dynamics of Innovation*. Oxford: Oxford University Press.

Nonaka, I. and G. von Krogh 2009. Perspective-tacit knowledge and knowledge conversion: controversy and advancement in organizational knowledge creation theory. *Organization Science* 20(3):635–52.

Norgan, N.G., A. Ferro-Luzzi and J.V.G.A. Durnin 1974. The energy and nutrient intake and the energy expenditure of 204 New Guinea adults. *Philosophical Transactions of the Royal Society of London, series B* 268:309–48.

Norman, D.A. 1988. *The Psychology of Everyday Things*. New York: Basic Books.

O'Connor, E. 2005. Embodied knowledge: the experience of meaning and the struggle towards proficiency in glassblowing. *Ethnography* 6(2):183–204.

——— 2006. Glassblowing tools: extending the body towards practical knowledge and informing a social world. *Qualitative Sociology* 29(2):177–93.

——— 2007. The centripetal force of expression: drawing embodied histories into glassblowing. *Qualitative Sociology Review* III(3):113–24.

——— n.d. Glassblowing: a case of explicating matter and practice through in situ ethnographic research. www.univ-paris1.fr/fileadmin/laboratoire_georges_friedmann/OConnor.pdf

O'Shea, M. 1996. *A Guide to the Snakes of Papua New Guinea*. Port Moresby: Independent Publishing Group.

Oakeshott, M. 1991. *Rationalism in Politics and Other Essays*. Indianapolis: Liberty Press.

Odell, G.H. 2001. Stone tool research at the end of the millennium: classification, function, and behavior. *Journal of Archaeological Research* 9:45–100.

Ohmagari, K. and F. Berkes 1997. Transmission of indigenous knowledge and bush skills among the Western James Bay Cree women of subarctic Canada. *Human Ecology* 25(2):197–222.

Ohtsuka, R. 1995. Carrying capacity and sustainable food production: the facts and prospects from Papua New Guinea. *Anthropological Science* 103(4):311–20.

Oliver, P. (ed.) 1969 *Shelter and Society*. London: Barrie and Rockliff the Cresset Press.

——— (ed.) 1971. *Shelter in Africa*. London: Barrie and Jenkins.

——— (ed.) 1997. *The Encyclopaedia of Vernacular Architecture of the World* (3 vols.). Cambridge: Cambridge University Press.

——— 2003 [1987]. *Dwellings: The Vernacular House Worldwide*. London: Phaidon Press.

Orme, B. 1981. *Anthropology for Archaeologists*. London: Duckworth.

Palladio, A. 2002 [1570] *The Four Books of Architecture*. (trans. R. Tavernor and R. Schofield) Cambridge, MA: MIT Press

Park, D.C. and C.-M. Huang 2010. Culture wires the brain: a cognitive neuroscience perspective. *Perspectives on Psychological Science* 5(4):391–400.

Pascual-Leone, A., A. Amedi, F. Fregni and L.B. Merabet 2005. The plastic human brain cortex. *Annual Review of Neuroscience* 28:377–401.

Pataki-Schweizer, K.J. 1980. *A New Guinea Landscape: Community, Space, and Time in the Eastern Highlands*. Seattle: University of Washington Press (Anthropological Studies in the Eastern Highlands of New Guinea IV).

Pels, P. 1998. The spirit of matter. On fetish, rarity, fact, and fancy. In P. Spyer (ed.) *Border fetishisms. Material objects in unstable spaces* (pp. 91–121). London: Routledge.

Pels, D., K. Hetherington and F. Vandenberghe 2002. The status of the object: performances, mediations, and techniques. *Theory, Culture and Society* 19(5/6):1–21.

Peltier, P. 2008 Men's houses, other people's houses. In S. Tcherkézoff and F. Douaire-Marsaudon (eds.), *The Changing South Pacific: Identities and Transformations* (pp. 63–84). Canberra: Australian National University E-Press.

Penniman, T.K. 1965. *A Hundred Years of Anthropology*. London: G. Duckworth.

Pereboom, D. 2001. *Living Without Free Will*. Cambridge: Cambridge University Press.

Pétrequin, P. and A.-M. Pétrequin 1993. *Ecologie d'un outil: la hache de pierre en Irian Jaya (Indonésie)*. París: CNRS (Monographie du CRA 12).

——— 2011. The twentieth-century polished stone axeheads of New Guinea: why study them? In V. Davis and M. Edmonds (eds.) *Stone Axe Studies III* (pp. 333–50). Oxford: Oxbow Books.

Piaget, J. 1950 [1947]. *The Psychology of Intelligence* (trans. M. Piercy and D.E. Berlyne). London: Routledge and Kegan Paul.

——— 1954 [1937]. *The Construction of Reality in the Child* (trans. M. Cook). New York: Basic Books.

Piccolpasso, C. 1980 [1556–9]. The Three Books of the Potter's Art (trans. R. Lightbown and A. Caiger-Smith). London: Scolar Press.

Piggott, S. 1989. *Ancient Britons and the Antiquarian Imagination. Ideas from the Renaissance to the Regency*. London: Thames and Hudson.

Pink, S. 2009. *Doing Sensory Ethnography*. London: Sage.

Plocki, Z. 1975. *Towards a Melanesian Style in Architecture*. Boroko: Institute of Papua New Guineas Studies (Occasional Paper 3).

Polanyi, M. 1958. *Personal Knowledge. Towards a Post-critical Philosophy.* Chicago: University of Chicago Press.

——— *1962. Personal Knowledge.* London: Routledge and Kegan Paul.

——— 1966. *The Tacit Dimension.* London: Routledge and Kegan Paul

——— 1968. Logic and psychology. *American Psychologist* 23(1):27–43.

——— 1969a. Knowing and being. In M. Grene (ed.) *Knowing and Being: Essays by Michael Polanyi* (pp. 123–37). London: Routledge and Kegan Paul.

——— 1969b. The logic of tacit inference. In M. Grene (ed.) *Knowing and Being: Essays by Michael Polanyi* (pp. 138–58). London: Routledge and Kegan Paul.

——— 1969c. Sense-giving and sense-reading. In M. Grene (ed.) *Knowing and Being: Essays by Michael Polanyi* (pp. 181–207). London: Routledge and Kegan Paul.

Polanyi, M. and H. Prosch 1975. *Meaning*. Chicago: University Of Chicago Press.

Pollard J. 2004. The art of decay and the transformation of substance. In C. Renfrew, C. Godsen and E. DeMarrais (eds.) *Substance, Memory, Display: Archaeology and Art* (pp. 47–62). Cambridge: McDonald Institute for Archaeological Research.

Popper, K.R. 1972 [1963]. *Conjectures and Refutations: The Growth of Scientific Knowledge*. London: Routledge and Kegan Paul.

Pospisil, L. 1963. *Kapauku Papuan Economy.* New Haven: Yale University Publications in Anthropology 67.

Post Courier 2011a. Embel denies K51m. *Post Courier*, 12 December 2011.

Post Courier 2011b. K1.2m NRL Bid funds queried. *Post Courier*, 21 December 2011.

Pothos, E.M. 2007. Theories of artificial grammar learning. *Psychological Bulletin* 133(2) 227–44.

Powell, R. (ed.) 1983. *Architecture and Identity*. Singapore: Concept Media.

Pratt, D.D. 1998. *Five Perspectives on Teaching in Adult and Higher Education*. Malabar, FL: Krieger Publishing Company.

Preda, A. 1999. The turn to things: arguments for a sociological theory of things. *The Sociological Quarterly* 40(2):347–66.

Preston Blier, S. 1987. *The Anatomy of Architecture: Ontology and Metaphor in Batammaliba Architectural Expression.* Cambridge: Cambridge University Press.

Pretty, G.L. 1969. Salvage ethnography in New Guinea: the South Australian Museum expedition to the Southern Highlands District, Papua, December 1968–January 1969. Report for the Wenner-Gren Foundation for Anthropological Research.

Prosch, H. 1986. *Michael Polanyi: A Critical Exposition*. Albany: SUNY Press.

Pugin, A.W.N. 1836. *Contrasts: Or, a Parallel Between the Noble Edifices of the Fourteenth and Fifteenth Centuries and Similar Buildings of the Present Day*. London: Author's Publication.

Quinn, N. and D. Holland 1987. Culture and cognition. In D. Holland and N. Quinn (eds.) *Cultural Models in Language and Thought* (pp. 3–40). Cambridge: Cambridge University Press.

Radcliffe-Brown, A.R. 1952. *Structure and Function in Primitive Society*. London: Cohen and West.

——— 1953. Letter to Lévi-Strauss. In S. Tax, L.C. Eiseley, I. Rouse and C.F. Voegelin (eds.) *An Appraisal of Anthropology Today* (p. 109). Chicago: University of Chicago Press.

Ragette, F. 2003. *Traditional Domestic Architecture of the Arab Region*. Fellbach: Edition Axel Menges.

Ramachandran, V.S. and W. Hirstein 1997. Three laws of qualia: what neurology tells us about the biological functions of consciousness. *Journal of Consciousness Studies* 4:429–57.

Rapoport, A. 1969. *House, Form and Culture.* Englewood Cliffs, NJ: Prentice-Hall.

Rappaport, R.A. 1968. *Pigs for the Ancestors: Ritual in the Ecology of a New Guinea people*. New Haven: Yale University Press.

Rathje, W. and C. Murphy 1992. *Rubbish! The Archaeology of Garbage*. New York: HarperCollins.

Reber A.S. 1967. Implicit learning of artificial grammars. *Journal of Verbal Learning and Verbal Behavior* 6:855–63.

——— 1993. *Implicit Learning and Tacit Knowledge: An Essay on the Cognitive Unconscious*. Oxford: Oxford University Press.

Reithofer, H. 2006. *The Python Spirit and the Cross: Becoming Christian in a Highland Community of Papua New Guinea* (Göttinger Studien zur Ethnologie bd 16). Berlin: LIT Verlag.

Renfrew, C. 2004. Towards a theory of material engagement. In E. DeMarrais, C. Gosden and C. Renfrew (eds.) *Rethinking Materiality: The Engagement of Mind with the Material World* (pp. 1–7). Cambridge: McDonald Institute for Archaeological Research.

——— 2007. *Prehistory: The Making of the Human Mind*. London: Weidenfeld and Nicolson.

Rensel, J. 1997. Introduction. In J. Rensel and M. Rodman (eds.) *Home in the Islands: Housing and Social Change in the Pacific* (pp. 7–26). Honolulu: Hawaii University Press.

Rensel, J. and M. Rodman (eds.) 1997. *Home in the Islands: Housing and Social Change in the Pacific* Honolulu: Hawaii University Press.

Reynolds, P.J. 1979. *Iron-age Farm: The Butser Experiment*. London: British Museum Publications.

Ribeiro, R. and H. Collins 2007. The bread-making machine: tacit knowledge and two types of action. *Organisation Studies* 28(9):1417–33.

Rich, A.C. 1977. *Of Woman Born: Motherhood as Experience and Institution.* London: Virago.

Richards, P. 1993. Cultivation: knowledge or performance? In M. Hobart (ed.) *An Anthropological Critique of Development: The Growth Ignorance* (pp. 61–78). London: Routledge.

Riggins. H. 1994. *The Socialness of Things: Essays on the Socio-semiotics of Objects.* Berlin: Mouton de Gruyter.

Riggs A.D., V.E.A. Russo and R.A. Martienssen 1996. *Epigenetic Mechanisms of Gene Regulation.* Plainview, NY: Cold Spring Harbor Laboratory Press.

Rilling, J.K. 2008. Neuroscientific approaches and applications within anthropology. *American Journal of Physical Anthropology: Yearbook of Physical Anthropology* Supplement 47(137):2–32.

Rio, K. 2008. The scale of materiality: objects between the social and the material. In H. Glørstad and L. Hedeager (eds.) *Six Essays on the Materiality of Society and Culture* (pp. 155–72). Lindome: Bricoleur Press.

Rizzolatti, G. and L. Craighero 2004. The mirror-neuron system. *Annual Review of Neuroscience* 27:169–92.

Roberts, J. 2006. Limits to communities of practice. *Journal of Management Studies* 43(3):623–39.

Robinson, T.M. 1991. *Heraclitus: Fragments. A Text and Translation with a Commentary by T.M. Robinson.* Toronto: University of Toronto Press.

Rock, I. 2010. *Build Your Own House: How to Plan, Manage and Build the Home of your Dreams.* Yeovil: J.H. Haynes and Co Ltd..

Rogoff, B. 1990. *Apprenticeship in Thinking: Cognitive Development in Social Context.* New York: Oxford University Press.

Rorty, R. 1991. *Objectivity, Relativism, and Truth: Philosophical Papers, vol. 1.* Cambridge: Cambridge University Press.

Rosaldo, R. 1986. From the door of his tent: the fieldworker and the inquisitor. In J. Clifford and G.E. Marcus (eds.) *Writing Culture: The Poetics and Politics of Ethnography* (pp. 77–97). Berkeley, CA: University of California Press.

Rose, W. 2001 [1937]. *The Village Carpenter.* Hertford: Stobart Davies.

Rowlands, M. 2004. Relating archaeology and anthropology. In J. Bintliff (ed.) *A Companion to Archaeology* (pp. 473–89). Oxford: Blackwell.

Rowlands, M. 2010. *The New Science of the Mind: From Extended Mind to Embodied Phenomenology.* Cambridge, MA: MIT Press.

Ruby J. 2000. *Picturing Culture: Explorations of Film and Anthropology.* Chicago: University of Chicago Press.

Rudofsky, B. 1987 [1964]. *Architecture without Architects: A Short Introduction to Non-pedigreed Architecture.* Albuquerque: University of New Mexico Press.

Ruff, W.M. and R.E. Ruff 1990. The village studies project for the recording of traditional architecture. In N. Lutkehaus, C. Kaufmann, W.E. Mitchell, D. Newton, L. Osmundsen and M. Schuster (eds.) *Sepik Heritage: Tradition and Change in Papua New Guinea* (pp. 568–86). Bathurst: Crawford House Press.

Rupert, R. 2004. Challenges to the hypothesis of extended cognition. *Journal of Philosophy* 101:389–428.

——— 2009. *Cognitive Systems and the Extended Mind.* Oxford: Oxford University Press.

Ruskin, J. 1849 *The Seven Lamps of Architecture.* London: Smith, Elder and Co..

——— 1907. *The Poetry of Architecture; or, The Architecture of the Nations of Europe Considered in its Association with Natural Scenery and National Character* (ed. F.C. Tilney). London: George Routledge and Sons.

Russell, B. 1946. *History of Western Philosophy.* London: George Allen and Unwin.

——— 1948. *Human Knowledge: Its Scope and Limits.* London: George Allen and Unwin.

Ryan, C. 2011. *Traditional Construction for a Sustainable Future.* London: Spon Press.

Ryan, D'A. 1959. Clan formation in the Mendi valley. *Oceania* 29:257–90.

——— 1961 Gift exchange in the Mendi valley. Ph.D. Thesis, Sydney University.

Rykwert, J. 1981. *On Adam's House in Paradise: The Idea of the Primitive Hut in Architectural History* (2nd edn.). Cambridge, MA: MIT Press.

Ryle, G. 1949. *The Concept of Mind.* London: Hutchinson's University Library.

——— 1971. *Collected Papers 2.* London: Hutchinson.

Sahlins, M. 2002. *Waiting for Foucault, Still.* Chicago: Prickly Paradigm Press.

Saini, B.S. 1965. An architect looks at New Guinea. *Architecture in Australia* 54(1):82–107.

——— 1966. Durability of bush materials. *South Pacific Bulletin* 16(1):21–5.

Salisbury, R. 1962. *From Stone to Steel: Economic Consequences of a Technological Change in New Guinea.* Melbourne: Melbourne University Press.

Santos-Granero, F. (ed.) 2013. *The Occult Life of Things: Native Amazonian Theories of Materiality and Personhood.* Tuscon: University of Arizona Press.

Sartre, J-P. 1962 [1936–7]. *Transcendence of the Ego: An Existentialist Theory of Consciousness* (trans. F. Williams and R. Kirkpatrick). New York: Noonday Press.

Scharmer, C.O. 2000. Organizing around not-yet-embodied knowledge. In G. von Krogh, I. Nonaka and T. Nichiguchi (eds.). *Knowledge Creation. A Source of Value* (pp. 13–60). Basingstoke: Macmillan Press.

Schefold, R. 1997 Anthropology. In P. Oliver (ed.) *The Encyclopaedia of Vernacular Architecture of the World* (vol. 1), pp. 6–8. Cambridge: Cambridge University Press.

Schieffelin, E.L. 1982. The *bau a* ceremonial hunting lodge: an alternative to initiation. In G. Herdt (ed.) *Rituals of Manhood: Male Initiation in Papua New Guinea* (pp. 155–200). Berkeley: University of California Press.

Schieffelin, E.L. and R. Crittenden (eds.) 1991. *Like People You See in a Dream.* Stanford: Stanford University Press.

Schiffer, M.B. with A.R. Miller 1999. *The Material Life of Human Beings: Artifacts, Behaviour and Communication.* London: Routledge.

Schleiermacher, F. 1998 [1838]. *Hermeneutics and Criticism* (ed. and trans. A. Bowie). Cambridge: Cambridge University Press.

Schneider, D.M. 1965. Some muddles in the models: or, how the system really works. In M. Banton (ed.) *The Relevance of Models of Social Anthropology* (pp. 25–85). London: Tavistock.

Schön, D. 1983. *The Reflective Practitioner: How Professionals Think in Action.* London: Temple Smith.

Scoditti, G.M.G. 2012. *Notes on the Cognitive Texture of an Oral Mind: Kitawa, a Melanesian Culture.* Wantage: Sean Kingston Publishing.

Searle, J.R. 1981. Minds, brains and programs. In D.R. Hofstadter and D.C. Dennett (eds.) *The mind's I: Fantasies and Reflections on Self and Soul* (pp. 353–72). New York: Basic Books.

——— 1983. *Intentionality: An Essay in the Philosophy of Mind.* Cambridge: Cambridge University Press.

Shankland, D. 2012. Introduction: archaeology and anthropology: divorce and partial reconciliation. In D. Shankland (ed.) *Archaeology and Anthropology: Past, Present and Future* (pp. 1–18). London: Berg (Association of Social Anthropologists Monograph 48).

Shennan, S.J. 2008. Evolution in archaeology. *Annual Review of Anthropology* 37:75–91.

Shore, C. and S. Wright 1999. Audit culture and anthropology: neo-liberalism in British higher education. *Journal of the Royal Anthropological Institute* 5:557–75.

Shott, M.J. 2003. Chaîne opératoire and reduction sequence. *Lithic Technology* 28(2):95–105.

Shott, M.J. and P. Sillitoe 2001. The mortality of things: correlates of use life in Wola material culture using age-at-census data. *Journal of Archaeological Method and Theory* 8(3):269–302.

——— 2004. Modelling use-life distributions in archaeology using New Guinea Wola ethnographic data. *American Antiquity* 69(2):339–55.

——— 2005. Use-life and curation in New Guinea experimental used flakes *Journal of Archaeological Science* 32:653–63.

Shotter, J. and A.M. Katz 1996. Articulating a practice from within the practice itself: establishing formative dialogues by the use of a 'social poetics'. *Concepts and Transformation* 1:213–37.

Shove, E., M. Watson, M. Hand and J. Ingram 2007. *The Design of Everyday Life.* Oxford: Berg.

Shunk, D.H. 2000. *Learning Theories: An Educational Perspective* (3rd edn.). Upper Saddle River, NJ: Prentice-Hall.

Siebert, S.F. 1993. The abundance and site preferences of rattan (*Calamus exilis* and *Calamus zollingeri*) in two Indonesian national parks. *Forest Ecology and Management* 59(1–2):105–13.

Sillitoe, P. 1978. Big men and war in New Guinea. *Man* 13:252–71.

——— 1979a. *Give and Take: Exchange in Wola Society.* Canberra: Australian National University Press.

——— 1979b. Man-eating women: fears of sexual pollution in the Papua New Guinea Highlands. *The Journal of the Polynesian Society* 88(1):77–97.

——— 1979c. Stone versus steel. *Mankind* 12(2):151–61.

——— 1980. The art of war: Wola shield designs. *Man* 15(3):483–501.

——— 1983. *Roots of the Earth: The Cultivation and Classification of Crops in the Papua New Guinea Highlands.* Manchester: Manchester University Press.

——— 1985. Divide and no one rules: the implications of sexual divisions of labour in the Papua New Guinea Highlands. *Man* 20:494–522.

——— 1988a. *Made in Niugini: Technology in the Highlands of Papua New Guinea.* London: British Museum Publications.

——— 1988b. From head-dresses to head-messages: the art of self-decoration in the highlands of Papua New Guinea. *Man* 23(2):298–318.

——— 1995. An ethnobotanical account of the plant resources of the Wola region, Southern Highlands Province, Papua New Guinea. *Journal of Ethnobiology* 15(2):201–35.

——— 1996. *A Place Against Time: Land and Environment in the Papua New Guinea Highlands.* Amsterdam: Harwood Academic.

——— 2002. After the 'affluent society': cost of living in the Papua New Guinea Highlands according to time and energy expenditure-income. *Journal of Biosocial Science* 34:433–61.

——— 2003. *Managing Animals in New Guinea: Preying the Game in the Highlands.* London: Routledge.

——— 2007. Counting on local knowledge. In P. Sillitoe (ed.) *Local Science vs Global Science: Approaches to Indigenous Knowledge in International Development* (pp. 257–78). Oxford: Berghahn.

——— 2010a. *From Land to Mouth: The Agricultural 'Economy' of the Wola of the New Guinea Highlands.* New Haven: Yale University Press.

——— 2010b. Trust in development: some implications of knowing in indigenous knowledge. *Journal of the Royal Anthropological Institute* (NS) 16:12–30.

——— 2012a. The anthropological imagination and British Iron Age society. In D. Shankland (ed.) *Archaeology and Anthropology: Past, Present and Future* (pp. 145–71). London: Berg (Association of Social Anthropologists Monograph 48).

——— 2012b. From participant-observation to participant-collaboration: some observations on participatory-cum-collaborative approaches. In R. Fardon, O. Harris, T.H.J. Marchand, M. Nuttall, C. Shore, V. Strang and R.A. Wilson (eds.) *The Sage Handbook of Social Anthropology* (pp. 183–200). London: Sage Publications.

——— 2015. The dialogue between indigenous studies and engaged anthropology: some first impressions. In P. Sillitoe (ed.) *The Collaborative Moment: The Implications of Indigenous Studies for an Engaged Anthropology* (pp. 1–29). Farnham: Ashgate.

——— n. d. The results of a survey of the composition of plant communities in the highlands of Papua New Guinea. www.dur.ac.uk/paul.sillitoe/veg.html.

Sillitoe, P. and A. Bicker 2004. Introduction: Hunting for theory, gathering ideology. In A. Bicker, P. Sillitoe and J. Pottier (eds.) *Development and Local Knowledge: New Approaches to Issues in Natural Resources Management, Conservation and Agriculture* (pp. 1–18). London: Routledge.

Sillitoe, P. and J. Sillitoe 2009. *'Grass-Clearing Man': A Factional Ethnography of Life in the New Guinea Highlands.* Long Grove, IL: Waveland Press.

Sillitoe, P., P. Stewart and A.J. Strathern 2002. *Horticulture in Papua New Guinea: Case Studies from the Southern and Western Highlands* (*Ethnology* Monograph 18). Pittsburgh: Pittsburgh University Anthropology Department.

Simondon G. 2005. *L'individuation a la lumiere des notions de forme et d'information.* Grenoble: Jérôme Millon.

Singleton, J. 1989. Japanese folkcraft pottery apprenticeship: cultural patterns of an educational institution. In M.W. Coy (ed.) *Apprenticeship: From Theory to Method and Back Again* (pp. 13–30). New York: SUNY Press.

Sinnett, P.F. and H.M. Whyte 1973. Epidemiological studies in a Highland population of New Guinea: environment, culture, and health status. *Human Ecology* 1(3):245–77.

Smelser, N.J. 1959. *Social Change in the Industrial Revolution.* London: Routledge and Kegan Paul.

Smith, A. 1993 [1776]. *An Inquiry into the Nature and Causes of the Wealth of Nations.* (abr. L. Dickey). Indianapolis: Hackett Publishing Co.

Smith, P. 1988. *Houses of the Welsh Countryside.* Aberystwyth: Royal Commission on the Ancient and Historical Monuments of Wales.

Smith, P.F. 2005. *Architecture in a Climate of Change: a Guide to Sustainable Design.* Oxford: Elsevier/Architectural Press.

Smith, R.M. 1980. The time that ethnography forgot. *Canberra Anthropology* 3(2):81–94.

Souvatzi, S. 2012. Space, place and architecture: a major meeting point between archaeology and social anthropology? In D. Shankland (ed.) *Archaeology and Anthropology: Past, Present and Future* (pp. 145–71). London: Berg (Association of Social Anthropologists Monograph 48).

Speake, J. (ed.) 2009. *The Oxford Dictionary of Proverbs.* Oxford: Oxford University Press.

Spector, T. 2012. *Identically Different: Why You Can Change Your Genes.* London: Weidenfeld and Nicolson.

Speiser, F. 1996 [1923]. House structure. In *Ethnology of Vanuatu: An Early 20th Century Study* (trans. D.Q. Stephenson) (pp. 99–114). Honolulu: University of Hawaii Press.

Spencer, H. 1904 [1862]. *First Principles. A System of Synthetic Philosophy*, vol 1. London: Williams and Norgate.

——— 1940 [1884]. *The Man versus The State.* Caldwell, ID: Caxton.

——— 1969 [1882–98]. *Principles of Sociology* (abr. and ed. S. Andreski). London: Macmillan.

Spradley, J.P. 1980. *Participant Observation.* Orlando, FL: Harcourt College Publishers.

Spriggs, M. (ed.) 1977. *Archaeology and Anthropology: Areas of Mutual Interest* (British Archaeological Survey Supplementary Series 19). Oxford: BAR Publishing.

——— 2008. Ethnographic parallels and the denial of history. *World Archaeology* 40:538–52.

Stadler, M.I. and P.A. Frensch (eds.) 1998. *Handbook of Implicit Learning.* London: Sage.

Stangroom, J. and J. Garvey 2007. *The Great Philosophers: From Socrates to Foucault.* London: Capella.

Stasch, R. 2011a. Korowai treehouses and the everyday representation of time, belonging, and death. *The Asia Pacific Journal of Anthropology* 12(3):327–47.

——— 2011b. The camera and the house: The semiotics of New Guinea "tree-houses" in global visual culture. *Comparative Studies in Society and History* 53(1):75–112.

Steensberg, A. 1980. *New Guinea Gardens: A Study of Husbandry with Parallels in Prehistoric Europe.* London: Academic Press.

Sternberg, R.J. 1999. What do you know about tacit knowledge? Making the tacit become explicit. In R.J. Sternberg and J.A. Horvath (eds.). *Tacit Knowledge in Professional Practice: Researcher and Practitioner Perspectives* (pp. 231–6). London: Lawrence Erlbaum Associates.

Stewart, P. and A.J. Strathern 1999. Female spirit cults as a window on gender relations in the Highlands of Papua New Guinea. *Journal of the Royal Anthropological Institute* 5(3):345–60.

Stocking, G.W. 1987. *Victorian Anthropology.* New York: Free Press.

Stoffregen, A.T. 2004. Breadth and limits of the affordance concept. *Ecological Psychology* 16(1):79–85.

Stone, G.D. 2007. Agricultural deskilling and the spread of genetically modified cotton in Warangal. *Current Anthropology* 48(1):67–103.

Stout, D. 2005. The social and cultural context of stone-knapping skill acquisition. In V. Roux and B. Bril (eds.) *Stone Knapping: The Necessary Conditions for a Uniquely Hominin Behaviour* (pp. 35–48). Cambridge: McDonald Institute for Archaeology.

Stout, D. and T. Chaminade 2007. The evolutionary neuroscience of tool making. *Neuropsychologia* 45:1091–100.

——— 2009. Making tools and making sense: complex intentional behaviour in human evolution. *Cambridge Archaeological Journal* 19:85–96.

——— 2012. Stone tools, language and the brain in human evolution. *Philosophical Transactions of the Royal Society. Series B* 367(1585):75–87.

Stout, D., N. Toth and K. Schick 2006. Acheulean toolmaking and hominin brain evolution: a pilot study using positron emission tomography. In N. Toth and K. Schick (eds.) *The Oldowan: Case Studies into the Earliest Stone Age* (pp. 321–31). Gosport, IN: Stone Age Institute Press.

Stout, D., N. Toth, K. Schick and T. Chaminade 2009. Neural correlates of Early Stone Age toolmaking: technology, language and cognition in human evolution. In C. Renfrew, C. Frith and L. Malafouris (eds.) *The Sapient Mind: Archaeology Meets Neuroscience* (pp. 1–19). Oxford: Oxford University Press.

Stout, D., N. Toth, K. Schick, J.C. Stout and G. Hutchins 2000. Stone tool-making and brain activation: positron emission tomography (PET) studies. *Journal of Archaeological Science* 27:1215–23.

Strathern, A.J. 1966. Despots and directors in the New Guinea highlands. *Man* (NS) 1(3):356–67.

——— 1972. *One Father, One Blood: Descent and Group Structure among the Melpa People.* London: Tavistock.

——— (ed.) 1982. *Inequality in New Guinea Highlands Society.* Cambridge: Cambridge University Press.

——— 1987. Social classes in Mount Hagen? The early evidence. *Ethnology* 26:245–60.

——— 1996. *Body Thoughts.* Ann Arbor: University of Michigan Press.

Strathern, A.J. and Stewart, P.J. 1998. Seeking personhood: anthropological accounts and local concepts in Mount Hagen, Papua New Guinea. *Oceania* 68(3):170–88.

Strathern, M. (ed.) 1987. *Dealing with Inequality: Analyzing Gender Relations in Melanesia and Beyond.* Cambridge: Cambridge University Press.

——— 1988. *The Gender of the Gift: Problems with Women and Problems with Society in Melanesia.* Berkeley, CA: University of California Press.

Strawson, G. 1986. *Freedom and Belief.* Oxford: Oxford University Press.

Sturt, G. 1923. *The Wheelwright's Shop.* Cambridge: Cambridge University Press.

Suggett, R.F. 2005. *Houses and History in the March of Wales: Radnorshire 1400–1800.* Aberystwyth: Royal Commission on the Ancient and Historical Monuments of Wales.

Suggett, R.F. and G. Stevenson 2010. *Introducing Houses of the Welsh Countryside / Cyflwyno Cartrefi Cefn Gwlad Cymru.* Aberystwyth: Royal Commission on the Ancient and Historical Monuments of Wales.

Takahashi C., J. Diedtichsen and S. Watt 2009. Integration of vision and haptics during tool use. *Journal of Vision* 9(6):3–13.

Tallis, R. 2011 *Aping Mankind: Neuromania, Darwinitis and the Misrepresentation of Humanity.* Durham: Acumen.

Tambiah, S.J. 1990. *Magic, Science, Religion and the Scope of Rationality.* Cambridge: Cambridge University Press.

Tarde, G. 1899 [1898]. *Social Laws – An Outline of Sociology* (trans. H.C. Warren). London: Macmillan Co.

Tehrani, J.J. and M. Collard 2009. On the relationship between inter-individual cultural transmission and population level cultural diversity: a case study of weaving in Iranian tribal populations. *Evolution and Human Behavior* 30(4):286–300.

Theophilus Presbyter 1979 [*c.* 1125]. *On Divers Arts* (trans. J.G. Hawthorne and C.S. Smith). New York: Dover Publications.

Thomas, J. 2012. Archaeology, anthropology, and material things. In D. Shankland (ed.) *Archaeology and Anthropology: Past, Present and Future* (pp. 219–32). London: Berg (Association of Social Anthropologists Monograph 48).

Thurnwald, R. 1916. Bánaro society: social organisation and kinship system of a tribe in the interior of New Guinea. *American Anthropological Association Memoirs* 3(4):251–391.

Tilley, C. 2004. *The Materiality of Stone.* Oxford: Berg.

——— 2006. Introduction. In C. Tilley, W. Keane, S. Kuechler, M. Rowlands and P. Spyer (eds.) *Handbook of Material Culture* (pp. 1–6). London: Sage.

——— 2007. Materiality in materials. *Archaeological Dialogues* 14(1):16–20.

Tilley, C., W. Keane, S. Kuechler, M. Rowlands and P. Spyer (eds.) 2006 *Handbook of Material Culture.* London: Sage.

Tisdale, K. 2001. Dissertation and distress in a cognitive apprenticeship in reading. *Reading Research and Instruction* 41:51–82.

Tostevin, G.B. 2011. Introduction. *Reduction Sequence, Chaîne Opératoire, and Other Methods: The Epistemologies of Different Approaches to Lithic Analysis.* Special Issue *PaleoAnthropology*:293–6.

Townsend, W.H. 1969. Stone and steel tool use in a New Guinea society. *Ethnology* 8:199–205.

Tsoukas, H. 2003. Do we really understand tacit knowledge? In M. Easterby-Smith and M.A. Lyles (eds.) *The Blackwell Handbook of Organizational Learning and Knowledge Management* (pp. 410–27). Oxford: Blackwell.

Turner, V. 1969. *The Ritual Process: Structure and Anti-structure.* London: Routledge and Kegan Paul.

Tylor, E.B. 1871. *Primitive Culture: Researches into the Development of Mythology, Philosophy, Religion, Language, Art and Custom* (2 vols). London: John Murray.

——— 1889. On a method of investigating the development of institutions: applied to laws of marriage and descent. *Journal of the Royal Anthropological Institute* 18:245–73.

Ulijaszek S.J. 1995. *Human Energetics in Biological Anthropology.* Cambridge: Cambridge University Press (Cambridge Studies in Biological Anthropology).

Upton, D. and J.M. Vlach 1986. *Common Places: Readings in American Vernacular Architecture.* Athens, GA: University of Georgia Press.

Usher, R., I. Bryant and R. Johnston 1997. *Adult Education and the Postmodern Challenge: Learning Beyond the Limits.* London: Routledge.

Vance, M.A. 1983. *Vernacular Architecture: A Bibliography.* Monticello, IL: Vance Bibliographies, Architecture series.

Varela, F.J., E. Thompson and E. Rosch. 1991. *The Embodied Mind: Cognitive Science and Human Experience.* Cambridge, MA: MIT Press.

Vayda, A.P. 2009. *Explaining human actions and environmental changes.* Lanham, MD: AltaMira Press.

Vellinga, M., P. Oliver and A. Bridge 2007. *Atlas of Vernacular Architecture of the World.* Abingdon: Routledge.

Verdon, M. 1982. Where have all their lineages gone? Cattle and descent among the Nuer. *American Anthropologist* 84(3):566–79.

Vitruvius Pollio, Marcus 1999. *Ten books on architecture* (trans. I.D. Rowland; comm. and illus. T. Noble Howe). Cambridge: Cambridge University Press.

Vogel, A.A. 1953. *Papuans and Pygmies*. London: Arthur Barker Ltd..

von Krogh, G. and J. Roos 1996. Imitation of knowledge: a sociology of knowledge perspective. In G. von Krogh and J. Roos (eds.) *Managing Knowledge. Perspectives on Cooperation and Competition* (pp. 32–54). London: Sage.

Vygotsky, L.S. 1978. *Mind in Society: The Development of Higher Psychological Processes*. Cambridge, MA: Harvard University Press.

Wackernagel, M. and W.E. Rees 1996. *Our Ecological Footprint: Reducing Human Impact on the Earth*. Gabriola Island, BC: New Society Publishers.

Wagner, R. 1967. *The Curse of Souw: Principles of Daribi Clan Definition and Alliance in New Guinea*. Chicago: University of Chicago Press.

——— 1981. *The Invention of Culture*. Chicago: University of Chicago Press.

Wagner, R.K. 1987. Tacit knowledge in everyday intelligent behavior. *Journal of Personality and Social Psychology* 52(6):1236–47.

Wagner, R.K. and R.J. Sternberg 1985. Practical intelligence in real-world pursuits: the role of tacit knowledge. *Journal of Personality and Social Psychology* 49(2):436–58.

——— 1986. Tacit knowledge and intelligence in the everyday world. In R.J. Sternberg and R.K. Wagner (eds.) *Practical Intelligence: Nature and Origins of Competence in the Everyday World* (pp. 51–83). Cambridge: Cambridge University Press.

Ward, C. 2012. *Evolutionary psychology and the problem of neural plasticity*. In K.S. Plaisance and T.A.C. Reydon (eds.) *Philosophy of Behavioral Biology* (pp. 235–54). Dordrecht: Springer.

Warnier J.-P. 1999. *Construire la culture materielle. L'homme qui pensait avec ses doigts*. Paris: Presses Universitaires de France.

Waterson, R. 1990. *The Living House: An Anthropology of Architecture in South-East Asia*. Singapore: Oxford University Press.

Watson, J.B. 1983. *Tairora Culture: Contingency and Pragmatism*. Seattle: Washington University Press.

Weiner, A.B. 1992. *Inalienable Possessions: The Paradox of Keeping-While-Giving*. Berkeley, CA: University of California Press.

Weiner, J.F. 1991. *The Empty Place: Poetry, Space and Being Among the Foi of Papua New Guinea*. Bloomington: Indiana University Press.

Wenger, E. 1998. *Communities of Practice: Learning, Meaning, and Identity*. Cambridge: Cambridge University Press.

Were, G. 2013. On the materials of mats: thinking through design in a Melanesian society. *Journal of the Royal Anthropological Institute* 19(3):581–99.

Wheeler, M. 2010. Minds, things and materiality. In L. Malafouris and C. Renfrew (eds.) *The Cognitive Life of Things: Recasting the Boundaries of the Mind* (pp. 29–37). Cambridge: McDonald Institute for Archaeological Research.

White, A.G. 1985. *Primitive Architecture: A Selected Bibliography*. Monticello, IL: Vance Bibliographies, Architecture series.

White, J.P. 1990. Review of *Made in Niugini*. *The Australian Journal of Anthropology* 1(1):71–3.

Whitehead, A.N. 1925. *Science in the Modern World*. New York: Macmillan.

——— 1938. *Modes of Thought*. New York: Macmillan.

Wiessner, P. 2002. The vines of complexity: egalitarian structures and the institutionalization of inequality among the Enga. *Current Anthropology* 43(2):233–69.

Williams, F.E. 1939. Report on the grasslanders. *Annual Report for the Territory of Papua, 1938–39*: 39–67.

Windelband, W. and G. Oakes 1980. History and natural science. *History and Theory* 19(2):165–8.

Wittgenstein, L. 1922 (1921). *Tractatus Logico-Philosophicus* (trans. C.K. Ogden with G.E. Moore, F.P. Ramsey and L. Wittgenstein). London: Routledge and Kegan Paul.

——— 1953. *Philosophical investigations* (trans. G.E.M. Anscombe). Oxford: Blackwell.

Wodehouse, L. 1980. *Indigenous Architecture Worldwide: A Guide to Information Sources*. Detroit: Gale Research.

Wohlt, P.B. 2004. Descent group composition and population pressure in a fringe Enga clan, Papua New Guinea. *Human Ecology* 32(2):137–62.

World Commission on Environment and Development 1987. *Our Common Future*. Oxford: Oxford University Press. [Often referred to as the Brundtland Report.]

Wright, E. (ed.) 2008. *The Case for Qualia*. Cambridge, MA: MIT Press.

Wylie, A. 1985. The reaction against analogy. *Advances in Archaeological Theory and Method* 8:63–111.

Young, M. 1993. The Kalauna house of secrets. In J.J. Fox (ed.) *Inside Austronesian houses: Perspectives on Domestic Designs for Living* (pp. 180–93). Canberra: Department of Anthropology in association with the Comparative Austronesian Project, Research School of Pacific Studies, Australian National University.

Zent, S. 2001. Acculturation and ethnobotanical knowledge loss among the Piaroa of Venezuela: demonstration of a quantitative method for the empirical study of TEK change. In L. Maffi (ed.) *On Biocultural Diversity: Linking Language, Knowledge, and the Environment* (pp. 190–211). Washington: Smithsonian Institution Press.

Zimmer, C. 2014. Secrets of the brain: new technologies are shedding light on biology's greatest unsolved mystery: how the brain really works. *National Geographic* 225(2):28–57.

INDEX

Page numbers in *italic* refer to captions to maps and plates. Page numbers with suffixes 'f', 'n' and 't' refer to figures, footnotes and tables respectively. Plant names refer to both plant species and derivative materials.

www.ingramcontent.com/pod-product-compliance
Lightning Source LLC
LaVergne TN
LVHW081946200926
848484LV00009B/2512
* 9 7 8 1 9 0 7 7 7 4 4 5 4 *